Lectures in Theoretical High Energy Physics

Lectures in Theoretical High Energy Physics

Edited by

H. H. ALY

Department of Physics,
American University of Beirut,
Lebanon

A WILEY-INTERSCIENCE PUBLICATION
John Wiley & Sons Ltd
London New York Sydney

First Published 1968 by John Wiley & Sons Ltd.

Library of Congress catalog card No. 68-56016

SBN 470 02520 4

Printed in Great Britain at the Pitman Press, Bath

Preface

In the last decade we have witnessed a substantial change in the understanding of the highly complex phenomena of elementary particles. The various aspects of high energy physics, since it usually means the physics of elementary particles, is so vast and the new findings are so frequent that it is becoming increasingly difficult to have one author cover all the aspects in one book of a reasonable size. This fact has indeed led to a number of specialized conferences, schools and series of seminars.

These gatherings of specialized physicists have proved to be highly profitable for a variety of reasons. Among them is the direct contact and discussions on a certain topic of common interest. It also helps looking at a certain problem from a different angle and even by using different calculational techniques, which would have been perhaps overlooked, had it not been for the individual discussions. There is also the very vital aspect of helping young and growing institutions like ours a great deal, by lecturing to graduate students and also have them exposed to the latest development in this rapidly changing field of high energy physics.

In this volume of lectures on various aspects of high energy physics, the goal was to cover some of the active subjects both by reviewing the field and also reporting some new results obtained by the speakers.

These thirteen chapters are specifically written in a style and language suitable to graduate students who are entering the field. But the book is also believed to have a value to scientists actively working in some of the fields covered here. It can therefore be used as a reference book. In all these chapters, the main emphasis was laid on the physics of the subjects under discussion.

We have been particularly happy to have all these visitors come to the American University of Beirut to lecture to us on topics selected by them.

These lectures were given in an informal atmosphere so as to encourage discussions and questions. It is then believed that only the texts of these papers should go on record.

These series of lectures were sponsored by a financial support of the School of Arts and Sciences of the American University of Beirut and also the Research Corporation.

In this connexion, it is a great pleasure to thank Professor E. T. Prothro, Dean of the School of Arts and Sciences for his help and encouragement. The generous support and considerable understanding of Professor A. B. Zahlan, Chairman of the Physics Department of the A.U.B., indeed made this book possible.

H. H. ALY

Beirut, Lebanon
June 1968

Contributing Authors

Aly, H. H.	Department of Physics, American University of Beirut, Beirut, Republic of Lebanon.
Hagen, C. R.	Department of Physics, University of Rochester, New York, U.S.A.
Lurie, D.	Dublin Institute for Advanced Study, Dublin, Ireland.
Moffat, J. W.	Department of Physics, University of Toronto, Toronto, Canada.
Muller, H. J. W.	Institut für Theoretische Physik, Universität München, 8 Munchen 13, Schellingstrasse 2–8, Germany.
Nilsson, J.	Institute for Theoretical Physics, Sven Hultin Gate, Goteberg, Sweden.
Pietschmann, H.	Institute for Theoretical Physics, University of Vienna, Austria.
Riazuddin	Institute of Physics, University of Islamabad, Rawalpindi, Pakistan.
Sarker, A. Q.	Institute of Physics, University of Islamabad, Rawalpindi, Pakistan.
Sudarshan, E. C. G.	Department of Physics, Syracuse University, New York, U.S.A.
Taylor, J. G.	Queen Mary College, University of London, Mile End Road, London E.1.
Uretsky, T. L.	High Energy Physics Division, Argonne National Laboratory, Illinois, U.S.A.
Warnock, R. L.	High Energy Physics Division, Argonne National Laboratory, Illinois, U.S.A., and Illinois Institute of Technology, Chicago, U.S.A.

Contents

Selected Topics in Current Algebra

H. PIETSCHMANN

1 Introduction

About five years ago, M. Gell-Mann in his paper on broken symmetries[1] suggested the study of commutation relations of internal symmetry operators. Taken between suitably chosen physical particle states, these commutation relations yield information on dynamical quantities such as form factors and the like. The suggestion was first taken up by the Vienna group in an extensive application of the isospin algebra[2]. When Fubini and Furlan showed how the method could be applied to non-conserved operators[3], Adler and Weisberger were able to explain the weak axial vector coupling constant renormalization[4]. For years this had been an open problem of elementary particle physics so that the striking success made the method immediately popular.

A flood of publications was the result of restless work in this field; the first big wave seems to have passed and we are already in possession of excellent review articles and lecture notes. Some of them attempt at completeness[5], others review the most important ideas[6]. However, in the author's opinion one very important introduction into the field is still missing: a presentation of the actual calculations and techniques which have to be used in order to arrive at the spectacular results. The student who is faced with the new field for the first time will find it most helpful, to say the least, if he can follow the computations in a few typical cases to the very end.

In these lectures, therefore, we pick two of the more important results of current algebra, the derivation of the Adler-Weisberger sum rule and the application of the 'soft pion method' to leptonic K-decays and present them in a hopefully easy and pedagogical way. The general ideas of the method will be discussed first, in section 2.

Natural units ($\hbar = c = 1$) are used and the notation follows closely that of reference 7.

2 The Method of Current Algebra in Elementary Particle Physics

It has been pointed out in the introduction that the method of current algebras essentially consists in extracting information on dynamical quantities, such as form factors or the like, from suitably chosen commutation relations. How this can be done is outlined in this section.

The method shall first be demonstrated here on a very simple example, the interaction of a proton with an external electro-magnetic field, described by an interaction Lagrangian

$$L_{\mathrm{I}} = e\bar{\psi}\gamma^{\mu}\psi A_{\mu}{}^{\mathrm{ext}} = ej^{\mu}A_{\mu}{}^{\mathrm{ext}} \tag{1}$$

The current-operator j^{μ} between 1-proton states takes on the well-known form[8]:

$$\langle p'|j^{\mu}(x)|p\rangle = \frac{1}{(2\pi)^3}\frac{M}{\sqrt{(p_0 p_0')}}\bar{u}(p')[\gamma^{\mu}F_1(q^2) + i\sigma^{\mu\nu}q_{\nu}F_2(q^2)]u(p)e^{iqx} \tag{2}$$

with

$$q = p' - p \tag{3}$$

where the Dirac-Formfactor $F_1(q^2)$ and the Pauli-Formfactor $F_2(q^2)$ are related to the spacial spread of the proton charge and its magnetic moment due to the cloud of strongly interacting virtual particles. $F_2(0)$ is just the anomalous magnetic moment of the proton times $\hbar/2Mc$ where M is the proton mass. It should be stressed that (2) directly follows from Lorentz invariance and current conservation.

The charge operator is defined by

$$Q = e\int d^3x j_0(x) = e\int d^3x \psi^{\dagger}(x)\psi(x) \tag{4}$$

An extra term involving the pion field operators has been omitted in (4), implying that the pions are thought to be bound states of the fermion-antifermion system but this assumption is not crucial here. From (2) and (4) it follows that the charge of a proton is just $eF_1(0)$:

$$\begin{aligned}\langle p'|Q|p\rangle &= e\frac{M}{p^0}\bar{u}(p)\gamma_0 u(p)F_1(0)\delta^{(3)}(\mathbf{p}' - \mathbf{p}) \\ &= eF_1(0)\delta^{(3)}(\mathbf{p}' - \mathbf{p})\end{aligned} \tag{5}$$

If strong interactions were switched off, one would have $F_1(q^2) \equiv 1$ and $F_2(q^2) \equiv 0$ for all q, thus the charge would just be e.

It is now possible to get information on $F_1(0)$ from the commutation relation of the charge with the proton field operators

$$[Q,\psi(x)] = -e\psi(x) \tag{6}$$

Though this commutation relation is evident from the property of the charge operator, it can be derived from (4), using the anticommutation relations of the proton field operators

$$[\psi(\mathbf{x},t),\psi^\dagger(\mathbf{x}',t)] = \delta^{(3)}(\mathbf{x}-\mathbf{x}') \tag{7}$$

The method to be described here contains essentially three steps:

1st step: Take the commutation relation between a suitable pair of states. Which states are suitable depends, of course, on the problem one wants to solve. Here, we shall take the vacuum to the left and a 1-proton state to the right of (6). The right-hand side of (6) is, of course, just the proton wave-function multiplied into the charge *e*.

2nd step: Insert a complete set of intermediate states in the bilinear part of the commutation relation

$$e\langle 0|\psi(x)|p\rangle = \Sigma\langle 0|\psi(x)|\alpha\rangle\,\langle\alpha|Q|p\rangle - \Sigma\langle 0|Q|\beta \times \beta|\psi(x)|p\rangle \tag{8}$$

3rd step: This is the most difficult one. It requires the manipulation of the matrix elements occurring in the sum over intermediate states. In general, it is useful to single out the 1-particle contribution to the sum. Some methods of handling the rest will be described below.

In the simple example under consideration the summation over intermediate states becomes trivial. This is because charge is conserved as a consequence of the continuity equation

$$\frac{\partial}{\partial x^\mu}\, j^\mu(x) = 0 \tag{9}$$

But because of $\dot{Q} = 0$, Q commutes with the Hamiltonian and is thus diagonal as long as there is no degeneracy in energy. Consequently, the summation over α in (8) has the 1-proton state as the only non-vanishing contribution and the sum over β disappears completely because the charge of the vacuum is zero. Using (5), it then follows that

$$e\langle 0|\psi(x)|p\rangle = \int d^3q\langle 0|\psi(x)|q\rangle\,\langle q|Q|p\rangle = eF_1(0)\langle 0|\psi(x)|p\rangle \tag{10}$$

or

$$F_1(0) = 1 \tag{11}$$

which is the well-known fact that the total charge is not changed by strong interactions as a consequence of current-conservation.

At this point the reader might wonder what would have changed if a quantized photon-field had been used in (1) instead of the external field A^{ext}. It is a well-known fact that vacuum-polarization phenomena do indeed change e and this change, in any actual computation, in fact turns out to be infinitely large. In this case we would, however, face a degeneracy in energy of the 1-proton state with any proton + n-photon state because their continuum starts at the proton mass. It is precisely due to the long range of electromagnetic interactions that the charge is 'renormalized'. Thus the correct expression of (11) is that the 'total charge is not changed by strong interactions'[9].

The wide field of fruitful application of the method of current algebras is that of internal symmetries. Let any internal symmetry be defined by the algebra of its generators I:

$$[I_\alpha, I_\beta] = c_{\alpha\beta}{}^\gamma I_\gamma \tag{12}$$

The reader who feels a bit uncertain on the grounds of higher symmetries may always keep in mind isospin, in which case $c_{\alpha\beta}{}^\gamma$ is simply $i\varepsilon_{\alpha\beta\gamma}$ and the I's are the 3 isospin operators. One can now proceed to define a set of currents

$$J_\mu{}^\alpha(x) = \bar{\psi}(x)\Gamma_\mu\Lambda^\alpha\psi(x) \tag{13}$$

where Γ_μ is, for instance, γ_μ or $\gamma_\mu\gamma_5$ or a combination of both and Λ^α is some representation of the I's in the space of the ψ's. If the ψ's are to be the nucleon spinors, say, the Λ^α's are simply the Pauli matrices in isospin space; in case the ψ's stand for quarks, the Λ^α's are the well-known 3 × 3 matrices of Gell-Mann[10].

In analogy to (4), one now defines 'generalized charges' by

$$Q_\alpha = GI_\alpha = G\int d^3x J_0{}^\alpha(x) \tag{14}$$

For those currents $J_\mu{}^\alpha$ which are not conserved, the corresponding Q_α depend on time explicitly. Suppose

$$\frac{\partial}{\partial x_\mu} J_\mu{}^\alpha(x) = a\phi_\alpha(x) \tag{15}$$

where $\phi_\alpha(x)$ is some non-vanishing operator. Upon integration over all space, one finds

$$\dot{Q}_\alpha(t) = \int d^3x \frac{\partial}{\partial t} J_0{}^\alpha(x) = a\int d^3x \phi_\alpha(x) \tag{16}$$

Multiplication of (12) by G^2 yields a commutation relation of the 'generalized charges'

$$[Q_\alpha, Q_\beta] = G^2 c_{\alpha\beta}{}^{\gamma} I_\gamma \tag{17}$$

In analogy to (5) one would like to have a relation

$$\langle p'|Q_\alpha|p\rangle = Gf_\alpha(0)\delta^{(3)}(\mathbf{p}' - \mathbf{p}) \equiv G_\alpha\delta^{(3)}(\mathbf{p}' - \mathbf{p}) \tag{18}$$

This is, however, no longer true in general. It shall be shown below that for non-conserved currents the matrix-element (18) explicitly depends on the external momentum **p** of the particles. But it will also be shown that one can always achieve a form similar to (18) by taking a certain limit of the momentum **p**. The form (18) shall thus be assumed for the moment to secure simplicity.

To be concrete, the generator on the right-hand side of (17) shall now be assumed to be diagonal; in the case of isospin I_3, the commutation relation then reads

$$G^2 c I_a = [Q^+, Q^-] \tag{19}$$

where c is some Clebsch-Gordan coefficient of the algebra. One now goes through the three steps described above. The first step is to sandwich the commutation relation between suitable states; 1-particle states, that is, in this case.

$$G^2 c\langle p'|I_a|p\rangle = \langle p'|Q^+Q^-|p\rangle - \langle p'|Q^-Q^+|p\rangle \tag{20}$$

Step 2 is just the insertion of a complete set of intermediate states on the bilinear side of the commutation relation. In step 3 one first singles out the 1-particle contribution to the complete set to obtain (recall equation 18):

$$G^2 c i_a \delta^{(3)}(\mathbf{p}' - \mathbf{p}) = G_a{}^2 \hat{c}^2 \delta^{(3)}(\mathbf{p}' - \mathbf{p}) + \\ + \sum_{\beta \neq q} [\langle p'|Q^+|\beta\rangle \langle\beta|Q^-|p\rangle - \langle p'|Q^-|\beta\rangle \langle\beta|Q^+|p\rangle] \tag{21}$$

where i_a is the eigenvalue of I_a to the state $|p\rangle$ and $\hat{c} = \sqrt{(ci_a)}$ is again some Clebsch-Gordan coefficient. The somewhat symbolic notation $\beta \neq q$ under the sum means that the 1-particle contribution is now excluded.

There are now two important questions to be answered. First, what is the size of the terms contributing to the sum in (21)? Notice that they represent the contribution to the change in the coupling constant due to interactions, that is to say the form factor at zero momentum transfer. To answer this question, one conveniently writes the total Hamiltonian as the sum of three contributions:

$$H = H_0 + fH_1 + eH_2 \tag{22}$$

where H_0 is the part symmetric under the internal symmetry group, H_1 is the symmetry-breaking term and H_2 is the electromagnetic interaction. We can distinguish three cases:

(*a*) The 'generalized charges' $Q^{\pm}$ commute with the total Hamiltonian. In this case the current is conserved and the sum in (21) vanishes. One thus arrives at a result similar to (11) (recall the discussion after equation 11).

(*b*) $Q^{\pm}$ commute with H_0 and H_1 but not with H_2, i.e.

$$\mathrm{i}\dot{Q}^{\pm} = [Q^{\pm},H] = e[Q^{\pm},H_2] \tag{23}$$

It will be shown presently that the sum is proportional to e^2.

(*c*) $Q^{\pm}$ commute neither with H_1 nor with H_2, i.e.

$$\mathrm{i}\dot{Q}^{\pm} = [Q^{\pm},H] = f[Q^{\pm},H_1] + 0(e) \tag{24}$$

The sum is now proportional to f^2.

The second question is a technical one: how to compute the sum. The most important method shall be discussed now; it can be used whenever one has information on $\dot{Q}^{\pm}$. Suppose $\dot{Q}^{\pm}$ were given, say

$$\dot{Q}^{\pm} = a\phi^{\pm} \tag{25}$$

The Heisenberg equation of motion then reads

$$\mathrm{i}a\phi^{\pm} = Q^{\pm}H - HQ^{\pm} \tag{26}$$

Taking it between states and recalling (25) gives

$$\langle\beta|\,Q^{\pm}|p\rangle = \mathrm{i}a\frac{\langle\beta|\phi^{\pm}|p\rangle}{E_p - E_\beta} \tag{27}$$

where E_α are the energies of the states. In case (*b*) described above, $\phi^{\pm}$ can be calculated directly, for one knows H_2 explicitly. In the more important case (*c*), the so-called hypothesis of partial conservation of axial vector currents (PCAC) provides a tool for many important applications via equations (15) and (16). This will be discussed in detail in connection with the Adler-Weisberger sum rule.

We have seen that in the case of strictly conserved currents, i.e. in the case of a rigorous internal symmetry, the sum in (21) vanishes. That is to say that only a degenerate set of 1-particle states contributes to the complete set of intermediate states. These particles are members of a certain multiplet of the symmetry group. They are said to 'saturate' the sum rule. If the symmetry is broken, there may be 'leakage' from the multiplet.

It remains to discuss the question of momentum dependence raised in connection with (18). As an example for a non-conserved current, let us take the strangeness changing weak vector current, denoted by $J_\mu{}^+(x)$. In terms of quark field operators p, n, λ this current would have the structure $(\bar{\lambda}p)$.

A matrix element of this current can be expressed through two form factors. On invariance arguments one has

$$\langle\pi^0(q')|J_\mu{}^+(x)|K^+(q)\rangle = \frac{1}{(2\pi)^3}\frac{1}{(4q_0q_0')^{1/2}}[(q+q')_\mu f_+(t) + \\ + (q-q')_\mu f_-(t)]\mathrm{e}^{\mathrm{i}x(q'-q)} \quad (28)$$

with

$$t = (q'-q)^2 \quad (29)$$

As usual, we define the 'generalized charge'

$$S^+ = \int \mathrm{d}^3x J_0{}^+(x) \quad (30)$$

and obtain at time $x_0 = 0$

$$\langle\pi^0(q')|S^+|K^+(q)\rangle = \delta^{(3)}(\mathbf{q}'-\mathbf{q})\frac{1}{(4\omega_\pi\omega_K)^{1/2}}[(\omega_\pi+\omega_K)f_+(t_0) + \\ + (\omega_K-\omega_\pi)f_-(t_0)] \quad (31)$$

where (μ being the pion mass):

$$\omega_\pi = (\mu^2+\mathbf{q}^2)^{1/2} \quad \omega_K = (m_K{}^2+\mathbf{q}^2)^{1/2} \quad t_0 = (\omega_K-\omega_\pi)^2 \quad (32)$$

Hence $\langle\pi^0(q')|S^+|K^+(q)\rangle$ depends on $|\mathbf{q}|$ and one actually gets a continuous set of sum rules, for all values of $|\mathbf{q}|$. It is now tempting to select the rest frame of the external particles as the most natural choice. However, in this system (31) becomes

$\mathbf{q}\to 0$:

$$\langle\pi^0(q')|S^+|K^+(q)\rangle = \delta^{(3)}(\mathbf{q}'-\mathbf{q})\frac{1}{(4\mu m_K)^{1/2}}[(\mu+m_K)f_+(\hat{t}) + \\ + (m_K-\mu)f_-(\hat{t})] \quad (33)$$

with

$$\hat{t} = (m_K-\mu)^2 \quad (34)$$

This is still a somewhat unhandy expression. On the other hand, for $|\mathbf{q}|\to\infty$ one obtains a form similar to (18), thus the wanted answer:

$$|\mathbf{q}|\to\infty: \ \langle\pi^0(q')|S^+|K^+(q)\rangle = \delta^{(3)}(\mathbf{q}'-\mathbf{q})f_+(0) \quad (35)$$

The question as to which value of $|\mathbf{q}|$ yields the 'best' sum rule will be taken up again when the Adler-Weisberger sum rule will be derived.

3 The Adler-Weisberger Sum Rule

It has already been pointed out that the hypothesis of partial conservation of the axial vector current (PCAC) will be the main source of information for the derivation of the Adler-Weisberger sum rule.

First, let us recall that the axial vector current cannot be conserved because if it were the pion could not decay. This is apparent from the most general form of the axial vector current matrix element between the 1-pion state and the vacuum

$$\langle 0|j_\mu{}^5(x)|\pi(q)\rangle = \frac{1}{(2\pi)^{3/2}}\frac{1}{\sqrt{(2q^0)}}\mu f(q^2)q_\mu e^{-iqx} \tag{36}$$

where the usual normalization factors have been taken out of the form factor $f(q^2)$ as well as the pion mass so that all form factors are dimensionless. But this form factor merely is a constant because of $q^2 = \mu^2$ where μ is the pion mass.

$$f(q^2) = f(\mu^2) \equiv f_\pi \tag{37}$$

Conservation of the current would require

$$\frac{\partial}{\partial x_\mu}\langle 0|j_\mu{}^5(x)|\pi(q)\rangle = \frac{-i}{(2\pi)^{3/2}}\frac{1}{\sqrt{(2q^0)}}\mu f_\pi \mu^2 e^{-iqx} = 0 \tag{38}$$

and thus $f_\pi = 0$. But the inverse pion lifetime is, of course, proportional to the square of the matrix element (36). More precisely, it is given by

$$\Gamma_\pi = G^2 f_\pi{}^2 \frac{\mu^3 m_\mu{}^2}{8\pi}\left(1 - \frac{m_\mu{}^2}{\mu^2}\right)^2 \tag{39}$$

(A detailed derivation of (39) can be found in reference 11.) From the experimental pion lifetime one finds $f = 0{\cdot}93$. If it were zero, the lifetime would be infinite. Hence the derivative of the axial vector current has to be a non-vanishing (pseudoscalar) operator. The most simple pseudoscalar operator which is available is the pion field operator itself. PCAC, in the form given by M. Gell-Mann and M. Levy[12], now states that the derivative of the axial vector current is simply proportional to the pion field operator. (For finer details see J. Bernstein, S. Fubini, M. Gell-Mann and W. Thirring[13].)

$$\frac{\partial}{\partial x_\mu} j_\mu{}^5(x) = a\phi(x) \tag{40}$$

In order to work out the proportionality constant a one writes down the most general form of the axial vector current matrix element between a proton and a neutron state, say:

$$\langle P(p_2)|j_\mu{}^{+5}(x)|N(p_1)\rangle = \frac{1}{(2\pi)^3}\frac{M}{\sqrt{(E_1E_2)}}\bar{u}(\mathbf{p}_2)[\gamma_\mu G_1(q^2) + q_\mu G_2(q^2)]\gamma_5 u(\mathbf{p}_1)e^{iqx} \tag{41}$$

Notice that for the axial vector current the 'magnetic moment' term changes into the 'induced pseudoscalar'[11]. Taking the derivative of (41) as well as the limit $q^2 = 0$ yields

$$\left\langle P(p_2)\left|\frac{\partial}{\partial x_\mu} j_\mu{}^{+5}(x)\right| N(p_1)\right\rangle = \frac{i}{(2\pi)^3}\frac{2M^2}{\sqrt{(E_1E_2)}}\bar{u}(\mathbf{p}_2)\gamma_5 u(\mathbf{p}_1)\lambda e^{-iqx} + 0(q^2) \tag{42}$$

where

$$\lambda \equiv G_1(0) \quad (= `-G_A/G_V') \tag{43}$$

and use has been made of the Dirac equation and the anticommutativity of γ_μ and γ_5.

To obtain the matrix element of the right-hand side of (40), one has to recall the equation of motion for the pion field operator:

$$(\Box + \mu^2)\phi^+(x) = \sqrt{2}gj_\pi{}^+(x) = \sqrt{2}g\bar{\psi}(x)\gamma_5\frac{\tau^+}{\sqrt{2}}\psi(x). \tag{44}$$

The matrix element of the γ_5 current is

$$\langle P(p_2)|j_\pi{}^+(x)|N(p_1)\rangle = \frac{1}{(2\pi)^3}\frac{M}{\sqrt{(E_1E_2)}}\bar{u}(p_2)\gamma_5 u(p_1)K(q^2)e^{iqx} \tag{45}$$

where the normalization of the form factor is

$$K(\mu^2) = 1 \tag{46}$$

From (44) and (45) it follows that for vanishing q^2:

$$\langle P(p_2)|\phi^+(x)|N(p_1)\rangle = \frac{1}{(2\pi)^3}\frac{M}{\sqrt{(E_1E_2)}}\frac{K(0)}{\mu^2}\sqrt{2}g\bar{u}(p_2)\gamma_5 u(p_1)e^{-iqx} + 0(q^2) \tag{47}$$

Comparison of (47), (42) and (40) yields the wanted result:

$$a = \frac{i\sqrt{2}M\mu^2\lambda}{gK(0)} \tag{48}$$

One can still go a step further: from invariance reasons it follows that

$$\langle 0|j_\mu{}^5(x)|\pi(q)\rangle = \frac{-iq_\mu}{\mu^2}\,a\langle 0|\phi(x)|\pi(q)\rangle = \frac{-ia}{\mu^2}\,q_\mu\,\frac{1}{(2\pi)^{3/2}}\,\frac{1}{(2q^0)^{1/2}}\,e^{-iqx} \quad (49)$$

Comparison with (36) and (37) yields

$$f_\pi = \frac{\sqrt{2}M\lambda}{gK(0)} \quad (50)$$

For $K(0) = 1$ (which is certainly only approximately valid) this is the so-called Goldberger-Treiman relation[14]. It predicts $f_\pi = 0{\cdot}83$ in rather good agreement with the experimental value.

It is now time to derive the Adler-Weisberger sum rule[4,15]. To this end, one has to assume the following commutation relation

$$2I_3 = [Q_a{}^+,\ Q_a{}^-] \quad (51)$$

where I_3 is the third component of the total isospin-operator, related to the weak vector current by (note that isospin-symmetry and CVC is assumed here):

$$I_3 = \int d^3x j_0{}^3(x) \quad (52)$$

and the 'generalized charges' $Q_a{}^\pm$ are sometimes called 'isotopic chiralities'. They are expressed through the axial vector current by

$$Q_a{}^\pm = \int d^3x j_0{}^{5\pm}(x) \quad (53)$$

The basic commutation relation (51) can be derived from canonical commutation relations together with (52) and (53) provided one assumes that the pion is a bound state in the baryon-antibaryon system rather than an elementary particle. In this case, no explicit pion term enters into the weak vector current, leaving complete symmetry between $j_\mu(x)$ and $j_\mu{}^5(x)$. If the pion is elementary and there is no scalar boson (called 'σ'), the basic commutator (51) does not follow. It is, however, true in the quark model of elementary particles. If one assumes an internal symmetry group to hold, the generators defined as the space-integral over the time component of the currents (13) will automatically obey the same commutation relations as the relevant matrices Λ^α of (13). This is a consequence of canonical equal time commutation relations between field operators. However, the commutation relation may still be true even if the symmetry is broken. This is one of the advantages of the current algebra approach over the direct application of the symmetry group.

As an illustration, let us assume a quark model with quarks of vanishing bare mass, interacting through a parity-conserving four-fermion interaction of the type V and/or A. With Γ_μ of (13) given by $\gamma_\mu(1 \pm \gamma_5)$, the sum and difference of vector and axial vector currents (positive and negative chiral projections) each form a group SU(3). Since positive chiral projections commute with negative chiral projections and vice versa, the underlying group is SU(3) × SU(3)[16]. Among others, it leads to the commutation relation (51). When the quarks acquire a mass, the axial vector currents are no longer conserved and thus the isotopic chiralities do not commute with the Hamiltonian, i.e. the symmetry is broken. The commutation relation (51), however, is still true.

We can now apply the rules set up in the previous section; the first step is to sandwich (51) between states, i.e. 1-proton states in this case. Since the proton is an eigenstate of isospin, one has

$$\langle P(p_2)|2I_3|P(p_1)\rangle = \delta^{(3)}(\mathbf{p}_2 - \mathbf{p}_1) \tag{54}$$

and hence from (51)

$$\delta^{(3)}(\mathbf{p}_2 - \mathbf{p}_1) = \langle P(p_2)|[Q_a{}^+, Q_a{}^-]|P(p_1)\rangle \tag{55}$$

From the complete set of intermediate states to be inserted on the right-hand side of (55), the 1-neutron state shall be singled out. Notice that it is the only 1-particle state which contributes.

$$\begin{aligned}\delta^{(3)}(\mathbf{p}_2 - \mathbf{p}_1) = &\sum_{\text{spins}} \int \mathrm{d}^3q \langle P(p_2)|\,Q_a{}^+|N(q)\rangle \,\langle N(q)|\,Q_a{}^-|P(p_1)\rangle\, + \\ &+ \sum_{\alpha \neq N} \langle P(p_2)|\,Q_a{}^+|\alpha_{\text{out}}\rangle \,\langle \alpha_{\text{out}}|\,Q_a{}^-|P(p_1)\rangle\, - \\ &- \sum_{\beta} \langle P(p_2)|\,Q_a{}^-|\beta_{\text{out}}\rangle \,\langle \beta_{\text{out}}|\,Q_a{}^+|P(p_1)\rangle\end{aligned} \tag{56}$$

In the second term of the commutator no 1-particle state contributes at all due to the isotopic selection rules of $Q_a{}^-$ and $Q_a{}^+$.

The space integral over the time-component of (41) gives the 1-particle matrix elements of $Q_a{}^+$:

$$\langle P(p_2)|\,Q_a{}^+|N(q)\rangle = \frac{M}{E}\,\bar{u}(\mathbf{p}_2)\gamma_0\gamma_5 u(\mathbf{q})\lambda\delta^{(3)}(\mathbf{p}_2 - \mathbf{q}) \tag{57}$$

where λ is defined by (43). A similar relation holds for the matrix element of $Q_a{}^-$, so that one has

$$\begin{aligned}&\sum_{\text{spins}} \int \mathrm{d}^3q \langle P(p_2)|\,Q_a{}^+|N(q)\rangle \,\langle N(q)|\,Q_a{}^-|P(p_1)\rangle = \\ &\quad = \frac{M^2}{E^2}\,\bar{u}(\mathbf{p}_1)\gamma_0\gamma_5\,\frac{M + \not{p}_1}{2M}\,\gamma_0\gamma_5 u(\mathbf{p}_1)\lambda^2\delta^{(3)}(\mathbf{p}_2 - \mathbf{p}_1) \\ &\quad = \frac{M}{2E^2}\,\lambda^2\delta^{(3)}(\mathbf{p}_2 - \mathbf{p}_1)\bar{u}(\mathbf{p}_1)(\gamma_0\not{p}_1\gamma_0 - M)u(\mathbf{p}_1)\end{aligned} \tag{58}$$

where the sum over spins has been converted into the energy projection operator by means of

$$\sum_{\text{spins}} u(\mathbf{p})\bar{u}(\mathbf{p}) = \frac{M + \not{p}}{2M} \tag{59}$$

and $E \equiv p^0 = (M^2 + \mathbf{p})^{1/2}$. In order to work out the remaining Dirac algebra one recalls the property of the Dirac spinors under parity transformations:

$$\gamma_0 u(\mathbf{p}) = u(-\mathbf{p}) \qquad \bar{u}(\mathbf{p})\gamma_0 = \bar{u}(-\mathbf{p}) \tag{60}$$

This gives

$$\begin{aligned}\bar{u}(\mathbf{p})\gamma_0 \not{p} \gamma_0 u(\mathbf{p}) &= \bar{u}(-\mathbf{p})\not{p}u(-\mathbf{p}) = 2Eu^\dagger u - \bar{u}(-\mathbf{p})\hat{p}u(-\mathbf{p}) \\ &= 2\frac{E^2}{M} - M\end{aligned} \tag{61}$$

where

$$\hat{p} = E\gamma_0 + \mathbf{p}\boldsymbol{\gamma} \qquad u^\dagger u = \frac{E}{M} \tag{62}$$

Inserting this result in (58) and the latter, in turn, in (56) yields

$$\begin{aligned}\delta^{(3)}(\mathbf{p}_2 - \mathbf{p}_1) = \lambda^2 \left(1 - \frac{M^2}{E^2}\right) \delta^{(3)}(\mathbf{p}_2 - \mathbf{p}_1) + \\ + \sum_{\alpha \neq N} \langle P(p_2)| Q_a{}^+ |\alpha_{\text{out}}\rangle \langle \alpha_{\text{out}}| Q_a{}^- |P(p_1)\rangle - \\ - \sum_\beta \langle P(p_2)| Q_a{}^- |\beta_{\text{out}}\rangle \langle \beta_{\text{out}}| Q_a{}^+ |P(p_1)\rangle\end{aligned} \tag{63}$$

At this point the various methods to derive the Adler-Weisberger sum rule depart. Following Fubini and Furlan, one would now have to take the limit $E \to \infty$. We will here show, however, that this is not necessary and that one can keep E finite throughout the calculations, thus deriving the whole continuous set of sum rules for varying E.

Here is the point where PCAC has to be used. Equations (40) and (48) together with (27) allow one to express matrix elements of $Q_a{}^\pm$ between an arbitrary state and a 1-proton state in the following way:

$$\langle \beta_{\text{out}}| Q_a{}^\pm |P(p)\rangle = \frac{\sqrt{2}M\mu^2\lambda}{gK(0)} \frac{1}{E_\beta - E_p} \langle \beta_{\text{out}}| \int d^3x \phi_\pi{}^\pm(x) |P(p)\rangle \tag{64}$$

The space-time dependence of the matrix elements of $\phi(x)$ is, as always, given by

$$\langle \beta_{\text{out}}|\phi_\pi{}^\pm(x)|P(p)\rangle = e^{ix(q_\beta - p)} \langle \beta_{\text{out}}|\phi_\pi{}^\pm(0)|P(p)\rangle \tag{65}$$

so that one has

$$\langle P(p_2)|\int d^3x\phi_\pi^{\pm}(x)|\alpha_{\text{out}}\rangle\langle\alpha_{\text{out}}|\int d^3x\phi_\pi^{\pm}(x)|P(p_1)\rangle =$$
$$= (2\pi)^3\delta^{(3)}(\mathbf{p}_2 - \mathbf{q}_\alpha)\delta^{(3)}(\mathbf{p}_2 - \mathbf{p}_1)|\langle P(p_1)|\phi_\pi^{\pm}(0)|\alpha_{\text{out}}\rangle|^2 \quad (66)$$

Dropping the common δ-function and dividing by λ^2, the sum rule thus becomes

$$\frac{1}{\lambda^2} = 1 - \frac{M^2}{E^2} - \sum_{\alpha\neq N}\frac{2M^2\mu^4}{g^2K^2(0)}(2\pi)^6\delta^{(3)}(\mathbf{p} - \mathbf{q}_\alpha)\frac{|\langle P(p)|\phi_\pi^{+}(0)|\alpha_{\text{out}}\rangle|^2}{(E_\alpha - E)^2} +$$
$$+ \text{ similar term with } \phi^+ \to \phi^- \quad (67)$$

E_α and $\mathbf{q}_\alpha$ are related to the 'effective mass' (total energy in the centre of momentum system) of the intermediate state by

$$E_\alpha{}^2 - \mathbf{q}_\alpha{}^2 = M_\alpha{}^2 \quad (68)$$

For every particular intermediate state α the sum in (67) means integration over all momenta and summation over all spins. From this, the integration over the centre of mass momentum can be singled out, leaving integration over relative ('internal') momenta only. One can then split off the normalization factors from the matrix elements as follows:

$$\langle P(p)|\phi_\pi^{+}(0)|\alpha_{\text{out}}\rangle = \frac{1}{(2\pi)^3}\left(\frac{MM_\alpha}{EE_\alpha}\right)^{\frac{1}{2}}F_\alpha{}^+ \quad (69)$$

where F^+ is a scalar under Lorentz transformations. In this way, (67) becomes

$$\frac{1}{\lambda^2} = 1 - \frac{M^2}{E^2} - \frac{2M^2\mu^4}{g^2K^2(0)}\int d^3q_\alpha\int_{M+\mu}^{\infty} dW\sum_{\substack{\alpha\\ \text{int}}}\delta(W - M_\alpha)\delta^{(3)}(\mathbf{p} - \mathbf{q}_\alpha)$$
$$\frac{MM_\alpha}{EE_\alpha}\frac{1}{(E - E)^2}[|F_\alpha{}^+|^2 - |F_\alpha{}^-|^2] \quad (70)$$

Here, an integration over an energy variable W has been artificially inserted. Note that $M + \mu$ is the lowest value M_α can take on. 'Int' under the summation symbol denotes the fact that the integration over the centre of mass momentum has been taken out. It is now convenient to define a new Lorentz invariant function by

$$K^{\pm}[W,(p - q_\alpha)^2] = \sum_{\substack{\alpha\\ \text{int}}}\delta(W - M_\alpha)|F_\alpha{}^{\pm}|^2 \quad (71)$$

W and $(p - q_\alpha)^2$ are the only invariants available, so that $K^\pm$ can only depend on them. The momentum δ-function together with (68) and $E^2 - \mathbf{p}^2 = M^2$ allow for the following replacement:

$$E_\alpha = (W^2 + E^2 - M^2)^{1/2} \equiv R \tag{72}$$

so that one ends up with

$$\frac{1}{\lambda^2} = 1 - \frac{M^2}{E^2} - \frac{2M^2\mu^4}{g^2K^2(0)} \int_{M+\mu}^{\infty} \mathrm{d}W \, \frac{MW}{ER} \, \frac{1}{(R-E)^2} \{K^+[W,(R-E)^2] - K^-[W,(R-E)^2]\} \tag{73}$$

Though the sum rule has now taken on a nice and explicit form, it remains to relate the term within the braces to experimentally accessible quantities. To achieve this, we now derive the total cross-section for an off-shell pion of mass k^2 scattered on a proton. For an arbitrary final state α, the scattering matrix element is given by

$$S_\alpha^\pm = \langle \alpha_{\text{out}} | P(p)\pi^\pm(k)_{\text{in}} \rangle \tag{74}$$

The pion field operator

$$\phi_\pi^\pm(x) = \int \mathrm{d}^3k [f_k(x)a^\pm(k) + f_k^*(x)a^{\mp\dagger}(k)] \tag{75}$$

where

$$f_k(x) = \frac{1}{(2\pi)^{3/2}} \frac{1}{\sqrt{(2k^0)}} \, \mathrm{e}^{-ikx} \tag{76}$$

shall be taken out of the 'in'-state with the help of

$$a^{\pm\dagger}(k) = -\mathrm{i} \int \mathrm{d}^3x f_k(x) \frac{\overleftrightarrow{\partial}}{\partial t} \phi_\pi^\mp(x) \tag{77}$$

where the differential operator with the double arrow is defined by

$$A \overleftrightarrow{\partial} B = A\partial B - (\partial A)B$$

$$|P(p)\pi^\pm(k)_{\text{in}}\rangle = \lim_{t\to-\infty} a^{\pm\dagger}(k)|P(p)\rangle = \lim_{t\to-\infty} -\mathrm{i}\int \mathrm{d}^3x f_k(x) \frac{\overleftrightarrow{\partial}}{\partial t} \phi_\pi^\mp(x)|P(p)\rangle$$

$$= \lim_{t\to+\infty} -\mathrm{i}\int \mathrm{d}^3x f_k(x) \frac{\overleftrightarrow{\partial}}{\partial t} \phi_\pi^\mp(x)|P(p)\rangle + \mathrm{i}\int_{-\infty}^{\infty} \mathrm{d}t \frac{\mathrm{d}}{\mathrm{d}t} \int \mathrm{d}^3x f_k(x) \frac{\overleftrightarrow{\partial}}{\partial t} \phi_\pi^\mp(x)|P(p)\rangle$$

$$= |P(p)\pi^\pm(k)_{\text{out}}\rangle + \mathrm{i}\int \mathrm{d}^4x [f_k(x)\ddot{\phi}_\pi^\mp(x) - \ddot{f}_k(x)\phi_\pi^\mp(x)]|P(p)\rangle \tag{78}$$

But since $f_k(x)$ obeys the Klein-Gordon equation, one has

$$\ddot{f}_k(x) = \Delta f_k(x) - \mu^2 f_k(x) \tag{79}$$

By partial integration one can throw the Laplace operator onto $\phi(x)$ to obtain

$$|P(p)\pi^{\pm}(k)_{\text{in}}\rangle = |P(p)\pi^{\pm}(k)_{\text{out}}\rangle + i\int d^4x[f_k(x)(\Box + \mu^2)\phi_\pi{}^{\mp}(x)]|P(p)\rangle \tag{80}$$

Inserting this in (74) and writing out the space-time dependence of the matrix element of $\phi_\pi{}^{\pm}(x)$ yields, after integration over d^4x:

$$S_\alpha{}^{\pm} = \delta_{\alpha,\text{in}} - i(2\pi)^{5/2}\frac{1}{\sqrt{(2k^0)}}\delta^{(4)}(k + p - q_\alpha)(k^2 - \mu^2)\langle\alpha_{\text{out}}|\phi^{\mp}(0)|P(p)\rangle \tag{81}$$

The relation of the S-matrix element to cross-section is most easily achieved via the definition of the invariant reaction matrix T[11].

$$S_\alpha{}^{\pm} = \delta_{\alpha,\text{in}} + i(2\pi)^{-1/2}(8E_\alpha k^0 E)^{-1/2}\delta^{(4)}(k + p - q_\alpha)T^{\pm} \tag{82}$$

The total cross-section is thereby related to $T^{\pm}$ by

$$\sigma_{\text{tot}}{}^{\pm} = \frac{\pi}{\text{flux}}\frac{1}{2k^0E}\int\frac{d^3q_\alpha}{2E_\alpha}\sum_{\substack{\alpha\\ \text{int}}}\delta^{(4)}(q_\alpha - k - p)|T_{fi}{}^{\pm}|^2 \tag{83}$$

In the centre of mass system the flux is given by

$$\text{flux} = \frac{|\mathbf{p}|}{k^0} + \frac{|\mathbf{p}|}{E} \tag{84}$$

where $\mathbf{p}$ is the centre of mass 3-momentum. In this system the following three relations hold

$$\begin{aligned} E^2 &= M^2 + \mathbf{p}^2 \\ k^2 &= k_0{}^2 - \mathbf{p}^2 \\ W &= q_\alpha{}^0 = k^0 + E \end{aligned} \tag{85}$$

so that one can express E, k_0 and $|\mathbf{p}|$ through W and k^2 in the following way:

$$\begin{aligned} E &= (M^2 + W^2 - k^2)/2W \\ k^0 &= (W^2 - M^2 + k^2)/2W \\ |\mathbf{p}| &= w(M^2, W^2, k^2)/2W \end{aligned} \tag{86}$$

where $w(M^2, W^2, k^2)$ is the famous totally symmetric function that occurs in every kinematical analysis

$$w(M^2, W^2, k^2) = [M^4 + W^4 + k^4 - 2(M^2W^2 + M^2k^2 + W^2k^2)]^{1/2} \tag{87}$$

Inserting (81) and (86) in (82) and (83) and recalling (71) yields

$$\sigma_{\text{tot}}^{\pm}(W, k^2) = 2\pi \frac{M(k^2 - \mu^2)^2}{w(M^2, W^2, k^2)} K^{\pm}(W, k^2) \tag{88}$$

Notice that this is a covariant equation which therefore holds in any arbitrary system.

For the pion-mass equal to $\Delta \equiv R - E$, one can now insert the total cross-section in the sum rule (73) to obtain

$$\frac{1}{\lambda^2} = 1 - \frac{M^2}{E^2} - \frac{M^2\mu^4}{g^2K^2(0)}\frac{1}{\pi}\int_{M+\mu}^{\infty} \mathrm{d}W \frac{Ww(W^2, M^2, \Delta^2)}{E(\Delta + E)\Delta^2(\Delta^2 - \mu^2)^2}[\sigma_{\text{tot}}^{+}(W, \Delta^2) - \sigma_{\text{tot}}^{-}(W, \Delta^2)] \tag{89}$$

This is the final form of the sum rule for finite E. It is really a continuous set of sum rules, depending on the external parameter E which can vary between M and ∞. But since $\Delta = R - E$ depends on W (recall 72), the integration over W is an integration over total cross-sections whose pion-mass variable changes with the integration variable; this is an unwanted feature. It can be dispensed with by taking the limit $E \to \infty$; the necessity of this limit appears now in a new light.

The limit $E \to \infty$ may be taken under the integral sign only if the integral converges uniformly at the upper limit of integration. Assuming this, one obtains with

$$E\Delta = \tfrac{1}{2}(W^2 - M^2) + 0(1/E) \tag{90}$$

the Adler-Weisberger sum rule

$$1 - \frac{1}{\lambda^2} = \frac{4M^2}{\pi g^2 K^2(0)} \int_{M+\mu}^{\infty} \mathrm{d}W \frac{W}{W^2 - M^2}[\sigma_0^{+}(W) - \sigma_0^{-}(W)] \tag{91}$$

Here, $\sigma_0^{\pm}(W) = \sigma_{\text{tot}}^{\pm}(W, 0)$ are the total cross-sections for zero mass pions. This and the factor $K^2(0)$ in front of the integral require still some analytic continuation to the mass-shell before numbers are obtained from (91). There are some minor uncertainties in this procedure and thus the values quoted by Adler and Weisberger differ slightly.

$$\text{Adler: } |\lambda| = 1{\cdot}24$$
$$\text{Weisberger: } |\lambda| = 1{\cdot}16 \tag{92}$$

This compares very favourably with the experimental value[17]

$$\lambda = 1{\cdot}198 \pm 0{\cdot}022 \tag{93}$$

It is worth mentioning that one of the nice features of (91) is that high energy values of $\sigma_0^{\pm}(W)$, which are not known, are suppressed by the weight factor in front of the bracket. The Pomeranchuk theorem guarantees that the term in the bracket vanishes at $W \to \infty$ so that the integral converges nicely.

Noticing the quantitative agreement of the calculation with the experimental value is not yet all; perhaps even more important is the qualitative understanding one derives from (91). Before the calculation of Adler and Weisberger, the weak axial vector coupling constant renormalization persisted to be infinite. But even when tricks with cut-offs were played, it tended to decrease the value of λ and no plausible argument could produce a λ larger than one. Now, we know that $\lambda > 1$ stems from the fact that in the low energy region the π^+p total cross-section is larger than the π^-p total cross-section, mainly due to the famous (3,3) resonance at 1236 MeV total energy. In fact, if we only insert this resonance in (91), we get $\lambda \approx 1{\cdot}44$ and it is the contribution from higher resonances that damps the value down to 1·2. This may shed some light on the SU(6) calculations which can be reproduced with a current algebra technique in which the intermediate states are restricted to the 56-representation[18] (i.e. the baryon octet plus resonance decuplet thus excluding higher resonances). The resulting value of $\lambda = 5/3$ is also too large.

From the sum rule (91), we can derive relations for the S-wave scattering lengths of pion nucleon scattering at threshold[19] if we make the additional assumption that the total cross-sections $\sigma_0^{\pm}(W)$ for scattering of zero-mass pions do not depend explicitly on the pion mass. With this assumption in mind, the derivative of (91) with respect to the pion-mass reads

$$\sigma_0^{+}(M + \mu) - \sigma_0^{-}(M + \mu) = 0 \tag{94}$$

This equation should at least hold approximately true for scattering of physical pions. Hence, for the S-wave scattering lengths a_{2I} for total isospin I one has

$$\tfrac{1}{9}(2a_1 + a_3)^2 - a_3^2 = 0 \tag{95}$$

which requires

$$\text{either} \quad a_1 - a_3 = 0 \quad \text{or} \quad a_1 + 2a_3 = 0 \tag{96}$$

Experimentally, the second condition is rather well satisfied[20]

$$a_1 + 2a_3 = (-0{\cdot}035 \pm 0{\cdot}012)\mu^{-1} \tag{97}$$

which has to be compared to the individual scattering lengths of order $0{\cdot}2\mu^{-1}$. The result (96) has also been derived by S. Weinberg with different methods[21].

4 Leptonic Kaon-Decays

The method of current algebra together with PCAC proved to be powerful in many ways. In this section, we discuss the possibility of relating to each other general matrix elements of the form $\langle\beta,\pi|J_\mu|\alpha\rangle$ and $\langle\beta|J_\mu|\alpha\rangle$ provided the extrapolation to zero mass and zero momentum of the pion is meaningful[23].

The method used here works again essentially in three steps.

1*st step:* By means of the ordinary reduction technique, the matrix element $\langle\beta,\pi|J_\mu|\alpha\rangle$ is turned into an integral over $\langle\beta|[\phi_\pi,J_\mu]|\alpha\rangle$.

2*nd step:* PCAC relates the latter to the derivative of $\langle\beta|[j_\nu,J_\mu]|\alpha\rangle$.

3*rd step:* Current algebra expresses the matrix element of the commutator by $\langle\beta|J_\mu|\alpha\rangle$ and the chain is completed.

In the application to leptonic Kaon-decays, $|\alpha\rangle$ will be a K^+ state of momentum q, say. J_μ has to be identified with $J_\mu{}^+(x)$ of equation (28). (Note that it contains both vector and axial vector.) We will only use integrated currents (recall equation 30) to avoid ambiguities arising from so-called Schwinger terms[22].

Let us now turn to the details of the computation. A matrix element of the form

$$\langle\beta,\pi^0(q')_{\text{out}}|S^+|K^+(q)\rangle = \langle\beta|a_{\text{out}}{}^0(q')S^+|K^+(q)\rangle \tag{98}$$

can be expressed through a retarded commutator if one subtracts

$$\langle\beta|S^+A_{\text{out}}{}^0(q')|K^+\rangle = 0 \tag{99}$$

so that

$$\langle\beta,\pi^0(q')_{\text{out}}|S^+|K^+(q)\rangle = \lim_{t\to\infty}\theta(t)\langle\beta|[a^0(q'),S^+]|K^+(q)\rangle \tag{100}$$

Using the hermitian conjugate of (77) and the procedure outlined in (78), one obtains

$$\langle\beta,\pi^0(q')_{\text{out}}|S^+|K^+(q)\rangle = \mathrm{i}\int \mathrm{d}^4x\,\frac{1}{(2\pi)^{3/2}}\,\frac{1}{(2q_0{}')^{1/2}}\,\mathrm{e}^{-q'x}(\Box+\mu^2)$$
$$\theta(x^0)\langle\beta|[\phi^0(x),S^+]|K^+(q)\rangle \tag{101}$$

This concludes the first of the three steps. Before we go on to use PCAC, we change the d'Alambertian operator into $-q'^2$ by means of a partial integration. Since it contains a partial integration with respect to time, surface terms cannot uncritically be neglected. This is immediately clear from (78). If we uncritically perform a partial integration with respect to time there and if we drop surface terms, we could prove that no scattering can ever take place. In the present case, however, these terms do not contribute; we will return to this point later on.

In order to get rid of the exponential which is unwanted in the relation, the limit $q_\mu' \to 0$ is taken in (101). This requires that the pion which is taken out from the outgoing state is unphysical with vanishing 4-momentum. It is for this reason that the method is called 'soft pion method'.

$$\lim_{q_\mu' \to 0} (2\pi)^{3/2}(2q_0')^{1/2}\langle\beta, \pi^0(q')_{\text{out}}|S^+|K^+(q)\rangle = i\mu^2 \int d^4x \, \theta(x^0)\langle\beta|[\phi^0(x), S^+]|K^+(q)\rangle \quad (102)$$

PCAC in the form of (40) and (48) can now be used.

$$\lim_{q' \to 0} (2\pi)^{3/2}(2q_0')^{1/2}\langle\beta, \pi^0(q')_{\text{out}}|S^+|K^+(q)\rangle = = \frac{gK(0)}{M\lambda} \int d^4x \theta(x^0) \frac{\partial}{\partial x_\mu} \langle\beta_{\text{out}}|[J_\mu^5(x), S^+]|K^+(q)\rangle \quad (103)$$

Note the difference of $\sqrt{2}$ to (48) because a neutral pion is now involved.

Once again, a partial integration with respect to time has to be performed in the integral

$$I = \int d^4x \theta(x^0) \frac{\partial}{\partial x_\mu} j_\mu^5(x) = \int d^4x \theta(t) \frac{\partial}{\partial t} j_0^5(x) \quad (104)$$

Keeping the surface terms, partial integration yields (recall 53):

$$I = \lim_{t \to \infty} \int d^3x \theta(t) j_0^5(x) = \int d^4x \delta(t) j_0^5(x) = Q_a(\infty) - Q_a(0) \quad (105)$$

A matrix element of $Q_a(t)$ has the following time dependence

$$\langle\alpha|Q_a(t)|\beta\rangle = \langle\alpha|Q_a(0)|\beta\rangle e^{it(E_\alpha - E_\beta)} \quad (106)$$

Hence at $t \to \infty$ the matrix element oscillates if $E_\alpha \neq E_\beta$ and can be dropped according to the usual rules of quantum mechanics. That this reasoning may be dangerous has been shown by Okubo[24]. A rigorous derivation of the same result has been given by B. Schroer[25].

With all this in mind, (103) becomes

$$\lim_{q' \to 0} (2\pi)^{3/2}(2q_0')^{1/2}\langle\beta, \pi^0(q')_{\text{out}}|S^+|K^+(q)\rangle = = -\frac{gK(0)}{M\lambda} \langle\beta_{\text{out}}|[Q_a^0, S^+]|K^+(q)\rangle \quad (107)$$

In the third step of the method, it remains to use the following integrated current commutator

$$[Q_a, S^+] = \tfrac{1}{2}S^+ \tag{108}$$

In the language of SU(3), this is the following commutator

$$[H_1, E_2] = \frac{1}{2\sqrt{3}}\, E_2 \tag{109}$$

but it should be stressed again that we do not assume SU(3) to be a strict symmetry. Insertion of (108) into (107) yields the basic equation

$$\lim_{q'\to 0} (2\pi)^{3/2}(2q_0')^{1/2}\langle \beta, \pi^0(q')_{\text{out}}|S^+|K^+(q)\rangle = $$

$$= -\frac{gK(0)}{2M\lambda}\langle \beta_{\text{out}}|S^+|K^+(q)\rangle \tag{110}$$

Specializing β to various states directly yields relations between measurable quantities.

First, let us put $\beta = 0$. The left-hand side of (110) is then given by (36), we now define f_K so that the integrated current matrix element becomes

$$\langle 0|S^+|K^+(q)\rangle = (2\pi)^{3/2}\frac{1}{(2q_0)^{1/2}} f_K m_K{}^2 \delta^{(3)}(\mathbf{q}) \tag{111}$$

f_K can be determined from the decay $K^+ \to \mu^+ + \nu_\mu$ via an equation entirely analogous to (39).

With these definitions, (110) becomes

$$f_+(0,0) + f_-(0,0) = -\frac{gK(0)}{2M} f_K m_K \tag{112}$$

This is the famous equation derived by Callan, Treiman and Mathur, Okubo, Pandit[23]. The double 0 argument of $f_\pm$ indicates that not only the 3-momentum transfer vanishes but also the pion mass is continued to 0. The equation works surprisingly well for physical values of $f_\pm(0)$ and $K(0) = 1$. From $K^+ \to \mu^+ + \nu_\mu$, one obtains

$$f_K = 0{\cdot}070 \pm 0{\cdot}001 \tag{113}$$

Insertion of the values[26]

$$f_-/f_+ = 0{\cdot}46 \pm 0{\cdot}27$$

$$|f_+| = 0{\cdot}16 \pm 0{\cdot}01 \tag{114}$$

in (112) yields

$$|f_K| = 0{\cdot}074 \pm 0{\cdot}014 \tag{115}$$

in exceedingly good agreement with the value (113).

Nowhere in the derivation of (110) did the particular form of the ingoing K^+ state enter. It is therefore possible to change it into *a* π^+, provided the commutator (108) is changed into

$$[Q_a{}^0, I^+] = Q_a{}^+ \tag{116}$$

But for the decay $\pi^+ \to \pi^0 + e^+ + \nu_e$, the conserved vector current theory determines the form factors to be

$$f_+{}^\pi(0) = \sqrt{2} \qquad f_-{}^\pi(0) = 0 \tag{117}$$

so that the corollary of (112) is simply the Goldberger-Treiman relation (50). Dividing (112) by (50) yields

$$f_+(0,0) + f_-(0,0) = \frac{m_K}{\sqrt{2}\mu}\frac{f_K}{f_\pi} \tag{118}$$

This beautiful equation clearly demonstrates the power of the current algebra method.

Of course, one can now proceed to put β equal to π^0 and relate the K_{e3} decay to $K^+ \to 2\pi + e^+ + \nu_e$. This matter is extensively discussed in the original literature[27] and we therefore rather turn to a comparison between the methods of Sections 3 and 4. To this end, the commutator (108) shall now be taken between a K^+ state to the right and the vacuum to the left.

$$\langle 0|[Q_a{}^0, S^+]|K^+(q)\rangle = \tfrac{1}{2}\langle 0|S^+|K^+(q)\rangle \tag{119}$$

Suppose, for the moment, that the sum rule 'is saturated by 1-particle states', i.e. that the contribution from many-particle intermediate states can be neglected.

$$\int d^3k \langle 0|Q^0|\pi^0(k)\rangle\langle\pi^0(k)|S^+|K^+(q)\rangle = \tfrac{1}{2}\langle 0|S^+|K^+(q)\rangle \tag{120}$$

Recalling (36), (53), and (67), (120) leads to

$$(m_K + \mu)f_+(0) + (m_K - \mu)f_-(0) = \frac{2m_K{}^2}{\mu}\frac{f_K}{f_\pi} \tag{121}$$

The main difference between (121) and (118) is that in (121) the form factors are now taken at the physical pion mass. In this respect, (121) is better than (118) but quantitatively it is much worse, showing that there is considerable leakage from the 1-particle states.

We have here demonstrated only two of the many applications of current algebra. Long-lasting open questions of theoretical physics have been answered and new fields are being explored. In the field of strong interactions, a theory is entirely lacking, so that the best result to be achieved at present is a correlation of various measurable quantities rather than a computation from a few basic constants such as masses and coupling constants. In this limited field of wishes, current algebra is a very useful tool.

Acknowledgement

Part of these lecture notes have been accumulated while the author was visiting the Institute of Theoretical Physics in Gothenburg; I am grateful to Prof. Jan Nilsson for the kind hospitality at Chalmers Institute of Technology.

References

1. Gell-Mann, M., *Phys. Rev.*, **125**, 1067 (1962).
2. Balachandran, A. P., and Pietschmann, H., *Nucl. Phys.*, **43**, 321 (1963); Biritz, H., and Pietschmann, H., *Proc. Siena Int. Conf. on Elementary Particles*, Vol. 1, 403 (1963); Kummer, W., Pietschmann, H., Balachandran, A. P., *Ann. Phys.* (*N.Y.*), **29**, 161 (1964); Pietschmann, H., *Acta Phys. Austriaca, Suppl. I*, **1964**, 92; Stremnitzer, H., *Acta Phys. Austriaca*, **20**, 39 (1965).
3. Fubini, S., and Furlan, G., *Physics*, **1**, 229 (1965).
4. Weisberger, W., *Phys. Rev. Letters*, **14**, 1047 (1965); Adler, S., *Phys. Rev. Letters*, **14**, 1051 (1965).
5. e.g. Renner, B., *Lectures on Current Algebras*, Rutherford Laboratory Report RHEL/R 126 (1966) (contains 219 references).
6. e.g. Moffat, J., *Acta Phys. Austriaca, Suppl. III*, **1966**, 113; or Bell, J. S., *Proc. of the* 1966 *CERN School of Physics, CERN* 66–29, Vol. 1 (1966).
7. Bjorken, J. D., and Drell, S. D., *Relativistic Quantum Mechanics*, McGraw-Hill Book Comp., New York, 1966.
8. e.g. Drell, S. D., and Zachariasen, F., *Electromagnetic Structure of Nucleons*, Oxford University Press, 1961.
9. The author is indebted to Klaiber, B., and Nilsson, J., for a discussion on this point.
10. e.g. Gell-Mann, M., and Ne'eman, Y., *The Eightfold Way*, Benjamin, New York, Amsterdam, 1964.
11. Nilsson, J., and Pietschmann, H., *An Introduction to Weak Interaction Physics*, McGraw-Hill Book Comp., New York (in preparation).
12. Gell-Mann, M., and Levy, M., *Nuovo Cimento*, **16**, 705 (1960).
13. Bernstein, J., Fubini, S., Gell-Mann, M., Thirring, W., *Nuovo Cimento*, **17**, 757 (1960).
14. Goldberger, M., and Treiman, S., *Phys. Rev.*, **110**, 1178 (1958).
15. Adler, S. L., *Phys. Rev.*, **140**, B736 (1965); Weisberger, W. I., *Phys. Rev.*, **143**, 1302 (1966).
16. For further details cf. Marshak, R. E., and Okubo, S., *Nuovo Cimento*, **19**, 1226 (1961) and reference 1.

17. Conforto, G., *Acta Phys. Hungarica*, **22,** 15 (1967).
18. Lee, B. W., *Phys. Rev. Letters*, **14,** 676 (1965).
19. Olesen, P., and Pietschmann, H., *Nuovo Cimento*, **49,** 673 (1967).
20. Samaranayake, V. K., and Woolcock, W. S., *Phys. Rev. Letters*, **15,** 936 (1965).
21. Weinberg, S., *Phys. Rev. Letters*, **17,** 616 (1966).
22. Schwinger, J., *Phys. Rev. Letters*, **3,** 296 (1959).
23. Callan, C. G., and Treiman, S. B., *Phys. Rev. Letters*, **16,** 153 (1966); Mathur, V. S., Okubo, S., Pandit, L. K., *Phys. Rev. Letters*, **16,** 371 (1966).
24. Okubo, S., *Nuovo Cimento*, **41A,** 586 (1966).
25. Schroer, B., *On the Asymptotic Behaviour of Generalized Charges* (preprint).
26. Trilling, H., *Proc. Int. Conf. Weak Int.*, *ANL*, 1965.

PC and *T* Violation

J. G. Taylor

1 Introduction

The discrete symmetries of parity or space reflection *P*, time reversal *T*, and charge conjugation *C* present an interesting puzzle. Both *P* and *C* were found to be violated by the weak interactions back in 1957[1], whilst more recently[2] it was found that *CP* was violated in the decay of the long-lived neutral meson K_2^0. The *CPT* theorem[1] then implies that *T* is violated in K_2^0 decay. Thus each of the discrete symmetry operations *P*, *C* and *T* are violated by at least part of the Hamiltonian describing the total interaction between elementary particles. These violations are, however, at least at first glance, quite dissimilar. The violation of *P* and *C* by the weak interactions is very well described by the 'maximally' violating interaction $(V - A)$[1]. The same notion of 'maximal' violation of *T* or *CP* in K_2^0 decay does not seem applicable, especially since the ratio of *CP*-violation to *CP*-conservation in K^0 decays is of order one in a thousand in the amplitudes[2]. Further there is no other experimental evidence which unequivocally shows a violation of *T* or *CP*[3]. There have been many attempts to understand this apparent difference between *P*, *C* and *CP* or *T*; this paper is a further contribution to this problem. We present here a number of different ways in which *T* or *PC*-violation could occur in a 'maximal' fashion. We cannot necessarily distinguish between the different models which we present for *T*-violation at present, but will have to wait for further experimental and theoretical developments.

The arrangement of the paper is as follows. In the next section the present experimental situation is briefly reviewed for evidence and limits on *CP* and *T*-violation in various processes. In the following section the problem of defining discrete symmetry operators such as *P*, *C* or *T* when they are not exact symmetries of the physical system is discussed. The basic problem here is one of lack of precise definition of these operators in such a situation. We make quite precise the operators *P*, *C* and *T* which we will use, though this definition is not necessarily unique. We then relate the manner in which the discrete operators can be regarded as symmetry operators to the concept of reciprocity, based on the relation between moduli of expectation values.

This idea of relating moduli of expectation values but not the expectation values themselves is very natural from an observational point of view. However, if the states considered may be chosen arbitrarily, then a theorem of Wigner[4] implies that the expectation values are also equal, so that a symmetry of the system would result. However we have to make quite clear what is meant by a symmetry. We take it here to mean the equality of S-matrix elements between pairs of states and their transformed states under the discrete operation under consideration. We may weaken this symmetry to be a reciprocity by only equating moduli of S-matrix elements between pairs of states and their transformed states. The Wigner theorem only applies to this case to remove the moduli signs in such equalities if S commutes with the discrete operation, which is the same thing as requiring the operation to be a symmetry in the first place. Thus in the contrary case, the phases that arise when the moduli signs are removed from these equalities are now a measure of the violation of the corresponding symmetry operation; we can define the notion of maximal violation of a symmetry if these phases are maximal. This use of reciprocity to describe the violations of C, P and T or their combinations is discussed in Section 4. There are various types of reciprocity which may be imposed; they are discussed and various of their consequences obtained[5].

A completely different model for explicit CP-violation is that due to the existence of a magnetic monopole when the monopole current is chosen to be an axial vector current[6,7]. Such a model has many difficulties, and we describe an attempt to remove some of them in Section 5. Our basic conclusion in this section is that magnetic monopoles may exist but with properties very different from those ascribed to them according to classical arguments, and that they are possible sources of CP-violation of the 'mismatch' kind discussed by T. D. Lee[8].

2 The Experimental Situation

We start off, then, with a brief survey of the experimental results on violations or limits of violation for each of the discrete symmetries P, C, T and their various pairwise combinations, and of PCT. The interpretation of the experimental results at this level is not affected by the difficulty in definition of P, C or T when they are violated, as mentioned earlier; we are considering the ratios of conserving and violating parts of amplitudes in a phenomenological manner. It is only when we wish to relate a particular theory of violation to the experimental results that care must be taken that the phenomenological discrete operators are correctly related to the more carefully defined discrete operators in that particular theory. Thus we can safely delay the precise definition of P, C and T till the next section, and take these operators to be the phenomenological ones:

(i) P acting on the wave function of a system of particles of given momenta and helicities changes the signs of all the momenta and helicities.
(ii) C acting on such a wave-function changes the signs of the charges of all the particles.
(iii) T acting on such a wave-function changes the momenta of all the particles, at the same time complex conjugating the wave-function.

We consider the different parts of the interaction Hamiltonian separately.

2.1 *Strong interactions*

P: Let the total amplitude A for a process be regarded as the sum of a parity conserving term A_s and a parity violating one, A_p: $A = A_s + A_p$. Then the observation[9] of circular polarization of the order of 10^{-4} in a γ-transition in Ta^{181} implies

$$|A_p/A_s| \sim 10^{-6}$$

which improves on the limit $|A_p/A_s| < 10^{-4}$ of Tanner[10], and is of the right order of magnitude to correspond to parity violation by the weak interaction only.

T: No T-violation has been observed in purely strong reactions. The best limit on the validity of T-invariance comes from experiments designed to test the principle of detailed balance, that is that the rates of $A + B \rightarrow C + D$ and the reverse reaction are equal. The limits from this are 2% in p-p scattering[11] and certain nuclear reactions[12,13], the best limit being 0·4% from the reactions $Mg^{24} + d \rightleftharpoons Mg^{25} + p$[14].

C: We may use the CPT theorem and the above experimental results on P, T to conclude that C is conserved to a few percent in strong interactions. We may alternatively test C directly by considering the relation between the energy distribution of particles and their antiparticles in high energy reactions in which they are produced in pairs. Thus the energy distribution of π^+ and π^- or of K^+ and K^- are required to be equal in $p\bar{p}$ annihilation; this is so to 1% for the pion reactions, 2% for the kaon case[15].

CPT: We may combine results on C, P, T conservation to conclude that CPT is conserved to within a few percent in strong interactions.

2.2 *Electromagnetic interactions*

T violation is particularly interesting in these interactions following the suggestions[16] that the CP-violation in K_2^0 decay occurs through electromagnetic violation.

P: From the experiment of Boehm and Kankeleit[9] we may conclude that P is conserved electromagnetically to one part in 10^4: if we write $A_p = \alpha A_p^{(el)}$ where α is the fine structure constant, then $|A_p^{el}/A_s| < 10^{-4}$.

C: There is no evidence that C is violated in electromagnetic interactions; the evidence that it is conserved for hadron-photon interactions is, however, weak. There are two types of tests for electromagnetic C-violation. The first involves determining the rate of a process which is forbidden by C conservation; the second involves the comparison of the distributions of particles and antiparticles produced in particles decaying through electromagnetic interactions. In the first class of tests are

$$\pi^0 \rightarrow 3\gamma \tag{1}$$

$$\eta^0 \rightarrow \pi^0 + e^+ + e^- \tag{2}$$

The limit on the rate of reaction (1) is expected to be[16] $\lesssim 10^{-6}$ of $\pi^0 \rightarrow 2\gamma$ from electromagnetic C-violation, whilst experimentally[17] this number is 10^{-4}; for reaction (2) the rate is expected to be[16] of the same order as that of $\eta^0 \rightarrow 2\gamma$ whilst experimentally[18]

$$\text{rate } (\eta^0 \rightarrow \pi^0 e^+ e^-)/(\text{rate } \eta^0 \rightarrow 2\gamma) < 10^{-2}$$

However, the theoretical prediction for reaction (2) depends on an arbitrary parameter, the mean square radius of the mixed charge distribution between η^0 and π^0, so it is difficult to determine any upper limit on the C-violation from (2).

Further reactions which only proceed due to electromagnetic C-violation are[16]

$$\phi^0 \rightarrow \omega^0 + \gamma$$

$$\phi^0 \rightarrow \rho^0 + \gamma$$

$$\omega^0 \rightarrow \rho^0 + \gamma$$

but experimental data on these reactions is rare. In the second class of tests are asymmetries in the $\pi^+ - \pi^-$ distributions in the decays

$$\eta^0 \rightarrow \pi^+\pi^-\pi^0 \tag{3}$$

$$\eta^0 \rightarrow \pi^+\pi^-\gamma \tag{4}$$

The results for (3) have a chequered history, but seem to give no asymmetry: the asymmetry for 10^4 events[19] is $(0{\cdot}3 \pm 1)\%$, smaller numbers of events give quite different results[20]. The results for reaction (4) also give no asymmetry[21].

The limits on *C*-violation for leptonic electromagnetic interactions are much stronger; the radiative corrections arising from the usual interaction $e\bar{\psi}_l\gamma_\mu\psi_l A_\mu$ are in extremely good agreement with experiment[22]. It is difficult to set simple limits on a possible *C*-violation in such interactions without choosing a very definite model of *C*-violation. However if we assume an additional *C*-violating interaction of strength e' between photons and leptons, which gives an added contribution of order $e'^2/\hbar c = a$ to the *g*-value of the electron, we need $a < {\cdot}005\alpha^2 \sim 10^{-6}$.

T: There is no indication of *T*-violation in lepton-electromagnetic interactions, and we may replace *C* by *T* in the remarks made in the previous paragraph. There is also a simple reason[16] why *T*-violation of hadrons will also be difficult to see. For a hadron of spin $\frac{1}{2}$ the electromagnetic form factor is

$$\langle N'|J_\mu(0)|N\rangle = \mathrm{i}e\bar{u}_{N'}[\gamma_\mu F_1 + \mathrm{i}(n_\mu' + n_\mu)F_2 + (n_\mu' - n_\mu)F_3]u_N \qquad (5)$$

where n, n' are the 4-momenta of the hadron in the states $|N\rangle$, $|N'\rangle$ respectively, and F_1, F_2, F_3 are functions of $(n - n')^2$ only. J_μ, the electromagnetic current, is assumed to be hermitian, which requires F_1, F_2 and F_3 all to be real. *T*-invariance requires F_1 and F_2 to be real, F_3 to be purely imaginary, so if F_3 is non-zero there is *T*-violation. However on the mass shell, current conservation requires F_3 to vanish, so that *T*-violation can only arise from off-mass shell effects. Thus in a nucleus any electromagnetic *T*-violation arising from F_3 being non-zero will be reduced by a factor of the order of (v/m_N), where v is the average velocity of a nucleon in the nucleus, m_N is the mass of a nucleon. This factor is of the order of 1/10 on average, so electromagnetic *T*-violation will never be more than a few percent arising from a single nucleon, unless a particular reaction is considered for which the *T*-conserving electromagnetic reaction is also reduced by a similar factor, or in which the off-mass shell reduction is removed.

The first alternative is being considered by N. Tanner[23], with the reaction

$$\gamma + O^{16} \rightleftharpoons \alpha + C^{12}$$

This reaction and its converse are first forbidden, so proceed by Coulomb scattering effects, which should be $0(\alpha^{\frac{1}{2}}) \sim 1/10$, in comparison with the *T*-violating effect $0(v/m_N) \sim 1/20$. Thus a possible 50% violation of detailed balance may occur, though this may be reduced by various numerical factors of 2π, etc.

The second alternative is involved in the suggestion of Barshay[24] to investigate the violation of detailed balance in

$$\gamma + d \rightleftharpoons n + p \tag{6}$$

For suitable γ-ray energy (290 MeV lab.) it is possible that the effective T-violation arises from the matrix element of the electromagnetic current J_μ between a nucleon and a (3-3)N^* resonance; the argument on reduction of T-violation effects related to (5) does not apply to such a matrix element, and even a 40% violation of detailed balance is possible.

Various theoretical estimates have been made of the amount of T-violation in

$$\gamma + N \rightarrow \pi + N$$

This has been done by putting in phenomenological T-violating terms in certain of the multipole amplitudes[25] or in γNN^* vertices[26]. The limits on the amount of allowed T-violation are not very strong, and T-violating phases of up to 20° seem to be allowed[25].

Detailed balance is not as strong as T-invariance, unless all possible reactions are investigated. We will discuss this situation in greater detail in the next section, but it is evident that tests of T-violation of a different nature should also be made. One of these is in the polarization of the Λ^0 in the electromagnetic decay

$$\Sigma^0 \rightarrow \Lambda^0 + e^+ + e^-$$

If p_+, p_-, p denote the 3-momenta of the e^+, e^-, Λ^0 respectively, and we define

$$\mathbf{n} = \mathbf{p}_\Lambda \times (\mathbf{p}_+ + \mathbf{p}_-), \quad \mathbf{n}' = \mathbf{p}_+ \times \mathbf{p}_-$$

and if in the decay $\Lambda^0 \rightarrow p + \pi^0$ we denote by $\mathbf{p}$ the 3-momentum of the proton, then T-violation corresponds to a non-zero value of $\langle \mathbf{p} \,.\, \mathbf{n}' \rangle$ or of $\langle \mathbf{p} \,.\, \mathbf{n} \rangle$, where $\langle\rangle$ denotes the weighted average. It is found that[27]

$$\langle \mathbf{p} \,.\, \mathbf{n}' \rangle = 0{\cdot}00 \pm 0{\cdot}03$$
$$\langle \mathbf{p} \,.\, \mathbf{n} \rangle = 0{\cdot}06 \pm 0{\cdot}03$$

2.3 *Weak interactions*

P and *C*: These are violated maximally, as is described by the $(V - A)$ current-current weak interaction[1].

PCT: It is a consequence of the *CPT* theorem[1] that stable particle and antiparticle masses are equal. It is known that $|m_{K^0} - m_{\bar{K}^0}| \sim 10^{-14} m_{K^0}$ so that if the K^0 and $\bar{K}^0$ were stable particles the allowed violation of *CPT* would be one part in 10^{14}. Due to the weak interactions the K^0 and $\bar{K}^0$ are unstable, so we may only conclude that *PCT*-violation due to the strong part of the Hamiltonian satisfies[28]

$$(PCT)_{\text{strong}} < 10^{-14}$$

whilst in the electromagnetic part the *PCT*-violation is

$$(PCT)_{\text{el}} < 10^{-12}$$

and in the weak part

$$(PCT)_{\text{weak}} < 10^{-8} \tag{7}$$

We note that this limit for the weak violation of *PCT* is only on the $\Delta S = 0$ part of the interaction $H_\omega{}^0$; nothing is known of the behaviour of the $\Delta S = 1$, $H_\omega{}^1$, part under *PCT* at the same level of accuracy. We may obtain

$$(PCT)_{\text{weak}} < 10^{-3}$$

for the $\Delta S = 1$ part of the weak interaction from the experimental value of the difference between the lifetimes τ^+ and τ^- of the K^+ and K^-,[29] which should be equal by the *PCT* theorem:

$$|\tau^+ - \tau^-| < 10^{-3}\tau^+$$

Similar limits also follow for the $\Delta S = 0$ part for the μ and π-meson lifetimes,

$$\text{for } \pi^+, \pi^-: |\tau^+ - \tau^-| < 3 \times 10^{-3}\tau^+$$

$$\text{and for } \mu^+, \mu^- \; |\tau^+ - \tau^-| < 10^{-3}\tau^+$$

Similarly *CPT* implies the equality of particle and antiparticle masses and of *g*-values for μ and μ^-, with experimental limits[28] for the μ-meson

$$g^+ - g^- \leqslant 6 \times 10^{-6}g^+$$

$$m^+ - m^- < 10^{-4}m^+$$

though the limits on *CPT*-violation of $H_\omega{}^0$ obtained from these does not reduce the limit (7).

CP: This is definitely violated in the decay of $K_2{}^0$ into $\pi^+\pi^-$ [21] and $\pi^0\pi^0$ [30] where it was found that the decay rates of $K_2{}^0$ in units of 10^6 s^{-1} are

$$K_2{}^0 \rightarrow \pi^+\pi^-: \quad 2 \times 10^{-2}$$

$$K_2{}^0 \rightarrow \pi^0\pi^0: \quad 5 \times 10^{-2}$$

There are numerous models which attempt to describe this violation of *CP*. The ones still consistent with the above experimental data are that the violation is due to *C*-violation in electromagnetic interactions[16] or that it is due to a *CP*-violating $\Delta I = 3/2$ effect in the $K \to 2\pi$ amplitude[31]. We will present further models in the following sections.

Another test of *CP* is that the ratio *R* of the rates of K_2^0 to $\pi^+ e^- \nu$ and $\pi^- e^+ \nu$ should be the same over the Dalitz plot[32]; experimentally the average value $\langle R \rangle$ of *R* over the Dalitz plot is $\langle R \rangle = 0{\cdot}93 \pm 0{\cdot}05$.

T: We may use the experimental limits on *CPT* and *CP*-violation to obtain a violation of *T* of at most one in 10^3 in $H_\omega^{\ 1}$. The definite *CP*-violation in $H_\omega^{\ 1}$ and the limit of *CPT*-violation in $H_\omega^{\ 1}$ is still consistent with *T*-invariance but *CPT*-violation in $H_\omega^{\ 1}$, though we have no definite model for this. A limit on *T*-violation in $H_\omega^{\ 1}$ is given[33] by the polarization of μ perpendicular to the decay plane in

$$K_2^0 \to \pi^- \mu^+ \nu$$

The value of this polarization is $-0{\cdot}05 \pm 0{\cdot}18$, consistent with no *T*-violation. The value of the relative phase between the β-decay coupling constants has also been measured[34] and found to be $\psi = 180 \pm 8°$, again consistent with *T*-invariance. There is also data on electric dipole moments of particles (which would be absent for *T*-conservation); there is no indication of any observable dipole moment, with the smallest upper limit being $2{\cdot}4 \times 10^{-20}$ for the neutron. On a rough estimate the possible dipole moments from various *T*-violating models are an order of magnitude smaller than this[36].

Finally it was thought for a time that there might be a violation of *T* in non-leptonic hyperon decays[5]. However recent data on the asymmetry parameters in Σ-decay have given far better agreement with the $\Delta I = \frac{1}{2}$ rule and also removed the possible *T*-violation. This is discussed briefly in Section 4.

3 Definition of *P*, *C* and *T* and Reciprocity

In the previous section we used the phenomenological definition of *P*, *C* and *T*. This definition is in terms of single particle wave-functions, and for example in the case of *P* corresponds to the simple operation of replacing the particle coordinates $\mathbf{r}$ by $-\mathbf{r}$. Parity violation corresponds to the statement that it is possible to observe a pseudoscalar under this operation of space inversion. But this must mean that the operator of space inversion cannot commute with the total Hamiltonian *H* for the system. This leads to the time dependence of *P*, if *P* is defined as the space-inversion operator $P(t)$ defined at a *particular* time *t* on any single particle wave-function:

$$[P(t)\psi](\mathbf{r},t) = \psi(-\mathbf{r},t)$$

Then $P(t) = e^{iHt}P(0)e^{-iHt}$ has a time dependence determined by the parity violating part of H. But since we have a time-dependent parity we cannot unambiguously use the phrase 'parity violating'. We have either to specify a particular time at which we have defined $P(t)$ and discussed its violation, or we have to consider a different parity operator which has no time dependence (and will not be the space inversion operator at all times).

It is possible to combine the above two possibilities by defining the parity operator as that acting on the incoming states, in other words $P(-\infty)$, or alternatively that acting on the outgoing states, $P(+\infty)$. There seems no reason to choose one or other of these values of t, nor indeed any finite value of t. It is customary to single out $t = 0$, and describe the amount of parity violation for $P(0)$, i.e. the commutator of H and $P(0)$; this does specify the amount of parity violation at all later times, though not in a manner which is most convenient for our discussion.

We will make a precise definition of P (and similarly of C and T) in terms of a complete set of eigenstates of the free part H_0 of the Hamiltonian H of the system. The definition will be the usual one that P applied to the single-particle states reverses particle momenta and helicities; this definition immediately extends to the many-particle eigenstates of H_0. Thus our operator P is defined on the whole of the space of physical states starting from

$$P|\mathbf{k},\lambda\rangle = \eta_P|-\mathbf{k},-\lambda\rangle \tag{8}$$

Similarly T and C are defined on single-particle states by

$$T|\mathbf{k},\lambda\rangle = \eta_T|-\mathbf{k},\lambda\rangle \tag{9}$$

$$C|\mathbf{k},\lambda\rangle = \eta_C|\mathbf{k},\lambda,C\rangle \tag{10}$$

where T is antiunitary in addition to (9), $\mathbf{k}$, λ denote particle momenta and helicities, $|C\rangle$ is the charge conjugate state to $|\rangle$, and η_P, η_C, η_T are arbitrary phases. We must remark here that the single-particle eigenstates of H_0 are taken to be the renormalized particle states, so that H_0 includes mass renormalization effects determined partly by the interactions entering H. We further remark that (8) and (9) make no reference to the definition of P, C or T as applied to unstable particles, such as the μ-meson. This lack of definition for such unstable particle states does not give rise to any further ambiguity in P, C or T in a theory in which unstable particles are regarded as resonances obtained from stable particles[37]; this is the case even in a theory in which field operators are introduced to describe the unstable particles. It is only if one wishes to introduce unstable particle states *per se* that definitions (8), (9), (10) will have to be extended, but then a precise definition should still be forthcoming in terms of these extended states, with a form very similar to (8), (9) and (10).

We wish to have P, C and T defined on a *complete* set of eigenstates of H_0. Since we are particularly interested in weak interactions we include in H_0 the free energies for neutrinos, *of both helicities*. It is usually stated that neutrinos only have one helicity, though it is more correct to say that the $(V - A)$ weak interaction only couples neutrinos of negative helicity to other particles; the positive helicity neutrinos are left non-interacting. Thus we define P, C and T on states composed of non-interacting hadrons, electrons, muons, photons and neutrinos of either helicity by obvious extensions of (8), (9) and (10) to product states; we expect such a set of states to form a complete set of states in the physical Hilbert space of states (neglecting bound states such as the deuteron, etc.).

We have thus achieved a precise definition of the discrete symmetry operators (DSO) P, C and T. We now wish to consider the relation between the DSO and the total Hamiltonian H or the S-matrix S. In particular we want to consider how we may characterize non-invariance of the system with Hamiltonian H, under the various DSO, in terms of S directly and avoid mention of H. In order to do that let V denote one of the DSO, and for any state $|A\rangle$ let $V|A\rangle = |A^v\rangle$ denote the V-transformed state derived from $|A\rangle$; V will be unitary (P,C) or antiunitary (T). For any states $|A\rangle$, $|B\rangle$.

$$\langle A|S|B\rangle = \langle A^v|VSV^{-1}|B^v\rangle \quad (V \text{ unitary}) \tag{11}$$

$$= \langle A^v|VSV^{-1}|B^v\rangle^* \quad (V \text{ antiunitary}) \tag{12}$$

Then evidently $[S,V] = 0$ iff (if and only if) $\langle A|S|B\rangle = \langle A^v|S|B^v\rangle$ for all states $|A\rangle$, $|B\rangle$ if V is unitary; if V is antiunitary this becomes

$$VSV^{-1} = S^+ \text{ iff } \langle A|S|B\rangle = \langle A^v|S|B^v\rangle^* \text{ for all } |A\rangle, |B\rangle$$

Thus V is an invariance operator or symmetry operator for the system iff V commutes with S (V unitary) or $VSV^{-1} = S^+$ (V antiunitary). Here we are using the term symmetry operator as one for which

$$\langle A|S|B\rangle = \langle A^v|S|B^v\rangle \quad (V \text{ unitary}) \tag{13}$$
$$= \langle A^v|S|B^v\rangle^* \quad (V \text{ antiunitary})$$

We may reduce the requirement (13) to the weaker statement between moduli:

$$|\langle A|S|B\rangle| = |\langle A^v|S|B^v\rangle| \tag{14}$$

If (14) is valid for any pairs of states $|A\rangle$, $|B\rangle$ of the system, we define V to be a *reciprocity* of the system. We note that Wigner's theorem[4] does not apply to (14) to enable one to remove the modulus signs from matrix elements of arbitrary operators, such as in (14). All we may conclude from (14) is that

$$\langle A|S|B\rangle = \langle A^v|S|B^v\rangle\, e^{i\phi_{AB}{}^v} \quad (V \text{ unitary}) \qquad (15)$$
$$= \langle A^v|S|B^v\rangle^*\, e^{i\phi_{AB}{}^v} \quad (V \text{ antiunitary})$$

where $\phi_{AB}{}^v$ is a real phase angle. If $\phi_{AB}{}^v = 0$ for arbitrary pairs of states $|A\rangle$, $|B\rangle$ then V is a symmetry; if $\phi_{AB}{}^v \neq 0$ for *some* pair $|A\rangle$, $|B\rangle$ then V is not a symmetry of the system. Thus the set of values of $\phi_{AB}{}^v$ for various pairs $|A\rangle$, $|B\rangle$ of states is a measure of the violation of the reciprocity V by the system.

We may even reduce the requirement (14) by limiting the possible pairs $|A\rangle$, $|B\rangle$ for which (14) is required to hold. Thus we may take only states of definite isotopic spin, or of definite particle type; we may even require (14) to hold only between states of the same orbital angular momentum in the centre-of-mass reference frame. We will define these relations and the corresponding phase angles $\phi_{AB}{}^v$ in the next section.

We note that we cannot expect (14) to be valid for states involving neutrinos when $V = P$, since S is the identity operator for positive helicity neutrinos and is certainly not the identity for the P-transformed negative helicity neutrinos. Thus we expect at least some restrictions on P-reciprocity.

We have now reached a position where we can specify violation of the various DSO in terms of certain reciprocity phase angles $\phi_{AB}{}^v$ which enter directly into observable quantities. Thus we have solved the problem we set out at the beginning of the section. Our solution in terms of the reciprocity phases $\phi_{AB}{}^v$ may not be the correct one, in the sense that for a given Hamiltonian the set of phases $\phi_{AB}{}^v$ which can be defined from (14) may not contain all the observable information derivable from *all* matrix elements. However, we may regard a weakened form of reciprocity for P, C or T as providing certain interesting predictions which test a general class of Hamiltonians. In order to determine what these predictions are we turn to precise definitions of the various weakened forms of reciprocity which were mentioned earlier.

4 Consequences of Reciprocity

We have already remarked that we cannot expect parity reciprocity to be valid for states involving neutrinos. Thus we limit our discussion of reciprocity for the DSO to strongly interacting particles only. We would like, if possible, to discuss violations of the DSO in such a manner that violations of P, C and T all appear as reciprocity phases $\phi_{AB}{}^{v}$, as in equation (15). For parity, if $|A\rangle$, $|B\rangle$ are eigenstates of orbital angular momentum with eigenvalues l_A, l_B,

$$\langle A|S|B\rangle = (-1)^{l_A+l_B}\langle A^P|S|B^P\rangle \tag{16}$$

Since a general state will in general be a linear combination of states with different orbital angular momenta it will not be possible for P-reciprocity to be valid for such states. We thus restrict P-reciprocity to be applied to states of given orbital angular momentum, and denote it by P_{rl}; when V denotes any one of P, C, or T and we require reciprocity between states of given orbital angular momentum and charge (isotopic spin) as in (15) we denote the condition by $V_{rl\alpha}(V_{rlI})$. This restricted form of reciprocity, depending as it does on a given reference frame, is not explicitly covariant. However this is the price we must be prepared to pay if we wish to discuss P and T-violation in the same way.

Reciprocity conditions have already been applied to the $K_2{}^0$ decay[38], in photo-pion reactions[25] and to some aspects of non-leptonic hyperon decays[39]. We will not consider the K-decays further here since our weakening of reciprocity by choosing states of given orbital angular momentum does not change the reciprocity conditions for spinless K-mesons already discussed in reference (38). We will consider here the non-leptonic hyperon decays, and discuss the various implications of $V_{rl\alpha}$ or V_{rlI} for them; many of our results may be extended to other processes not involving neutrinos.

Explicitly for the decay of a hyperon to a hyperon plus a meson, $N' \rightarrow N + \pi$, we postulate the following reciprocity relations among the partial wave amplitudes

$$M_l{}^{\text{out}}(\alpha) = \langle(\pi N)_{l,\alpha}{}^{\text{out}}|S|N'\rangle,\ M_l{}^{\text{in}}(\alpha) = \langle(\pi N)_{l\alpha}{}^{\text{in}}|S|N'\rangle$$

(where α is a label for a definite particle state or a definite isospin state of the πN system and 'out' or 'in' labels are with respect to the strong-interaction part of the Hamiltonian H).

$$\langle[(\pi N)_{l\alpha}{}^{\text{out}}]^T|S^+|(N')^T\rangle^*\, e^{i\phi_l{}^T(\alpha)} = M_l{}^{\text{out}}(\alpha) \tag{17}$$

$$\langle[(\pi N)_{l\alpha}{}^{\text{out}}]^{PCT}|S^+|(N')^{PCT}\rangle^*\, e^{i\phi_l{}^{PCT}(\alpha)} = M_l{}^{\text{out}}(\alpha) \tag{18}$$

$$\langle[(\pi N)_{l\alpha}{}^{\text{in}}]^T|S^+|(N')^T\rangle^*\, e^{i\phi_l{}^T(\alpha)} = M_l{}^{\text{in}}(\alpha) \tag{17'}$$

$$\langle[(\pi N)_{l\alpha}{}^{\text{in}}]^{PCT}|S^+|(N')^{PCT}\rangle^*\, e^{i\phi_l{}^{PCT}(\alpha)} = M_l{}^{\text{in}}(\alpha) \tag{18'}$$

$$\langle[(\pi N)_{l\alpha}{}^{\text{out}}]^C|S|(N')^C\rangle\, e^{i\phi_l{}^C(\alpha)} = M_l{}^{\text{out}}(\alpha) \tag{19}$$

$$\langle[(\pi N)_{l\alpha}{}^{\text{in}}]^C|S|(N')^C\rangle\, e^{i\phi_l{}^C(\alpha)} = M_l{}^{\text{in}}(\alpha) \tag{19'}$$

$$\langle[(\pi N)_{l\alpha}{}^{\text{out}}]^P|S|(N')^P\rangle\, e^{i\phi_l{}^{P(\alpha)}} = M_l{}^{\text{out}}(\alpha) \tag{20}$$

$$\langle[(\pi N)_{l\alpha}{}^{\text{in}}]^P|S|(N')^P\rangle\, e^{i\phi_l{}^{P(\alpha)}} = M_l{}^{\text{in}}(\alpha) \tag{20'}$$

The *T*, *C*, *P* and *PCT*-transformed states are defined by

$$\left.\begin{aligned}
|[(\pi N)_{l,\alpha,\hat{p}_\pi,\hat{S}_N}{}^{\text{out}}]^T\rangle &= \eta_T|(\pi N)_{l,\alpha,-\hat{p}_\pi,-\hat{S}_N}{}^{\text{in}}\rangle \\
|[(\pi N)_{l,\alpha,\hat{p}_\pi,\hat{S}_N}{}^{\text{out}}]^{PCT}\rangle &= \eta_{PCT}(-1)^l|(\overline{\pi N})_{l,\alpha,\hat{p}_\pi,\hat{S}_N}{}^{\text{in}}\rangle \\
|[(\pi N)_{l,\alpha,\hat{p}_\pi,\hat{S}_N}{}^{\text{out}}]^C\rangle &= \eta_C|(\overline{\pi N})_{l,\alpha,\hat{p}_\pi,\hat{S}_N}{}^{\text{out}}\rangle \\
|[(\pi N)_{l,\alpha,\hat{p}_\pi,\hat{S}_N}{}^{\text{out}}]^P\rangle &= \eta_P(-1)^l|(\pi N)_{l,\alpha,-\hat{p}_\pi,-\hat{S}_N}{}^{\text{out}}\rangle
\end{aligned}\right\} \tag{21}$$

and $\hat{p}_\pi$, $\hat{S}_N$ are the centre of mass momentum and nucleon polarization unit vectors, $\eta_{PCT} = \eta_P\eta_C\eta_T$, and η_P, η_C, η_T are arbitrary phases.

We will first see that (17) and (17′) are equivalent, as are (20) and (20′), basically because *P* and *T* do not change the nature of the particles involved; this is not true of *C*, so that the equality of the *C* and *PCT*-reciprocity phases in (18) and (18′), (19) and (19′) implies relations between particle and antiparticle decay rates.

We discuss first the equivalence between (17) and (17′). We have

$$M_l{}^{\text{out}}(a) = \sum_I C_I(a)a_l(I)\, e^{i\delta_l(I)} \tag{22}$$

where $C_I(a)$ is the appropriate Clebsch-Gordon coefficient, $a = \alpha$ or I, $a_l(I) = \langle(\pi N)^{\text{standing}}|S|N'\rangle$, $\delta_l(I)$ is the πN phase shift in the *I*-spin channel *I* with angular momentum *l*, and we assume that the strong interaction part of *H* is invariant under each of *P*, *C*, *T*. Then $T_{\text{r}la}$ of (17) implies

$$\sum_I C_I(a)a_l(I)\, e^{i\delta_l(I)} = \sum_I C_I(a)a_l{}^*(I)\, e^{i\delta_l(I)+i\phi_l{}^T(a)}(-1)^l \tag{23}$$

We have not included the spinor term $\bar{u}_N\Gamma_l u_{N'}$ with $\Gamma_0 = 1$, $\Gamma_1 = \boldsymbol{\sigma}\cdot\hat{\mathbf{p}}_\pi$ since these automatically cancel on either side of (23) except for the term $(-1)^l$. If there is just one value of *I* contributing in (22) (as in Λ decay with the $\Delta I = \frac{1}{2}$ rule) then we may conclude immediately from (23) that

$$\phi_l{}^T(a) = 2\theta_l + \pi l \quad (\text{mod. } 2\pi) \tag{24}$$

where $a_l = |a_l|e^{i\phi_l}$; $\phi_l{}^T(a)$ is thus independent of the 'in' or 'out' labels and (17′) follows. If there are two values of *I*, $I = \frac{1}{2}$ and $I = \frac{3}{2}$, then from (23) we need

$$\sum_{I,I'} C_I(a)C_{I'}(a)a_l{}^*(I)a_l(I')\sin[\delta_l(I) - \delta_l(I')] = 0$$

Since in general $\delta_l(\frac{1}{2}) \neq \delta_l(\frac{3}{2})$ then

$$a_l(\tfrac{1}{2})a_l{}^*(\tfrac{3}{2}) = a_l(\tfrac{3}{2})a_l{}^*(\tfrac{1}{2}) \tag{25}$$

and (24) again follows, so completing the proof of equivalence between (17) and (17′).

The equivalence between (20) and (20′) follows similarly, though now with the identity

$$a_l(\tfrac{1}{2})a_l(\tfrac{3}{2}) = a_l(\tfrac{1}{2})a_l(\tfrac{3}{2})$$

replacing (25).

Our condition (19) and (19′), with the same reciprocity phase for both equations, may be shown to give rise to non-trivial relations between amplitudes for particles and amplitudes for antiparticles. Thus from (19):

$$\sum_I C_I(a)a_l(I)\,\mathrm{e}^{\mathrm{i}\delta_l(I)} = \sum_I \bar{C}_I(a)\bar{a}_l(I)\,\mathrm{e}^{+\mathrm{i}\delta_l(I)}\,\mathrm{e}^{\mathrm{i}\phi_l{}^C(a)} \tag{26}$$

and from (19′) we obtain

$$\sum_I C_I(a)a_l(I)\,\mathrm{e}^{-\mathrm{i}\delta_l(I)} = \sum_I \bar{C}_I(a)\bar{a}_l(I)\,\mathrm{e}^{-\mathrm{i}\delta_l(I)}\,\mathrm{e}^{\mathrm{i}\phi_l{}^C(a)} \tag{27}$$

where $\bar{a}_l(I)$ is the amplitude similar to $a_l(I)$ but involving antiparticles in place of particles. Combining (26) and (27) as before,

$$\sum_{I,I'} C_I(a)\bar{C}_{I'}(a)a_l(I)\bar{a}_l(I')\sin[\delta_l(I) - \delta_l(I')] = 0 \tag{28}$$

For only one I-spin amplitude (28) becomes an identity; in the case of two values for $I, I' = \frac{1}{2}, \frac{3}{2}$, we have again $\delta_l(\frac{1}{2}) \neq \delta_l(\frac{3}{2})$, so

$$a_l(\tfrac{1}{2})\bar{a}_l(\tfrac{3}{2})C_{\frac{1}{2}}(a)\bar{C}_{\frac{3}{2}}(a) = a_l(\tfrac{3}{2})\bar{a}_l(\tfrac{1}{2})C_{\frac{3}{2}}(a)\bar{C}_{\frac{1}{2}}(a) \tag{29}$$

An example of one of the equations of (29) is for $\Sigma^+ \to p\pi^0$, where we take $a = \alpha$ (given particle states), so that $C_{\frac{1}{2}} = -\bar{C}_{\frac{1}{2}}$, $C_{\frac{3}{2}} = +\bar{C}_{\frac{3}{2}}$ and we obtain

$$a_l(\tfrac{1}{2})/\bar{a}_l(\tfrac{1}{2}) = -a_l(\tfrac{3}{2})/\bar{a}_l(\tfrac{3}{2}) \tag{30}$$

If we use (18) and (18′) we obtain in place of (29) the equation

$$a_l(\tfrac{1}{2})\bar{a}_l{}^*(\tfrac{3}{2})C_{\frac{1}{2}}(a)\bar{C}_{\frac{3}{2}}(a) = a_l(\tfrac{3}{2})\bar{a}_l{}^*(\tfrac{1}{2})C_{\frac{3}{2}}(a)\bar{C}_{\frac{1}{2}}(a) \tag{31}$$

We may rewrite (30) and its similar relation from (31) respectively as

$$\theta_l(\tfrac{1}{2}) - \theta_l(\tfrac{3}{2}) = \bar{\theta}_l(\tfrac{1}{2}) - \bar{\theta}_l(\tfrac{3}{2}) + \pi \quad (\text{mod. } 2\pi)$$
$$\theta_l(\tfrac{1}{2}) - \theta_l(\tfrac{3}{2}) = -\bar{\theta}_l(\tfrac{1}{2}) + \bar{\theta}_l(\tfrac{3}{2}) + \pi \quad (\text{mod. } 2\pi)$$

which may be tested by experiment.

We note that it is the strong final-state interactions among the hadrons which helps to bring about the simple relation (24) between the T-reciprocity phases $\phi_l{}^T(a)$ and the phases of the weak transition matrix elements.

If $\phi_l{}^T(\alpha)$ is independent of l we obtain T-reciprocity for the weak transition matrix element to the particle state α. Conversely we may separate a reciprocity relation for the total weak transition matrix element into separate ones for the separate partial wave amplitudes. Thus $T_{\mathrm{r}l}$ is equivalent to $T_{\mathrm{r}l\alpha}$ iff $\phi_l{}^T(\alpha)$ is independent of l (where $T_{\mathrm{r}\alpha}$ is T-reciprocity between given particle states). In the case of $T_{\mathrm{r}lI}$, $C_I(a) = \delta_{Ia}$, where a denotes the I-spin of the particular final state being considered, so that in (27) the strong phase shifts cancel exactly and (24) follows. As before $T_{\mathrm{r}I}$ is equivalent to $T_{\mathrm{r}lI}$ iff $\phi_l{}^T(I)$ is independent of l.

If $\delta_l(\frac{1}{2}) \neq \delta_l(\frac{3}{2})$, then evidently $T_{\mathrm{r}l\alpha}$ is a stronger condition than $T_{\mathrm{r}lI}$ and is equivalent to it iff $\phi_l{}^T(I)$ is independent of I. In the case of $T_{\mathrm{r}l\alpha}$, (24) imposes the condition that $\theta_l(\frac{1}{2}) = \theta_l(\frac{3}{2})$ while this does not follow from $T_{\mathrm{r}lI}$ in general; in fact only $T_{\mathrm{r}l\alpha}$ leads to a reduction in the number of independent phase angles.

There are numerous equalities for particles and antiparticles which we must be careful about if we weaken PCT to PCT-reciprocity[40]; in particular, the equality of masses and lifetimes of particles and antiparticles follows from PCT. The experimental situation as discussed in Section 2 agrees with this equality, so we wish to derive these equalities from PCT reciprocity.

We consider first the total lifetimes, again for hyperon non-leptonic decays. $(PCT)_{l\alpha}$ reciprocity yields for the antiparticle decay lifetime

$$\begin{aligned}\sum_\alpha |\bar{M}_\alpha|^2 &= \sum_\alpha [|\bar{M}_{s\alpha}{}^{\mathrm{in}}|^2 + |\bar{M}_{p\alpha}{}^{\mathrm{in}}|^2] \\ &= \sum_{\alpha,I,I'} (C_I C_{I'})_\alpha \{ e^{i[\delta_s(I) - \delta_s(I')]} M_s{}^{\mathrm{out}*}(I) M_s{}^{\mathrm{out}}(I') \\ &\qquad + e^{i[\delta_p(I) - \delta_p(I')]} M_p{}^{\mathrm{out}*}(I) M_p{}^{\mathrm{out}}(I\,)\}\end{aligned}$$

where the summation also involves integration over final momenta. Since $M_l(I)$ is independent of α and $\sum_\alpha (C_I C_{I'})_\alpha = \delta_{II'}$, we have

$$\begin{aligned}\sum_\alpha |\bar{M}_\alpha|^2 &= \sum_I [|M_s{}^{\mathrm{out}}(I)|^2 + |M_p{}^{\mathrm{out}}(I)|^2] \\ &= \sum_\alpha |M_\alpha|^2\end{aligned}$$

so proving the equality between particle and antiparticle lifetimes.

In order to prove equality of masses of particles and antiparticles we postulate PCT-reciprocity for the diagonal matrix elements of the S-matrix, and work to lowest order in the weak interaction. Then for an eigenstate $|\psi\rangle$ of H representing a particle of spin $\frac{1}{2}$ at rest,

$$\langle\psi|H|\psi\rangle = \langle\psi^{PCT}|H|\psi^{PCT}\rangle^* \, e^{i\phi^{PCT}} \tag{32}$$

Then
$$e^{i\phi^{PCT}} = \langle\psi|H|\psi\rangle / \langle\psi^C|H|\psi^C\rangle \tag{33}$$

where (33) follows from (32) in lowest order in the T-violating part of H, assuming that it is a part of the PCT-violating part of H. But from the hermitian character of H we deduce from (33) that $\phi^{PCT} = 0$ (mod. 2π), so proving the equality of the masses of particle and antiparticle.

We may also deduce equalities of partial decay rates of particle and antiparticle. Thus from $C_{rl\alpha}$ we have, again for hyperon non-leptonic decay,

$$\begin{aligned}|M_\alpha|^2 &= |M_s^{\text{out}} + M_p^{\text{out}}|^2 = |\bar{M}_s^{\text{out}}\, e^{i\phi_s^C} + \bar{M}_p^{\text{out}}\, e^{i\phi_p^C}|^2 \\ &= |\bar{M}_s^{\text{out}}|^2 + |\bar{M}_p^{\text{out}}|^2 + 2\text{Re}[\bar{M}_s^{\text{out}}\bar{M}_p^{\text{out}*}\, e^{i(\phi_s^C - \phi_p^C)}] \\ &= |\bar{M}_\alpha|^2 \text{ iff } \phi_s^C = \phi_p^C\end{aligned}$$

Therefore the interference term implies the inequality of the partial differential decay rates of particle and antiparticle, while the integrated partial decay rates will be equal, since the interference term does not contribute to the integral over angles. Thus a difference in the partial integrated decay rates of, for example, $\Sigma^+ \to p\pi^0$ and $\bar{\Sigma}^+ \to \bar{p}\pi^0$ would imply a breakdown of $C_{rl\alpha}$[41].

We remark that for the integrated partial decay rates $(PCT)_{rl\alpha}$ and $P_{rl\alpha}$ alone are not sufficient to prove equality, since (18) and (20) only imply $|M_l^{\text{out}}|^2 = |\bar{M}_l^{\text{in}}|^2$, $|M_l^{\text{in}}|^2 = |\bar{M}_l^{\text{out}}|^2$. We need also to have $T_{rl\alpha}$ and rotation invariance, so giving the relation

$$M_l^{\text{out}} = M_l^{\text{in}*}\, e^{i\phi_l^T} \text{ or } |M_l^{\text{out}}| = M_l^{\text{in}}|$$

It is possible to obtain relations between the various reciprocity phases; thus if we assume $P_{rl\alpha}$, $C_{rl\alpha}$, $T_{rl\alpha}$ and $(PCT)_{rl\alpha}$ in hyperon non-leptonic decays we obtain[5]

$$\phi_l^T(\alpha) = \phi_l^C(\alpha) + \phi_l^P(\alpha) + \phi_l^{PCT}(\alpha) \quad (\text{mod. } 2\pi) \tag{34}$$

A similar result holds if we replace α by I in the above.

We also remark that we have no evident mechanism for violation of PCT, since this requires either going completely outside a Hamiltonian basis, or assuming a non-hermitian Hamiltonian. Either of these possibilities has grave difficulties associated with it.

We conclude that the reciprocity conditions discussed above provide a framework in which to discuss violation of the DSO's in a unified fashion for the strongly interacting particles. The experimental situation is that, other than for the case of $K_2^0 \to 2\pi$ decay discussed in reference (38), all reciprocity phases for T-violation are consistent with zero, while for all processes the reciprocity phases for P-violation are consistent with 0 or π and for PCT-violation are consistent with zero (hence giving the C-violating phases as 0 or π, by equation 34).

In particular in the case of non-leptonic Σ-decay, for the asymmetry parameters α^+, α^-, α^0 for $\Sigma^+ \to n\pi^+$, $\Sigma^- \to n\pi^-$, $\Sigma^+ \to p\pi^0$, we have from the SU(6) prediction that the S-wave amplitude for $\Sigma^+ \to n\pi^+$ vanishes[42], the $\Delta I = \frac{1}{2}$ rule, and $T_{rl\alpha}$, that[5]

$$\alpha^+ = 0$$

$$\alpha^- = \pm 2\sqrt{2}|a_s^{(0)}(\tfrac{1}{2})a_p^{(0)}(\tfrac{3}{2})|(2|a_s^{(0)}(\tfrac{1}{2})|^2 + |a_p^{(0)}(\tfrac{3}{2})|^2)^{-1} \cos \tfrac{1}{2}(\phi_s^{(0)T} - \phi_p^{(0)T})$$

$$\alpha^0 = 2[\pm\sqrt{2}|a_s^{(0)}(\tfrac{1}{2})a_p^{(0)}(\tfrac{3}{2})| \pm |a_s^{(0)}(\tfrac{1}{2})a_p^{(0)}(\tfrac{1}{2})|] \times \times (3|a_s^{(0)}(\tfrac{1}{2})|^2 + |a_p^{(0)}(\tfrac{3}{2})|^2)^{-1} \cos \tfrac{1}{2}(\phi_s^{(0)T} - \phi_p^{(0)T})$$

Experimentally $\alpha^+ = \alpha^- \approx 0$, $\alpha^0 \sim 1$[42,43], so that $a_p^{(0)}(\frac{3}{2}) = 0$ and $\alpha^0 = 2x/(3x^2 + \frac{1}{3}) \cos \frac{1}{2}(\phi_s^{(0)T} - \phi_p^{(0)T})$, with $x = |a_s^{(0)}(\frac{1}{2})|/|a_p^{(0)}(\frac{1}{2})|$. The only possibility of having $\alpha^{(0)} = 1$ is to take $x = \frac{1}{3}$, $\phi_s^{(0)T} = \phi_p^{(0)T}$, which is consistent with T-conservation and with the branching ratio of $\Sigma^+ \to p\pi^0$ to $\Sigma^+ \to n\pi^+$. As in the case of Λ or Ξ non-leptonic decays the asymmetry parameter β for each of these processes is equal to the corresponding parameter α multiplied by $\tan[\frac{1}{2}(\phi_s^T - \phi_p^T)]$ for the relevant channel, and since all the values of β are consistent with zero[44] then all these processes are not only consistent with $T_{rl\alpha}$ reciprocity but with T-invariance; the case of maximal T-violation due to some T-reciprocity phases being 180° is definitely excluded in these processes.

We remark finally that except for our remark on the equality of particle and antiparticle masses the results of this section are not restricted to lowest order in the DSO-violating part of the Hamiltonian. Thus our results also apply if the electromagnetic interactions violate C or T in a manner consistent with one or other of the reciprocity conditions embodied in (17) to (21) and their extensions to other processes. It has not been shown that any DSO-violating Hamiltonian may satisfy such reciprocity conditions; this is an interesting question, but the lack of an answer to it should not prevent us from determining the experimental consequences of such conditions. This is what we have done here, albeit in a very preliminary fashion.

5 Magnetic Monopole Theory of *CP*-Violation

The magnetic monopole was introduced by Dirac[45] to enable the electric and magnetic field strengths **E**, **H** to enter Maxwell's equations symmetrically. This symmetry is achieved by introducing the monopole current $\hat{j}_\mu(x)$, as well as the charge current $j_\mu(x)$, so that Maxwell's equations become

$$\partial_\mu F_{\mu\nu} = j_\nu,\ \partial_\mu \hat{F}_{\mu\nu} = \hat{j}_\nu \tag{35}$$

where $F_{\mu\nu}$ is the usual antisymmetric field tensor with components (**E**,**H**). We take the source of j_μ and $\hat{j}_\mu$ to be spin-$\frac{1}{2}$ particles with fields ψ, χ respectively. Then $j_\mu = e\bar{\psi}\gamma_\mu\psi$ is the usual expression for the conserved electric current. There are two simple possibilities in constructing $\hat{j}_\mu$ from χ*:

$$\hat{j}_\mu = g\bar{\chi}\gamma_\mu\chi \tag{36}$$

or

$$\hat{j}_\mu = ig\bar{\chi}\gamma_\mu\gamma_5\chi \tag{37}$$

We note that the vector current (36) is conserved for any monopole mass, while the axial vector current (37) is not, but will only be conserved for zero monopole mass; further, the vector current separately violates T and P-invariance, while the axial vector current conserves P and violates T. We may see this as follows: the Maxwell equations (35) are

$$\begin{aligned} \nabla \cdot \mathbf{E} &= j_0 \\ \dot{\mathbf{E}} - \nabla \times \mathbf{H} &= -\mathbf{j} \\ \nabla \cdot \mathbf{H} &= \hat{j}_0 \\ \dot{\mathbf{H}} + \nabla \times \mathbf{E} &= -\hat{\mathbf{j}} \end{aligned}$$

For the vector current (36) under P, $\mathbf{E} \rightarrow -\mathbf{E}$, $\mathbf{H} \rightarrow \mathbf{H}$, $j_0 \rightarrow j_0$, $\mathbf{j} \rightarrow -\mathbf{j}$, $\hat{j}_0 \rightarrow \hat{j}_0$, $\hat{\mathbf{j}} \rightarrow -\hat{\mathbf{j}}$, so that P is violated. Under T, $\mathbf{E} \rightarrow \mathbf{E}$, $\mathbf{H} \rightarrow -\mathbf{H}$, $j_0 \rightarrow j_0$, $\hat{j}_0 \rightarrow \hat{j}_0$, $+\mathbf{j} \rightarrow -\mathbf{j}$ and $\hat{\mathbf{j}} \rightarrow -\hat{\mathbf{j}}$, so again T is violated. In the case of the axial vector current (37), under P, $\hat{j}_0 \rightarrow -\hat{j}_0$ and $\hat{\mathbf{j}} \rightarrow \hat{\mathbf{j}}$, and under T, $\hat{j}_0 \rightarrow -\hat{j}_0$, $\hat{\mathbf{j}} \rightarrow \hat{\mathbf{j}}$, so (38) violates T but conserves P. We described the good evidence in Section 2 for the conservation of P in electromagnetic interactions (though equation 38 with 36 is invariant under the combined operation of parity *and* monopole conjugation). We also wish to be able to consider particles which carry *both* electric and magnetic charge, which would not be possible if we choose the vector current to be the monopole current. So for this and the previous reason we take the axial vector current model (37) with which to generate CP-violation.

We are thus faced with the problem of quantizing (37) and (38), and then calculating the amount of CP-violation to be expected from such a model. In order to proceed further it is necessary to consider the problem

* We use hermitian γ_0 and γ_5, anti-hermitian γ_1, γ_2, γ_3, with $[\gamma_\mu,\gamma_\nu]_+ = 2g_{\mu\nu}$, $g_{00} = +1$, $g_{11} = g_{22} = g_{33} = -1$, $g_{ij} = 0 (i \neq j)$, $\gamma_5{}^2 = 1$, and $\bar{\psi} = \psi^+\gamma_0$.

of the monopole mass. For the axial vector current will only be conserved for zero monopole mass; if we wish to start with a non-zero bare monopole mass it will be necessary to add a term to the axial current so as to ensure its conservation. A possible term for this is the non-local field[46] $2\,mig\,(\Box^2)^{-1}\partial_\mu\bar{\chi}\gamma_5\chi$. However this non-local term will explicitly introduce a pseudoscalar massless meson, which would be coupled with strength g to the monopole. Since we will later find the Dirac quantization condition (DQC)[45] that $eg \sim 1$, then we expect this boson to be strongly produced. There is no experimental evidence for such a particle; there is, of course, no experimental evidence for a magnetic monopole either. However once we have introduced one unseen particle it would seem better to avoid dragging in any more; moreover there are reasons which we shall discuss shortly which might enable the invisibility of the monopole to be explained, but similar reasons cannot be set up for a pseudoscalar massless boson with electromagnetic strength coupling to protons (through virtual monopole pairs).

An alternative possibility is to start with zero bare mass for the monopoles and introduce monopole mass by spontaneously breaking the γ_5-symmetry, using the axial vector current (37). We expect the matrix element of j_μ, after quantization, when taken between single monopole states to have the form

$$\langle p'|j_\mu(0)|p\rangle = \bar{u}(p')X_\mu(p,p')u(p) \tag{39}$$

where

$$X_\mu(p,p') = F_1(q^2)[\gamma_\mu\gamma_5 + 2m\gamma_5 q_\mu(q^2)^{-1}] + i(p_\mu' + p_\mu)\gamma_5 F_2(q^2) \tag{40}$$

with $q = p - p'$, and m is the physical monopole mass. We have used P-conservation in (40), and current conservation to require the pole term in q^2. This pole term will again correspond to a pseudoscalar massless boson, the Goldstone boson, arising from the spontaneous breaking of a continuous symmetry[47]. It is not necessary that such a boson be present—in the case of electrodynamics with zero bare mass for the electron the solution with non-zero physical mass has a non-interacting Goldstone boson[48]. In a similar manner it may be that $F_1(0) \equiv 0$ for all values of e and g, so that the Goldstone boson is non-interacting in our case also. Alternatively it may be necessary to impose the condition $F_1(0) = 0$ on e and g. This condition, together with the DQC may thus give unique values for e and g (after renormalization).

We now may use the fact that j_μ is hermitian to deduce that both F_1 and F_2 must be real. Invariance under T requires F_1 to be real, F_2 to be purely imaginary. Since our axial vector current (37) does not conserve T, we see that the effects of T or CP-violation are given by F_2. We thus have to calculate F_2. Thus our programme is to

(*a*) Quantize (35) and (37) (which we will call the axial vector monopole theory, or AVM for short).
(*b*) Determine under which conditions the Goldstone boson is absent.
(*c*) Determine the value of the CP-violating vertex function $F_2(q^2)$, or at least $F_2(0)$.

At this stage of the work it is certainly not possible to say that the CP-violating effects will be large, due to the DQC. Indeed we might suspect that the condition $F_1(0) = 0$ will require cancellations, and $F_2(0)$ may be small, compared, say, to the monopole strength g. Thus we cannot rule out this model for CP-violation due to large CP-violating effects, as has been done already[6].

In order to proceed with the programme outlined above, let us perform (*a*), the quantization, following the method of Schwinger[49]. It was actually the vector-monopole theory (VM) which was quantized in the latter reference, though it is straightforward to adapt the discussion given there to the AVM. The DQC $eg = n\hbar c$ ($\frac{1}{2}n\hbar c$ if ψ and χ are identical) is required for AVM, as for VM, in order to preserve Lorentz invariance and ensure unobservability of the direction of the singular lines (strings) leading from the magnetic monopoles away to infinity.

Following reference (49), we choose the energy density for the AVM as

$$T^{00} = \tfrac{1}{2}(\mathbf{E}^2 + \mathbf{H}^2) + \bar{\psi}\boldsymbol{\gamma}\cdot(-i\boldsymbol{\nabla} - e\mathbf{A}^T - e\mathbf{A}_g)\psi + m_e\bar{\psi}\psi + \bar{\chi}\boldsymbol{\gamma}\cdot(-i\boldsymbol{\nabla} - g\gamma_5\mathbf{B}^T - \gamma_5 g\mathbf{B}_l)\chi \tag{41}$$

where $\mathbf{E} = \mathbf{E}^T - \boldsymbol{\nabla}\phi$, $\mathbf{H} = \mathbf{H}^T - \boldsymbol{\nabla}\hat{\phi}$, $\phi(\mathbf{x}) = \int d^3\mathbf{y}\mathscr{D}(\mathbf{x}-\mathbf{y})j_0(\mathbf{y})$, $\hat{\phi}(\mathbf{y}) = \int d^3\mathbf{y}\mathscr{D}(\mathbf{x}-\mathbf{y})\hat{j}_0(\mathbf{y})$

with $\mathbf{H}^T = \boldsymbol{\nabla}\times\mathbf{A}^T$, $\mathbf{E}^T = -\boldsymbol{\nabla}\times\mathbf{B}^T$ being the transverse components of $\mathbf{H}$ and $\mathbf{E}$ respectively, and $\mathscr{D}(x) = (4\pi|\mathbf{x}|)^{-1}$; further $A_g(\mathbf{x}) = \int d^3\mathbf{y}\mathbf{a}(\mathbf{x}-\mathbf{y})\hat{j}_0(\mathbf{y})$, $\mathbf{B}_e(x) = -\int d^3\mathbf{y}\mathbf{a}(\mathbf{x}-\mathbf{y})j_0(\mathbf{y})$, and $\mathbf{a}(\mathbf{x})$ satisfies the equation

$$\boldsymbol{\nabla}\mathscr{D}(\mathbf{x}) = -\boldsymbol{\nabla}\times\mathbf{a} - \tfrac{1}{2}\mathbf{n}(\mathbf{n}\cdot\mathbf{x}/|\mathbf{x}|)\delta_{\mathbf{n}}(\mathbf{x}) \tag{42}$$

In (42) $\mathbf{n}$ is a unit vector and $\delta_{\mathbf{n}}(\mathbf{x})$ is the δ-function in the 2-dimensional plane orthogonal to $\mathbf{n}$. The quantization of the system is now achieved by postulating the non-zero equal-time commutation and anticommutation relations:

$$\begin{aligned} &i[A_k{}^T(\mathbf{x}), B_l{}^T(\mathbf{x}')]_- = \Sigma_{klm}\partial_m\mathscr{D}(\mathbf{x}-\mathbf{x}') \\ &[\psi_i{}^+(\mathbf{x}), \psi_j(\mathbf{x}')]_+ = [\chi_i{}^+(\mathbf{x}), \chi_j(\mathbf{x}')]_- = \delta_{ij}\delta^3(\mathbf{x}-\mathbf{x}') \end{aligned} \tag{43}$$

We may satisfy the commutation relation between the potentials $\mathbf{A}^T$ and $\mathbf{B}^T$ in (43) by the choice

$$A_l{}^T(\mathbf{x}) = (2\pi)^{-3/2}\int d^3\mathbf{k}(2|\mathbf{k}|)^{-\frac{1}{2}}[\sum_{\lambda=1,2} e_l{}^{(\lambda)}a^{(\lambda)}(\mathbf{k})\, e^{i\mathbf{k}\cdot\mathbf{x}} + \text{hermitian conjugate}]$$

$$B_l{}^T(\mathbf{x}) = (2\pi)^{-3/2}\int d^3\mathbf{k}(2|\mathbf{k}|)^{-\frac{1}{2}}[\sum_{\lambda=1,2} (\hat{\mathbf{k}} \times \mathbf{e}^{(\lambda)})_l a^{(\lambda)}(k)\, e^{i\mathbf{k}\cdot\mathbf{x}} + \text{hermitian conjugate}] \quad (44)$$

where $\mathbf{e}^{(1)}$, $\mathbf{e}^{(2)}$, $\hat{\mathbf{k}}$ form an orthogonal system of unit vectors in 3-dimensional space, with $\hat{\mathbf{k}} = \mathbf{k}/|\mathbf{k}|$, and $[a^{(\lambda)}(\mathbf{k}),\ a^{(\mu)+}(\mathbf{k}')]_- = \delta_{\lambda\mu}\delta^3(\mathbf{k}-\mathbf{k}')$, $[a^{(\lambda)}(\mathbf{k}), a^{(\mu)}(\mathbf{k}')] = [a^{(\lambda)}(\mathbf{k})^+, a^{(\mu)}(\mathbf{k}')^+] = 0$. The anticommutation relations between the spinor fields ψ and χ are the usual ones.

We may now set up the interaction representation picture in which the fields develop according to the free Hamiltonian density

$$H_0 = \tfrac{1}{2}(\mathbf{E}^{T^2} + \mathbf{H}^{T^2}) + \bar{\psi}\boldsymbol{\gamma}\cdot(-i\nabla)\psi + m_e\bar{\psi}\psi + \bar{\chi}\boldsymbol{\gamma}(-i\nabla)\chi \quad (45)$$

and the state vectors according to the interaction Hamiltonian density

$$H_1 = -e\bar{\psi}\boldsymbol{\gamma}\cdot(\mathbf{A}^T + \mathbf{A}_g)\psi - g\bar{\chi}\boldsymbol{\gamma}\gamma_5\cdot(\mathbf{B}^T + \mathbf{B}_e)\chi + H_c \quad (46)$$

where $\int H_c d^3\mathbf{x}$ is the sum of the Coulomb energies of the electric charges and the monopoles:

$$\int H_c(\mathbf{x})d^3\mathbf{x} = \int d^3\mathbf{x}d^3\mathbf{y}\mathscr{D}(\mathbf{x}-\mathbf{y})[j_0(\mathbf{x})j_0(\mathbf{y}) + \hat{j}_0(\mathbf{x})\hat{j}_0(\mathbf{y})] \quad (47)$$

We may now evaluate S-matrix elements between eigenstates of H_0 in the usual fashion. We may express the Feynman rules which result, by suitable combination of the Coulomb electric and magnetic terms, as follows:

(*a*) internal spinor lines are described by the usual Feynman propagators for the ψ and χ fields.

(*b*) internal photon lines are of two types:

(i) those joining vertices x, y with both electric currents $j_\mu(x)$, $j_\nu(y)$, or magnetic currents $\hat{j}_\mu(x)$, $\hat{j}_\nu(y)$, and are then $g_{\mu\nu}D_F(x-y)$.

(ii) those joining vertices x, y with one vertex having an electric current $j_\mu(x)$, the other having a magnetic current $\hat{j}_\nu(y)$; the contribution from these vertices is covariantly expressed as $\varepsilon_{\mu\nu\lambda\sigma}\partial_\lambda n_\sigma\ \text{sgn}[n(x-y)]\partial^{-1}D_F(x-y)$ where n_μ is a time-like vector with $n^2 = 1$, and $\partial = \partial_\mu n_\mu$.

(*c*) external spinor or photon lines have the usual form.

Such a set of rules gives covariant and 'string'-independent results provided that the Dirac quantization condition holds:

$$eg = \tfrac{1}{2}\hbar cn \tag{48}$$

where n is an integer, as has been shown by the general discussion of Schwinger[49].

We may now use these Feynman rules to proceed to step (*b*) of our programme. The method of doing this is to set up the Bethe-Salpeter equation for monopole pair scattering, using the single photon exchange contribution, and taking internal monopole propagators with mass non-zero. The condition $F_1(0) = 0$ may then be imposed on this equation.

We will discuss this problem along the lines of reference[48] elsewhere. We will turn here to a discussion of (*c*), or at least to the effect on the photon propagator of the *CP*-violation caused by the monopole current. In particular we wish to consider the corrections to the photon propagator joining two electric currents. These corrections will play a role in giving *T*-violating effects in electron-photon interactions, so in atomic physics. We noted in Section 2 that the allowed upper limit on such effects are several orders of magnitude smaller than that in hadronic physics; it has already been noted[6] that such atomic physics effects should be large if the Dirac quantization condition (48) is satisfied. This is only true for a massless monopole; if we take a massive monopole of large mass m we may estimate the correction to the photon propagator due to virtual monopole-antimonopole pairs in the following manner. We take the correction terms to be just a sum of self-energy bubbles, so that the corrected propagator is

$$D_{F\mu\nu}' = D_{F\mu\nu} + G_{\mu\nu} \tag{49}$$

where $D_{F\mu\nu} = g_{\mu\nu}/p^2$, and

$$G_{\mu\nu} = \sum_{n=1,m=0}^{\infty} [H(\Pi^5 H)^n (\Pi D_F)^m]_{\mu\nu} \tag{50}$$

where $H_{\mu\nu} = \Sigma_{\mu\nu\sigma\lambda} p_\sigma n_\lambda \bar{D}(p)$, $\bar{D}(p)$ being the Fourier transform of $\text{sgn}(nx)\partial^{-1} D_F(x)$, and Π, Π^5 are the self-energies for the ψ and χ field respectively. We may count the resulting powers of p^2 in (49) and (50), and find that $G_{\mu\nu}$ behaves for large p^2 as $(p^2)^{-1}$, and has a pole at $p^2 = 0$. We are only interested in the term in $G_{\mu\nu}$ proportional to $g_{\mu\nu}$, due to current conservation. If we take from this term the pole term $(p^2)^{-1}$ and add it to

$D_{F\mu\nu}$ in (49) and then perform a wave-function renormalization, we are left with an added correction to the photon propagator which we may write approximately as proportional to $(p^2 - m^2)^{-1}$ (since it vanishes as $(p^2)^{-1}$ as $p^2 \to \infty$, and will also vanish as $m \to \infty$); the constant of proportionality is expected to be of order of unity, since even though $g^2/\hbar c > 137/4$, from the DQC, we have the term $\Pi^5 H(1 - \Pi^5 H)^{-1}$, which will be of order unity. The lower limit on m allowed by experiment when the photon propagator is $[1/p^2 + 1/(p^2 - m^2)]$ is 1 BeV[50], which is not very high; it may still be possible to identify monopoles with one or other of the hadrons or their constituent quarks[51].

Thus we may conclude that it may be possible to ascribe *CP*-violation to the presence of magnetic monopoles described by the axial vector current (37). We hope to evaluate the resulting amount of *CP*-violation in $K_{20} \to 2\pi$ decay elsewhere; there are also many other interesting problems, in particular the nature of the broken γ_5-symmetry. We may specifically ask if this breaking produces a mass for the complete propagator joining two $\hat{\jmath}_\mu$-vertices. If not, then it does not seem possible to relate monopoles to hadrons; if it does we may attempt to relate this massive field to axial vector and pseudoscalar mesons. A similar effect may arise in the complete propagator joining an electric and a magnetic current. We hope to return to these and other questions elsewhere.

We note that our monopole theory as presented above has a very different behaviour than that of a classical monopole. Indeed since the matrix elements of γ_5 and $\gamma_5\gamma_\mu$ between a static (massive) monopole spinor is zero, the static charge of a monopole is zero. Thus the monopole field is not an inverse square field.

We note finally that we may regard our theory as invariant under $PC_{m\gamma}$ but not under PC_m or PC_γ, where C_γ is the electric charge conjugation operator, C_m is the magnetic charge conjugation operator, and $C_{m\gamma}$ is the joint electric and magnetic charge conjugation operator. In the approximation $g = 0$, this is the usual PC_γ-invariance; in the approximation $e = 0$ this is PC_m-invariance. It is only the interaction term which produces PC_γ or PC_m non-invariance, and so gives a theory of 'mismatch' of charge conjugation operators of the kind discussed in general by T. D. Lee[8].

Acknowledgement

The author would like to thank Professor A. Zahlan for the hospitality of the Department of Physics, American University of Beirut, where part of this work was done.

References

1. See, for instance the discussion and references in Gasiorowicz, S., *Elementary Particle Physics*, Wiley and Sons, New York, 1966, especially Part IV.
2. Christenson, J. H., Cronin, J. W., Fitch, V. L., and Turlay, R., *Phys. Rev. Letters*, **13**, 138 (1964).
3. We discuss the experimental situation in more detail in the next section.
4. Wigner, E. P., *Group Theory and its Application to the Quantum Mechanics of Atomic Spectra*, Academic Press, 1959, Appendix to Chap. 20.
5. This work was done in collaboration with Simon, J., *Reciprocity and Non-Leptonic Hyperon Decays*, Rutgers University preprint (unpublished).
6. Salam, A., *Phys. Lett.*, **22**, 683 (1966).
7. Taylor, J. G., *A Non-Classical Theory of Magnetic Monopoles, Phys. Rev. Letters*, **18**, 713 (1967).
8. Lee, T. D., *Phys. Rev.*, **140B**, 959, 967 (1965).
9. Boehm, F., and Kankeleit, E., *Phys. Rev. Letters*, **14**, 312 (1965).
10. Tanner, N., *Phys. Rev.*, **107**, 1203 (1957).
11. Abashien, A., and Hafner, E. M., *Phys. Rev. Letters*, **1**, 255 (1958).
12. Rosen, L., and Brollay, J. E., *Phys. Rev. Letters*, **2**, 98 (1959).
13. Bodansky, D., and others, *Phys. Rev. Letters*, **2**, 101 (1959).
14. Bodansky, D., and others, *Phys. Rev. Letters*, **17**, 589 (1966).
15. Baltay, C., and others, *Phys. Rev. Letters*, **15**, 591 (1965).
16. Bernstein, J., Feinberg, G., and Lee, T. D., *Phys. Rev.*, **139B**, 1650 (1965); Barshay, S., *Phys. Lett.*, **17**, 78 (1965).
17. Cline, D., and Dowd, R. M., *Phys. Rev. Letters*, **14**, 530 (1965).
18. Crawford, F. S., and Price, Le Roy R., *Phys. Rev. Letters*, **15**, 123 (1965).
19. Cnops, A. M., and others, *Phys. Lett.*, **22**, 546 (1966).
20. Larribe, A., and others, *Phys. Lett.*, **23**, 600 (1966).
21. Fortney, L. R., and others, Berkeley Conf., 1966; Baltay, C., and others, *Phys. Rev. Letters*, **16**, 1224 (1966).
21. Bowen, R. A., and others, *Phys. Lett.*, **24**, (1967); Crawford, F. S., and Price, L. R., *Phys. Rev. Letters*, **16**, 333 (1966).
22. See, for example, the discussion of Pipkin, F. M., 'Nucleon form-factors and evidence for the validity of quantum electrodynamics at small distances,' *Oxford Conference on Elementary Particles*, September, 1965.
23. Tanner, N., private communication.
24. Barshay, S., *Phys. Rev. Letters*, **17**, 49 (1966).
25. Christ, N., and Lee, T. D., *Phys. Rev.*, **148**, 1520 (1966).
26. Hadjioannou, F. T., Iliopoulos, J., and Mennessier, G., *Nuovo Cimento*, **46**, 281 (1966).
27. Glasser, R. G., and others, *Phys. Rev. Letters*, **17**, 603 (1966). On the basis of SU(3) it is expected that $\langle \mathbf{p} . \mathbf{n} \rangle = 0$.
28. Lee, T. D., *Weak Interactions and Questions of C, P, T Non-Invariance*, Rutherford Lab. preprint (unpublished).
29. Farley and others, *Nuovo Cimento*, **25**, 281 (1966).
30. Gaillard, J.-M., and others, *Phys. Rev. Letters*, **18**, 20 (1967); Cronin, J., Kunz, P., Risk, W., and Wheeler, P., *Phys. Rev. Letters*, **18**, 25 (1967).
31. Truong, T. N., *Phys. Rev. Letters*, **13**, 358 (1964). See also Barshay, S., *Phys. Rev.*, **149**, 1229 (1966).

32. Verhay, L. J., and others, *Phys. Rev. Letters*, **17**, 669 (1966), and Young, K. K., Longo, M. J., Helland, J. A., *Phys. Rev. Letters*, **18**, 806 (1967).
33. Abranes, R. J., and others, *Phys. Rev. Letters*, **17**, 606 (1966).
34. Burgy, M. T., and others, *Phys. Rev. Letters*, **1**, 324 (1958).
35. Gibson, W. M., *Nuovo Cimento*, **25**, 882 (1966).
36. See, for example, the report *CP-Violation* by Prentki, J., at the Oxford Conference on Elementary Particles, 1965.
37. Compare with the remarks in Lee, T. D. and Wick, G. C., *Space Inversion, Time Reversal and Other Discrete Symmetries in Local Field Theories*, Columbia University preprint (unpublished).
38. Patil, S. H., Tomogowa, Y., and Yao, Y-P, *Phys. Rev.*, **142**, 1041 (1966).
39. Lee, T. D., unpublished lecture notes (1965).
40. Lee, T. D., Oehme, R., and Yang, C. N., *Phys. Rev.*, **106**, 340 (1957); Lüders, G., and Zumino, B., *Phys. Rev.*, **106**, 345 (1957).
41. See also Okubo, S., *Phys. Rev.*, **109**, 984 (1958).
42. Rosen, S. P., and Pakuasa, S., *Phys. Rev. Letters*, **13**, 773 (1964).
43. Bazin, M., and others, *Phys. Rev.*, **140B**, 1358 (1965).
44. Cronin, J. W., and Overseth, O. E., *Phys. Rev.*, **129**, 1795 (1963), and *KP interactions at* 2·24 *BeV/C*, Appendix VII, BNL Preprint 9542, C-58 (unpublished).
45. Dirac, P. A. M., *Proc. Roy. Soc.* (*London*), **A133**, 60 (1931); *Phys. Rev.*, **74**, 817 (1948).
46. Private communication from Simon, J.
47. Goldstone, J., *Nuovo Cimento*, **19**, 154 (1961); Goldstone, J., Salam, A., and Weinberg, S., *Phys. Rev.*, **127**, 965 (1962). For recent discussions and references see Higgs, P. W., *Phys. Rev.*, **145**, 1156 (1966).
48. Willey, R. S., *Phys. Rev.*, **153**, 1364 (1967).
49. Schwinger, J., *Phys. Rev.*, **144**, 1087 (1966).
50. Pipkin, F. M., 'Nucleon form-factors and evidence for the validity of quantum electrodynamics at small distances,' *Proc. Oxford Conf. on Elementary Particles*, 1965.
51. Taylor, J. G., *Quarks and Magnetic Monopoles*, Oxford University, unpublished (1967).

Bootstraps and Their Field Quantization

J. G. TAYLOR

1 Introduction

It is the purpose of this paper to present a quantum field theory approach to bootstraps. By a bootstrap we mean here a dynamical system which contains no elementary particles but only composite ones; these composite particles 'bootstrap' each other in the sense that they bind each other, to produce themselves as composites, through forces which arise from exchange of themselves. Such a bootstrap approach has the attractive possibility of determining the values of all coupling constants and masses in terms of one basic mass, these values being unique[1].

By a quantum field theory approach we mean here a description of the dynamical system by means of a set of field equations satisfying certain properties, those of Lorentz invariance, locality, and possession of a Hamiltonian with a non-negative spectrum being the most important[2]. Such an approach is essentially an off-the-mass shell approach to a dynamical system. Whilst it may in principle be possible to describe a system entirely by means of its on-mass shell properties, such a possibility does not yet appear to have been effectively realized. The use of analytic properties of S-matrix elements has helped very much in such an on-mass shell discussion; this use has not enabled production and annihilation processes to be handled with any great success. We use field theory as an alternative to the on-mass shell S-matrix approach because it enables many-particle intermediate states to be gathered together at once and dealt with by operator techniques. This advantage is partly illusory, since it is difficult to unscramble the information contained in the operator equations. However we will see that we can deduce certain interesting results, by suitable techniques, from the field theoretic form of bootstrapped systems. It is the purpose of this paper to present some of these results.

In order to describe these results we review briefly in the next section a field theoretic approach to composite particles[3]. We then extend this to a bootstrapped system in the following section, such a system having the wave function renormalization constants of all the particles set equal to zero. In order to obtain any results for such a system it is necessary to understand its quantization. This problem is discussed in Section 3 by various methods, and a satisfactory understanding obtained. This is then used to determine which possible systems can bootstrap each other. A complete answer to this question is obtained for non-derivative systems: there are none. For derivative systems the same answer cannot be obtained; however we deduce the existence of certain conserved quantum numbers for some systems of this type. We discuss these and their relation to bootstrapped symmetries briefly in the final section.

2 Composite Particles in Quantum Field Theory

Any local field theory describing a system of particles commences with bare or non-interacting particles and clothes them to become physical particles by means of the interaction term in the Hamiltonian. Thus there will be two single particle states, $|\text{ bare }\rangle$ and $|\text{ physical }\rangle$ for each type of particle.

If we define the constant $Z = |\langle \text{bare} \mid \text{physical}\rangle|^2$, being the probability of observing a bare particle in a physical particle state, then Z is the particle's wave function renormalization constant. We may set $Z = 0$ as corresponding to the absence of a bare particle state, so that provided we add the further condition that there is a physical particle present, we may consider this particle as being a composite of the other particles present. We denote these two conditions ($Z = 0$ *and* the existence of a physical particle) as the compositeness condition (for this particle).

There has been a great deal of discussion in the literature[3] of what further conditions need be imposed in order that the particle is composite; the discussion in the last paragraph would imply that further conditions are not required (though may be so for a particular model). Before we discuss this situation, we will see briefly how the compositeness conditions described above lead to a certain type of field equation.

Let us consider two elementary particles described by renormalized field operators $\phi_i (i = 1,2)$, with Lagrangian density

$$L = \sum_{i=1,2} L_0(\phi_i)[1 - (1 - Z_i)] + \tfrac{1}{2} \sum_{i=1,2} Z_i \delta m_i^2 \phi_i^2 + g_0 \phi_1 \phi_2^2 Z_1^{\frac{1}{2}} Z_2 \qquad (1)$$

where $L_0(\phi_i) = -\frac{1}{2}m_i^2\phi_i^2 - \frac{1}{2}(\partial_\mu\phi_i)^2$; the masses m_i are the renormalized masses of the particles. The field equation for ϕ_1 arising from (1) is

$$(\Box + m_1^2)\phi_1 = (1 - Z)(\Box + m_1^2)\phi_1 + Z_1\delta m_1^2\phi_1 + g_0 Z_1^{\frac{1}{2}} Z_2 \phi_2^2 \quad (2)$$

If we consider the limit of (2) as $Z_1 \to 0$ we obtain

$$\phi_1 = \lambda\phi_2^2 \quad (3)$$

where

$$\lambda = -\lim_{Z_1 \to 0} g_0 Z_1^{-\frac{1}{2}} Z_2 (\delta m_1^2)^{-1} = -\lim_{Z_1 \to 0} (Z_v g_r / \delta m_1^2 Z_1) \quad (4)$$

and $g_r = Z_v^{-1} Z_1^{\frac{1}{2}} Z_2 g_0$ is the renormalized coupling constant, Z_v is the vertex function renormalization constant defined by the requirement that the complete vertex function is equal to g_r on the mass shell. Thus provided that λ is finite the field equation (2) reduces to the algebraic equation (3) in the limit $Z_1 \to 0$. Since (3) is equivalent to the composite field operator of Zimmermann[4] we have obtained a composite theory in this limit for particle 1.

There are divergence difficulties in this discussion, especially since we do not expect the product on the right hand side of (3) to be well defined. We meet the divergence problem by introducing a cut-off Λ, so that Z_1^{-1} is in general finite, and choose m_r and g_r so that $Z_1 = 0$. We then let $\Lambda \to \infty$, keeping $Z_1 = 0$ and m_1 fixed. If a limit exists with λ finite and non-zero then a composite particle pole will occur at mass m_1, as can be seen from the analysis of the related Green's functions[3]; if it does not exist for any choice of m_r then we cannot obtain such a composite theory. We thus regard the presence of a composite particle as independent of the existence of divergences, but depending on the dynamics of the system.

We may reverse the argument presented above going from (2) to (3), so that a composite particle may be made as similar to an elementary particle as possible, and we may thus achieve the equivalence between a composite particle and an elementary particle with its wave function renormalization constant set equal to zero.

There has been the further suggestion[3] that it is also necessary to have $Z_v = 0$ in order to have a composite particle. Such a condition appears to lead to the vanishing of the composite vertex function off the mass shell, and makes an off-mass shell theory given by (3) vanish, so that a contradiction seems to arise. The origin of this difficulty was pointed out earlier[5] in the remark that we may rewrite (3) as a field equation provided we have a momentum dependent coupling constant:

$$(\Box + m_1{}^2)\phi_1 = \lambda(\Box + m_1{}^2)\phi_2{}^2$$

or

$$(-p^2 + m_1{}^2)\tilde{\phi}_1 = \lambda(-p^2 + m_1{}^2)\,\tilde{\phi}_2{}^2 \tag{5}$$

where $\tilde{\phi}$ is the Fourier transform of ϕ. Then $\tilde{\lambda} = \lambda(-p^2 + m_1{}^2)$ is the effective coupling constant in (5), and it vanishes exactly on the mass shell $p^2 = m_1{}^2$, as is required in order that the inhomogeneous term in the vertex function equation be zero there and the equation reduce to an eigenvalue equation for the mass. However, if this momentum dependence is not recognized, but a constant behaviour is assumed for the coupling constant, then this constant must be zero in order that the homogeneous vertex equation have a solution on the mass shell; off the mass shell this solution must vanish identically.

It is thus necessary to be very careful as to the way in which (3) is to be regarded as the limit of (2): either (3) is to be forced into the mould of an elementary particle theory with momentum independent renormalization constants, or it is to be treated in a more general fashion, with the understanding that (3) is the limit of (2) when the compositeness condition is applied. To spell out the difference between these two approaches even more than we did in the last paragraph, we remark that the first approach, requiring that Z_v be a constant *even in the limit* $Z_1 \to 0$ requires $Z_v = 0$, and no non-trivial off-mass shell description of the composite particle can occur. On the other hand, if we let Z_v become momentum dependent at $Z_1 = 0$, a non-trivial off-mass shell discussion of the composite particle may then be given, provided $Z_v = 0$ on the mass shell.

We choose the second approach here. Thus we describe the composite particle ϕ_1 by the field operator $\lambda\phi_2{}^2$ as in (3); the quantization of ϕ_1 is thus achieved by quantizing ϕ_2 in the standard fashion (with interaction attractive enough to produce a bound state of mass m_1). We may now use the techniques of off-mass shell and operator methods directly in this field quantization approach. We will see in the next section that this will give us certain interesting results very rapidly.

We have remarked that (3) may be regarded as the limit of (2) as $Z_1 \to 0$. It is natural to remark here that the canonical commutation relation

$$[\dot{\phi}_1(\mathbf{x},t), \phi_1(\mathbf{y},t)]_- = -\mathrm{i}Z_1{}^{-1}\delta^3(\mathbf{x} - \mathbf{y}) \tag{6}$$

becomes poorly defined as $Z_1 \to 0$. However we may use (3) in the limit

$Z_1 = 0$, so that (6), with the canonical commutation relations for the field ϕ_2, becomes in that limit,

$$[\phi_1(\mathbf{x},t), \dot{\phi}_1(\mathbf{y},t)]_- = -4iZ_2^{-1}\lambda\phi_1(\mathbf{x},t)\delta^3(\mathbf{x}-\mathbf{y}) \tag{7}$$

Since (7) is well defined, we have avoided the appearance of Z_1^{-1} in (6). We might indeed recognize the limit of Z_1^{-1} on the right of (6) as the operator $4\lambda\phi_1$ as $Z_1 \to 0$.

3 Bootstraps in Quantum Field Theory

We now bootstrap a set of particles by requiring that each of them be regarded as a composite of some of the others (including itself) in the field theoretic sense of equation (3). Thus each composite particle is described by a field operator ϕ_i which is some suitable polynomial function of the set of fields ϕ_j, $1 \leqslant j \leqslant N$ (where N is the number of distinct particles being bootstrapped). We thus have

$$\phi_i = F_i(\phi_1, \ldots, \phi_N) \tag{8}$$

for $1 \leqslant i \leqslant N$, where each function F_i is a polynomial in the fields and possibly their derivatives. We need to add to (8) the specific quantization conditions contained in the canonical commutation relations. However these relations can no longer be made well-defined in the composite limit $Z_i \to 0$ of elementary particles by the same method as was used in (7) if there are *no* elementary particles.

There is an alternative way to approach this problem[6]. We may perform a different canonical transformation on the unrenormalized field quantities than that of the renormalization transformation, and avoid the factor Z^{-1} on the right of (6) completely. We treat the problem of the interaction between unrenormalized fields ψ_0 and ϕ_0 with Hamiltonian

$$H = \tfrac{1}{2}[\pi_0^2 + (\nabla\psi_0)^2 + m^2\psi_0^2] - \tfrac{1}{2}\delta m^2\psi_0^2 + g_0\psi_0\phi_0^2 + H_0(\phi_0) \tag{9}$$

where $H_0(\phi_0)$ is the free Hamiltonian of the field ϕ_0. We perform the canonical transformation

$$\begin{aligned} \psi_0 &\to Z^{-\frac{1}{2}}\psi_0 = \psi_1 \\ \pi_0 &\to Z^{-\frac{1}{2}}\pi_0 = \pi_1 \qquad \text{(instead of } \pi_0 \to Z^{\frac{1}{2}}\pi_0 \text{ arising in the} \\ H &\to Z^{-1}H = K \qquad \text{renormalization transformation)} \end{aligned} \tag{10}$$

where Z is the wave function renormalization constant for the field ψ_0. We then see that K in the new variables ψ_1, π_1, is

$$K = \tfrac{1}{2}[\pi_1^2 + (\nabla\psi_1)^2 + m^2\psi_1^2] - \tfrac{1}{2}\delta m^2\psi_1^2 + Z^{-\frac{1}{2}}g_0\psi_1\phi_0^2 + Z^{-1}H_0(\phi_0) \tag{11}$$

We quantize (11) by means of the canonical variables π_1, ψ_1, where ψ_1 will satisfy the canonical commutation relations (6) without a factor Z^{-1}. The basic problem to be tackled now is to discuss the limit of (11) as $Z \to 0$. The objects of interest will be the Green's functions constructed by means of the field ψ_1; since Z^{-1} does not enter the commutation relation for ψ_1 with $\dot{\psi}_1$ then we cannot expect ψ_1 to be the renormalized field. Thus while we have apparently removed the factor Z^{-1} on the right of (6) we have still to deal with the renormalization problem; the physical Green's functions are to be expected to depend on the renormalized field which has the Z^{-1} factor in the right of (6). It may still be possible to describe the quantization by this approach; we will instead turn to the 'sum over paths' quantization method, since we can use it immediately to obtain results on allowed bootstrap systems.

This quantization process gives that for any system with Lagrangian density $L[\phi]$, depending on fields ϕ, then the Green's functions of the system are proportional to

$$\int F(\phi) \exp\{i \int L[\phi] d^4x\} d\mu(\phi) \tag{12}$$

where $F(\phi)$ is a suitable polynomial in the fields and $d\mu(\phi)$ is the measure defined on cylinder sets (finite dimensional subsets) as the product of Lebesgue measures and then extended to the real line. We may define μ by lattice space methods[7] by dividing space-time into a finite number N of cells each of volume less than some small quantity ε. Representative points x_k are then taken in the kth cell and the function ϕ replaced by the N numbers $\phi(x_1), \ldots, \phi(x_N)$, and $d\mu(\phi) = \prod_{i=1}^{N} d\phi(x_i)$. The integral (12) is then evaluated, and finally ε is taken to zero.

In the case of a system such as (2), it is known that this method gives the same result as the method of canonical quantization, using the canonical commutation relations (6). We may now quantize (3) by this method, thus avoiding the difficulty of the right hand side of (6) as $Z_1 \to 0$. This method of quantization cannot give different results for the Green's functions from that related to (11), since for $Z_1 \neq 0$ the results are identical.

It is immediate that the Green's functions for a system with tensor fields of type (8) only (with no derivatives) will have trivial Green's functions in coordinate space equal to the symmetrized product of δ-functions of pairs of coordinate differences. Thus the system possesses no particle solutions.

We have to be more careful when spinor fields are present, since the lattice space integration in this case cannot automatically take account of the anticommutativity of the integration variables. We define the integration over the spinor fields as an integration over lattice space which is then antisymmetrized jointly with respect to space and spin variables of the anticommuting fields in $F(\phi)$ of (12). When we consider Green's functions of an equal number of spin $\frac{1}{2}$ particles and antiparticles, then only commuting quantities arise at each lattice point[7], and lattice space methods will be applicable. Again the system possesses no particle solutions.

We have thus given a proof of the conjecture made[3] that algebraic equations of the form of (12) without derivatives have no non-trivial solutions, i.e. solutions which may be used to describe particles, with the propagators having real poles in momentum space. It will always be necessary to have a derivative coupling in order to have a non-trivial bootstrap. If we relate the presence of derivatives to the need for subtractions in dispersion relations, we see that our result implies that arbitrary (subtraction) constants may be necessary to obtain non-trivial bootstrap solutions. It may be possible to avoid such constants by a suitable non-perturbative high energy behaviour for the corresponding Green's functions, but this is a very difficult problem which we do not wish to go into here[9].

We see then that the only way to obtain a non-trivial bootstrap will be to use a derivative coupling in the functions F_i on the right of (8). It is not possible to use the lattice space quantization method in this case since contributions from separate lattice points are not separated in the integral, so cannot be integrated over. In order to deduce any general properties for such cases, we have to consider the structure of the system in more detail. It is still possible to use the 'sum over paths' quantization to set up a field theoretic form of (8) and investigate some of its properties. We turn to this in the next section.

4 Bootstrapped Symmetries and Conserved Quantum Numbers

There has been a great deal of work recently attempting to derive the internal hadron symmetries of SU(3), SU(6), etc. by means of bootstrap arguments[10]. The arguments in the previous section have shown that *no* such symmetries can arise from non-derivative couplings, when all intermediate states are taken into account in a crossing-symmetric Lorentz invariant manner by means of a field theoretic description. It may be possible to give an alternative description which possesses those properties without using field theory (either explicitly or implicitly); the successes of quantum electrodynamics and current algebras lead at least to the conclusion that a field theory description is extremely useful, and has not yet been displaced.

Thus in order to bootstrap internal symmetries it is necessary to consider derivative couplings. Such couplings are very suggestive; indeed the simplest Yukawa coupling giving a non-trivial meson-nucleon bootstrap arises from the interaction $\bar{\psi}\gamma_8\gamma_u\psi.\ \partial_\mu\phi$, and the resulting bootstrap equations are

$$\begin{aligned} \psi &= \lambda\gamma_5\gamma_\mu\psi\partial_\mu\phi \\ \phi &= \lambda^1\partial_\mu(\bar{\psi}\gamma_5\gamma_\mu\psi) \end{aligned} \tag{13}$$

We see that the second of equations (13) is indeed the PCAC model of the pseudoscalar meson. It is certainly not possible to cancel factors in (13), nor is it possible to puantize (13) by a function space integration and obtain explicit Green's functions.

It may however be possible to obtain general results on the structure of a field theoretic solution to (13), if such a solution exists. In particular we may recognize the possible existence of additively conserved quantum numbers due to the 'kinks' which may arise in the solutions to (13). In particular a solution of (13) must satisfy

$$(1 - \lambda\gamma_5\gamma_\mu\partial_\mu\phi)\psi = 0 \tag{13a}$$

so that at points where $\psi = 0$ we need

$$(\partial_u\phi^2) = -\lambda^{-2} \tag{14}$$

Thus neglecting points where $\psi = 0$ we need the derivatives of ϕ lying on the non-linear manifold (14). This manifold is doubly connected for $\lambda^2 > 0$, and its third homotopy group is equal to the additive group N of the integers; there will thus be an additively conserved integer for each state, the integer corresponding essentially to the number of times that the vector $\partial_\mu\phi$ rotates round the 'hole' in the manifold (14). By the vector $\partial_\mu\phi$ is meant the set of eigenvalues of the vector operator ϕ at a fixed time but at various points of space which are used to describe the state of the system at this time.

We note that since λ is real in order that we have a unitary time development, then it will always be possible to choose $\lambda^2 > 0$. It is necessary to consider what other possible bootstraps may arise which also allow kinks. The restrictions on the interaction term, in order that the Lagrangian equations do not possess characteristics which depend on the solution, has already been discussed in the case of elementary particles[3];

the result is that only terms linear in derivatives of tensor fields are allowed. If we extend this result to the bootstrap situation, this means that a scalar particle cannot bootstrap itself even through derivative coupling, since this is not allowed, whilst this is possible in the manner of (13) or its generalizations when spinors are allowed. Thus our restriction on the interaction term is not very strong, in that it allows bootstraps from any interaction Lagrangian $F(\bar{\psi},\psi)(\bar{\psi}\gamma_5\gamma_\mu\psi)\partial_\mu\phi$, for any invariant function of $\bar{\psi}$ and ψ. Such an interaction would give a generalized PCAC relation

$$\phi = \lambda^1\partial_\mu[F(\bar{\psi},\psi)(\bar{\psi}\gamma_5\gamma_\mu\psi)] \tag{15}$$

while the other equation of (13) becomes

$$\psi = \lambda[(\partial F/\partial\bar{\psi})(\bar{\psi}\gamma_5\gamma_\mu\psi) + F(\bar{\psi},\psi)\gamma_5\gamma_\mu\psi]\partial_\mu\phi \tag{16}$$

The corresponding generalization of (14) is more difficult to analyse now; for example in the case of $F = a\bar{\psi}\psi$, then (14) becomes

$$[1 - \lambda a(\bar{\psi}\gamma_5\gamma_\mu\psi)\partial_\mu\phi]^2 = -a^2(\partial_\mu\phi)^2 \tag{17}$$

which depends on the self-consistent solution ψ of (15) and (16). Then (17) will have

$$(\partial\phi)^2 < 0$$

so that the manifold of values of ϕ will have a non-trivial third homotopy group. Thus kinks will be expected to arise in this case, with at least one additively conserved quantum number.

It is possible to conjecture that such a quantum number will correspond to an internal symmetry of the system. Since it is necessary to break such an exact conservation law for all known cases except for conservation of nucleon number or lepton number (which are not usually regarded as generators of intrinsic symmetry groups) it is necessary to give a mechanism for breaking the conservation of the kink number. It may be possible to do this by considering the bootstrap to be only partial—the wave function renormalization constants of the hadrons are all much less than unity, so in a first approximation are zero; we may then obtain a non-trivial self-consistent solution to the bootstrap equations with possible conserved internal quantum numbers. The rigorous conservation of these quantum numbers is then broken by going to the first order approximation in the wave function renormalization constants. Whilst there are no results on this approach as yet, it may be a possible method of bootstrapping broken symmetries; it is evidently very closely related to strong coupling theory.

References

1. The numerical successes and difficulties of the bootstrap approach to elementary particles is well described in the article 'Bootstraps,' by Diu, B., report at the Ecole Internationale de la Physique des Particules Elementaries, Herceg Novi, Jugoslavia, 1966.
2. A set of postulates for local quantum field theories is discussed by Taylor, J. G., *Nuovo Cimento, Suppl.*, **1,** 857 (1964), paper I.
3. A more complete discussion is contained in the paper by Broido, M. M., and Taylor, J. G., *Phys. Rev.*, **147,** 993 (1966), as well as in references cited there.
4. Zimmermann, W., *Nuovo Cimento*, **10,** 597 (1958).
5. Taylor, J. G., *Nuovo Cimento, Suppl.*, **1,** 857 (1964), paper VI.
6. This was suggested by Broido, M. M., (*private communication*).
7. See Heber, C., 'Lattice space method in functional integration,' *Winter School in Theoretical Physics*, Karpacz, Poland, 1967, Univ. of Wroclaw press.
8. This is discussed more fully in Heber, G., and Kühnel, A., *Z. Naturforsch.*, **19a,** 1245 (1964), or Symanzik, K., *Z. Naturforsch.*, **9a,** 809 (1954), Appendix.
9. Certain results on this problem have been described by Taylor, J. G., *Nuovo Cimento, Suppl.*, **1,** 857 (1964), paper IV, and in case of scattering by a singular potential, by Aly, H. H., and Taylor, J. G., *A New Field Theory Approach to Singular Potentials*, American University of Beirut preprint (1967).
10. See, for example, *High Energy Physics and Elementary Particles*, Book III, Trieste, 1965.
11. As discussed by Finkelstein, D., *J. Math. Phys.*, **7,** 1218 (1966).

The Discrete Symmetries *P*, *C* and *T*

J. NILSSON

1 Introduction

In recent years there has been a revival of the interest in the discrete symmetries *P* (space reflection), *C* (charge conjugation), *T* (time reversal) and their various combinations. There are several reasons for this turn of events. The first and most immediate motivation for a reexamination of these invariance principles is the fact that some, if not all, seem to be violated; the main difference being the level at which the violations occur.

It all started in 1956 when Lee and Yang suggested that space reflection is not a valid symmetry for weak interaction processes. Experiments in nuclear β decay and muon decay confirmed that this suggestion was correct. It was realized at about the same time that weak interactions are not only *P*-violating but also *C*-violating. Some sort of balance in the laws of physics was restored by the assumption by Landau and others that although *P* and *C* are not separately violated, the combined *CP* is still a symmetry operation respected by all interactions. Recent experiments on K_2^0 decays have cast reasonable doubts also on the validity of the *CP* symmetry. Since the operation *CPT* seems to be a true invariance of any reasonable theory (the *CPT* theorem) a *CP*-violation necessarily implies that *T*-invariance is also violated.

Before 1956 it was generally believed that all the symmetries mentioned here were absolute symmetries of nature. At present we are in the midst of drastic changes in our understanding of nature and none of the symmetries, with the possible exception of *CPT*, may survive this development. At the same time we have been forced to take a good look at the present formulation of the discrete symmetries with an open mind to possible redefinitions. Clearly this state of affairs places the field of invariance principles in the centre of interest.

The second reason for the present surge of interest in symmetries including the discrete symmetries is the fact that we still lack a dynamical theory for elementary particle processes. It has been known for a long time that invariance arguments can be used to derive predictions, which can be subjected to experimental tests, and the deductions do not require knowledge about the detailed form of the interaction. Hence, symmetry principles can be verified or falsified at a stage when we still lack the complete theory. In this way they provide very powerful tools to bring order in our empirical findings with regard to particle processes while we still search for the appropriate theory.

2 The Basic Formalism and Invariance Principles

To establish our notations and the general conventions to which we shall adhere, we begin with a brief review of the basic formalism. In this context we shall also introduce the concept of symmetry and derive some of the immediate implications which follow from invariance principles. For a more complete discussion of these questions we refer to standard textbooks[1].

2.1 *The Hilbert space of physical state vectors*

An experiment in physics consists of two parts: (i) the preparation of the initial state of the physical system which is under study, and (ii) performing measurements on the system at a later time. It is the aim of theory to account for or 'explain' the correlations one may observe between different preparations and the observations which one may make at a later time.

The preparation of the initial state corresponds to performing a series of measurements on the physical system to define its properties. After the preparation the system is represented by a state vector $|\alpha(\Omega)\rangle$, where α denotes the outcome of the various measurements. Thus each state vector is labelled by the results of measurements. The preparation extends over a finite domain Ω in space and time, and the state vectors will in general depend on Ω as implied by the notation used.

The conventional formulation of quantum theory, on which we base our description, can be summarized in the following way. The manifold H of all possible state vectors form a linear vector space or, more precisely, a separable Hilbert space. Each observable, that is measurable, property A of a physical system is represented by a linear hermitian operator A acting on H. The (real) eigenvalues a_i of A are the only possible results of an exact, sharp measurement of the property A. The maximal amount of information one may obtain for a physical system corresponds to *a complete set of compatible* (commuting) *measurements*, that is measurements which do not perturb or interfere with each other. If the set is complete, then the system is represented by a state vector which is unique apart from an arbitrary overall phase factor. Such a state vector is labelled by all the eigenvalues $a, a', \ldots$ of the observables $A, A', \ldots$ in the complete set. If the set of measurements is not complete, then one distinguishes between two cases: (i) *Pure states*; the representing state vector is a linear superposition of unique state vectors so that the phase relations between the various components are known. Clearly the case of a unique state vector is a special case of a pure state. (ii) *Mixed states*; the representing state vector is a statistical superposition of unique state vectors so that the phase relations between the components are unknown. The two cases are physically distinguishable by the fact that for pure states one may observe interference effects which are absent if one is dealing with mixed states.

Returning to our discussion of an experiment we note that some time after the preparation the system is once more subjected to measurements, which constitute the second phase of the experiment. The interaction between the physical system and the measuring apparatus will affect the latter and the registration of how it is affected is the outcome of the experiment. As a result of this second set of measurements one has new information about the physical system, hence it must be represented by a new state vector $|\beta(\Omega')\rangle$, the final state. This new state vector will depend on the space-time region Ω' in which the measurements were performed. The properties of the initial state at the time and place of the latter measurements is given by the probability distribution over all possible outcomes of these measurements. The fundamental principle of quantum mechanics now asserts that this probability distribution $w(\beta,\alpha)$ is given by

$$w(\beta, \alpha) = |\langle\beta(\Omega')|\alpha(\Omega)\rangle|^2 \tag{1}$$

2.2 *The S-operator*

In elementary particle experiments there are specific features we shall take advantage of. We will be concerned with decay processes and scattering processes. These have in common that for times much larger than some characteristic reaction time all particles are free, i.e. non-interacting, and we can label the states with single particle labels such as the individual momenta and spins. However, we have previously recognized that the state vectors in general depend on the space-time region for the labelling measurements. For an elementary particle process the 'asymptotic' regions, where the particles are free, correspond to $t \rightarrow -\infty$ for the initial state and $t \to +\infty$ for the final outgoing state. With this in mind we write (1) in the following way

$$w(\beta, \alpha) = |\langle \beta; \text{out}|\alpha; \text{in}\rangle|^2 \tag{2}$$

where 'in' and 'out' refer to asymptotic free particle states for $t \to -\infty$ and $t \to +\infty$ respectively.

For a free particle of mass m and spin s we may take as a complete set of measurements a determination of its 3-momentum $\mathbf{p}$ and its helicity λ, but other choices are possible and in some cases more convenient. For many applications the 'plane wave' states $|m, s; \mathbf{p}, \lambda\rangle$ are useful, however, since in most experiments one actually measures at least the momenta of the particles. Of course, one cannot determine the momentum exactly since it is a continuous quantity, but one can determine it within an arbitrarily small but finite interval, and the corresponding state vector will then be an appropriate superposition of plane wave states. If the momentum spread is small enough the idealization of a plane wave state is often relevant and suitable to facilitate computations. The price one has to pay for this convenience is sometimes a lack of mathematical rigour in the formalism.

With regard to 2-particle states and many-particle states the most obvious extension is to form simple product states of the form

$$|\mathbf{p}, \mathbf{p}', \lambda, \lambda'\rangle \equiv |\mathbf{p}, \lambda\rangle|\mathbf{p}', \lambda'\rangle \tag{3}$$

In some cases the analysis of a process is more transparent if one uses other basis vectors corresponding to a different choice of a complete set of observables. For example, it is often instructive to analyse the process in terms of angular momentum. The transformation from basis (3) to an angular momentum basis is straightforward but involves tools which we shall not develop here. A brief discussion of these questions is given in Appendix 2.

For the no-particle state, i.e. the vacuum state, and the single-particle states there is no distinction between in-states and out-states since nothing can happen in those cases. For states of two or more particles a distinction is necessary, however, since our in-state may undergo scattering, etc., so that

$$|\alpha; \text{in}\rangle \neq |\alpha; \text{out}\rangle$$

The set of all free particle in-states, as well as the set of all free particle out-states, each form a complete set of basis vectors in the physical Hilbert space H. Thus the set of in-states and the set of out-states provide us with two different choices of basis vectors and, consequently, there must exist a unitary operator S which connects the two sets, that is

$$|\alpha; \text{in}\rangle = S|\alpha; \text{out}\rangle \tag{4}$$

for all α. The unitarity of the S-operator implies that

$$SS^\dagger = S^\dagger S = 1$$

or

$$S^\dagger = S^{-1}$$

Introducing the S-operator of (4) we may now write

$$|\alpha; \text{in}\rangle = \sum_\beta |\beta; \text{out}\rangle\langle\beta; \text{out}|S|\alpha; \text{out}\rangle = \sum_\beta |\beta; \text{out}\rangle S_{\beta\alpha} \tag{5}$$

with

$$S_{\beta\alpha} = \langle\beta; \text{out}|S|\alpha; \text{out}\rangle \tag{6}$$

From (5) we conclude that given the in-state $|\alpha; \text{in}\rangle$ the matrix elements $S_{\beta\alpha}$ represent the probability amplitudes for this in-state to appear in the channel $|\beta; \text{out}\rangle$ at large times. This is consistent with (2) which may now be rewritten in the following way

$$w(\beta, \alpha) = |\langle\beta; \text{out}\ |S|\alpha; \text{out}\rangle|^2 = |S_{\beta\alpha}|^2 \tag{7}$$

From this expression for the transition probability it is a trivial matter to derive the expressions for cross-sections and decay rates. We omit the derivations and just quote the results[2] for the following two types of processes

$$a \rightarrow b_1 + b_2 + \ldots + b_n \tag{8}$$

and

$$a + b \rightarrow c_1 + c_2 + \ldots + c_n \tag{9}$$

For the decay rate λ of process (8) one finds

$$\lambda = \frac{1}{(2\pi)^{3n-4}} \cdot \frac{1}{2E_a} \sum_{\text{polarizations}} \int \frac{d^3p_1}{2E_1} \cdots \frac{d^3p_n}{2E_n} \delta^4(P_f - P_i)|\langle f|T|i\rangle|^2 \tag{10}$$

where the transition matrix element $\langle f|T|i\rangle$ is defined by

$$\langle f|S|i\rangle = \delta_{fi} + \mathrm{i}(2\pi)^4\delta^4(P_f - P_i) \cdot N \cdot \langle f|T|i\rangle \tag{11}$$

and N is a normalization factor defined by

$$N = \prod_{\substack{\text{initial}\\ \text{particles}}} \frac{1}{\sqrt{(2VE_i)}} \prod_{\substack{\text{final}\\ \text{particles}}} \frac{1}{\sqrt{(2VE_f)}} \tag{12}$$

For convenience a finite normalization volume V is used. The cross section σ for process (9) is similarly given by

$$\sigma = \frac{1}{(2\pi)^{3n-4}} \cdot \frac{1}{4E_a E_b v_{\text{in}}} \cdot \sum_{\substack{\text{polari-}\\\text{zations}}} \int \frac{d^3p_1}{2E_1} \cdots \frac{d^3p_n}{2E_n} \, \delta^4(P_f - P_i) |\langle f|T|i\rangle|^2 \quad (13)$$

where v_{in} is the relative velocity of the incoming particle with respect to the target particle. If the initial particles carry spin, then one has to take the appropriate spin averages over the initial state.

We conclude this section by noting that all physical information is contained in the S-operator or, alternatively, the T-operator. A discussion of invariance principles, which is the main subject of these lectures, can then clearly be restricted to a discussion of symmetry properties of these operators.

2.3 *Invariance principles*

An invariance principle or symmetry operation of a physical system is a one-to-one correspondence which assigns to each physically realizable state $|\alpha\rangle$ another state $|\alpha'\rangle$ such that all transition probabilities are preserved, that is

$$w(\beta', \alpha') = w(\beta, \alpha) \quad (14)$$

or

$$|\langle\beta'|\alpha'\rangle|^2 = |\langle\beta|\alpha\rangle|^2$$

Similarly there is a correspondence between the observables in the two alternative descriptions. It has been shown by Wigner that if the mapping $|\alpha\rangle \to |\alpha'\rangle$ is to satisfy (14) then it is realized by means of a unitary or an antiunitary operator. We summarize the essential features of the two alternatives.

(i) *Unitary symmetry operations;*

$$\begin{aligned} &|\alpha\rangle \to |\alpha'\rangle = U|\alpha\rangle \equiv |U\alpha\rangle \\ &U^\dagger = U^{-1} \\ &\langle\beta|U^\dagger|\alpha\rangle = \langle U\beta|\alpha\rangle = \langle\alpha|U|\beta\rangle^* \end{aligned} \quad (15)$$

For an arbitrary operator Ω one obtains

$$\langle\beta|\Omega|\alpha\rangle = \langle\beta|U^{-1}U\,\Omega\,U^{-1}U|\alpha\rangle = \langle\beta'|\Omega'|\alpha'\rangle \quad (16)$$

where

$$\Omega' = U\,\Omega\,U^{-1}$$

This is the transformation law for an operator of the theory. If the operator Ω is invariant under the symmetry operation U, then $\Omega' = \Omega$ so that

$$\Omega = U\,\Omega\,U^{-1}; \qquad [\Omega, U] = 0 \tag{17}$$

and

$$\langle\beta'|\Omega|\alpha'\rangle = \langle\beta|\Omega|\alpha\rangle \tag{18}$$

The last equation implies that the matrix elements of Ω are the same in the two descriptions.

(ii) *Antiunitarity symmetry operations;*

$$\begin{aligned} &|\alpha\rangle \to |\alpha'\rangle = A|\alpha\rangle \equiv |A\alpha\rangle \\ &A^{\dagger} = A^{-1} \\ &A\lambda|\alpha\rangle = \lambda^{*}A|\alpha\rangle \\ &\langle\beta|A^{\dagger}|\alpha\rangle = \langle A\beta|\alpha\rangle^{*} = \langle\alpha|A|\beta\rangle \end{aligned} \tag{19}$$

For an arbitrary operator Ω one obtains

$$\begin{aligned} \langle\beta|\Omega|\alpha\rangle = \langle\beta|A^{-1}A\,\Omega\,A^{-1}A|\alpha\rangle &= \langle\beta'|(A\,\Omega\,A^{-1})|\alpha'\rangle^{*} \\ &= \langle\alpha'|(A\,\Omega\,A^{-1})^{\dagger}|\beta'\rangle = \langle\alpha'|\Omega'|\beta'\rangle \end{aligned} \tag{20}$$

where

$$\Omega' = (A\,\Omega\,A^{-1})^{\dagger}$$

This is the transformation law for Ω under an antiunitary symmetry transformation. More precisely, (20) implies that the operator Ω' taken between the states $|\alpha'\rangle$ and $|\beta'\rangle$ yields the same result as the operator Ω taken between the states $|\beta\rangle$ and $|\alpha\rangle$. If Ω is invariant under the transformation, then

$$\Omega^{\dagger} = A\,\Omega\,A^{-1} \tag{21}$$

which is to be compared to (17) for the unitary case.

After this rather formal introduction of the symmetry concept it is instructive to consider an explicit example. We choose the principle of Lorentz invariance, that is the invariance under proper, orthochronous Lorentz transformations (Poincaré transformations). This principle asserts that the laws of physics are invariant under Poincaré transformations which relate the space time coordinates of the same physical event as it is observed from two different inertial frames. From the previous discussion we know that this implies that there exists a unitary operator relating the state vectors and the observables which represent the physical system in the two descriptions. The case of an antiunitary correspondence is clearly inapplicable for transformations that can be reached continuously from the identity transformation, since the identity operator is unitary.

This formulation of the relativity principle is usually called *the passive formulation*. It is a rule which relates the descriptions of the same physical system with reference to two different coordinate frames. However, one may adopt an alternative point of view, which is known as *the active formulation* of the relativity principle. Instead of considering two different observers, in relative motion or otherwise related by a Poincaré transformation, one may consider one and the same observer as he observes two physical events, which are related by the same Poincaré transformation as the two observers in the passive formulation. The active formulation implies that if $|\alpha\rangle$ is a possible state of a system, then $|\alpha'\rangle = U|\alpha\rangle$, where U represents an arbitrary Poincaré transformation, is also a possible state of the system as seen by the same observer. Thus, in the active formulation the relativity principle relates the observations one observer may make on two different physical systems, which are identical apart from a relative Poincaré transformation.

All the transformations of the Poincaré group are of a geometrical nature and it is experimentally feasible to test the relativity principle both in the active and the passive formulation and to establish their equivalence. For other transformations, such as inversions (i.e. time reversal), we can in effect only test the invariance principle formulated in the active manner, since we cannot realize the 'inverted' observer. Nevertheless, in the passive manner one may always formulate hypothetical invariance principles involving such 'pathological' coordinate transformation, but they will not be physically meaningful unless they represent a true invariance operation. If not they will merely be mathematical operations. Correspondingly, in the active formulation one finds that the operations, which are not symmetry operations, in some cases transform physically realizable states into states which do not occur in nature. This means that the corresponding state vectors lie outside the physical Hilbert space. Mathematically, this means that the symmetry operator cannot be unitary or antiunitary since it leads out of the space. A well-known example of this phenomenon occurs in connection with the space reflection operation and the neutrino particle for which only one of the two conceivable helicity states is realized in nature. Of course, one may enlarge the space of state vectors to include also these non-physical states so that one may define the corresponding symmetry operators. However, this is, as previously stated, only a mathematical construction and we can attach no physical meaning to it.

Since we are ultimately interested in experimental tests of the invariance principles, we shall consistently adopt the active formulation of all symmetry operations.

2.4 *Invariance properties of the S-operator*

The basic requirement on any symmetry operation was expressed in (14). Recalling relation (7), we see that a symmetry principle can be expressed as an equality between the squared modulus of certain S-matrix elements. Thus, in the active formulation *symmetry operators yield relations between the appropriate S-matrix elements corresponding to different experimental situations.* From this we may then derive relations between different measurable quantities and these predictions can be tested in experiments. It is important to realize that no further information about the S-operator beyond the assumed symmetry property is needed to make such predictions. It is then clear that invariance principles provide useful and powerful tools in the study of elementary particle processes.

From the previous section we know that the invariance of the S-operator implies

$$S = USU^{-1} \tag{22}$$

if it is a unitary operation, and

$$S^{\dagger} = ASA^{-1} \tag{23}$$

if it is an antiunitary operation. We will encounter symmetry principles of both types. In those cases where we have an explicit expression for the S-operator, (22) and (23) permit us to examine invariance properties of the theory directly. This is the case, for example, in many field-theoretic models. Assuming that we know the Hamiltonian density and that it has the form

$$H(x) = H_0(x) + H'(x) \tag{24}$$

where $H_0(x)$ and $H'(x)$ refer to the free part and the interaction part respectively, standard methods yield the following perturbation expression for the S-operator:

$$S = 1 + \sum_{n=1}^{\infty} \frac{(-\mathrm{i})^n}{n!} \int \mathrm{d}^4x_1 \ldots \mathrm{d}^4x_n T[H'(x_1) \ldots H'(x_n)] \tag{25}$$

where T is the conventional time-ordering symbol. From (25) it is clear that in this case invariance of the S-operator follows from the following conditions

$$H'(x) = UH'(x)U^{-1} \tag{26}$$

if it is a unitary symmetry operator, and

$$H'(x) = AH'(x)A^{-1} \tag{27}$$

if the symmetry operation is represented by an antiunitary operator. In (27) we have used the fact that $H'(x)$ is hermitian.

To any unitary operator U_i one may define a hermitian operator H_i by the relation

$$U_i = \exp[\mathrm{i}\, H_i] \tag{28}$$

For continuous symmetry groups the operators H_i are closely related to the infinitesimal generators of the group. Since the operators H_i are hermitian, they may represent observables and their eigenvalues may be used as quantum numbers, which are conserved provided the operators U_i represent invariance operations. To see this consider the process $|\alpha\rangle \rightarrow |\beta\rangle$. We assume

(i) that H_i is related to a symmetry operation U_i by (28), that is $[H_i, S] = 0$;
(ii) $|\alpha\rangle$ and $|\beta\rangle$ are eigenvectors of H_i corresponding to the eigenvalues h_α and h_β respectively.

From this we obtain

$$h_\alpha\langle\beta|S|\alpha\rangle = \langle\beta|SH_i|\alpha\rangle = \langle\beta|H_iS|\alpha\rangle = h_\beta\langle\beta|S|\alpha\rangle \tag{29}$$

or

$$(h_\alpha - h_\beta)\langle\beta|S|\alpha\rangle = 0 \tag{30}$$

and hence

$$\langle\beta|S|\alpha\rangle \neq 0 \Rightarrow h_\alpha = h_\beta \tag{31}$$

This result implies that the initial and the final states are eigenstates of H_i corresponding to the same eigenvalue; the quantum number h is conserved. Invariance principles may in this way give rise to conservation laws. We reach the same result for Hamiltonian theories by considering the Heisenberg equation for an observable H_i, which does not depend on the time coordinate explicity. In that case the Heisenberg equation reads

$$\frac{\mathrm{d}H_i}{\mathrm{d}t} = \mathrm{i}\,[H, H_i] \tag{32}$$

where H denotes the Hamiltonian of the system. If H_i is conserved, i.e. its time derivative vanishes, then H_i and H commute and we may deduce the conservation law for h.

If we apply these results to the case of Poincaré invariance for the S-operator we find that

$$[P_\mu, S] = 0$$
$$[J_{\mu\nu}, S] = 0 \tag{33}$$

where P_μ and $J_{\mu\nu}$ are the hermitian generators of the Poincaré group. It follows from the first relation above that the matrix elements of S between states of definite values of P_μ must have the following diagonal form

$$\langle p_\mu', \alpha'|S|p_\mu, \alpha\rangle = \delta^4(p_\mu' - p_\mu)S(p_\mu, \alpha', \alpha) \tag{34}$$

where α' and α are the additional quantum numbers used to specify the states. If the state vectors are so normalized that the S-matrix elements are relativistic invariants, then the function $S(p_\mu, \alpha', \alpha)$ must also be an invariant function constructed out of the variables p_μ, α' and α. The four-dimensional δ-function in (34) is precisely the δ-function extracted in the definition of the T-matrix in (11). In that definition a normalization factor N was taken out to make the remainder a relativistically invariant function for spinless particles. For particles with spin the spin average has the same property.

2.5 *Final state interactions*

So far we have not taken full advantage of the unitarity relation for the S-operator. In this section we shall exploit this property and deduce some important relations for the T-matrix elements. These relations will turn out to be very useful in the later discussions of time reversal and *CPT*-invariance. They shed light on a complication called final state interactions which plague some particle processes as we shall see later.

Consider a process whereby an initial state $|\alpha\rangle$ goes over into a final state $|\beta\rangle$. We have given the probability $w(\beta, \alpha)$ for this to occur in (7)

$$w(\beta, \alpha) = |\langle\beta|S|\alpha\rangle|^2 = \langle\alpha|S^\dagger|\beta\rangle\langle\beta|S|\alpha\rangle \tag{35}$$

Conservation of probability implies that the total probability that the state $|\alpha\rangle$ at the end is found in some channel $|\beta\rangle$ must be unity, that is

$$\sum_\beta w(\beta, \alpha) = 1 \tag{36}$$

or

$$\sum_\beta \langle\alpha|S^\dagger|\beta\rangle\langle\beta|S|\alpha\rangle = \langle\alpha|S^\dagger S|\alpha\rangle = 1 \tag{37}$$

If this is to hold for any state $|\alpha\rangle$ then it follows that

$$S^\dagger S = 1 \tag{38}$$

or

$$S^\dagger = S^{-1} \tag{39}$$

and this is the unitarity condition for the S-operator.

To rewrite the unitarity condition (38) in terms of the T-operator we put

$$S = 1 + \mathrm{i}\,T \tag{40}$$

This definition of the transition operator differs from the previous one of (11) by a trivial normalization factor, which is irrelevant for the present discussion. Inserting (40) into (38) we find

$$\mathrm{i}\,(T^{\dagger} - T) = T^{\dagger}T \tag{41}$$

If we take the matrix element of this operator relation between the states $|\alpha\rangle$ and $|\beta\rangle$ we obtain

$$\mathrm{i}\,[\langle\beta|T^{\dagger}|\alpha\rangle - \langle\beta|T|\alpha\rangle] = \langle\beta|T^{\dagger}T|\alpha\rangle \tag{42}$$

Introducing a complete set of intermediate states on the right-hand side yields

$$\mathrm{i}\,[\langle\beta|T^{\dagger}|\alpha\rangle - \langle\beta|T|\alpha\rangle] = \sum_{\gamma}\langle\beta|T^{\dagger}|\gamma\rangle\langle\gamma|T|\alpha\rangle \tag{43}$$

or

$$\mathrm{i}\,[T_{\alpha\beta}{}^{*} - T_{\beta\alpha}] = \sum_{\gamma} T_{\gamma\beta}{}^{*}T_{\gamma\alpha} \tag{44}$$

Of course, the summation on the right-hand side is subjected to restrictions imposed by conservation laws and other symmetry conditions. In special cases there are no possible intermediate states $|\gamma\rangle$, and we deduce from (44)

$$\langle\alpha|T|\beta\rangle^{*} = \langle\beta|T|\alpha\rangle \tag{45}$$

In other cases the right-hand side of (44) is very small because possible intermediate states can only be reached by means of weak or electromagnetic interactions. In those cases it is sometimes possible to evaluate the corrections to the approximate relation (45).

The physical content of these results is most easily described by considering a specific process where the right-hand side of (44) cannot be neglected. Such a process is the normal decay of a Λ-particle

$$\Lambda \rightarrow p + \pi^{-}$$

If we neglect all possible intermediate states, then the process is described by diagram (a) of figure 1. However, the proton and the π-meson in the final state may rescatter, a strong process described by diagram (b) of figure 1. This intermediate state will give a non-negligible contribution to the sum in (43). Due to the rescattering the momenta of the proton and the π-meson will change. Of course, there are other intermediate states which contribute to this process. Collectively, these effects are called final state interactions for obvious reasons. As mentioned before these

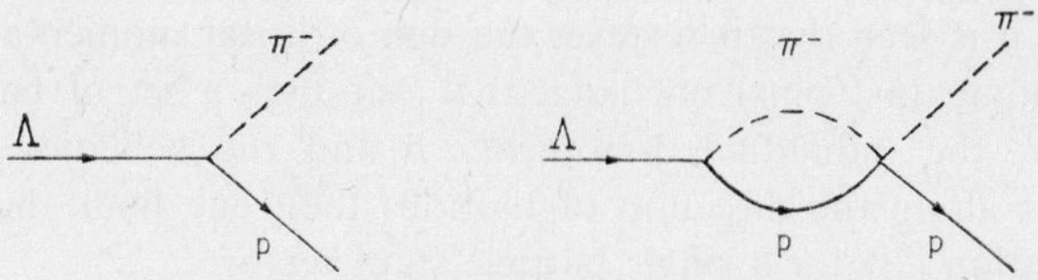

FIGURE 1 Decay diagrams for the Λ-particle

effects will be of importance for the considerations related to time reversal invariance and we shall then make use of the results of this section.

3 Space Reflection and Parity

3.1 *Definition and action on state vectors*

The space reflection transformation is a classical concept defined by the coordinate transformation

$$(x_0, \mathbf{x}) \to (x_0', \mathbf{x}') = (x_0, -\mathbf{x}) \tag{46}$$

It corresponds to a description of the same physical event in a right-handed or a left-handed, respectively, coordinate frame. In quantum theory the transformation is implemented by a unitary operator P. Since the transformation is of classical origin we know its action on various observables. For example, it must satisfy the following conditions

$$\begin{aligned} Px_\mu P^{-1} &= \varepsilon(\mu)x_\mu \\ PP_\mu P^{-1} &= \varepsilon(\mu)P_\mu \\ PJ_{\mu\nu}P^{-1} &= \varepsilon(\mu)\varepsilon(\nu)J_{\mu\nu} \end{aligned} \tag{47}$$

where x_μ is the coordinate operator, P_μ the momentum operator and $J_{\mu\nu}$ the angular momentum operator (more precisely, $J_{\mu\nu}$ for μ, $\nu \neq 0$ is an angular momentum operator while $J_{\mu 0} = -J_{0\mu}$ is a generator of a pure Lorentz transformation). We have further introduced the symbol $\varepsilon(\mu)$ which is defined as $\varepsilon(0) = 1$ and $\varepsilon(k) = -1$ for $k = 1, 2, 3$. In adopting the active formulation we must recall that the space reflection invariance is not generally valid (see further discussion below). As a consequence, P acting on a physical state vector may in some cases lead out of the physical Hilbert space. However, only the weak interactions seem to violate P-invariance, and for many important processes the influence of the weak interactions is negligible. One may then neglect their contributions to the S-matrix element, which will then satisfy the requirements imposed by P-invariance.

For later applications we need to know the action of P on state vectors. From (47) it is seen that it reverses the sign of linear momenta but leaves angular momenta (spins) unchanged. If one uses a set of basis vectors labelled by the individual 3-momenta $\bar{p}$ and the helicities λ (= spin component along the direction of motion) then one finds the following transformation law for a single particle state vector

$$P|\mathbf{p}, \lambda\rangle = \eta_p \exp(-\mathrm{i}\,\pi s)|-\mathbf{p}, -\lambda\rangle \tag{48}$$

where η_p is the intrinsic parity of the state and s is its invariant spin. For a precise definition of the state vector $|\mathbf{p}, \lambda\rangle$ we refer to Appendix 2. In Appendix 2 we have collected the definitions of all the different sets of basis vectors which will be used in these lectures. Also some of their main properties are summarized there.

It follows from (48) that P^2 acting on any state vector will map it onto itself, hence P^2 must be a multiple of the identity and one may choose the undetermined phase factor in such a way that P is hermitian and unitary, that is

$$P = P^\dagger = P^{-1} \tag{49}$$

With this choice of phase the eigenvalues of P are restricted to ± 1. We further note that in the ordinary coordinate space a space reflection P followed by a rotation $R(2n\pi)$ around an arbitrary axis yields the same result as P alone, provided n is an integer. Thus if

$$P'(n) = PR(2n\pi);\ n \text{ integer}$$

then either one of P and P' may serve as the space reflection transformation. Consider now an eigenstate $|\eta\rangle$ of P for which

$$P|\eta\rangle = \eta|\eta\rangle$$

If $|\eta\rangle$ is a state of integer spin one finds

$$P'(n)|\eta\rangle = \eta|\eta\rangle$$

Similarly, for a state of half-integer spin one obtains

$$P'(n)|\eta\rangle = \eta(-1)^n|\eta\rangle$$

since the half-integer spin representations of the rotation group are double-valued. As a consequence of this one may only define relative parities for fermions. In passing we note finally that one could also have identified P^2 with a rotation $R(2n\pi)$. For n odd, one would then obtain the eigenvalues ± 1 for P^2 with the minus sign referring only to states of half-integer spin. If this is done then the physical space reflection operator may have the eigenvalues $\pm \mathrm{i}$ and ± 1. We mention this since attempts have been made to exploit this additional freedom in the definition of the space reflection transformation[3]. Since nothing of interest has come out of these attempts, however, there is at present no reason to consider this possibility further.

Returning to the transformation properties of the various sets of basis vectors we list the ones we shall make use of. For single-particle states we have

$$P|\mathbf{p}, \lambda\rangle = \eta_p \exp(-\mathrm{i}\,\pi s)|-\mathbf{p}, -\lambda\rangle$$
$$P|j, m, p, \lambda\rangle = \eta_p(-1)^{j-s}|j, m, p, -\lambda\rangle \tag{50}$$

Similarly for 2-particle states one finds

$$P|\mathbf{P} = 0; J, M, p; \lambda_1, \lambda_2\rangle$$
$$= \eta_p(1)\eta_p(2)(-1)^{J-s_1-s_2}|\mathbf{P} = 0; J, M, p; -\lambda_1, -\lambda_2\rangle \tag{51}$$
$$P|\mathbf{P} = 0; J, M, p; l, \sigma\rangle = \eta_p(1)\eta_p(2)(-1)^l|\mathbf{P} = 0; J, M, p; l, \sigma\rangle$$

In this last set of basis vectors one has first coupled the two individual spins $\mathbf{s}_1$ and $\mathbf{s}_2$ to a resultant spin $\boldsymbol{\sigma}$, which is then coupled to the orbital angular momentum $\mathbf{l}$ to yield a total angular momentum (J, M). It is seen from (51) that these vectors are particularly useful for space reflection considerations since they are eigenvectors of P. Finally, in some applications we shall also encounter 3-particle states. We shall then always make use of basis vectors in which the relative orbital angular momentum $\mathbf{l}$ of particle 1 and 2 is coupled to the orbital angular momentum $\mathbf{L}$ of particle 3 (with reference to the centre of mass of particles 1 and 2) to a total orbital angular momentum (see figure 2). Similarly the spins $\mathbf{s}_1$ and $\mathbf{s}_2$ are coupled

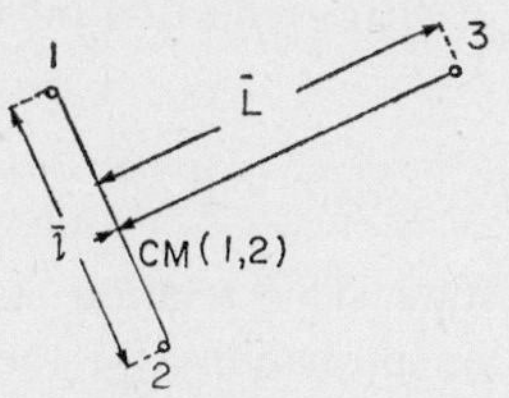

FIGURE 2 The coupling of orbital angular momenta in a 3-particle state

to $\boldsymbol{\sigma}$ and this in turn is coupled to $\mathbf{s}_3$. Finally the total orbital angular momentum and the total spin $\boldsymbol{\Sigma}$ are coupled to yield a total angular momentum (J, M). The corresponding basis vectors are denoted

$|\mathbf{P} = 0; J, M; l, \sigma; L, \Sigma\rangle$. They are eigenvectors of P as indicated by the transformation law

$$P|\mathbf{P} = 0; J, M; l, \sigma; L, \Sigma\rangle$$
$$= \eta_P(1)\eta_P(2)\eta_P(3)(-1)^{l+L}|\mathbf{P} = 0; J, M; l, \sigma; L, \Sigma\rangle \tag{52}$$

From the transformation rules for the state vectors it is a straightforward task to deduce the transformation rules which creation and destruction operators must obey. From this one may then obtain the transformation rules for the field operators of a second-quantized theory. Although we shall not make use of field-theoretic concepts we list the transformation rules of the fields $\phi(x)$, $\psi(x)$ and $A_\mu(x)$, describing particles of spin 0, $\frac{1}{2}$ and 1 respectively. For spin 1 we only consider the case of a massless field (the electromagnetic field).

$$\begin{aligned} P\phi(x_0, \mathbf{x})P^{-1} &= \eta_P\phi(x_0, -\mathbf{x}) \\ P\psi(x_0, \mathbf{x})P^{-1} &= \eta_P\gamma_0\psi(x_0, -\mathbf{x}) \\ P\bar{\psi}(x_0, \mathbf{x})P^{-1} &= \eta_P{}^*\bar{\psi}(x_0, -\mathbf{x})\gamma_0 \\ PA_\mu(x_0, \mathbf{x})P^{-1} &= \varepsilon(\mu)\eta_P A_\mu(x_0, -\mathbf{x}) \end{aligned} \tag{53}$$

A spin-0 field for which $\eta_P = +1$ is called a *scalar field*. If $\eta_P = -1$ the field is said to be a *pseudoscalar field*.

In passing we note that off-hand there is no compelling reason for choosing the intrinsic parity η_P of (48) to be the same for a particle and its antiparticle. In fact, for fermions and antifermions one must choose the relative intrinsic parity to be odd in order that the field equations should remain invariant. Hence, for a fermion/antifermion state one has

$$\eta_P(1)\eta_P(2) = -1 \tag{54}$$

3.2 *Some applications of the space reflection symmetry*

Under the assumption that the S-operator is space reflection invariant we have from (22)

$$\langle\beta; \text{out}|S|\alpha; \text{out}\rangle = \langle P\beta; \text{out}|S|P\alpha; \text{out}\rangle \tag{55}$$

where $|P\alpha\rangle$ refers to the space-reflected state. This relation states that the matrix element and hence the transition probability for the process $|\alpha\rangle \to |\beta\rangle$ is the same as that of the process $|P\alpha\rangle \to |P\beta\rangle$. From this one may in many important cases derive selection rules or restrictions on the possible form of the matrix elements. We will demonstrate this in a few examples below.

(i) *The* $\Theta-\tau$ *puzzle*. The original suggestion of Lee and Yang, that the weak interactions do not respect space reflection invariance, was made to resolve the so-called $\Theta - \tau$ puzzle. One had observed what one believed to be two different particles Θ and τ. They had the same mass and were distinguished solely by their different decay modes

$$\Theta^+ \to \pi^+ + \pi^0$$
$$\tau^+ \to \pi^+ + \pi^+ + \pi^-$$

If parity is a good quantum number in these decays, then we will demonstrate below that the relative intrinsic parity of the two particles must be odd. Now we know that parity is violated in these decays and that the two particles in fact are the same (the *K*-meson). To establish the difficulty which one was faced with before the parity violation was discovered, let us assume that both particles have spin zero (the argument can easily be generalized to higher spins).

For the Θ-particle decay the total angular momentum l of the final state is zero due to angular momentum conservation. Thus the parity of the final state is

$$\eta = \eta(\pi^+)\eta(\pi^0)(-1)^0 = +1$$

since the π-mesons are pseudoscalar particles. Parity conservation then requires that $\eta_\Theta = +1$.

In the τ decay we denote the relative orbital angular momentum of the $2\pi^+$ system by l and the orbital angular momentum of the π^- relative to the centre of mass of the $2\pi^+$ system by L. If the spin of the τ is zero, then angular momentum conservation requires $\mathbf{l} + \mathbf{L} = 0$ or $l = L$. The total parity of the final state is then $(-1)^3(-1)^{l+L} = -1$, and conservation of parity requires that $\eta_\tau = -1$. Thus the Θ and the τ seem to have opposite parity. The complete discussion of this problem is found in reference[4].

(ii) *The* β-*decay of polarized* Co^{60}. The first actual verification of Lee and Yang's hypothesis that parity is not conserved in weak processes was obtained from the study of the decay[5]

$$Co^{60} \to Ni^{60} + e^- + \nu_e$$

The Co^{60} nucleus has the spin-parity assignment 5^+ in its ground state and it decays into an excited state of Ni^{60} with 4^+. Since Co^{60} has non-vanishing spin one may polarize a sample of Co^{60} nuclei. In this way one defines a direction in space. Consider now the two experimental configurations of figure 3. Clearly the two configurations are related by a space reflection transformation in the active sense. From (55) we know that if the theory is invariant under P, then the transition probability for the two decays must be the same, a prediction one can easily test experimentally. Thus, one should investigate if in a sample of polarized Co^{60} nuclei the number of electrons emitted in the forward direction with regard to $\langle \boldsymbol{\sigma}_N \rangle$ is the same as the number emitted in the backward direction. In the

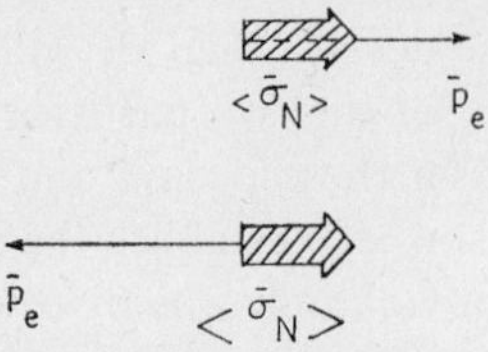

FIGURE 3 Decay configuration in the β-decay of polarized Co^{60} nuclei

experiment by Wu and others[5] one found that the electrons were preferentially emitted in the backward direction, from which one then could conclude that P in fact is violated.

We may express the result in a slightly different way. In an experiment, where one observes (i) $\langle \boldsymbol{\sigma}_N \rangle$ and (ii) the momentum $\mathbf{p}_e$ of the emitted electron but no other quantities, the most general form of the transition rate consistent with rotational invariance is given by

$$\frac{\mathrm{d}\lambda}{\mathrm{d}(\cos\theta)} = A(1 + \alpha\langle \boldsymbol{\sigma}_N \rangle \cdot \hat{\mathbf{p}}_e) \tag{56}$$

where θ is the angle between the two vectors $\langle \boldsymbol{\sigma}_N \rangle$ and $\hat{\mathbf{p}}_e$. Since the second term changes sign under a space reflection it is clear that P is a valid symmetry only if $\alpha = 0$. Experimentally one found $\alpha_{\text{exp}} \approx -v/c$, where v is the electron velocity.

We note in passing that an observed asymmetry is evidence for P-violation. On the other hand the absence of a P-violating effect may always be accidental and cannot be taken as conclusive evidence for P conservation.

(iii) *Space reflection and the process* $\nu_\mu + n \rightarrow \mu^- + p$. In many important applications invariance arguments are used to restrict the possible form of matrix elements. We shall demonstrate how the method works by considering the process $\nu_\mu + n \rightarrow \mu^- + p$ in relation to the various discrete transformations. The process is due to the weak interactions and it has been investigated in connection with the discovery that there exist two kinds of neutrinos, namely ν_μ and ν_e.

Two configurations related by a space reflection are depicted in figure 4. Since parity is not conserved by the weak interaction we cannot deduce an equality between the two matrix elements. Instead we shall demonstrate

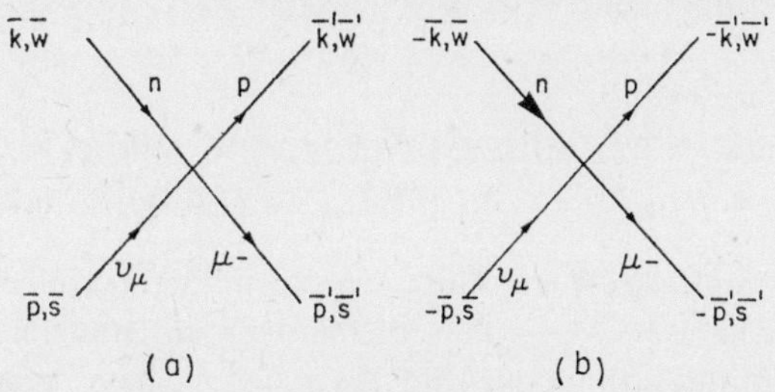

FIGURE 4 Two configurations of the process $\nu_\mu + n \to \mu^- + p$ related by a space reflection

the difference between (a) and (b) within the conventional *V-A* theory of the weak interactions. This is a field-theoretic model with an interaction described by the Hamiltonian density

$$H'(x) = \frac{G}{\sqrt{2}} J_\mu^\dagger(x) J^\mu(x) \tag{57}$$

where $G \approx 10^{-5}/m_p^2$ is the weak interaction coupling constant and the current J_μ may be decomposed into

$$J_\mu(x) = J_\mu^l(x) + J_\mu^h(x) \tag{58}$$

corresponding to a leptonic part and a hadronic part (involving strongly interacting particles $\equiv$ hadrons). This separation is motivated by the fact that leptons do not interact strongly. For that reason the matrix element of the leptonic current has a very simple structure.

If we treat the problem to lowest order in the perturbation expansion we obtain for the process (a):

$$\begin{aligned}\langle p, \mu^-|S|n, \nu_\mu\rangle &= -\mathrm{i}\langle p, \mu^-|\textstyle\int \mathrm{d}^4x H'(x)|n, \nu_\mu\rangle \\ &= -\frac{\mathrm{i}G}{\sqrt{2}}(2\pi)^4\delta^4(P_f - P_i)\langle p|J_\mu^\dagger(0)|n\rangle l^\mu \end{aligned}\tag{59}$$

with[6]

$$l^\mu \equiv \langle \mu^-|J^\mu(0)|\nu_\mu\rangle = \bar{u}(\mathbf{p}', \mathbf{s}')\gamma^\mu(1+\gamma_5)u(\mathbf{p}, \mathbf{s}) \tag{60}$$

In (60) we have omitted some trivial normalization factors which are irrelevant for the present discussion. For the hadronic matrix element $\langle p|J_\mu^\dagger(0)|n\rangle$ we cannot neglect the structure effects due to the strong interactions. However, we know that the matrix element shall transform as a four-vector under proper Lorentz transformations. Using this fact and the properties of the γ-matrices and the spinors $u(\mathbf{p}, \mathbf{s})$ one can easily show that the most general form is given by

$$\begin{aligned} h_\mu &\equiv \langle p|J_\mu{}^\dagger(0)|n\rangle \\ &= \bar{u}(\mathbf{k}', \mathbf{w}')\{F_1(q^2)\gamma_\mu + \mathrm{i}\, F_2(q^2)\sigma_{\mu\nu}q^\nu + F_3(q^2)q_\mu + \\ &\quad + [G_1(q^2)\gamma_\mu + \mathrm{i}\, G_2(q^2)\sigma_{\mu\nu}q^\nu + G_3(q^2)q_\mu]\gamma_5\}u(\mathbf{k}, \mathbf{w}) \end{aligned} \tag{61}$$

where $F_i(q^2)$ and $G_i(q^2)$ are unknown form-factors depending on the variable q^2, where $q = k' - k$ is the momentum transfer. These form-factors are complex functions and the i factors have been inserted for convenience. We shall see later that with the conventions of (61), time reversal invariance will permit us to choose the form factors to be real functions. The corresponding expressions $l_\mu(P)$ and $h_\mu(P)$ for the process (b) are similarly given by

$$l^\mu(P) = \bar{u}(-\mathbf{p}', \mathbf{s}')\gamma^\mu(1 + \gamma_5)u(-\mathbf{p}, \mathbf{s}) \tag{62}$$

and

$$\begin{aligned} h_\mu(P) = \bar{u}(-\mathbf{k}', \mathbf{w}')\{&F_1(q^2)\gamma_\mu + \mathrm{i}\varepsilon(\nu)F_2(q^2)\sigma_{\mu\nu}q^\nu + \varepsilon(\mu)F_3(q^2)q_\mu + \\ &+ [G_1(q^2)\gamma_\mu + \mathrm{i}\varepsilon(\nu)G_2(q^2)\sigma_{\mu\nu}q^\nu + \varepsilon(\mu)G_3(q^2)q_\mu]\gamma_5\}u(-\mathbf{k}, \mathbf{w}) \end{aligned} \tag{63}$$

The factors $\varepsilon(\mu)$ take care of the change in sign for the space components of the momentum transfer under the space reflection. To compare the matrix elements for the two processes we make use of the following relations

$$\begin{aligned} u(-\mathbf{k}, \mathbf{w}) &= \gamma_0 u(\mathbf{k}, \mathbf{w}) \\ \bar{u}(-\mathbf{k}, \mathbf{w}) &= \bar{u}(\mathbf{k}, \mathbf{w})\gamma_0 \end{aligned} \tag{64}$$

Inserting these in (62) and (63) we obtain

$$\begin{aligned} l^\mu(P) &= \bar{u}(\mathbf{p}',\mathbf{s}')\gamma_0\gamma^\mu(1 + \gamma_5)\gamma_0 u(\mathbf{p}, \mathbf{s}) \\ &= \varepsilon(\mu)\bar{u}(\mathbf{p}', \mathbf{s}')\gamma^\mu(1 - \gamma_5)u(\mathbf{p}, \mathbf{s}) \end{aligned} \tag{65}$$

and

$$\begin{aligned} h_\mu(P) = \varepsilon(\mu)\bar{u}(\mathbf{k}', \mathbf{w}')\{&F_1(q^2)\gamma_\mu + \mathrm{i}F_2(q^2)\sigma_{\mu\nu}q^\nu + F_3(q^2)q_\mu - \\ &- [G_1(q^2)\gamma_\mu + \mathrm{i}G_2(q^2)\sigma_{\mu\nu}q^\nu + G_3(q^2)q_\mu]\gamma_5\}u(\mathbf{k}, \mathbf{w}) \end{aligned} \tag{66}$$

In forming the product $h_\mu(P)l^\mu(P)$ we see that the factors $\varepsilon(\mu)$ cancel out. Comparison between l_μ and $l_\mu(P)$ and between h_μ and $h_\mu(P)$ shows that all terms containing γ_5 have changed sign. Since an overall change of sign for the S-matrix element does not affect the transition probability, we find that the cross-sections for the processes (a) and (b) differ because of the *simultaneous* presence of one term with γ_5 and one without γ_5 in l_μ and h_μ separately. This gives rise to changes in the relative sign between the different terms in the product $h_\mu l^\mu$ as compared to $h_\mu(P)l^\mu(P)$. If we introduce the symbolic notation

$$h_\mu l^\mu \equiv \phi(F_i, G_i; \gamma_5) \tag{67}$$

then the result of this section may be expressed as follows:

$$h_\mu(P)l^\mu(P) = \phi(F_i, G_i; -\gamma_5) \tag{68}$$

To establish the difference between the transition rates of (a) and (b) experimentally one must design experiments which depend on the change in the relative signs due to a space reflection transformation. This is equivalent to saying that one must study pseudoscalar quantities such as the correlation $\boldsymbol{\sigma}_\mu, \boldsymbol{p}_\mu$, that is the longitudinal polarization of the muon.

3.3 *Tests of space reflection invariance*

In this section we will give a brief summary of the present status of space reflection invariance with regard to the strong, electromagnetic and weak interactions. For a more complete discussion of the experiments we refer to the original papers.

(i) *Parity conservation in strong interaction processes.* There is so far no experimental evidence for parity violations in strong processes. Since any strong process involves also electromagnetic and weak corrections the precision by which one may establish parity conservation for the strong interactions is limited to the level at which these corrections show up. As we shall see below there is strong evidence for parity conservation also for the electromagnetic interactions, and consequently one may push the limit for parity violating effects down to the level of the weak corrections.

(1) Parity mixing in the energy levels of nuclei. The energy levels of nuclei are eigenstates of the Hamiltonian H. If space reflection invariance holds, then we have seen that $[H, P] = 0$ and one may classify the eigenstates of H also with respect to their parity. If space reflection invariance does not hold, then H will in general induce transitions between states of different parity. The fact that one can classify the nuclear levels according to parity, as is done in nuclear spectroscopy, is thus by itself an indication that parity cannot be violated in any significant way.

On the basis of the parity assignments for nuclear levels, one may now study the restrictions on various nuclear reactions imposed by space reflection invariance. A suitable reaction, which has been studied[7], is

$$Li^{6*} \rightarrow He^4 + d \tag{69}$$

where Li^{6*} refers to the 3·56 Mev level with the spin-parity assignment 0^+. Space reflection invariance forbids this break-up process. To see this we first note that the spin and parity of He^4 and d are found to be 0^+ and 1^+ respectively. If further angular momentum conservation is taken into account, one finds that the final state must have odd parity and the reaction can only take place if parity conservation does not hold. On the other hand, if parity is violated by the strong nuclear forces then there may be a small admixture of negative parity in the Li^{6*} level, that is, the state will be a superposition of state vectors with opposite parity. If the negative parity component has an amplitude F as compared to the 0^+ component, then one may estimate $|F|^2$ by looking for the rate of the break-up process (69).

Experimentally the break-up of Li^{6*} presents severe difficulties. These are circumvented by considering the inverse process for which the selection rule obviously works equally well. In practice one then looks for the reaction.

$$He^4 + d \rightarrow Li^{6*} \rightarrow Li^6 + \gamma \tag{70}$$

where the final state is reached via the resonant intermediate state Li^{6*}. If the intermediate state is unstable under the break-up process then there is a partial width $\Gamma_{d\alpha}$ of the corresponding level, and this width enters in the cross-section for the process (70) as a measure of the probability that the initial state may reach the intermediate state. A small width $\Gamma_{d\alpha}$ clearly corresponds to a small chance for the process to occur. Experimentally one has determined an upper limit for the width of 0·2 ev from which one may conclude that $|F|^2 < 10^{-7}$.

(2) Circular polarization in nuclear γ-transitions. If parity is strictly conserved, then the γ-rays emitted following the bombardment of an unpolarized target by unpolarized particles can show no circular polarization. Since circular polarization is a correlation of the type $(\boldsymbol{\sigma} \cdot \mathbf{k})$ it is a result of an interference between parity conserving and parity non-conserving parts in the interaction. It is then proportional to the amplitude F introduced in the previous example. Several experiments of this general type have been reported. We shall only mention one of these with implications for the presence of weak parity-violating forces between the nucleons in the nuclei. In the experiment[8] one measured the γ-transitions of 482 kev

in Ta^{181}. The transition is between a $5/2^+$ state and a $7/2^+$ state and it may occur as an $M1$ or $E2$ emission if parity is conserved. If parity is not conserved there may be admixtures of $5/2^-$ and $7/2^-$ states and also the $E1$ matrix element may be non-vanishing. A circular polarization may then be the result of interference between the $E1$ and $M1$ matrix elements. One observed a circular polarization $P = -(2{\cdot}0 \pm 0{\cdot}4) \times 10^{-4}$ from which one concluded that there are parity non-conserving nuclear forces. Expressed in terms of the amplitude F one concludes from this experiment that $|F| \approx 10^{-6}$. On the basis of the V-A theory with a current x current interaction one expects an effect of about this size, and the result was interpreted as a verification of this prediction. More recent experiments show no effect and the result of reference[8] is at present in doubt[9].*

(3) Angular correlation of γ-emission from polarized nuclei. Unless parity conservation breaks down the emission of particles from an isolated nuclear level cannot display odd powers of cos θ, where θ is the angle between the initial polarization direction and the momentum of the emitted particle. An experiment of this type is the decay of polarized Cd^{114*} through γ-emission[10]

$$n + Cd^{113} \rightarrow Cd^{114*} \rightarrow Cd^{114} + \gamma \tag{71}$$

By using polarized neutrons they obtain polarized Cd^{114*}. Spin and parity assignments are 1^+ and 0^+ for Cd^{114*} and Cd^{114} respectively, so that is must be an $M1$ emission if parity is conserved. If parity is not conserved also $E1$ radiation is possible and there will be a term proportional to cos θ in the angular correlation. From the experiment one obtained $E1/M1 = (4 \pm 8) \times 10^{-4}$ corresponding to $|F|^2 < 2 \times 10^{-9}$, consistent with no parity violation.

(4) The electric dipole moment of the neutron. A very stringent but less unambiguous test of P-invariance for the strong interactions is the measurement of the electric dipole moment of the neutron. At present the upper limit is set by

$$\frac{\mu_n}{e} < (0{\cdot}1 \pm 2{\cdot}4) \times 10^{-20}\,\text{cm}$$

but more accurate measurements are in progress. Space reflection invariance requires it to vanish. However, also from time reversal invariance

* Note added in proof.
More accurate measurements have been performed and indicate a small P-violating effect of the order $|F| \approx 10^{-7}$. cf. L. Wolfenstein, *Proc. Heidelberg Int. Conf. on Elementary Particles*, North-Holland Publishing Company, 1968.

one arrives at the same conclusion. Therefore, the result cannot be attributed to P-invariance with any certainty.

(ii) *Parity conservation in electromagnetic processes.* All the experiments quoted in connection with the strong interactions clearly have bearing also on the electromagnetic interactions, although the accuracy in this case is rather poor. Much more accurate evidence for parity conservation in electromagnetic interactions is obtained from optical spectroscopy of atomic levels. The evidence in the literature suggests that impurities of parities in the energy levels of atoms or in their electromagnetic transitions is less than about 10^{-7} in intensity, that is, the parity-violating amplitude must be less than 10^{-3} of the parity-conserving amplitude.

(iii) *Parity conservation and the weak interactions.* In many of the applications of space reflection invariance we used examples of weak processes, and it was found that in fact they violate this invariance principle strongly. In the conventional V-A theory the parity violation is maximal as implied by the fact that vector and axial vector parts are of equal strength. This theory seems to account for all known weak processes quite accurately and parity-violating effects have been established in a wide variety of reactions.

4 Charge Conjugation

The classical theory of electromagnetism can be derived from (i) Coulomb's law for the force between charges and (ii) the theory of special relativity. In Coulomb's law there is full symmetry between positive and negative charges; equal charges repel each other and unequal attract each other. Since the theory of special relativity only concerns space-time concepts, which are detached from the concept of charge, it is clear that the classical theory of electromagnetism also exhibits a basic symmetry between the two kinds of charges.

Later, when the relativistic quantum theory was developed, it was found that to each particle there must exist an antiparticle with the same space-time properties as the original one. Electromagnetism was incorporated in analogy with the classical theory and consistency then required that a particle and its antiparticle must have opposite charges. In this way one was led to consider the symmetry of charge more appropriately as a particle-antiparticle symmetry. For historical reasons one retained the name charge conjugation symmetry, although particle-antiparticle symmetry is a more accurate name for it now.

We may remark already at this point that invariance under charge conjugation C most probably holds for strong and electromagnetic interactions (see Section 4.3 below), while the weak interactions violate it. This situation thus equals the situation for P-invariance previously discussed and it seems that the appropriate particle-antiparticle transformation is the combined operation CP. To this we shall return in Section 5.

4.1 *Definition and action on state vectors*

It has already been stated that particles and antiparticles have the same space-time properties. This requires that the operator representing the transformation must commute with all the generators of space-time transformations, that is, the Poincaré group extended to include also the inversions (P and T). On the other hand, charge, baryon number and lepton number have opposite sign for particle and antiparticle, and the corresponding operators in Hilbert space must then anticommute with the charge conjugation operator C. Thus we may classify the various operators with regard to their commutation rules with C; *operators of the first class* commute with C and *operators of the second class* anticommute with C.

We now define the action of the charge conjugation operator on a single-particle state $|\alpha; \mathbf{k}, \lambda\rangle$ by the relation

$$|\alpha; \mathbf{k}, \lambda\rangle \rightarrow |\alpha'; \mathbf{k}', \lambda'\rangle = C|\alpha; \mathbf{k}, \lambda\rangle = \eta_c| -\alpha; \mathbf{k}, \lambda\rangle \tag{72}$$

where α denotes all the labels corresponding to observables of the second class (charge, etc.) and η_c is a phase factor in order that C be a unitary operator. Applying a second charge conjugation operator one is led back to the original state, and by choosing the phase factor η_c suit the one may identify C^2 with the identity operator. With this choiceably eigenvalues of C are ± 1. Clearly, eigenstates of C can those states only be, for which all charges α are zero. For example, only those states which have $Q = 0$, $B = 0$ and $L = 0$ can be eigenstates of C. These conditions express the obvious fact that only if the particle is identical with its antiparticle can the corresponding single particle state be an eigenstate of C. In the examples we shall encounter there are three particles which satisfy these conditions; the photon, the π^0 meson and the η meson.

To determine the charge parity η_c for the photon we recall that the whole concept of charge conjugation originated from the electromagnetic interaction which exhibited invariance under the transformation. The interaction is due to the coupling of the electromagnetic current to the electromagnetic field. Under a C transformation the current changes sign and, hence, also the electromagnetic field must change sign in order that the interaction be invariant. Since the photon is the elementary quantum of the electromagnetic field we conclude

$$C|\gamma\rangle = -|\gamma\rangle \tag{73}$$

Since space-time properties are not affected by the C transformation we may extend (73) to an arbitrary n-photon state

$$C|n\gamma\rangle = (-1)^n|n\gamma\rangle \tag{74}$$

Similar considerations for the π^0 meson and the η meson yield

$$C|\pi^0\rangle = |\pi^0\rangle \tag{75}$$

$$C|\eta\rangle = |\eta\rangle \tag{76}$$

and correspondingly for the n-particle states involving neutral mesons.

We next proceed to the 2-particle states such as e^+e^-, $p\bar{p}$, etc., since these may be eigenstates of C, and we shall here only consider such states. Then the so-called *generalized Pauli principle* applies. To specify the content of this principle we regard a particle and its antiparticle as two different states of the same particle. The two states differ by opposite sign for all the quantum numbers (charges) corresponding to observables of the second class. The generalized Pauli principle now asserts that any fermion-antifermion state is antisymmetric under the transformation which exchanges the two particles, while a boson-antiboson state is symmetric under the same transformation. In field theory this principle expresses the fact that creation and destruction operators satisfy (i) commutation relations for bosons and (ii) anticommutation relations for fermions. If we denote the spin and coordinate exchange operator by $A(\sigma, r)$, then we may write the principle in the following way

$$CA(\sigma, r)|\alpha, \bar{\alpha}\rangle = (-1)^{2s}|\alpha, \bar{\alpha}\rangle \tag{77}$$

Here α and $\bar{\alpha}$ denote any particle and its antiparticle, respectively, of spin s. Thus any particle-antiparticle state which is an eigenstate of $A(\sigma, s)$ is simultaneously an eigenstate of C by (77). We have previously introduced the 2-particle basis vectors $|\mathbf{P} = 0; J, M, p; l, \sigma\rangle$ which are eigenstates of the parity operator. These states are also eigenvectors of $A(\sigma, r)$ and a closer examination yields

$$A(\sigma, r)|\mathbf{P} = 0; J, M, p; l, \sigma\rangle = (-1)^{l+\sigma-2s}|\mathbf{P} = 0; J, M, p; l, \sigma\rangle \tag{78}$$

and hence

$$C|\mathbf{P} = 0; J, M, p; l, \sigma\rangle = (-1)^{l+\sigma}|\mathbf{P} = 0; J, M, p; l, \sigma\rangle \tag{79}$$

This result will be most important for applications later. Relation (79) clearly holds even if one does not restrict oneself to states with vanishing total momentum.

Once the transformation rules for state vectors are given we may deduce the corresponding rules for field operators. Just as in the case of the space reflection transformation we quote only the results for fields of spin 0, $\frac{1}{2}$ and 1.

$$\begin{aligned} C\phi(x)C^{-1} &= \eta_c\phi^{\dagger}(x) \\ C\psi(x)C^{-1} &= \eta_c\mathscr{C}\bar{\psi}^T(x) \\ C\bar{\psi}(x)C^{-1} &= -\eta_c{}^*\psi^T(x)\mathscr{C}^{-1} \\ CA_\mu(x)C^{-1} &= -\eta_c A_\mu(x) \end{aligned} \tag{80}$$

where $\mathscr{C}$ is a 4×4 matrix, whose properties are given in (92) below.

4.2 *Some applications of charge conjugation invariance*

Invariance of the S-operator under charge conjugation implies

$$\langle\beta|S|\alpha\rangle = \langle C\beta|S|C\alpha\rangle \tag{81}$$

If the states $|\alpha\rangle$ and $|\beta\rangle$ are eigenstates of C with the eigenvalues η_α and η_β, then (81) may be written

$$\langle\beta|S|\alpha\rangle = \eta_\alpha\eta_\beta\langle\beta|S|\alpha\rangle \tag{82}$$

so that only if $\eta_\alpha\eta_\beta = +1$ do we obtain a non-vanishing matrix element. Under these circumstances C-invariance provides us with a selection rule. The more general case of (81) always leads to relations between matrix elements imposing restrictions on the possible form for these matrix elements. We shall see below how these remarks apply in a few cases.

(i) *The decay* $\pi^0 \rightarrow n\gamma$. The more restrictive relation (82) applies only if the two states $|\alpha\rangle$ and $|\beta\rangle$ are eigenstates of C. We have previously seen that a necessary but not always sufficient condition for a state to be an eigenstate of C is that it is neutral ($Q = 0$) and we have found that both π^0 states and photon states in fact are eigenstates of C. From (74) and (75) we know that $\eta_c(\pi^0) = +1$ and $\eta_c(\gamma) = -1$ and hence

$$\langle n\gamma|S|\pi^0\rangle = (-1)^n\langle n\gamma|S|\pi^0\rangle \tag{83}$$

For $\langle n\gamma|S|\pi^0\rangle \neq 0$ we then obtain $n = 0$ mod. 2, that is, a π^0 can only decay into an even number of photons if the interaction is C-invariant. Clearly the same conclusion holds for the decay of the η-meson into photons.

(ii) *The decay of positronium into photons*. Positronium is a bound state of (e^+e^-). Provided positronium is in a definite (l, σ) state it will be an eigenstate of C with the eigenvalue $(-1)^{l+\sigma}$. This follows directly from (79). For the decay

$$(e^+e^-)_{l,\sigma} \rightarrow n\gamma \tag{84}$$

we obtain the following selection rule

$$(-1)^{l+\sigma} = (-1)^n \tag{85}$$

or $l + \sigma = n$ mod. 2. From this selection rule we find (conventional spectroscopic notation is used for the positronium states)

$$\begin{aligned} {}^1S_0 &\rightarrow 2\gamma;\ l = \sigma = 0;\ n = 2 \\ &\not\rightarrow 3\gamma;\ l = \sigma = 0;\ n = 3 \\ {}^3S_1 &\not\rightarrow 2\gamma;\ l = 0;\ \sigma = 1;\ n = 2 \\ &\rightarrow 3\gamma;\ l = 0;\ \sigma = 1;\ n = 3 \end{aligned} \tag{86}$$

etc.

The $p\bar{p}$ annihilation into π^0's is, of course, completely analogous to the positronium decay and one obtains equivalent selection rules.

(iii) *The decay* $\eta \rightarrow \pi^+ + \pi^- + \pi^0$. This decay is of electromagnetic nature, since it does not conserve G-parity, a concept we shall not be concerned with, however. The decay has been proposed as a suitable process to test C-invariance for the electromagnetic interactions. As pointed out by Bernstein, Feinberg and Lee[11], many previous tests of C-invariance with regard to the electromagnetic interactions are ambiguous.

If the interaction responsible for the η decay is C-invariant, then we obtain from (81)

$$\langle \pi^+(\mathbf{k}_+), \pi^-(\mathbf{k}_-), \pi^0(\mathbf{k})|S|\eta\rangle = \eta_c\langle \pi^-(\mathbf{k}_+), \pi^+(\mathbf{k}_-), \pi^0(\mathbf{k})|S|\eta\rangle \tag{87}$$

where η_c is a phase factor which, however, is irrelevant for the present discussion. This relation implies that in η decay the probability of finding a π^+ with momentum $\mathbf{k}_+$ and a π^- with momentum $\mathbf{k}_-$ is the same as finding a π^+ with momentum $\mathbf{k}_-$ and a π^- with momentum $\mathbf{k}_+$. This in turn implies that the energy spectra for π^+ and π^- in η decay should be the same (the energy spectrum of π^+ is obtained by integrating the decay rate over all angles and the energies of the π^- and the π^0, etc.). It also implies that the charged π-meson with the largest energy in η decays should be a π^+ as often as a π^-. An unbalance in this respect clearly violates C-invariance. The experimental results seem to favour no C-violation for this process but further clarification would be desirable (see below).

(iv) *Charge conjugation and the process* $\nu_\mu + n \to \mu^- + p$. We have previously studied this process in Section 3.2 in connection with the space reflection transformation. In this section we will discuss it with regard to the charge conjugation transformation and as before we work within the framework of the *V-A* theory.

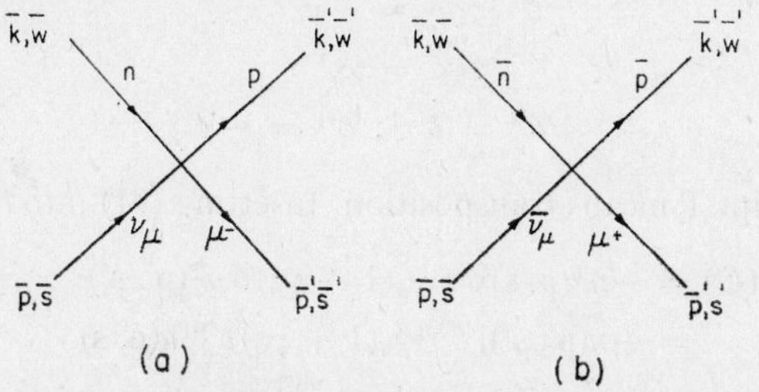

FIGURE 5 Two processes (a) $\nu_\mu + n \to \mu^- + p$ and (b) $\bar{\nu}_\mu + \bar{n} \to \mu^+ + \bar{p}$ which are related by a charge conjugation transformation

In figure 5 the diagrams for the process $\nu_\mu + n \to \mu^- + p$ and the charge conjugate process $\bar{\nu}_\mu + \bar{n} \to \mu^+ + \bar{p}$ are given. Since the C-transformation does not affect the space-time properties of a state there is in this case no change in the momenta and the spins. The matrix element of the process (a) is given by (59), (60) and (61). To compute the matrix element of process (b) it must be realized that although the *V-A* Hamiltonian $H'(x)$ of (57) is hermitian each term by itself does not have this property, that is the currents of (58) are not hermitian. The relevant part of the S-operator for the process (b) is in fact the hermitian conjugate of the part describing the process (a)*, and the matrix element is given by

$$\langle \bar{p}, \mu^+ | S | \bar{n}, \bar{\nu}_\mu \rangle = -\frac{iG}{\sqrt{2}} (2\pi)^4 \delta^4(P_f - P_i) \langle \bar{p} | J_\mu(0) | \bar{n} \rangle l^\mu(C) \quad (88)$$

where

$$l^\mu(C) \equiv \bar{v}(\mathbf{p}, \mathbf{s}) \gamma^\mu (1 + \gamma_5) v(\mathbf{p}', \mathbf{s}') \quad (89)$$

and

$$\begin{aligned} h_\mu(C) &\equiv \langle \bar{p} | J_\mu(0) | \bar{n} \rangle = \langle \bar{n} | J_\mu{}^\dagger(0) | \bar{p} \rangle^* \\ &= \{\bar{v}(\mathbf{k}', \mathbf{w}')[F_1(q^2)\gamma_\mu - iF_2(q^2)\sigma_{\mu\nu}q^\nu - iF_3(q^2)q_\mu + \quad (90) \\ &\quad + G_1(q^2)\gamma_\mu\gamma_5 - iG_2(q^2)\sigma_{\mu\nu}q^\nu\gamma_5 - G_3(q^2)q_\mu\gamma_5] v(\mathbf{k}, \mathbf{w})\}^* \end{aligned}$$

* Note that $\langle p | J_\mu{}^\dagger(0) | n \rangle = \langle \bar{p} | C J_\mu{}^\dagger(0) C^{-1} | \bar{n} \rangle$ and $C J_\mu{}^\dagger(0) C^{-1} = J_\mu(0)$ if we leave out overall phase factors. A similar relation holds for the lepton part.

This last relation is most easily derived if one makes explicit use of an effective current operator in terms of field operators[12]. To compare the matrix elements of the two processes we make use of the following properties of the Dirac spinors

$$\begin{aligned} v(\mathbf{p}, \mathbf{s}) &= \mathscr{C}\bar{u}^{\mathrm{T}}(\mathbf{p}, \mathbf{s}) \\ \bar{v}(\mathbf{p}, \mathbf{s}) &= -u^{\mathrm{T}}(\mathbf{p}, \mathbf{s})\mathscr{C}^{-1} \end{aligned} \tag{91}$$

with the matrix $\mathscr{C}$ defined by

$$\begin{aligned} \mathscr{C}^{-1}\gamma_\mu\mathscr{C} &= -\gamma_\mu{}^{\mathrm{T}} \\ \mathscr{C}^{-1}\gamma_5\mathscr{C} &= \gamma_5{}^{\mathrm{T}} \\ \mathscr{C}^\dagger = \mathscr{C}^{-1}&; \ \mathscr{C}^{\mathrm{T}} = -\mathscr{C} \end{aligned} \tag{92}$$

The superscript T means transposition. Inserting (91) into $l^\mu(C)$ we obtain

$$\begin{aligned} l_\mu(C) &= -u^{\mathrm{T}}(\mathbf{p}, \mathbf{s})\mathscr{C}^{-1}\gamma_\mu(1 + \gamma_5)\mathscr{C}\bar{u}^{\mathrm{T}}(\mathbf{p}', \mathbf{s}') \\ &= -\bar{u}(\mathbf{p}', \mathbf{s}')[\mathscr{C}^{-1}\gamma_\mu(1 + \gamma_5)\mathscr{C}]^{\mathrm{T}}u(\mathbf{p}, \mathbf{s}) \\ &= \bar{u}(\mathbf{p}', \mathbf{s}')[\gamma_\mu{}^{\mathrm{T}}(1 + \gamma_5)^{\mathrm{T}}]^{\mathrm{T}}u(\mathbf{p}, \mathbf{s}) \\ &= \bar{u}(\mathbf{p}', \mathbf{s}')[\gamma_\mu(1 - \gamma_5)u(\mathbf{p}, \mathbf{s}) \end{aligned} \tag{93}$$

and similarly for $h_\mu(C)$

$$\begin{aligned} h_\mu(C) = \bar{u}(\mathbf{k}', \mathbf{w}')\{&F_1{}^*(q^2)\gamma_\mu + \mathrm{i}F_2{}^*(q^2)\sigma_{\mu\nu}q^\nu + F_3{}^*(q^2)q_\mu - \\ &- [G_1{}^*(q^2)\gamma_\mu + \mathrm{i}G_2{}^*(q^2)\sigma_{\mu\nu}q^\nu + G_3{}^*(q^2)q_\mu]\gamma_5\}u(\mathbf{k}, \mathbf{w}) \end{aligned} \tag{94}$$

With the symbolic notations of (67) we may write this result in the following way

$$h_\mu(C)l^\mu(C) = \phi(F_i{}^*, G_i{}^*; -\gamma_5) \tag{95}$$

Once more we find that it is the simultaneous presence of terms with γ_5 and without γ_5 in both l_μ and h_μ that makes it impossible to achieve C-invariance. Any quantity which depends on the interference between the two kind of terms will change sign when we pass from the process (a) to the process (b).

4.3 *Tests of charge conjugation invariance*

We proceed to discuss some of the more stringent tests of C-invariance for the three types of interactions.

(i) *C-invariance in strong interaction processes.* There is so far no evidence for C-violation in strong interaction processes. The most accurate test seems to be annihilation experiments of $p\bar{p}$ at rest. It is believed that the annihilation takes place from an *S*-state ($l = 0$)[13]. From (79) it then follows that the initial state is an eigenstate of C with $\eta_c = +1$ for $\sigma = 0$ (antiparallel spins) and $\eta_c = -1$ for $\sigma = 1$ (parallel spins). Thus, from angular momentum conservation, we conclude that the $p\bar{p}$ state with $\eta_c = +1$ will go over into an overall *S*-state and the state with $\eta_c = -1$ into a *P*-state. Therefore, if one performs the angular integrations over the final states, there will be no interference between the contributions from the two possible initial $p\bar{p}$ states, and the momentum distribution for the annihilation products must be invariant under interchange of positive and negative particles. In the experimental test of this prediction[14] one has examined for example the following annihilation channels

$$\begin{aligned} p + \bar{p} &\rightarrow \pi^+ + \pi^- + \pi^0 \\ &\rightarrow \pi^+ + \pi^+ + \pi^- + \pi^- \\ &\rightarrow \pi^+ + \pi^- + x^0 \qquad\qquad (m_{x^0} > m_{\pi^0}) \\ &\rightarrow K^+ + \pi^- + \bar{K}^0 \\ &\rightarrow K^- + \pi^+ + K^0 \\ &\rightarrow K^+ + \pi^- + \pi^0 + \bar{K}^0 \\ &\rightarrow K^- + \pi^+ + K^0 + \pi^0 \end{aligned}$$

In the *K*-channels one compared energy and momentum distributions of the K^+ and the K^-, etc. If α is the ratio of the *C*-violating amplitude to the *C*-conserving amplitude, then one obtained

$$|\alpha_\pi| < 0{\cdot}01$$
$$|\alpha_K| < 0{\cdot}03$$

In passing we note that in the absence of final state interactions the same result follows from *CPT* invariance. However, the final state interactions are of strong nature and a *C*-violation is thus expected to give observable effects. This remark touches upon a very important question, namely, a specific prediction may follow from several alternative invariance principles and it may be impossible to interpret a verification of the prediction as evidence for any one of them separately.

(ii) *C-invariance in electromagnetic processes.* In many cases of electromagnetic processes one or several of the participating particles also interact strongly, and tests of *C*-invariance in such cases obviously have bearing also on the question of *C*-invariance for the strong interactions although indirectly.

(1) The decay $\pi^0 \to 3\gamma$. We have previously found that this decay is strictly forbidden if *C*-invariance holds. The dominant decay mode of the π^0 is clearly $\pi^0 \to 2\gamma$ and an upper limit on the branching ratio of the two processes has been determined[15]

$$B = \frac{\lambda(\pi^0 \to 3\gamma)}{\lambda(\pi^0 \to 2\gamma)} \lesssim 5 \times 10^{-6}$$

(2) The decay $\eta \to \pi^+ + \pi^- + \pi^0$. Denote by N_+ the number of η decays where $E(\pi^+) > E(\pi^-)$ and by N_- the number of events with $E(\pi^-) > E(\pi^+)$. We have previously shown that *C*-invariance implies that $N_+ = N_-$. One has determined the asymmetry parameter *A*, where

$$A = \frac{N_+ - N_-}{N_+ + N_-}$$

in a number of experiments[16] and the results are listed in Table 1. It is

TABLE 1

Investigator	Number of decays	*A* per cent
Baltay and others	1351	7·2 ± 2·8
Rutherford–Saclay	705	−6 ± 4
CERN	10665	0·3 ± 1·0

seen that the Cern result rests on the best statistics. In view of the fundamental character of this experiment it is certainly desirable to investigate the matter further.

(3) The decay $\eta \to \pi^+ + \pi^- + \gamma$. The previous discussion of $\eta \to 3\pi$ applies equally well to this process, and one may look for an asymmetry in the energy distribution for the two charged π mesons as evidence for a *C*-violation. If we define the asymmetry parameter *A* in the same way as before it is found[17] that

$$A = (1{\cdot}5 \pm 2{\cdot}5) \times 10^{-2}$$

which is consistent with *C*-conservation for this electromagnetic process.

(4) The decay $\eta \to \pi^0 + e^+ + e^-$. This is clearly an electromagnetic process and to lowest order in the electromagnetic coupling it is described

by the diagram of figure 6. The relevant part of the transition matrix element is $\langle \pi^0 | J_\mu^{EM}(0) | \eta^0 \rangle$ where $J_\mu^{EM}(0)$ is the electromagnetic current operator. We have previously given the transformation properties of the

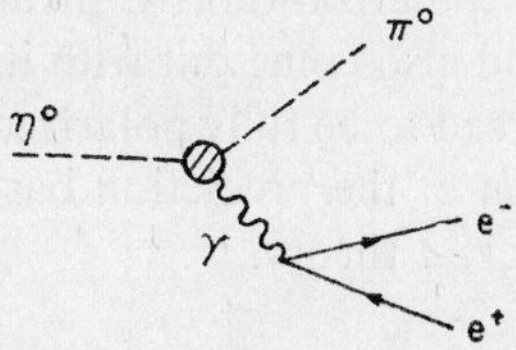

FIGURE 6 The decay $\eta \to \pi^0 + e^+ + e^-$

two state vectors, which appear in this matrix element, and of the current operator. With those relations in mind we find

$$\begin{aligned}\langle \pi^0 | J_\mu^{EM}(0) | \eta^0 \rangle &= \langle \pi^0 | C^{-1} C J_\mu^{EM}(0) C^{-1} C | \eta^0 \rangle \\ &= \langle \pi^0 | C J_\mu^{EM}(0) C^{-1} | \eta^0 \rangle = -\langle \pi^0 | J_\mu^{EM}(0) | \eta^0 \rangle\end{aligned}$$

and we conclude that this matrix element must vanish unless C is violated. Of course, if the electromagnetic current operator has a component which is even under C[11], then (i) C-invariance is violated by the electromagnetic interactions and (ii) the decay $\eta \to \pi^0 + e^+ + e^-$ is no longer forbidden to this order in the coupling constant. Experimentally[18] one has determined the branching ratio for this decay with the following result

$$B \equiv \frac{\lambda(\eta \to \pi^0 + e^+ + e^-)}{\lambda(\eta \to \pi^0 + \pi^+ + \pi^-)} < 0{\cdot}0045$$

which supports the assumption of C-invariance.

It should be noted that this process is allowed by higher order terms in the perturbation expansion even if C-invariance is an exact symmetry. For example, the two-photon exchange diagram does not vanish due to C-invariance. Therefore, one expects that the process may occur but it must be substantially suppressed if C-invariance holds.

(*iii*) *C-invariance of the weak interactions.* The weak interactions violate C-invariance strongly. There are many verifications of this and in the conventional V-A theory the violation is maximal. We will only mention one experiment which clearly demonstrates the C-violation in weak processes.

Consider the processes $\pi^+ \to \mu^+ + \nu_\mu$ and $\pi^- \to \mu^- + \bar{\nu}_\mu$. They are clearly related by a C-transformation. It is experimentally established that in the first one of these decays the μ^+ is emitted fully polarized with the spin antiparallel to its momentum. Since a C-transformation does not affect space-time properties (such as the polarization), C-invariance implies that the μ^- in the second decay should also come out with its spin antiparallel to its momentum. It is found instead to be fully polarized with its spin parallel to its momentum in violation of the prediction based on C-invariance, but in full agreement with the V-A theory.

5 The *CP* Transformation

In the previous discussion the P and C transformations were treated separately. Reviewing the present experimental situation, it was concluded that with regard to the strong and the electromagnetic interactions we have no evidence for violations of separate P and C-invariance. For the weak interactions, non-invariance under the separate P and C transformations is firmly established. In fact, in almost all weak processes which have been observed so far the P and C-violating effects are maximal or nearly maximal. The few exceptions to this general rule are all results of accidental cancellations. In view of these observations the question immediately arises whether the weak interactions are invariant under the combined operation CP, and we shall devote this chapter to that question. For obvious reasons we restrict the discussion to weak processes.

Since we have already given the transformation properties for state vectors under separate P and C transformations, it is a trivial matter to derive the corresponding rules for the operation CP, and we proceed directly to a discussion of some relevant applications.

5.1 *Some applications of the CP symmetry*

From (22) we obtain the following condition on the S-matrix elements, provided the S-operator is invariant under the CP transformation.

$$\langle\beta|S|\alpha\rangle = \langle CP\beta|S|CP\alpha\rangle \tag{96}$$

If we are dealing with states $|\alpha\rangle$ and $|\beta\rangle$, which are eigenstates of CP, then (96) leads to a selection rule. Since the charge conjugation is involved it is clear that the eigenstates of CP must have $Q = B = Y = 0$, and this restricts the class of processes for which CP-invariance imposes absolute selection rules. On the other hand, it should be borne in mind that absolute selection rules generally provide us with the best opportunities to test invariance principles to a high degree of accuracy. This will be evident in the case of *CP*-invariance below.

(i) *The decays* $K^0 \to 2\pi$ and $K^0 \to 3\pi$. For the present discussion *we assume* that *invariance under the CP transformation* is a rigorous symmetry law respected by all interactions. Under this assumption we shall derive absolute selection rules which can be and actually have been subjected to experimental tests.

Neutral *K*-mesons are obtained by associated production for which the strong (electromagnetic) interactions are responsible. These interactions have in common that they conserve the additive quantum number of hypercharge Y or alternatively the strangeness S, where $Y = B + S$. The implication of associated production is that the neutral *K*-mesons obtained in this way at the moment of creation represent eigenstates of hypercharge. There are two such states denoted K^0 and $\bar{K}^0$ for which $Y = +1$ and $Y = -1$ respectively. Obviously Y is an observable of the second class with regard to C and neither K^0 nor $\bar{K}^0$ by itself can be an eigenstate of C or CP. However, with our previous phase convention for the space reflection transformation we obtain for a single-particle state of a neutral *K*-meson *at rest*

$$P|K^0\rangle = -|K^0\rangle$$
$$P|\bar{K}^0\rangle = -|\bar{K}^0\rangle \tag{97}$$

and we may choose the phases for the C transformation such that

$$C|K^0\rangle = -|\bar{K}^0\rangle$$
$$C|\bar{K}^0\rangle = -|K^0\rangle \tag{98}$$

and hence*

$$CP|K^0\rangle = |\bar{K}^0\rangle$$
$$CP|\bar{K}^0\rangle = |K^0\rangle \tag{99}$$

On the basis of (99) we may now define the following eigenstates of CP

$$|K_1{}^0\rangle = \frac{1}{\sqrt{2}}(|K^0\rangle + |\bar{K}^0\rangle)$$

$$|K_2{}^0\rangle = \frac{-\,i}{\sqrt{2}}(|K^0\rangle - |\bar{K}^0\rangle) \tag{100}$$

* With our phase convention for the time reversal transformation T (cf. Section 6) one finds $CPT|K^0\rangle = |\bar{K}^0\rangle$, which is the conventional definition of the relation between the two neutral *K*-mesons. With this latter definition no restriction to the rest system is needed.

corresponding to the eigenvalues +1 respectively −1. Of course, the states $|K^0\rangle$ and $|\bar{K}^0\rangle$ may just as well be regarded as coherent superpositions of $|K_1{}^0\rangle$ and $|K_2{}^0\rangle$. From (100) one obtains

$$|K^0\rangle = \frac{1}{\sqrt{2}}(|K_1{}^0\rangle + \mathrm{i}|K_2{}^0\rangle)$$

$$|\bar{K}^0\rangle = \frac{1}{\sqrt{2}}(|K_1{}^0\rangle - \mathrm{i}|K_2{}^0\rangle)$$

We have previously stated that the strong and the electromagnetic interactions conserve hypercharge. As a consequence there will be no transitions of the type $|K^0\rangle \rightleftarrows |\bar{K}^0\rangle$ due to these interactions. However, the decays of the neutral K-mesons are all due primarily to the weak interactions and these do not conserve hypercharge. This means that due to weak interactions the transitions $|K^0\rangle \rightleftarrows |\bar{K}^0\rangle$ occur and although we create pure states $|K^0\rangle$ and $|\bar{K}^0\rangle$ these will develop into coherent mixtures of the type

$$|\phi\rangle = \alpha(t)|K^0\rangle + \beta(t)|\bar{K}^0\rangle \tag{102}$$

For the discussion of the decays of neutral K-mesons, the states $|K^0\rangle$ and $|\bar{K}^0\rangle$ are clearly not appropriate since they do not permit us to identify from which state the decay occurred. Thus the states $|K^0\rangle$ and $|\bar{K}^0\rangle$ will not exhibit exponential decay laws and they will not be characterized by a unique lifetime. If we instead consider the states $|K_1{}^0\rangle$ and $|K_2{}^0\rangle$ and if CP is a *rigorously valid symmetry*, then we know that the transitions $|K_1{}^0\rangle \rightleftarrows |K_2{}^0\rangle$ are rigorously forbidden and a $|K_1{}^0\rangle$ state respectively a $|K_2{}^0\rangle$ state will retain its identity until the moment of decay. These states will then decay exponentially with unique lifetimes. One concludes that in this case the decay of K^0 will be characterized by two lifetimes corresponding to the two components $|K_1{}^0\rangle$ and $|K_2{}^0\rangle$ and similarly for $\bar{K}^0$. These conclusions are verified by observation and it turns out that

$$\tau(K_1{}^0) = (0{\cdot}87 \pm 0{\cdot}01) \times 10^{-10}\mathrm{s}$$

$$\tau(K_2{}^0) = (5{\cdot}68 \pm 0{\cdot}26) \times 10^{-8}\mathrm{s}$$

that is, the lifetimes differ by almost three orders of magnitude. Beyond this, the decay modes of the states $|K_1{}^0\rangle$ and $|K_2{}^0\rangle$ are different and we shall in this context restrict our discussion to the decay modes with two or three π-mesons in the final state.

Consider first the decay of a neutral K-meson into two π-mesons; $\pi^+\pi^-$ or $\pi^0\pi^0$. In the rest system of the K-meson, $J = 0$ for the initial state, and angular momentum conservation then implies that $J = l = 0$

also in the final state. From (51) and (79) we conclude that since $l = 0$ we have

$$CP|\pi^+, \pi^-\rangle = |\pi^+, \pi^-\rangle$$
$$CP|2\pi^0\rangle = |2\pi^0\rangle \tag{103}$$

To the extent that CP is a good quantum number we then find that the decay

$$K_2{}^0 \not\rightarrow 2\pi$$

is forbidden, while

$$K_1{}^0 \rightarrow 2\pi$$

can occur.

For the decay into three π-mesons we conclude that $J = 0$ implies $L = l$. The parity of any such state is odd and hence

$$CP = -1 \quad \text{for } 3\pi^0$$
$$CP = (-1)^{l+1} \text{ for } \pi^+ + \pi^- + \pi^0$$

where l is the relative orbital angular momentum of the $\pi^+\pi^-$ pair. We conclude that

$$K_1{}^0 \not\rightarrow 3\pi^0$$

is forbidden, while

$$K_2{}^0 \rightarrow 3\pi^0$$

is allowed. Similarly

$$K_1{}^0 \rightarrow \pi^+ + \pi^- + \pi^0$$

is allowed if $l = 1, 3, 5, \ldots$, and

$$K_2{}^0 \rightarrow \pi^+ + \pi^- + \pi^0$$

if $l = 0, 2, 4, \ldots$, while for other l-values these decays are forbidden by CP-invariance.

We shall not pursue this subject further at this point but we shall return to it in Section 5.2 since these decays have played a very important role in recent investigations regarding CP-invariance.

(ii) *The decays* $\pi^+ \rightarrow \mu^+ + \nu_\mu$ *and* $\pi^- \rightarrow \mu^- + \bar{\nu}_\mu$. In Section 4.3 we examined the experimental results with regard to the muon polarization for the decays $\pi^+ \rightarrow \mu^+ + \nu_\mu$ and $\pi^- \rightarrow \mu^- + \bar{\nu}_\mu$. Experimentally it is known that the muon is fully polarized in both cases but the μ^+ has the helicity (= spin component along the momentum vector) -1 while the helicity of the μ^- in the second process is $+1$. This was found to violate C-invariance, since the C transformation relates the matrix elements for the two processes provided the helicity of the μ^+ is the same as that of the μ^-. However, the added space reflection in the CP transformation changes the sign of the helicity and there is no contradiction between the observations and CP-invariance. This is demonstrated in more detail in figure 7.

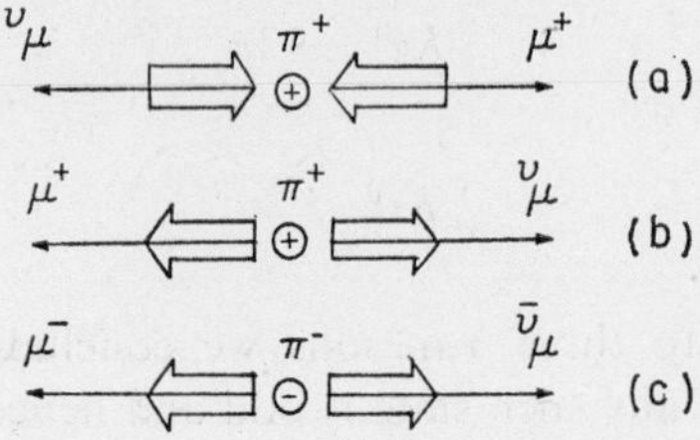

FIGURE 7 Decay configuration for the process $\pi^+ \to \mu^+ + \nu_\mu$ (a), the space reflected configuration (b) and the CP transformed configuration (c)

(iii) *The CP transformation and the process* $\nu_\mu + n \to \mu^- + p$. We have previously considered this process in relation to the P and the C transformations. As before, we write the relevant S-matrix element for this process in the following symbolic way

$$h_\mu l^\mu = (F_i, G_i; \gamma_5) \tag{104}$$

We have found the following relations for the P and the C transformed processes

$$h^\mu(P) l_\mu(P) = \phi(F_i, G_i; -\gamma_5) \tag{105}$$

and

$$h^\mu(C) l_\mu(C) = \phi(F_i{}^*, G_i{}^*; -\gamma_5) \tag{106}$$

For the combined operation CP we then obtain

$$h^\mu(CP) l_\mu(CP) = \phi(F_i{}^*, G_i{}^*; \gamma_5) \tag{107}$$

and we conclude that CP-invariance can hold only if all the form-factors F_i and G_i ($i = 1, 2, 3$) are real. This prediction is verified by experiments* although the accuracy is not sufficient to rule out small CP-violating effects since a small but non-vanishing relative phase factor for the form-factors is very difficult to detect. We shall return to these questions later in the discussion of time reversal invariance.

* The same form-factors appear in the matrix elements for other processes such as the beta-decay of neutrons, and the experimental tests refer to other processes than the one considered here.

5.2 *Experimental tests of CP-invariance in weak processes*

fter the discovery of charge conjugation and parity violations in weak processes the possibility of symmetry under *CP* was suggested[19] as a remedy for the apparent asymmetry between world and antiworld. The *CP* symmetry is closely related to the symmetry under time reversal through the *CPT* theorem (cf. Section 7) and many of the early tests refer to *T*-invariance rather than to *CP*-invariance directly.

We have seen in the previous section that the implications of *CP* symmetry fall into two categories: (i) absolute selection rules, and (ii) reality properties for form-factors. Predictions of the second kind are mostly derived from *T*-invariance and we shall discuss them in Section 6. They all have in common that small deviations from $T\,(\sim CP)$ invariance are very difficult to detect and this is the reason why no *CP*-violating effects were discovered until one made use of an absolute selection rule of the type (i) above. We shall restrict the discussion of this section to the fundamental experiment by Christenson and others[20] in which one found a small but non-vanishing branching ratio for the decay $K_2{}^0 \rightarrow \pi^+ + \pi^-$. We have previously seen that this decay is absolutely forbidden if *CP* is conserved. Since *CP* and *T*-invariance have been very dear to theorists, there have been a large number of proposals made to account for the observations and still retain *CP*-invariance. Many of them have been ruled out on experimental grounds already and at the present time the simplest explanation is that in fact *CP* is violated, and we shall take this attitude. With *CP* violated the relevant decay states are not the $K_1{}^0$ and $K_2{}^0$ states previously introduced but some other linear combinations of K^0 and $\bar{K}^0$. These new states are usually denoted $K_L{}^0$ and $K_S{}^0$ referring to the long-living and the short-living components. Since the *CP*-violation is small (see below), one finds $K_S{}^0 \approx K_1{}^0$ and $K_L{}^0 \approx K_2{}^0$. A measure for the violation is given by the parameter η_{+-} and η_{oo} which are defined in the following way:

$$\eta_{+-} = \frac{A(K_L{}^0 \rightarrow \pi^+ + \pi^-)}{A(K_S{}^0 \rightarrow \pi^+ + \pi^-)} \tag{108}$$

$$\eta_{oo} = \frac{A(K_L{}^0 \rightarrow \pi^o + \pi^o)}{A(K_S{}^0 \rightarrow \pi^o + \pi^o)} \tag{109}$$

where *A* stands for the transition amplitude. Experimentally one has found[21]

$$|\eta_{+-}| = (1{\cdot}94 \pm 0{\cdot}09) \times 10^{-3}$$
$$\arg \eta_{+-} = (84 \pm 17)^0$$
$$|\eta_{oo}| = (4{\cdot}9 \pm 0{\cdot}5) \times 10^{-3}$$

We shall not pursue this subject further but refer to the extensive discussion of the neutral K-mesons in reference[22].

6 Time Reversal

6.1 *Definition and action on state vectors*

The time reversal transformation is defined as the coordinate transformation

$$(x_0, \mathbf{x}) \rightarrow (x_0', \mathbf{x}') = (-x_0, \mathbf{x}) \tag{110}$$

on the physical space-time. In quantum theory it is implemented by an operator T. Since the basic transformation involves classical concepts, we immediately know its action on various observables related to space and time. It must satisfy the following conditions

$$\begin{aligned} Tx_\mu T^{-1} &= -\varepsilon(\mu)x_\mu \\ TP_\mu T^{-1} &= \varepsilon(\mu)P_\mu \\ TJ_{\mu\nu}T^{-1} &= -\varepsilon(\mu)\varepsilon(\nu)J_{\mu\nu} \end{aligned} \tag{111}$$

provided that x_μ, P_μ and $J_{\mu\nu}$ are chosen to be hermitian operators. It can now easily be shown that the relations (111) are consistent with the commutation rules for the Poincaré algebra only if T is an antiunitary operator. To see this we recall the commutation rule for an angular momentum operator J_{kl} and P_k. In our metric it reads

$$[J_{kl}, P_k] = \mathrm{i}P_l \tag{112}$$

Since both J_{kl} and P_k change sign under time reversal the left-hand side is unchanged, while the right-hand side would change sign if T is a linear operator. However, if T is antilinear, then there is an extra change of sign on the right-hand side due to the factor i (cf. equation 19). Thus, to preserve the basic commutation rules T must be chosen to be an antiunitary operator.

Since T is an antiunitary operator, it cannot represent an observable. However, the operator T^2 is a linear operator. It is further clear that performing two consecutive time reversal transformations restores the original situation and hence T^2 must commute with all observables. It is, therefore, a multiple of the unit operator

$$T^2 = \lambda I \tag{113}$$

or

$$T = \lambda T^{-1} = \lambda T^\dagger \tag{114}$$

By hermitian conjugation of (114) we find

$$T^{\dagger} = \lambda^{*} T \tag{115}$$

which inserted in (114) yields $|\lambda|^2 = 1$, that is, λ is a phase factor. If we multiply by T from left and right in (114) and further make use of the antilinearity of T, then we deduce

$$\lambda^{*} I = T^2 = \lambda I \tag{116}$$

and thus $\lambda = \pm 1$. Since T^2 commutes with the S-operator it follows that its eigenvalue λ is a constant of motion; a state with $\lambda = +1$ can never develop into a state with $\lambda = -1$ if it is isolated. There is a profound difference between this result for T^2 and the previous result in Section 3.1 that for the space reflection operator P a suitable phase convention will yield $P^2 = I$. This latter result does not yield a selection rule. The selection rule which one obtains from T^2 is called a superselection rule[23] to mark its distinction from ordinary selection rules. *A superselection rule occurs whenever there is an observable* (like T^2) *which is strictly conserved and which commutes with all other observables* so that any physical state necessarily is an eigenstate of the operator. In contrast, an ordinary selection rule is obtained if there is a strictly conserved observable, which does not commute with all other observables. Physically, superselection rules respond to the fact that there are physical quantities which always have definite values for any realizable physical state, and these values remain constant as long as the system is not subjected to external interactions. Mathematically, the existence of a superselection rule leads to a decomposition of the Hilbert space into incoherent subspaces. In the case of T^2 one obtains

$$H = H_{+} \oplus H_{-}$$

where H_{+} is spanned by vectors with $\lambda = 1$ and H_{-} by vectors with $\lambda = -1$. It is easily seen that no observables can have non-vanishing matrix elements between states which belong to different incoherent subspaces, since T^2 commutes with all observables. Thus, if Ω is an arbitrary observable then one finds

$$\begin{aligned}\langle \lambda = +1 | \Omega | \lambda = -1 \rangle &= \langle + | (T^2)^{\dagger} T^2 \Omega | - \rangle \\ &= \langle + | (T^2)^{\dagger} \Omega T^2 | - \rangle = -\langle + | \Omega | - \rangle\end{aligned} \tag{117}$$

from which it follows that the matrix element must vanish. What has been stated here for T^2 clearly holds for any operator which yields a superselection rule. The result of (117) may be phrased differently: there is no way to determine the relative phase between state vectors which belong to different incoherent subspaces. We shall not pursue the subject further, but note that the notion of an observable sometimes is used incorrectly for quantities which are not measurable and, therefore, not observable. For example, in order to assign a charge parity to the photon we made reference to the form of the electromagnetic interaction and concluded that the photon has odd charge parity. This is just a useful but not unique convention and, hence, the charge parity is not measurable. As a consequence, C does not represent an observable. This is an important observation since the electric charge operator usually is assumed to yield a superselection rule but the charge operator does not commute with C as it would be required to do if C was an observable. Also, C clearly has non-vanishing matrix elements between states of different charge.

We proceed to discuss the action of T on various 1- and 2-particle state vectors. Consider first the plane wave state $|\mathbf{p}, \lambda\rangle \equiv |\phi, \theta, p; \lambda\rangle$. With the same phase convention as before one finds

$$T|\phi, \theta, p; \lambda\rangle = \exp[-\mathrm{i}\pi\lambda]|\phi + \pi, \pi - \theta, p; \lambda\rangle \tag{118}$$

corresponding to a change in sign for the 3-momentum but no change in the helicity. Similarly for the angular momentum states one easily obtains

$$T|j, m, p, \lambda\rangle = (-1)^{j-m}|j, -m, p, \lambda\rangle \tag{119}$$

For the 2-particle state vectors, which we have previously considered in Section 3.1, we just quote the following results

$$T|\mathbf{P} = 0; J, M, p; \lambda_1, \lambda_2\rangle = (-1)^{J-M}|\mathbf{P} = 0; J, -M, p; \lambda_1, \lambda_2\rangle \tag{120}$$

and

$$T|\mathbf{P} = 0; J, M, p; l, \sigma\rangle = (-1)^{J-M}|\mathbf{P} = 0; J, -M, p; l, \sigma\rangle \tag{121}$$

In the same fashion one may then construct many-particle states and investigate their properties under time reversal.

From equations (118) to (121) it is immediately seen that T^2 will have the eigenvalue $+1$ for integer J and -1 for half-integer J so that the corresponding state vectors belong to H_+ for integer J and to H_- for half-integer J. One can show that for the general case the eigenvalues of T^2 are given by

$$T^2 = (-1)^{2J} \tag{122}$$

in accord with the special cases considered above.

From these transformation laws for the state vectors one may deduce the following transformation properties for the fields describing particles of spin 0, $\frac{1}{2}$ and 1 respectively:

$$
\begin{aligned}
\phi(x_0, \mathbf{x}) &\rightarrow \phi'(-x_0, \mathbf{x}) \equiv [T\phi(-x_0, \mathbf{x})T^{-1}]^\dagger = \eta_T\phi^\dagger(x_0, \mathbf{x}) \\
\psi(x_0, \mathbf{x}) &\rightarrow \psi'(-x_0, \mathbf{x}) \equiv [T\psi(-x_0, \mathbf{x})T^{-1}]^\dagger = \eta_T\bar{\psi}(x_0, \mathbf{x})\mathscr{T} \\
\bar{\psi}(x_0, \mathbf{x}) &\rightarrow \bar{\psi}'(-x_0, \mathbf{x}) \equiv [T\bar{\psi}(-x_0, \mathbf{x})T^{-1}]^\dagger = \eta_T^*\mathscr{T}^{-1}\psi(x_0, \mathbf{x}) \qquad (123) \\
A_\mu(x_0, \mathbf{x}) &\rightarrow A_\mu'(-x_0, \mathbf{x}) \equiv [TA_\mu(-x_0, \mathbf{x})T^{-1}]^\dagger = \eta_T\varepsilon(\mu)A_\mu(x_0, \mathbf{x})
\end{aligned}
$$

where $\mathscr{T}$ is a 4×4 matrix, whose properties are specified in Appendix 1.

6.2 *Some applications of time reversal invariance*

From equation (23) it follows that invariance of the S-operator under the time reversal transformation implies the following relation

$$\langle\beta|S|\alpha\rangle = \langle T\alpha|S|T\beta\rangle \qquad (124)$$

where $|T\alpha\rangle \equiv T|\alpha\rangle$. We have previously seen that the action of T on a vector $|\alpha\rangle$ implies a change in sign for 3-momenta while the helicities remain unchanged. The explicit form of $|T\alpha\rangle$ beyond this depends on the choice of basis vectors. We may express (124) in terms of the transition operator t* and it then reads

$$\langle\beta|t|\alpha\rangle = \langle T\beta|t^\dagger|T\alpha\rangle^* = \langle T\alpha|t|T\beta\rangle \qquad (125)$$

In Section 2.5 we have shown that if final state interactions are negligible, then (45) holds and one obtains from (125)

$$\langle\beta|t|\alpha\rangle = \langle T\beta|t|T\alpha\rangle^* \qquad (126)$$

This relation implies equality for the transition amplitudes when all 3-momenta are reversed while helicities are unchanged corresponding to a reversal of spins. In the applications below we will demonstrate what results one may obtain from (125) and (126) when they are applicable.

(i) *Reciprocity relations.* Consider the following two processes

$$
\begin{aligned}
a + b &\rightarrow c + d \\
c + d &\rightarrow a + b
\end{aligned}
$$

at the same centre of mass energy W. They are related by an exchange of the initial and the final states. For convenience we choose to describe

* To avoid confusion in the notations we shall denote the transition operator by t rather than by T as one conventionally does (cf. Sections 2.2 and 2.5).

them in their respective centre of mass frames and we denote the cross sections by σ and σ' respectively. Since we work in the centre of mass frame we need only give $\mathbf{p}_a$ and $\mathbf{p}_c$ to specify the kinematics. Alternatively we may give the centre of mass total energy $W = E_a + E_b = E_c + E_d$ and the scattering angles for, say, the particle c. From (13) we may now derive the differential cross-section for the processes. For the first process we have

$$\frac{d\sigma}{dw_c} = \frac{1}{(2\pi)^2} \frac{1}{4E_aE_bv_{\text{in}}} \int dE_c \frac{p_c}{2} \int \frac{d^3p_d}{2E_d} \delta^3(\mathbf{p}_c + \mathbf{p}_d) \times$$

$$\times\ \delta(E_a + E_b - E_c - E_d)|\langle \mathbf{p}_c, \lambda_c; \mathbf{p}_d, \lambda_d|t|\mathbf{p}_a, \lambda_a; -\mathbf{p}_a, \lambda_b\rangle|^2 \quad (127)$$

where dw_c is the solid angle element within which particle c emerges. For $v_{\text{in}} = |\mathbf{v}_a - \mathbf{v}_b|$ we obtain

$$v_{\text{in}} = v_a + v_b = p_a \cdot \frac{E_a + E_b}{E_a \cdot E_b} = p_a \frac{W}{E_a \cdot E_b} \quad (128)$$

The integration in $\mathbf{p}_d$ can be performed by means of the 3-dimensional δ-function. This fixes $\mathbf{p}_d$ and, therefore, E_d may be expressed as a function of E_c. This must be taken into account when the remaining integration is carried out by means of the remaining δ-function. If this is done correctly one finds

$$\frac{d\sigma}{dw_c} = \frac{1}{(8\pi)^2} \cdot \frac{p_c}{p_a} \cdot |\langle \mathbf{p}_c, \lambda_c; -\mathbf{p}_c, \lambda_d|t|\mathbf{p}_a, \lambda_a; -\mathbf{p}_a, \lambda_b\rangle|^2 \quad (129)$$

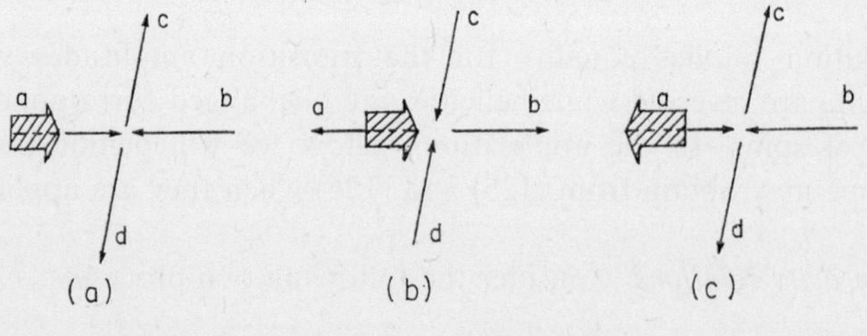

FIGURE 8 Momentum and spin orientations for a process (a), the reverse process (b) and the process obtained from (b) by a time reversal transformation (c)

Similarly for the second process one obtains (see figure 8b)

$$\frac{d\sigma'}{dw_a} = \frac{1}{(8\pi)^2} \cdot \frac{p_a}{p_c} \cdot |\langle -\mathbf{p}_a, -\lambda_a; \mathbf{p}_a, -\lambda_b|t|-\mathbf{p}_c, -\lambda_c; \mathbf{p}_c, -\lambda_d\rangle|^2 \quad (130)$$

We have chosen a configuration in (130) which is the reverse of the first process as demonstrated in figures 8a and 8b. If invariance under time reversal is assumed we may apply (125) to obtain

$$|\langle -\mathbf{p}_a, -\lambda_a; \mathbf{p}_a, -\lambda_b|t|-\mathbf{p}_c, -\lambda_c; \mathbf{p}_c, -\lambda_d\rangle|^2$$
$$= |\langle \mathbf{p}_c, -\lambda_c; -\mathbf{p}_c, -\lambda_d|t|\mathbf{p}_a, -\lambda_a; -\mathbf{p}_a, -\lambda_b\rangle|^2 \quad (131)$$

The right-hand side of (131) is almost the same as the matrix element which occurs in (129). The difference is demonstrated in figure 8c. It corresponds to a change of the spin directions. This difference vanishes if we sum over possible final helicities and average over the initial helicities. In this way we obtain the following equality from equations (129) to (131)

$$\frac{\mathrm{d}\sigma(W)}{\mathrm{d}w_a} = \frac{p_c^2}{p_a^2}\cdot\frac{(2S_c+1)(2S_d+1)}{(2S_a+1)(2S_b+1)}\cdot\frac{\mathrm{d}\sigma'(W)}{\mathrm{d}w_c} \quad (132)$$

which is our final result. Relations of the type (132) are known as *reciprocity relations.*

Of course, if instead of considering the reversed state (b) we had considered states with the opposite helicities, then a time reversal transformation would have taken us back to (a). Thus, from time reversal invariance one may deduce relations between a process and the reversed process with all spins reversed.

Reciprocity relations have been used to test time reversal invariance (cf. Section 6.3). One has also used a relation of the type (132) to determine the spin of the charged π-meson from the reactions[24]

$$p + p \rightleftarrows \pi^+ + d$$

(ii) *The decay* $K^+ \to \pi^0 + \mu^+ + \nu_\mu$. This decay has attracted attention recently in connection with possible tests of time reversal invariance for the weak interactions. We shall return to this aspect of the decay below, but for the time being *we assume invariance under the time reversal transformation* and note that no strong final state interactions can occur and hence one may presumably neglect the final state interactions. This makes relation (126) applicable and we shall use it. First, however, a few remarks about the structure of the relevant S-matrix element.

Since it is a weak process it is described by the Hamiltonian (57). If the weak interactions are treated to first order in the perturbation expansion, then we may write the relevant S-matrix element

$$\langle \pi^0, \mu^+, \nu_\mu|S|K^+\rangle = -\frac{\mathrm{i}G}{\sqrt{2}}(2\pi)^4\,\delta^4(P_f - P_i) \times \langle \pi^0|J_\mu(0)|K^+\rangle l^\mu \quad (133)$$

As before, we have omitted some trivial normalization factors and

$$l_\mu \equiv \bar{u}_\nu(\mathbf{q}, \mathbf{w})\gamma_\mu(1 + \gamma_5)\nu_\mu(\mathbf{q}', \mathbf{w}') \tag{134}$$

From Lorentz invariance we may deduce the following form for the hadronic matrix element

$$L_\mu \equiv \langle \pi^0 | J_\mu(0) | K^+ \rangle = f_+(p_K + p_\pi)_\mu + f_-(p_K - p_\pi)_\mu \tag{135}$$

where f_+ and f_- are unknown complex-valued functions of the only relativistic invariant one may form out of p_K and p_π, namely $(p_K{}^\mu p_{\pi\mu})$. To apply the result of (126) we must consider the same S-operator sandwiched between the time reversed states, that is, with the sign for 3-momenta and spins reversed. More precisely, from (126) and (133) we obtain

$$L_\mu l^\mu = [L_\mu(T) l^\mu(T)]^* \tag{136}$$

where the asterisk denotes complex conjugation. The factors $L_\mu(T)$ and $l^\mu(T)$ are defined by

$$L_\mu(T) \equiv \langle T\pi^0 | J_\mu(0) | T(K^+) \rangle = \varepsilon(\mu)[f_+(p_K + p_\pi)_\mu + f_-(p_K - p_\pi)_\mu] \tag{137}$$

and

$$l_\mu(T) = \bar{u}_\nu(-\mathbf{q}, -\mathbf{w})\gamma_\mu(1 + \gamma_5)(-\mathbf{q}', -\mathbf{w}') \tag{138}$$

The last equation may be rewritten in the following way

$$l_\mu{}^*(T) = \varepsilon(\mu)\bar{u}_\nu(\mathbf{q}, \mathbf{w})\gamma_\mu(1 + \gamma_5)(\mathbf{q}', \mathbf{w}') \tag{139}$$

or

$$l_\mu{}^*(T) = \varepsilon(\mu) l_\mu \tag{140}$$

In the scalar product $[L_\mu(T)l^\mu(T)]^*$ the factors $\varepsilon(\mu)$ cancel out and from (136) it then follows that the form-factors f_+ and f_- must be real functions. We have omitted some overall phase-factors which are unmeasurable, however, and hence irrelevant. A more careful statement of the condition placed upon the form-factors in order to satisfy time reversal invariance is, therefore, that f_+ and f_- have the same phase which we may put equal to zero without changing the content of the theory.

For processes like the one just considered, where final state interactions are negligible, the simple argument with vectors, previously used in connection with space reflection invariance, again holds. For example, in a $K_{\mu 3}$ decay one may determine (i) the 3-momentum of the π-meson, (ii) the 3-momentum of the μ^+ and (iii) the polarization of the μ^+. One may then determine whether there exists in a sample of $K_{\mu 3}$ decays a correlation of the type $\alpha(\mathbf{p}_\pi \times \mathbf{p}_\mu) \cdot \langle \boldsymbol{\sigma}_\mu \rangle$. If $\alpha \neq 0$ one observes a transverse polarization of the muon, that is, a polarization perpendicular to the plane defined by $\mathbf{p}_\pi$ and $\mathbf{p}_\mu$. Since a term of this type changes sign under time reversal, any value of α different from zero is evidence for a violation of time reversal invariance. A more detailed calculation based on the expression (133) for the matrix element shows that α is proportional to $\mathrm{Im}(f_+{}^*f_-)$, which vanishes if f_+ and f_- have the same phase. Thus $\alpha = 0$ implies that the form-factors must have the same phase in agreement with our previous findings. Experimentally there is so far no evidence for a transverse polarization of the muon but the experimental errors are still relatively large[25].

If final state interactions are non-negligible (strong or electromagnetic), then the simple argument with vectors fails since the momenta and the spins of the particles in the final state may change as a result of rescattering. By the same token, if the corresponding scattering phase shifts are small for some reason, then the corrections to the vector argument are small. To demonstrate the role played by the final state interactions we shall next consider the process $\Sigma^- \to n + \pi^-$. We shall not deal with a T-violating effect since the corresponding correlations are rather complicated and hence the computational work is more extensive. Instead we choose a P-violating correlation where final state interactions are important and T-invariance imposes restrictions.

(iii) *The decay* $\Sigma^- \to n + \pi^-$. We have chosen this process since no charge exchange scattering can occur in the final state and we do not have to invoke isospin in the discussion. An analogous process is $\Lambda \to p + \pi^-$. In that case there are contributions from two isospin channels and the analysis will be somewhat more complicated.

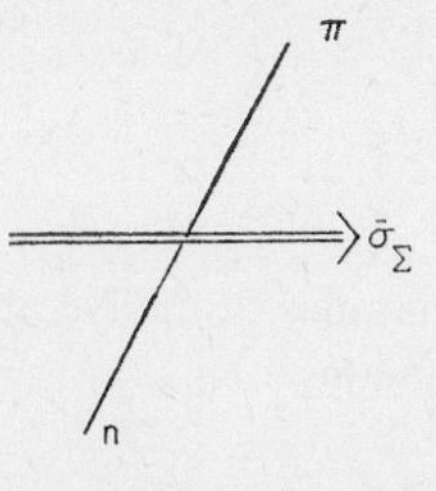

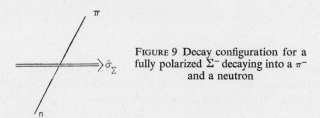

FIGURE 9 Decay configuration for a fully polarized Σ^- decaying into a π^- and a neutron

The Σ^--hyperons are produced by associated production. In general the Σ particles have a non-vanishing polarization due to final state interactions in the production process. For simplicity we consider the case of fully polarized particles whereby we may omit a discussion of the production process and we can focus our attention on the decay. In the rest system of the Σ, the decay configuration is given by figure 9. Angular momentum conservation requires that $J = M = \frac{1}{2}$ in the final state (the z-axis along the spin vector of the Σ particle). If we anticipate a P-violation in this weak decay, then the $n\pi$-system may be in an S-state or in a P-state. The separation with regard to orbital angular momentum is convenient in this analysis. We have (cf. Appendix 2)

$$\text{Initial state: } |j, m, p, \lambda\rangle = |\tfrac{1}{2}, \tfrac{1}{2}\rangle$$

$$\text{Final state: } |\mathbf{P} = 0; \phi, \theta, p; \lambda_1, \lambda_2\rangle = |\mathbf{P} = 0; \phi, \theta, p; \pm\tfrac{1}{2}, 0\rangle$$

corresponding to the two possible helicities for the final neutron. The decay rate λ is given by (cf. equation (10))

$$\frac{d\lambda}{d(\cos\theta)} = \text{const.} \times \sum_{s=\pm\frac{1}{2}} |\langle 0; \phi, \theta, p; s, 0|t|\tfrac{1}{2}, \tfrac{1}{2}\rangle|^2 \tag{141}$$

The two possible final states can be expressed in terms of an orbital angular momentum basis $|\mathbf{P} = 0; J, M, p; l, \sigma\rangle$ by the relations

$$|\mathbf{P} = 0; \phi, \theta, p; \tfrac{1}{2}, 0\rangle = |0; \tfrac{1}{2}, \tfrac{1}{2}, p; 0, \tfrac{1}{2}\rangle - \frac{1}{\sqrt{(4\pi)}} \cos\theta |0; \tfrac{1}{2}, \tfrac{1}{2}, p; 1, \tfrac{1}{2}\rangle$$

$$|\mathbf{P} = 0; \phi, \theta, p; -\tfrac{1}{2}, 0\rangle = -\frac{1}{\sqrt{(4\pi)}} \sin\theta \exp(-i\phi) |0; \tfrac{1}{2}, \tfrac{1}{2}, p; 1, \tfrac{1}{2}\rangle \tag{142}$$

so that

$$\frac{d\lambda}{d(\cos\theta)} = \text{const.} \times \{|A_s + A_p \cos\theta|^2 + |A_p \sin\theta|^2\} \tag{143}$$

with

$$A_s = \langle 0; \tfrac{1}{2}, \tfrac{1}{2}, p; 0, \tfrac{1}{2}|t|\tfrac{1}{2}, \tfrac{1}{2}\rangle$$

$$A_p = -\frac{1}{\sqrt{(4\pi)}} \langle 0; \tfrac{1}{2}, \tfrac{1}{2}, p; 1, \tfrac{1}{2}|t|\tfrac{1}{2}, \tfrac{1}{2}\rangle \tag{144}$$

corresponding to the S and P wave amplitudes. Finally, absorbing a factor $[|A_s|^2 + |A_p|^2]$ in the constant, we obtain

$$\frac{d\lambda}{d(\cos\theta)} = \text{const.} \times (1 + \alpha\cos\theta) \tag{145}$$

with

$$\alpha = \frac{2\text{Re}(A_s A_p^*)}{|A_s|^2 + |A_p|^2} \tag{146}$$

We next proceed to show that with time reversal invariance the phase angles for the two complex amplitudes A_s and A_p are given by the S and P

wave scattering phase shifts for $\pi^- n$ scattering. In the absence of final state interactions both amplitudes are real. Treating the S and P waves separately we employ the following simplified notation

$$\langle n, \pi; L|t|\Sigma\rangle \equiv \langle 0; \tfrac{1}{2}, \tfrac{1}{2}, p; L, \tfrac{1}{2}|t|\tfrac{1}{2}, \tfrac{1}{2}\rangle \tag{147}$$

Assuming time reversal invariance we obtain from (125)

$$\langle n, \pi; L|t|\Sigma\rangle = \langle T(n, \pi; L)|t^\dagger|T(\Sigma)\rangle^* \tag{148}$$

From (119) and (121) we find

$$\langle T(n, \pi; L)t^\dagger|T(\Sigma)\rangle = \langle 0; \tfrac{1}{2}, -\tfrac{1}{2}, p; L, \tfrac{1}{2}|t^\dagger|\tfrac{1}{2}, -\tfrac{1}{2}\rangle \tag{149}$$

and further using rotational invariance we conclude

$$\langle T(n, \pi; L)|t^\dagger|T(\Sigma)\rangle = \langle n, \pi; L|t^\dagger|\Sigma\rangle \tag{150}$$

so that (148) reads

$$\langle n, \pi; L|t|\Sigma\rangle = \langle n, \pi; L|t^\dagger|\Sigma\rangle^* \tag{151}$$

If final state interactions were negligible we would use (126) to deduce that the matrix elements in (151) are real. In this case, however, we must employ the complete expression (42) and we obtain

$$\langle n, \pi; L|t^\dagger|\Sigma\rangle^* = \langle n, \pi; L|t|\Sigma\rangle^* + \mathrm{i}\langle n, \pi; L|t^\dagger t|\Sigma\rangle^* \tag{152}$$

Thus, from (151) and (152) we arrive at

$$\langle n, \pi; L|t|\Sigma\rangle = \langle n, \pi; L|t|\Sigma\rangle^* + \mathrm{i}\sum_\gamma\langle n, \pi; L|t^\dagger|\gamma\rangle^*\langle\gamma|t|\Sigma\rangle^* \tag{153}$$

where a complete set of *physical* states $|\gamma\rangle$ has been introduced. Considering for a moment only the last term in (153), it is clear that each factor is a matrix element for a physical process (note that t is the full T-operator, which in principle describes all physical processes). We only have to consider those intermediate states $|\gamma\rangle$ for which the order of the last term in (153) is the same as the other two terms. This is the case, for example, if $|\gamma\rangle = |n, \pi; L\rangle$ in which case the first factor corresponds to $n\pi$ scattering in the appropriate channel L. This is a strong process, while the second factor in this case describes Σ^- decay into $(n\pi^-)$. A closer study immediately reveals that at the relevant energy, which is below the inelastic threshold, all other possible intermediate states involve additional electromagnetic and/or weak processes and hence we can neglect those terms. In this approximation we obtain

$$\langle n, \pi; L|t|\Sigma\rangle = \langle n, \pi; L|t|\Sigma\rangle^* + \\ + \mathrm{i}\langle n, \pi; L|t^\dagger|n, \pi; L\rangle^*\langle n, \pi; L|t|\Sigma\rangle^* \tag{154}$$

It follows from parity conservation for the strong interactions that the part of the S-operator which describes $n\pi$ scattering is diagonal in the orbital angular momentum basis which we have chosen. More precisely, one has

$$S|n, \pi; L\rangle = \exp(2i\delta_L)|n, \pi; L\rangle \tag{155}$$

where δ_L is the scattering phase shift in the channel L. From this we obtain

$$t|n, \pi; L\rangle = -i[\exp(2i\delta_L)-1]|n, \pi; L\rangle \tag{156}$$

since $S = 1 + it$. Inserting this into (154) we find

$$\langle n, \pi; L|t|\Sigma\rangle = \exp(2i\,\delta_L)\langle n, \pi; L|t|\Sigma\rangle^* \tag{157}$$

and thus

$$\arg[\langle n, \pi; L|t|\Sigma\rangle] = \delta_L \tag{158}$$

which is the result claimed before. With this result we can now write the asymmetry parameter α of (146) in its final form

$$\alpha = \frac{2|A_s| \cdot |A_p|}{|A_s|^2 + |A_p|^2} \cos(\delta_s - \delta_p) \tag{159}$$

and the presence of the cosine factor is here a result of final state interactions. Clearly δ_s and δ_p refer to the phase shifts taken at a centre of mass energy equal to the energy release in the Σ decay.

(iv) *Time reversal and the process* $\nu_\mu + n \rightarrow \mu^- - p$. We return to this process which serves as our test ground for invariance arguments. The

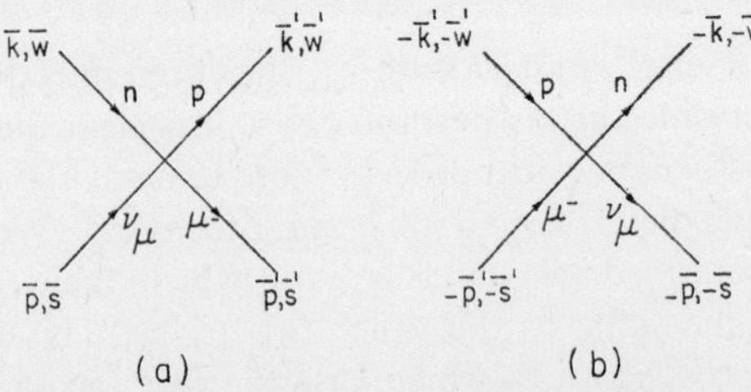

FIGURE 10 Two processes (a) $\nu_\mu + n \rightarrow \mu^- + p$ and (b) $\mu^- + p \rightarrow \nu_\mu + n$ related by a time reversal transformation

two configurations corresponding to the process $\nu_\mu + n \rightarrow \mu^- + p$ and the time-reversed process are given in figure 10. We have given the explicit form of the matrix element 10a in Section 3.2, and we proceed directly to the matrix element 10b. It is given by

$$\langle T(\nu_\mu, n)|S|T(\mu^-, p)\rangle = -\frac{iG}{\sqrt{2}}(2\pi)^4\delta^4(P_f - P_i)h_\mu(T)l^\mu(T) \tag{160}$$

where

$$h_\mu(T) \equiv \langle T(n)|J_\mu(0)|T(p)\rangle = \langle T(p)|J_\mu{}^\dagger(0)|T(n)\rangle^* \tag{161}$$

and

$$l^\mu(T) \equiv \langle T(\nu_\mu)|J_\mu{}^\dagger(0)|T(\mu^-)\rangle = \langle T(\mu^-)|J_\mu(0)|T(\nu_\mu)\rangle^* \tag{162}$$

In (161) and (162) we have taken notice of the fact that the hermitian conjugate parts of the currents contribute to the time reversed process, since initial and final states are exchanged. We next express the matrix elements in terms of elementary spinors

$$\begin{aligned} h_\mu{}^*(T) = \bar{u}(-\mathbf{k}', -\mathbf{w}')\{F_1(q^2)\gamma_\mu + \mathrm{i}\varepsilon(\nu)F_2(q^2)\sigma_{\mu\nu}q^\nu + \\ + \varepsilon(\mu)F_3(q^2)q_\mu + [G_1(q^2)\gamma_\mu + \mathrm{i}\varepsilon(\nu)G_2(q^2)\sigma_{\mu\nu}q^\nu + \\ + \varepsilon(\mu)G_3(q^2)q_\mu]\gamma_5\}u(-\mathbf{k}, -\mathbf{w}) \end{aligned} \tag{163}$$

To rewrite this matrix element we make use of the following relations:

$$\begin{aligned} u(-\mathbf{k}, -\mathbf{w}) &= \mathscr{T}\bar{u}^T(\mathbf{k}, \mathbf{w}) \\ \bar{u}(-\mathbf{k}, -\mathbf{w}) &= u^T(\mathbf{k}, \mathbf{w})\mathscr{T}^{-1} \end{aligned} \tag{164}$$

where the 4 × 4 matrix $\mathscr{T}$ has the following properties

$$\mathscr{T}^{-1}\gamma_\mu\mathscr{T} = \varepsilon(\mu)\gamma_\mu{}^T \tag{165}$$

Introducing (164) into (163) one finds

$$\begin{aligned} h_\mu(T) = \varepsilon(\mu)\bar{u}(\mathbf{k}', \mathbf{w}')\{F_1{}^*(q^2)\gamma_\mu + \mathrm{i}F_2{}^*(q^2)\sigma_{\mu\nu}q^\nu + F_3{}^*(q^2)q_\mu + \\ + [G_1{}^*(q^2)\gamma_\mu + \mathrm{i}G_2{}^*(q^2)\sigma_{\mu\nu}q^\nu + G_3{}^*(q^2)q_\mu]\gamma_5\}u(\mathbf{k}, \mathbf{w}) \end{aligned} \tag{166}$$

and similarly

$$l^\mu(T) = \varepsilon(\mu)\bar{u}(\mathbf{p}', \mathbf{s}')\gamma^\mu(1 + \gamma_5)u(\mathbf{p}, \mathbf{s}) \tag{167}$$

If we write

$$h_\mu l^\mu \equiv \phi(F_i, G_i; \gamma_5)$$

then we obtain from (166) and (167)

$$h_\mu(T)l^\mu(T) = \phi(F_i{}^*, G_i{}^*, \gamma_5) \tag{168}$$

T-invariance implies that

$$h_\mu l^\mu = h_\mu(T)l^\mu(T) \tag{169}$$

and hence all form-factors must be real (have the same phase). We had anticipated this result when we introduced the factors i in the definitions of the form-factors in Section 3.

It is of interest to note that the restrictions imposed on the matrix element (160) by time reversal invariance are exactly the same as those previously found to follow from *CP*-invariance (cf. equation 107). This is an indication of the close relationship between *CP* and *T*, a subject we shall return to in Section 7.

6.3 *The present status of time reversal invariance in strong, electromagnetic and weak processes*

In the last couple of years considerable attention has been devoted to the question of *T*-invariance in various processes, and we shall briefly review the most important findings which we classify according to the type of interaction responsible for the process.

(i) *Time reversal invariance in strong interaction processes.* The most stringent tests make use of reciprocity relations applied to nuclear reactions. For example, one has considered the processes[26] Mg^{24} (α, p) Al^{27} and Mg^{24} (d, p) Mg^{25} and one has found no disagreement with reciprocity. The upper limit on the T non-invariant amplitude obtained from these experiments is less than 1–2 $\times$ 10^{-3} times the T-conserving amplitude.

Another less accurate test of T-invariance in strong processes refers to a polarization–asymmetry equality in nucleon-nucleon scattering. If a beam of unpolarized protons hits an unpolarized target of protons one will find that the scattered protons are partially polarized along the normal to the scattering plane, which is defined by the momenta $\mathbf{p}_i$ and $\mathbf{p}_f$ of the incident and the scattered proton respectively (cf. figure 11). Since this

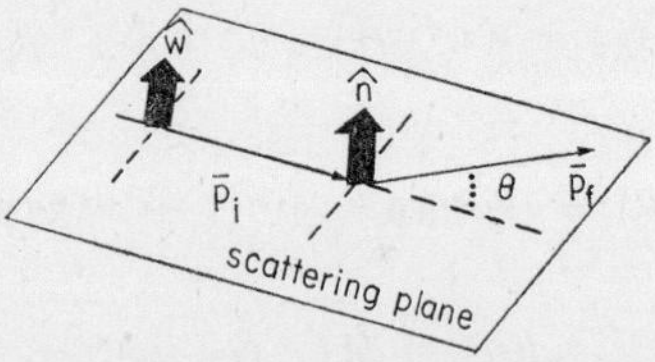

FIGURE 11 A configuration in scattering of a polarized proton on an unpolarized proton target

corresponds to a correlation of the type $\langle \boldsymbol{\sigma} \rangle \cdot (\mathbf{p}_i \times \mathbf{p}_f)$ it is clear that if T-invariance holds, then this polarization is the result of final state interactions. Furthermore, there can be no polarization in the scattering plane unless space reflection invariance is violated. We denote by $P(\theta)$ the polarization of a proton which has been scattered at an angle θ. Next consider the scattering of fully polarized protons on a target of unpolarized protons. In general there will be an azimuthal distribution of the form

$$N(\theta) = \text{const.} \times [1 + \alpha(\theta)\hat{\mathbf{w}} \cdot \hat{\mathbf{n}}] \tag{170}$$

where $\hat{\mathbf{w}}$ is a unit vector along the polarization vector of the incident protons and $\hat{\mathbf{n}} = (\mathbf{p}_i \times \mathbf{p}_f)/|\mathbf{p}_i \times \mathbf{p}_f|$. Equation (170) implies a left-right asymmetry corresponding to whether $\hat{\mathbf{w}}$ and $\hat{\mathbf{n}}$ are parallel or antiparallel. It can be shown that time reversal invariance implies that

$$P(\theta) = \alpha(\Theta) \tag{171}$$

Experimentally one has investigated these effects for $p\bar{p}$ scattering at an energy of 210 MeV and for $\theta = 30^{\circ}$ in the centre of mass frame[27]. One found

$$P(\theta) - \alpha(\theta) = -0{\cdot}014 \pm 0{\cdot}014$$

with $P(\theta)$ of the order of 25%. This places the upper limit on the T non-conserving amplitude at a few percent of the T-conserving amplitude. Much more extensive measurements to test the relation (171) at higher energies are in progress.

(ii) *Time reversal invariance in electromagnetic processes.* The test of T-invariance in electromagnetic processes are few and often ambiguous with regard to the interpretation. We shall return to some of these questions in Section 8.

(1) The electric dipole moment of the neutron. We have previously noted that a non-vanishing electric dipole moment for the neutron violates *both* T and P-invariance. The observed upper limit of the dipole moment sets the upper limit for the P or/and T-violating amplitude at $10^{-6} - 10^{-7}$ times invariant amplitude.

(2) Quantum electrodynamics. One may view the excellent agreement between theory and experiment in quantum electrodynamics as evidence for T-invariance since this invariance is explicitly present in the theory. Some caution in using this argument is necessary, however, as pointed out by Lee and others[11]. Many of the consequences of T-invariance follow also from the hermicity of the electromagnetic current operator and the fact that it is conserved.

(3) The decay $\Sigma^0 \rightarrow \Lambda^0 + e^+ + e^-$. The electromagnetic decay

$$\Sigma^0 \rightarrow \Lambda^0 + e^+ + e^-$$

has recently been investigated in the search for T-violations among electromagnetic processes[28]. It is of particular interest since the photon is virtual, and hence current conservation does not disguise a T-violation (cf. Section 8). One has looked for a polarization of the Λ-hyperon perpendicular to the decay plane defined by the vectors $\mathbf{p}_\Lambda$ and $\mathbf{q} = (\hat{\mathbf{p}}_{e^+} + \hat{\mathbf{p}}_{e^-})$ that is, a correlation term

$$\langle \boldsymbol{\sigma}_\Lambda \rangle \cdot (\mathbf{p} \times \mathbf{q})$$

A net polarization of the Λ can be detected as an asymmetry in the decay of the Λ-hyperon

$$\Lambda \rightarrow p + \pi^-$$

Experimentally this asymmetry was found to be

$$\langle \cos \theta \rangle_{\text{exp}} = 0{\cdot}048 \pm 0{\cdot}026$$

where $\cos \theta = \hat{\mathbf{p}}_p \cdot \hat{\mathbf{N}}$, $\mathbf{N} = \mathbf{p}_\Lambda \times \mathbf{q}$. The result indicates a small polarization for the Λ but it seems that the result is somewhat uncertain.

(4) Reciprocity relations, etc. Experiments are under way for testing *T*-invariance in the form of reciprocity for the processes

$$\gamma + d \rightleftarrows p + n$$

which clearly are of electromagnetic nature. Furthermore, one is looking for a correlation of the type $\langle \boldsymbol{\sigma} \rangle \cdot (\mathbf{p}_i \times \mathbf{p}_f)$ in the scattering of electrons on polarized nuclei. Since the final state interactions are of electromagnetic character one expects such a correlation also in the absence of *T*-violation, but then it should be of the order of α (= fine structure constant) and hence quite small.

(iii) *Time reversal invariance in weak processes.* There is a fair number of experimental tests of *T*-invariance for weak processes. They all have in common that they determine relative phase angles which are restricted by *T*-invariance. In general, the accuracy of such experiments is rather poor and no stringent tests are then possible. Therefore, it is not surprising that one has found no direct evidence for *T*-violations so far. There is, of course, indirect evidence for a small *T*-violation from the observation of the decay $K_2{}^0 \rightarrow 2\pi$. This decay is strictly forbidden by *CP*-conservation and, hence, its occurrence implies a *CP*-violation. From the *CPT* theorem (cf. next section) we then deduce that also *T* must be violated in this weak decay. We have previously discussed the $K_2{}^0 \rightarrow 2\pi$ decay and in this section we restrict ourselves to direct tests of *T*-invariance.

(1) *T*-violating effects in the β-decay of the neutron. A test of *T*-invariance is provided by a measurement of the electron-neutrino angular correlation in the decay of polarized neutrons

$$n \rightarrow p + e^- + \nu_e$$

The observed correlation is of the form $\langle \boldsymbol{\sigma}_n \rangle \cdot (\mathbf{p}_e \times \mathbf{p}_\nu)$. Invoking energy-momentum conservation one may rewrite this correlation in the following way (the neutrons decay at rest)

$$\langle \boldsymbol{\sigma}_n \rangle \cdot (\mathbf{p}_e \times \mathbf{p}_\nu) = -\mathbf{p}_{\text{recoil}} \cdot (\langle \boldsymbol{\sigma}_n \rangle \times \mathbf{p}_e) \tag{172}$$

where $\mathbf{p}_{\text{recoil}}$ is the 3-momentum of the recoiling proton. Thus, the presence of a term like (172) is established if one finds a non-vanishing up-down asymmetry for recoiling protons with regard to the plane defined by $\langle \boldsymbol{\sigma}_n \rangle$ and $\mathbf{p}_e$. In terms of the form-factors F_i and G_i previously introduced for the matrix element of the hadron current, the dominant contribution to this correlation is proportional to $\text{Im}[F_1(0)G_1^*(0)]$. We have already established that T-invariance requires the form-factors to be real and hence this term vanishes if time reversal is a good symmetry. The experimental result implies that the phase angle difference between these two form-factors is less than 8° (mod. π)[20].

(2) The transverse polarization in $K_{\mu 3}$ decay. This is a weak, strangeness-changing decay for which we have already discussed the implications of time reversal invariance. Experimentally one looks for a net polarization of the emitted muon perpendicular to the plane defined by $\mathbf{p}_\pi$ and $\mathbf{p}_\mu$, that is the relevant correlation is $\langle \boldsymbol{\sigma}_\mu \rangle \cdot (\mathbf{p}_\pi \times \mathbf{p}_\mu)$. Since there are no strong or electromagnetic (to order α) final state interactions, such a correlation is forbidden if T-invariance holds. Experimentally one has determined the transverse polarization $P_\perp$ of the muon to be[30].

$$P_\perp = 0{\cdot}007 \pm 0{\cdot}016$$

which is consistent with T-invariance.

(3) The proton polarization in the decay of polarized Λ-particles. In the decay

$$\Lambda \to p + \pi^-$$

of polarized Λ-hyperons one has measured the polarization of the emitted proton. It can be shown[31] that the polarization of the proton is given by

$$\langle \boldsymbol{\sigma}_p \rangle = \text{const.} \times \{-(\alpha - \mathbf{p}_p \cdot \mathbf{P})\mathbf{p}_p + \beta \mathbf{p}_p \times \mathbf{P} + \gamma[(\mathbf{p}_p \times \mathbf{P}) \times \mathbf{p}_p]\} \tag{173}$$

where the parameters α, β and γ are given in terms of S and P wave transition amplitudes (cf. the discussion of $\Sigma \to n + \pi^-$ in Section 6.2), and $\mathbf{P}$ is the polarization of the Λ-hyperon. Clearly the polarization perpendicular to the plane defined by $\mathbf{p}_p$ and $\mathbf{P}$ is a correlation which is non-invariant under T and hence $\beta = 0$ if (i) T-invariance holds and

(ii) one can neglect final state interactions. In the presence of final state interactions this term will be proportional to sin $(\delta_s - \delta_p)$ and one may determine the phase shifts $(\delta_s - \delta_p)$ for $p\pi^-$ scattering at the centre of mass energy of 37 MeV. Experimentally, one finds from the decay[32]

$$(\delta_s - \delta_p) = (7 \pm 8)^\circ$$

which should be compared to the result obtained from $p\pi^-$ scattering

$$(\delta_s - \delta_p) = (6{\cdot}5 \pm 0{\cdot}5)^\circ$$

7 The *CPT* Transformation

Since we have treated the three discrete transformations *C*, *P* and *T*, it is a straightforward task to investigate the physical implications of invariance under the combined operation *CPT*. For two reasons we shall pay special attention to this transformation: (i) It turns out that the *CPT* transformation by itself is of much more fundamental character than any one of its constituents and (ii) it imposes less stringent conditions on the theory than separate invariance under *C*, *P* and *T* and in that way it is a sort of minimal programme to have *CPT*-invariance.

The fundamental importance of the *CPT* transformation is a consequence of the so-called *CPT* theorem, which we shall briefly state omitting a proof since it goes beyond the scope of these lectures. The main content of the theorem is that any reasonable theory is *CPT*-invariant. In the next subsection we shall qualify this statement making it more precise. It is clear, however, that the depth of the theorem makes it particularly important to test the implications one may derive from *CPT*-invariance, and we shall discuss those aspects in some detail. In view of the very weak assumptions on which the *CPT* theorem is based it is not surprising that one finds less stringent conditions on a theory by invoking *CPT*-invariance than, say, from separate *C*, *P* or *T*-invariance. In fact our previous results with regard to the process $\nu_\mu + n \rightarrow \mu^- + p$ illustrates this statement, since it is immediately seen that *CPT*-invariance imposes no restrictions whatsoever on the form of the matrix element. We shall later see that this is to be expected, since we have assumed Lorentz invariance in deriving the most general form for the matrix element, and Lorentz invariance is the major ingredient of the *CPT* theorem.

7.1 *The CPT theorem*

So far in these lectures we have made no reference to local field theory although local field theory in many ways provides the most natural frame for a description of elementary particles. The *CPT* theorem emerges in local field theory from very general assumptions. It essentially asserts that any local Lagrangian field theory is invariant under the combined operation *CPT*, taken in any order whatever, provided the theory is invariant under proper orthochronous Lorentz transformations. For a proof or a more detailed discussion of this remarkable theorem we refer to[33].

Since the assumptions underlying the *CPT* theorem are so weak it is generally assumed that *CPT*-invariance is an absolute symmetry principle. If that is the case, and there is so far no indication to the contrary, then, say, tests of *CP*-invariance have immediate implications for the question of *T*-invariance. This was briefly mentioned in connection with the $K \to 2\pi$ decay. It was shown that the existence of this decay mode violates *CP*-invariance and we stated that as a consequence of *CPT*-invariance it is at the same time evidence for *T* non-invariance. In the same fashion the *CPT*-invariance offers opportunities to test *C*, *P* or *T*-invariance indirectly in cases where direct tests are inaccurate or beyond present technique.

7.2 *The present status of CPT-invariance*

Since the *CPT* transformation involves a time reversal operation, it is clearly represented by an antiunitary operator for which we shall use the notation Θ. Its action on state vectors can be read off from their transformation properties under separate *C*, *P* and *T* transformation. For the moment it suffices to note that a single-particle state of momentum **p** and helicity λ, under *CPT* transforms into an antiparticle state of momentum **p** and helicity $-\lambda$, that is the spin direction is reversed. Furthermore, an in-state is turned into an out-state.

Consider the process $|i; \text{in}\rangle \to |f; \text{out}\rangle$ where i and f stand for the arbitrary initial and final states characterized by the momenta and the helicities of the individual particles. The corresponding states with the particles replaced by their antiparticles will be denoted $|\bar{i}; \text{in}\rangle$ and $|\bar{f}; \text{out}\rangle$. Finally we denote by $|\tilde{i}; \text{in}\rangle$ and $|\tilde{f}; \text{out}\rangle$ the original states with the opposite signs for the helicities. Next consider a hermitian operator Ω which is assumed to be *CPT*-invariant, that is

$$\Theta \, \Omega \, \Theta = \Omega \tag{174}$$

For the matrix elements of Ω one then finds the following relation

$$\langle \bar{f}; \text{out}|\Omega|\bar{i}; \text{in}\rangle = \langle \bar{f}; \text{out}|\Theta^{-1}\Theta \, \Omega \, \Theta^{-1}\Theta|\bar{i}; \text{in}\rangle = \langle \tilde{f}; \text{in}|\Omega| \tilde{i}; \text{out}\rangle^* \tag{175}$$

We may further introduce the S-operator to rewrite (175) in the following way:

$$\langle \bar{f}; \text{out}|\Omega|\bar{\imath}; \text{in}\rangle = \langle \tilde{\bar{f}}; \text{out}|S^{-1}\,\Omega\, S|\tilde{\bar{\imath}}; \text{in}\rangle^* \tag{176}$$

If the states $|\tilde{\bar{f}}; \text{out}\rangle$ and $|\tilde{\bar{\imath}}; \text{in}\rangle$ are chosen to be eigenstates of the total Hamiltonian and hence of the S-operator, then we obtain

$$\langle \bar{f}; \text{out}|\Omega|\bar{\imath}; \text{in}\rangle = \exp\,[2i(\delta_f + \delta_i)]\langle \tilde{\bar{f}}; \text{out}|\Omega|\tilde{\bar{\imath}}; \text{in}\rangle^* \tag{177}$$

which relates the matrix elements of the hermitian operator Ω between particle states and the corresponding antiparticle states respectively. From this very important relation we may deduce a number of predictions which may be subjected to experimental tests.

(i) *Equality for the mass of a particle and its antiparticle.* We choose the states $|i; \text{in}\rangle$ and $|f; \text{out}\rangle$ to represent a single particle at rest, hence there is no distinction between in and out labels in this case. If we further consider the case with $\Omega = H$, where H is the total Hamiltonian (the energy operator), then (177) reads

$$\bar{m}\delta_{\bar{\imath}\bar{f}} = m^*\delta_{\tilde{\bar{\imath}}\tilde{\bar{f}}} \tag{178}$$

where $\bar{m}$ and m represent the rest masses of the antiparticle respectively the particle. Since the energy operator is hermitian and m represents a diagonal matrix element of H, it follows that m is real and we conclude that

$$\bar{m} = m \tag{179}$$

or the mass of a particle equals the mass of its antiparticle. The same result clearly follows also from the more restrictive assumption of C-invariance. Since C-invariance is not generally respected, it is an important observation that the very weak assumption of CPT-invariance is sufficient.

Experimentally, one has determined the masses for particles and antiparticles as a test of CPT-invariance, and some of the most accurate results are summarized in Table 2 below. The most precise test of CPT-

TABLE 2

The mass ratio for some particle-antiparticle systems

Particle	$\frac{m^+}{m^-} - 1$	Reference
$\mu^\pm$	10^{-4}	(34)
$\pi^\pm$	$0{\cdot}002 \pm 0{\cdot}002$	(35)
$K^\pm$	$0{\cdot}00 \pm 0{\cdot}01$	(35)

invariance refers to the measured mass difference between the two neutral K-mesons, namely $K_L{}^0$ and $K_S{}^0$. If we define M and $\bar{M}$ to be the diagonal matrix elements of the mass operator in the basis K^0 and $\bar{K}^0$, then *CPT*-invariance of the mass (energy) operator implies

$$M = \langle K^0|H|K^0\rangle = \langle \bar{K}^0|H|\bar{K}^0\rangle \equiv \bar{M} \tag{180}$$

From the measured mass difference $\Delta m \equiv m(K_L{}^0) - m(K_S{}^0)$ one may deduce that (180) holds to the accuracy $|\Delta m/m(K)| \simeq 10^{-14}$. To the extent that a decomposition $H = H_{st} + H_{em} + H_{wk}$ (according to the type of interaction) is meaningful, this result implies that the ratio of the *CPT*-violating amplitude and the *CPT*-conserving amplitude is less than 10^{-14} for H_{st}, less than 10^{-12} for H_{em} and less than 10^{-8} for the $\Delta S = 0$ part of H_{wk}.

(ii) *Equality for the lifetime of a particle and its antiparticle.* We next consider a state $|i; \text{in}\rangle$ corresponding to particle which can only decay as a result of the weak interactions. With regard to the strong and the electromagnetic interactions it is a steady state. Further, to start with we only consider the case where the *final state interactions are negligible*, so that (177) for $\Omega = H_{wk}$ takes the form

$$\langle f; \text{out}|H_{wk}|i; \text{in}\rangle = \langle \tilde{\bar{f}}; \text{out}|H_{wk}|\tilde{\bar{\imath}}\text{: in}\rangle^* \tag{181}$$

Inserting this result into expression (10) for the partial decay rate one immediately obtains $\bar{\lambda} = \tilde{\lambda}$, where $\bar{\lambda}$ refers to the antiparticle channel and $\tilde{\lambda}$ to the particle channel with all spins inverted. If we sum over possible spin orientations and hence deal with partial decay rates corresponding to no spin measurements in the initial or the final state, then one obtains

$$\lambda(i \to f) = \lambda(\bar{\imath} \to \bar{f}) \tag{182}$$

Among processes for which this result holds we just mention the K_{l2} and the K_{l3} decays. Due to (182) we conclude that for these processes the energy spectra and partial decay rates (after spin summation) are equal for the K^+ decay and the K^- decay.

$$\begin{aligned} \lambda(K^+ \to \mu^+ + \nu_\mu) &= \lambda(K^- \to \mu^- + \bar{\nu}_\mu) \\ \lambda(K^+ \to \pi^0 + \mu^+ + \nu_\mu) &= \lambda(K^- \to \pi^0 + \mu^- + \bar{\nu}_\mu) \end{aligned} \tag{183}$$

These predictions of *CPT*-invariance have so far not been tested to any degree of accuracy. For a more complete discussion of these decays in relation to *CPT*-invariance, we refer to the article by Lee and Wu[22].

In *the case of non-negligible final state interactions*, (181) is replaced by

$$\langle \bar{f}; \text{out}|H_{\text{wk}}|\bar{\imath}; \text{in}\rangle = \exp(2i\delta_f)\langle \tilde{f}; \text{out}|H_{\text{wk}}|\tilde{\imath}; \text{in}\rangle^* \tag{184}$$

provided $|\bar{f}; \text{out}\rangle$ is an eigenstate of the strong (and electromagnetic) interactions. In the more general case, when this is not so, one may always expand the final state in terms of such eigenstates to obtain a sum of terms with different phase factors on the right-hand side. The implications of these last remarks are most easily explained in terms of an example. Consider the decays

$$\begin{aligned} \Lambda &\rightarrow p + \pi^- \\ &\rightarrow n + \pi^0 \end{aligned}$$

which both occur due to the weak interactions. Since both particles in the final states are hadrons we may not neglect the final state interactions. As a matter of fact, an event which is identified as a decay $\Lambda \rightarrow p + \pi^-$ may instrinsically be a $\Lambda \rightarrow n + \pi^0$ decay in which there is a charge exchange scattering in the final state. Somehow one must first disentangle the final states. This is done by introducing eigenstates of the strong interactions. Since the strong interactions conserve isospin and parity we must deal with states of definite I and L. For the πN system there are two isospin channels with $I = 1/2$ and $I = 3/2$. The weak decay process is not parity-conserving and hence the final state can have $L = 0$ and $L = 1$. Neglecting the electromagnetic interactions there will be four eigenstates of the strong interactions involved in this process and there will be four phase shifts $\delta_{2J}(L, I)$ to consider, namely $\delta_1(0, 1/2)$, $\delta_1(0, 3/2)$, $\delta_1(1, 1/2)$ and $\delta_1(1, 3/2)$. We shall not carry the analysis any further[36] but note that the presence of several terms with different phase factors in (184) does not permit us to deduce equalities between the partial decay rates via particle conjugate channels. For example, it does *not* follow from *CPT*-invariance that $B = \bar{B}$, where B and $\bar{B}$ are defined by

$$B = \frac{\lambda(\Lambda \rightarrow n + \pi^0)}{\lambda(\Lambda \rightarrow p + \pi^-)}$$

$$\bar{B} = \frac{\lambda(\bar{\Lambda} \rightarrow \bar{n} + \pi^0)}{\lambda(\bar{\Lambda} \rightarrow \bar{p} + \pi^+)}$$

However, with the additional assumption of C or T-invariance this equality holds[37].

In some cases it so happens that the final state always is an eigenstate of the strong interaction S-operator due to various independent selection rules. In that case the phase factor in (184) does not lead to any complications. As an example of this we consider the 2-particle conjugate processes

$$K^{\pm} \to \pi^{\pm} + \pi^{o}$$

It follows from the generalized Pauli principle (Bose statistics) that the final 2π state must be an eigenstate of isospin with $I = 2$. Furthermore, angular momentum conservation requires that $L = 0$ (S-state). Since isospin and parity is conserved by the strong interactions, there is no other possible final state which can be reached by a strong final state interaction. The 2π final state is in this case an eigenstate of the strong interaction S-operator. Neglecting the influence of electromagnetic interactions one then obtains from CPT-invariance the following equality

$$\lambda(K^{+} \to \pi^{+} + \pi^{o}) = \lambda(K^{-} \to \pi^{-} + \pi^{o}) \tag{185}$$

The corrections due to electromagnetic interactions are expected to be of the order 10^{-4}. In the same fashion one can deduce the following relation from CPT-invariance (neglecting electromagnetic corrections)

$$\begin{aligned} &\lambda(K^{+} \to \pi^{+} + \pi^{+} + \pi^{-}) + \lambda(K^{+} \to \pi^{+} + \pi^{o} + \pi^{o}) = \\ &= \lambda(K^{-} \to \pi^{-} + \pi^{-} + \pi^{+}) + \lambda(K^{-} \to \pi^{-} + \pi^{o} + \pi^{o}) \end{aligned} \tag{186}$$

with the corrections expected to be of the same order as in the previous case. If one further assumes that there is no admixture of $I = 3$ in the final state (due to weak interaction selection rules), then one obtains more restrictive results, namely

$$\lambda(K^{+} \to \pi^{+} + \pi^{+} + \pi^{-}) = \lambda(K^{-} \to \pi^{-} + \pi^{-} + \pi^{+}) \tag{187}$$

and

$$\lambda(K^{+} \to \pi^{+} + \pi^{o} + \pi^{o}) = \lambda(K^{-} \to \pi^{-} + \pi^{o} + \pi^{o}) \tag{188}$$

A more extensive study of the consequences of CPT-invariance for the K decays is found in the article by Lee and Wu to which we have previously referred[22].

So far we have been considering partial decay rates corresponding to specific decay channels. If we sum over all possible channels to obtain the total decay rate (the lifetime) then we may always analyse the final state in terms of eigenstates of the non-weak S-operator and since in this case the partial rates are equal it follows that the same holds true for the total decay rate. Thus we conclude that the lifetime of a particle equals the lifetime of its antiparticle if the theory is CPT-invariant. This prediction has been examined experimentally and we list the most accurate results below in Table 3, which is taken from reference[25].

TABLE 3

The ratio of the lifetime for some particles and their antiparticles

Particles	$\frac{\tau_+}{\tau_-} - 1$
μ^+, μ^-	$0{\cdot}000 \pm 0{\cdot}001$
π^+, π^-	$0{\cdot}0018 \pm 0{\cdot}0040$
K^+, K^-	$-0{\cdot}0009 \pm 0{\cdot}0008$

(iii) *Equality for the anomalous magnetic moment of a particle and its antiparticle.* We have previously discussed the weak process $\nu_\mu + n \rightarrow \mu^- + p$ and introduced the concept of weak form-factors in order to obtain explicit expressions for the matrix elements of the weak hadronic current. Similarly, for the electromagnetic interactions of hadrons one needs explicit expressions for matrix elements of the electromagnetic current $J_\mu{}^{\mathrm{EM}}(x)$. Consider for example the scattering of a proton in an external field. To lowest order in the electromagnetic interactions this process is described by the diagram of figure 12, and the relevant part

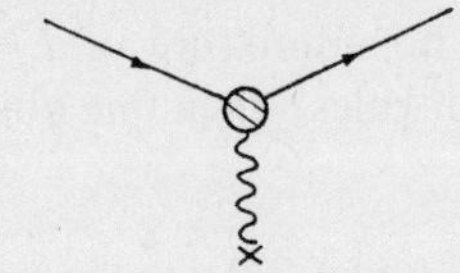

FIGURE 12 Proton scattering in an external electromagnetic field

of the matrix element for this process is $\langle p'|J_\mu{}^{\mathrm{EM}}(0)|p\rangle$. From Lorentz invariance and current conservation it is easily shown that the most general form for this matrix element is given by (some irrelevant normalization factors have been omitted)

$$\langle f; \mathrm{out}|J_\mu{}^{\mathrm{EM}}(0)|i; \mathrm{in}\rangle \equiv \langle \mathbf{p}', \mathbf{s}'|J_\mu{}^{\mathrm{EM}}(0)|\mathbf{p}, \mathbf{s}\rangle =$$
$$= e\bar{u}(\mathbf{p}', \mathbf{s}')[F_1(q^2)\gamma_\mu + \mathrm{i}F_2(q^2)\sigma_{\mu\nu}q^\nu]u(\mathbf{p}, \mathbf{s}) \qquad (189)$$

with $q = p' - p$. Considering the static limit, that is $q \rightarrow 0$, one easily identifies $eF_1(0)$ as the proton charge and $eF_2(0)$ as the anomalous magnetic moment of the proton. Since the charge of the proton equals the charge of the electron, it follows that $F_1(0) = 1$. From the fact that $J_\mu{}^{\mathrm{EM}}(0)$ is a hermitian operator it follows that $F_1(q^2)$ and $F_2(q^2)$ are real functions.

To investigate the consequences of *CPT*-invariance of the electromagnetic interaction, we first note the transformation property for the current operator in order that this condition be satisfied

$$\Theta J_\mu{}^{\mathrm{EM}}(0)\Theta^{-1} = -J_\mu{}^{\mathrm{EM}}(0) \tag{190}$$

From this we obtain the analogue of (177)

$$\langle \bar{f};\, \mathrm{out}|J_\mu{}^{\mathrm{EM}}(0)|\bar{\imath};\, \mathrm{in}\rangle = -\langle \tilde{f};\, \mathrm{out}|J_\mu{}^{\mathrm{EM}}(0)|\tilde{\imath};\, \mathrm{in}\rangle^* \tag{191}$$

The right-hand side may be evaluated from (189) with the result

$$\langle \bar{f};\, \mathrm{out}|J_\mu{}^{\mathrm{EM}}(0)|\bar{\imath};\, \mathrm{in}\rangle = -e\{\bar{u}(\mathbf{p}', -\mathbf{s}')[F_1(q^2)\gamma_\mu + \mathrm{i}F_2(q^2)\sigma_{\mu\nu}q^\nu]u(\mathbf{p}, -\mathbf{s})\}^* \tag{192}$$

If we make use of the following spinor relations (cf. Appendix 1)

$$\begin{aligned} u(\mathbf{p}, -\mathbf{s}) &= \gamma_0 \mathscr{T} \bar{u}^{-\mathrm{T}}(\mathbf{p}, \mathbf{s}) \\ \mathbf{u}(\mathbf{p}, -\mathbf{s}) &= u^{\mathrm{T}}(\mathbf{p}, \mathbf{s})\mathscr{T}^{-1}\gamma_0 \end{aligned} \tag{193}$$

we obtain

$$\langle \bar{f};\, \mathrm{out}|J_\mu{}^{\mathrm{EM}}(0)|\bar{\imath};\, \mathrm{in}\rangle = -e\bar{u}(\mathbf{p}', \mathbf{s}')[F_1(q^2)\gamma_\mu + \mathrm{i}F_2(q^2)\sigma_{\mu\nu}q^\nu]u(\mathbf{p}, \mathbf{s}) \tag{194}$$

The change in sign as compared to (189) just reflects the opposite charge for a particle and its antiparticle. From (189) and (194) we conclude that *CPT*-invariance implies equality for the particle and the antiparticle form-factors. In particular, their anomalous magnetic moments should be the same. This last prediction has been tested experimentally[25] (see Table 4).

TABLE 4

The gyromagnetic ratios for electrons and muons

Particles	$\frac{1}{2}(g_+ - g_-)$
e^+, e^-	$(1{\cdot}5 \pm 2)\alpha^2/\pi^2$
μ^+, μ^-	$(0 \pm 1{\cdot}5)\alpha^2/\pi^2$

In summary, we conclude that there is very strong evidence for *CPT*-invariance for the strong, the electromagnetic and the strangeness-conserving non-leptonic weak interactions. Also the purely weak processes lend rather strong support for *CPT*-invariance, while the semi-leptonic and the strangeness-changing weak interactions remain to be investigated with higher precision. It should be noted also that in general only some aspects of *CPT*-invariance have been examined. In no case do we have precision measurements directly checking the relation (177) without summation over spin directions. It would certainly be worthwhile to test the equality of the matrix elements for particle conjugate processes with the spins reversed as given by (177).

8 Concluding Remarks

We summarize the present experimental status of the discrete symmetries C, P, T or CP and CPT with regard to the strong, electromagnetic and weak interactions in Table 5. The figures quoted refer to the upper limit

TABLE 5

Upper limits for symmetry-violating amplitudes in the different types of interactions

Symmetry	Strong interactions	Electromagnetic interactions	Weak interactions
CPT	10^{-14}	10^{-12}	$\Delta S = 0$, non-lep. 10^{-8} other $\sim 10^{-3}$
$T \sim CP$	10^{-3}	non-leptonic?	violated; $\eta_{+-} \simeq 2 \times 10^{-3}$
P	10^{-6}	10^{-4}	violated
C	10^{-2}	non-leptonic 10^{-1}	violated

for the symmetry-violating amplitudes. The electromagnetic interactions of leptons are to a high degree of accuracy C and T-invariant, and these symmetries are incorporated in quantum electrodynamics.

We end this presentation of the discrete symmetries with some brief remarks concerning two important aspects of symmetry arguments, namely (i) a specific prediction may follow from more than one invariance principle and its verification cannot be uniquely attributed to any one of them, and (ii) the whole concept of symmetry when symmetry-violating interactions are present.

8.1 *Predictions which follow from more than one invariance principle*

On various occasions we have encountered predictions which can be derived from more than one principle. This is, for example, the case with the vanishing electric dipole moment for the neutron, which is a consequence of C or T-invariance. A violation of this prediction implies a violation of *both* C and T-invariance. A verification of the prediction lends support to both invariance principles but cannot be taken as conclusive evidence for either.

For this reason it is extremely important to clarify under what assumptions a certain prediction may be derived, and if there is more than one alternative. This inherent weakness of symmetry arguments was once again pointed out very clearly by Bernstein, Feinberg and Lee[11] who noted that very many predictions of *C* and *T*-invariance in electromagnetic processes can also be derived from other equally fundamental principles. As an example we consider the electromagnetic form-factors of a nucleon. In Section 7.2 we introduced the concept invoking Lorentz invariance and current conservation. If we only assume Lorentz invariance, then (189) is replaced by

$$\begin{aligned}\langle \mathbf{p}', \mathbf{s}'|J_\mu{}^{\mathrm{EM}}(0)|\mathbf{p}, \mathbf{s}\rangle &= \\ &= e\bar{u}(\mathbf{p}', \mathbf{s}')[F_1(q^2)\gamma_\mu + \mathrm{i}F_2(q^2)\sigma_{\mu\nu}q^\nu + F_3(q^2)q_\mu]u(\mathbf{p}, \mathbf{s}) \qquad (195)\end{aligned}$$

Current conservation implies

$$\langle \mathbf{p}', \mathbf{s}'|q^\mu J_\mu{}^{\mathrm{EM}}(0)|\mathbf{p}, \mathbf{s}\rangle = 0 \qquad (196)$$

and it follows from gauge invariance. Noting that the first term in (195) may be written

$$\bar{u}(\mathbf{p}', \mathbf{s}')\gamma_\mu q^\mu u(\mathbf{p}, \mathbf{s}) = \bar{u}(\mathbf{p}', \mathbf{s}')[\gamma_\mu p'^\mu - \gamma_\mu p^\mu]u(\mathbf{p}, \mathbf{s}) \qquad (197)$$

it follows from the Dirac equation for the two spinors that this term vanishes. In the second term we obtain $\sigma_{\mu\nu}q^\mu q^\nu$. Since $\sigma_{\mu\nu}$ is antisymmetric in μ and ν while $q^\mu q^\nu$ is symmetric, this term also vanishes. Thus, from condition (196) we conclude

$$q^2 F_3(q^2) = 0 \qquad (198)$$

or

$$F_3(q^2) = 0 \qquad (199)$$

except possibly for $q^2 = 0$. This result was anticipated in (189). However, one reaches the same conclusion by invoking (i) that $J_\mu{}^{\mathrm{EM}}(0)$ is hermitian and (ii) that under time reversal $J_\mu{}^{\mathrm{EM}}(0) \rightarrow \varepsilon(\mu)J_\mu{}^{\mathrm{EM}}(0)$. From (i) we obtain

$$\langle \mathbf{p}', \mathbf{s}'|J_\mu{}^{\mathrm{EM}}(0)|\mathbf{p}, \mathbf{s}\rangle = \langle \mathbf{p}, \mathbf{s}|J_\mu{}^{\mathrm{EM}}(0)|\mathbf{p}', \mathbf{s}'\rangle^* \qquad (200)$$

or

$$\begin{aligned}\bar{u}(\mathbf{p}', \mathbf{s}')[F_1\gamma_\mu + \mathrm{i}F_2\sigma_{\mu\nu}q^\nu + F_3 q_\mu]u(\mathbf{p}, \mathbf{s}) &= \\ &= \{\bar{u}(\mathbf{p}, \mathbf{s})[F_1\gamma_\mu - \mathrm{i}F_2\sigma_{\mu\nu}q^\nu - F_3 q_\mu]u(\mathbf{p}', \mathbf{s}')\}^* \\ &= u^\dagger(\mathbf{p}', \mathbf{s}')[F_1{}^*\gamma_\mu{}^\dagger + \mathrm{i}F_2{}^*\sigma_{\mu\nu}{}^\dagger q^\nu - F_3{}^* q_\mu]\gamma_0 u(\mathbf{p}, \mathbf{s}) \\ &= \bar{u}(\mathbf{p}', \mathbf{s}')[F_1{}^*\gamma_\mu + \mathrm{i}F_2{}^*\sigma_{\mu\nu}q^\nu - F_3{}^* q_\mu]u(\mathbf{p}, \mathbf{s}) \qquad (201)\end{aligned}$$

We conclude that hermiticity requires that F_1 and F_2 are real functions and F_3 a purely imaginary function. With regard to the condition (ii), we have previously considered the case of a weak current. The computations are in this case very similar and we shall not repeat them. The result is that the condition

$$\langle \mathbf{p}', \mathbf{s}' | J_\mu^{\mathrm{EM}}(0) | \mathbf{p}, \mathbf{s} \rangle = \varepsilon(\mu) \langle -\mathbf{p}', -\mathbf{s}' | J_\mu^{\mathrm{EM}}(0) | -\mathbf{p}, -\mathbf{s} \rangle^* \qquad (202)$$

implies that F_1, F_2 and F_3 must all be real functions. In order to satisfy both conditions we must then choose $F_3 = 0$, which is the same result as (199).

We shall not pursue the subject further, but we conclude that a verification of a prediction based on a symmetry argument never constitutes an absolute proof of the invariance principle, since (i) the same result may follow from an alternative set of principles as discussed above, or (ii) there may be an accidental cancellation, for example of dynamical origin. An example of this latter possibility is known from the parity-violating asymmetry parameters α in Σ decay. For the Σ_+^+ ($\Sigma^+ \to n + \pi^+$) and Σ_-^- ($\Sigma^- \to n + \pi^-$) decays the asymmetry parameters vanish while for Σ_0^+ ($\Sigma^+ \to p + \pi^0$) it is large. Off-hand one would say that the first two decays are parity-conserving while the third one is parity-violating. For the time being these observations can only be understood as accidental cancellations, although they neatly conform with the $\Delta I = \frac{1}{2}$ rule.

Finally, it should also be stated that although the fundamental laws of physics may satisfy certain symmetry properties, the same symmetry properties are reflected in the experimental results only for isolated systems with no 'external fields' present. For example, if we neglect the very small CP-violation, it was found that the two neutral K-mesons K_1^0 and K_2^0 retain their identity until the moment of decay since CP conservation forbids the transitions $K_1^0 \rightleftarrows K_2^0$. How then can one explain that a beam of K_2^0 regenerates K_1^0 mesons copiously when it passes through matter even though all interactions are CP-invariant to a high degree of accuracy? The reason is simply that the piece of matter through which the beam passes is not CP-invariant and it constitutes an 'external field'. Quite naturally, then, the regeneration effect depends on the difference between the scattering amplitudes for KN and $\bar{K}N$ scattering. In the same spirit it was first attempted to save the CP-invariance in the $K_L^0 \to 2\pi$ decay by invoking an external field to account for the apparent CP-violation. This possibility has lately been ruled out on experimental grounds.

8.2 *The concept of symmetry in the presence of symmetry-violating interactions*

We have consistently defined the various symmetry operations by giving their action on certain state vectors. In those cases where there is a classical analogue, we have chosen the definitions in such a way that the quantum-mechanical operators transform in the same way as their classical counterparts. This procedure may seem very straightforward, but in realistic applications there are some inherent difficulties. For one thing, the problem of how to treat unstable states has not been tackled more than superficially, and a completely satisfactory way to do that is not yet known[38]. The other complication arises in the context of approximate symmetries. We have seen that our definition of a space reflection operator requires a Hilbert space which is larger than the physical Hilbert space ($\equiv$ the space of physical state vectors) since occasionally non-physical vectors appear (e.g. the neutrino states with positive helicity). In a formal approach strictly based on group theory it is difficult to give an exact meaning to the concept of an approximate symmetry, and yet the concept has turned out to be most useful in elementary particle physics. In a Hamiltonian approach approximate symmetries emerge in a rather natural way.

Suppose that we can write the total Hamiltonian of a physical system in the following way

$$H = H_{\text{free}} + H_{\text{st}} + H_{\text{em}} + H_{\text{wk}} \tag{203}$$

where the strong, the electromagnetic and the weak parts depend on the corresponding coupling constants. Suppose further that one may define some limiting procedure so that, say, the weak and the electromagnetic part of the Hamiltonian vanish. In this limit, if it exists, it is conceivable that we obtain a model which exhibits full C and P symmetry. This means that the corresponding quantum mechanical operators would not contain any parts, which would lead out of the physical Hilbert space when they act on a physical state vector. On the other hand, in this limit the Hamiltonian will in general be invariant under entirely new transformations which render the C and P transformations ambiguous and some of the conventional state labels are no longer relevant. In the example considered above this happens for the electric charge, since it requires the presence of electromagnetism for an operational definition. Since the strong forces are charge independent one arrives at a model which is invariant under the isospin group. Under these circumstances a space reflection or a charge conjugation may just as well contain isospin transformations. Without the charge label we would retain transformation properties for observables analogous to the classical ones for P, and the C transformation would, for example, transform a p state into an $\bar{n}$ state, etc. In this way one faces the possibility of different C, P and T transformations corresponding to the limits which we conventionally refer to as strong, electromagnetic and weak interactions[39]. We shall not develop this subject further, but it is clear that once a symmetry is found not to hold rigorously we must approach the subject with an open mind for entirely new possibilities with drastic changes in old concepts. This is, of course, an observation which is particularly pertinent in these times when old symmetries repeatedly fail under closer scrutiny.

Appendix 1

Brief Review of the Dirac Equation

Relativistic particles of spin $\frac{1}{2}$ are described by a 4-component spinor wave-function $\psi(x)$. For free particles this wave-function satisfies the following equation of motion—*the Dirac equation*

$$\left(i\gamma_\mu \frac{\delta}{\delta x_\mu} - m\right)\psi(x) = 0 \tag{1}$$

The 4×4 matrices γ_μ ($\mu = 0, 1, 2, 3$) obey the anticommutation rules

$$\gamma_\mu\gamma_\nu + \gamma_\nu\gamma_\mu = 2g_{\mu\nu} \tag{2}$$

where $g_{\mu\nu}$ is the metric tensor, which we take to be

$$(g_{\mu\nu}) = \begin{pmatrix} 1 & 0 & 0 & 0 \\ 0 & -1 & 0 & 0 \\ 0 & 0 & -1 & 0 \\ 0 & 0 & 0 & 1 \end{pmatrix} \tag{3}$$

We shall use the following explicit representation of the γ-matrices

$$\gamma_0 = \begin{pmatrix} I & 0 \\ 0 & -I \end{pmatrix}; \quad \gamma_k = \begin{pmatrix} 0 & \sigma_k \\ -\sigma_k & 0 \end{pmatrix} \tag{4}$$

where I denotes the 2×2 unit matrix and σ_k ($k = 1, 2, 3$) is a Pauli matrix

$$\sigma_1 = \begin{pmatrix} 0 & 1 \\ 1 & 0 \end{pmatrix}; \quad \sigma_2 = \begin{pmatrix} 0 & -i \\ i & 0 \end{pmatrix}; \quad \sigma_3 = \begin{pmatrix} 1 & 0 \\ 0 & -1 \end{pmatrix} \tag{5}$$

With these conventions it is easily shown that γ_0 is hermitian and γ_k antihermitian. These properties are summarized by the relation

$$\gamma_\mu{}^\dagger = \gamma_0\gamma_\mu\gamma_0 \tag{6}$$

If the adjoint wave-function $\bar{\psi}(x)$ is defined by

$$\bar{\psi}(x) = \psi^{\dagger}(x)\gamma_0 \tag{7}$$

then it follows from (1) that in the case of free particles it satisfies the equation

$$i\frac{\delta}{\delta x_{\mu}}\bar{\psi}(x)\gamma_{\mu} + m\bar{\psi}(x) = 0 \tag{8}$$

The following two matrices occur frequently in the Dirac theory

$$\sigma_{\mu\nu} = \frac{1}{2i}(\gamma_{\mu}\gamma_{\nu} - \gamma_{\nu}\gamma_{\mu})$$

$$\gamma_5 = i\gamma_0\gamma_1\gamma_2\gamma_3 \tag{9}$$

The matrix $\sigma_{\mu\nu}$ satisfies the relation

$$\sigma_{\mu\nu}{}^{\dagger} = \gamma_0\sigma_{\mu\nu}\gamma_0 \tag{10}$$

while γ_5 is hermitian. Furthermore, the matrix γ_5 anticommutes with the four γ-matrices γ_{μ}

$$\gamma_{\mu}\gamma_5 + \gamma_5\gamma_{\mu} = 0 \tag{11}$$

In the standard representation (4), γ_5 is given by

$$\gamma_5 = \begin{pmatrix} 0 & -I \\ -I & 0 \end{pmatrix} \tag{12}$$

The Dirac equation (1) is satisfied by plane wave solutions of the form

$$\psi(x) = u(\mathbf{p})\exp[-\mathrm{i}p_{\mu}x^{\mu}]$$

$$\psi(x) = v(\mathbf{p})\exp[\mathrm{i}p_{\mu}x^{\mu}] \tag{13}$$

provided the coordinate–independent spinors $u(\mathbf{p})$ and $v(\mathbf{p})$ satisfy the following set of equations

$$(\gamma_{\mu}p^{\mu} - m)u(\mathbf{p}) = 0$$

$$(\gamma_{\mu}p^{\mu} + m)v(\mathbf{p}) = 0 \tag{14}$$

The solutions with $u(\mathbf{p})$ are identified with the particle solutions, while those with $v(\mathbf{p})$ describe antiparticles. It is easily seen that equations (14) have two independent solutions, each corresponding to two possible orientations of the spin. With the standard representation (4) the solutions are given by

$$u_i(\mathbf{p}) = \left(\frac{E+m}{2m}\right)^{\frac{1}{2}} \begin{pmatrix} \chi_i \\ \dfrac{\boldsymbol{\sigma}\cdot\mathbf{p}}{E+m}\chi_i \end{pmatrix} \tag{15}$$

$$v_i(\mathbf{p}) = \left(\frac{E+m}{2m}\right)^{\frac{1}{2}} \begin{pmatrix} -\dfrac{\boldsymbol{\sigma}\cdot\mathbf{p}}{E+m}\xi_i \\ \xi_i \end{pmatrix} \tag{16}$$

with $i = 1, 2$ and

$$\chi_1 = \begin{pmatrix} 1 \\ 0 \end{pmatrix} = -\xi_2; \quad \chi_2 = \begin{pmatrix} 0 \\ 1 \end{pmatrix} = \xi_1 \tag{17}$$

The normalization has been chosen in such a way that the following orthonormality conditions hold

$$\begin{aligned} \bar{u}_i(\mathbf{p})u_j(\mathbf{p}) &= \delta_{ij} \\ \bar{v}_i(\mathbf{p})v_j(\mathbf{p}) &= -\delta_{ij} \\ \bar{u}_i(\mathbf{p})v_j(\mathbf{p}) &= \bar{v}_i(\mathbf{p})u_j(\mathbf{p}) = 0 \end{aligned} \tag{18}$$

The adjoint spinors $\bar{u}(\mathbf{p})$ and $\bar{v}(\mathbf{p})$ are defined by

$$\begin{aligned} \bar{u}(\mathbf{p}) &= u^\dagger(\mathbf{p})\gamma_0 \\ \bar{v}(\mathbf{p}) &= v^\dagger(\mathbf{p})\gamma_0 \end{aligned} \tag{19}$$

and they satisfy equations analogous to (14)

$$\begin{aligned} \bar{u}(\mathbf{p})(\gamma_\mu p^\mu - m) &= 0 \\ \bar{v}(\mathbf{p})(\gamma_\mu p^\mu + m) &= 0 \end{aligned} \tag{20}$$

In the discussions of charge conjugation and time reversal two more matrices $\mathscr{C}$ and $\mathscr{T}$ are introduced. They have the following properties

$$\begin{aligned} \mathscr{C}^{-1}\gamma_\mu\mathscr{C} &= -\gamma_\mu{}^{\mathrm{T}} \\ \mathscr{C}^{-1}\gamma_5\mathscr{C} &= \gamma_5{}^{\mathrm{T}} \\ \mathscr{C}^\dagger &= \mathscr{C}^{-1} \\ \mathscr{C}^{\mathrm{T}} &= -\mathscr{C} \end{aligned} \tag{21}$$

and

$$\begin{aligned}\mathscr{T}^{-1}\gamma_\mu\mathscr{T} &= \varepsilon(\mu)\gamma_\mu{}^{\mathrm{T}}\\ \mathscr{T}^{-1}\gamma_5\mathscr{T} &= -\gamma_5{}^{\mathrm{T}}\\ \mathscr{T}^{\dagger} &= \mathscr{T}^{-1}\\ \mathscr{T}^{\mathrm{T}} &= -\mathscr{T}\end{aligned}\tag{22}$$

By direct inspection it is seen that

$$\mathscr{C} = -i\gamma_0\gamma_2 \tag{23}$$

and

$$\mathscr{T} = i\gamma_2\gamma_5 \tag{24}$$

satisfy these conditions. In the standard representation (4) they are given by

$$\begin{aligned}\mathscr{C} &= -i\begin{pmatrix}0 & \sigma_2\\ \sigma_2 & 0\end{pmatrix}\\ \mathscr{T} &= i\begin{pmatrix}-\sigma_2 & 0\\ 0 & \sigma_2\end{pmatrix}\end{aligned}\tag{25}$$

For applications regarding P, C and T we note the following relations, which can be obtained, for example, by direct inspection from (15) and (16), with the explicit representation (4) for the γ-matrices

$$\begin{aligned}u(-\mathbf{p}, \pm\mathbf{s}) &= \gamma_0 u(\mathbf{p}, \pm\mathbf{s})\\ v(-\mathbf{p}, \pm\mathbf{s}) &= -\gamma_0 v(\mathbf{p}, \pm\mathbf{s})\end{aligned}\tag{26}$$

with

$$\begin{aligned}u(\mathbf{p}, \mathbf{s}) &\equiv u_1(\mathbf{p})\\ u(\mathbf{p}, -\mathbf{s}) &\equiv u_2(\mathbf{p})\end{aligned}\tag{27}$$

etc. Similarly

$$\begin{aligned}v(\mathbf{p}, \mathbf{s}) &= \mathscr{C}\bar{u}^{\mathrm{T}}(\mathbf{p}, \mathbf{s})\\ \bar{v}(\mathbf{p}, \mathbf{s}) &= -u^{\mathrm{T}}(\mathbf{p}, \mathbf{s})\mathscr{C}^{-1}\end{aligned}\tag{28}$$

and

$$\begin{aligned}u(-\mathbf{p}, -\mathbf{s}) &= \mathscr{T}\bar{u}^{\mathrm{T}}(\mathbf{p}, \mathbf{s})\\ \bar{u}(-\mathbf{p}, -\mathbf{s}) &= u^{\mathrm{T}}(\mathbf{p}, \mathbf{s})\mathscr{T}^{-1}\end{aligned}\tag{29}$$

Appendix 2

Free Particle States

The basis vectors of the physical Hilbert space are labelled by the eigenvalues of a complete set of commuting observables, and any state vector representing a physical system can always be expanded in such a basis. Corresponding to different sets of commuting observables one obtains different sets of basis vectors, which are related by unitary transformations (Clebsch-Gordan type expansions). It is often important to know the explicit form of these transformations for the sets which are most frequently used in physics. It is also very important that a consistent set of phase conventions is used. Since there are many different conventions used in the literature, special care must be exercised in comparing results from different authors.

There are essentially two types of basis vectors which are used in particle physics; (i) plane wave states of definite momentum and (ii) angular momentum states of various kinds. In this appendix we shall define the sets which have been used in this series of lectures and give the relations between them. For a more complete discussion of these questions we refer to reference 1. We shall treat single-particle states, 2-particle states, etc., separately since the complexity of labelling the state vectors increases with increasing number of particles. We shall further restrict the discussion to free particle states for which single-particle labels can be used.

The classification of relativistic particle states is based on the Poincaré group, that is, the group of proper orthochronous Lorentz transformations. Within this scheme the kinematics of a free particle of mass m and spin s is related to a unitary irreducible representation (UIR) of the Poincaré group. These UIR's are characterized by the value of the two invariants (Casimir operators) $P^{\mu}P_{\mu}$ and $W^{\mu}W_{\mu}$ which in turn are related to the rest mass and the spin of the particle

$$P^{\mu}P_{\mu} = m^2$$

$$W^{\mu}W_{\mu} = -m^2 s(s+1)$$

with

$$W^{\mu} = \tfrac{1}{2}\varepsilon^{\mu\nu\rho\sigma} J_{\nu\rho} P_{\sigma}$$

and $\varepsilon^{0123} = -\varepsilon_{0123} = -1$. More precisely, the manifold of state vectors representing the possible states which the particle can occupy form a representation space of the UIR (m, s) of the Poincaré group. Alternatively, given the state vector representing any possible state of the particle, all the other physical state vectors representing the same particle (in different states of motion) are obtained by means of a Lorentz transformation acting on the original state vector. This property provides us with the means to construct the state vectors representing a particle from some standard state vector. We shall outline the procedure below but first we summarize the basic properties of the Poincaré group and its Lie algebra spanned by the infinitesimal generators P_{μ} and $J_{\mu\nu}$.

(i) $(a, \Lambda) : x^{\mu} = \Lambda_{\nu}{}^{\mu} x'^{\nu} + a^{\mu}$

(ii) $(a_1, \Lambda_1)(a_2, \Lambda_2) = (a_1 + \Lambda_1\, a_2, \Lambda_1\, \Lambda_2)$

(iii) $[P^{\mu}, P^{\nu}] = 0$

$[J^{\mu\nu}, P^{\rho}] = -\mathrm{i}(g^{\mu\rho}P^{\nu} - g^{\nu\rho}P^{\mu})$

$[J^{\mu\nu}, J^{\rho\sigma}] = -\mathrm{i}(g^{\mu\rho}J^{\nu\sigma} + g^{\nu\sigma}J^{\mu\rho} - g^{\mu\sigma}J^{\nu\rho} - g^{\nu\rho}J^{\mu\sigma})$

(iv) If we define a new set of generators by

$$J^{kl} = \sum_{m} \varepsilon^{klm} J^{m}; \quad k, l, m = 1, 2, 3$$

$$J^{ok} = -J^{ko} = K^{k}$$

then the commutation rules read

$$[P^{\mu}, P^{\nu}] = 0$$
$$[J^{k}, P^{o}] = 0$$
$$[J^{k}, P^{l}] = \mathrm{i}\, \varepsilon^{klm} P^{m}$$
$$[J^{k}, J^{l}] = \mathrm{i}\, \varepsilon^{klm} J^{m}$$
$$[K^{k}, P^{l}] = -\mathrm{i}\, \delta_{kl} P^{o}$$
$$[J^{k}, K^{l}] = \mathrm{i}\, \varepsilon^{klm} K^{m}$$
$$[K^{k}, K^{l}] = -\mathrm{i}\, \varepsilon^{klm} J^{m}$$
$$[K^{k}, P^{o}] = -\mathrm{i}\, P^{k}$$

1. *Single-particle states*

Consider a particle of mass $m > 0$ and spin s at rest. The state of this particle is completely specified by the spin component m_s along the

z-axis, since $P^\mu P_\mu$, $W^\mu W_\mu$, $\mathbf{P}$ and J^3 form a complete set of commuting observables. In the rest system the angular momentum operator $\mathbf{J}$ is identical with the spin operator, for which we shall use the notation $\mathbf{S}$. Altogether there are $(2s + 1)$ linearly independent states corresponding to $m_s = s, s - 1, \ldots, -s$. We denote the state vectors by $|\mathbf{p} = 0, m_s\rangle$ and we suppress the labels m and s. The transformations between the different states are accomplished with the raising and lowering operators $J_\pm = (J^1 \pm \mathrm{i}\, J^2)$ which satisfy the following relations

$$J_\pm|\mathbf{p} = 0, m_s\rangle = [(s \mp m_s)(s \pm m_s + 1)]^{\frac{1}{2}}|\mathbf{p} = 0, m_s \pm 1\rangle \qquad (1)$$

while J^3 is diagonal

$$J^3|\mathbf{p} = 0, m_s\rangle = m_s|\mathbf{p} = 0, m_s\rangle \qquad (2)$$

A finite rotation $R(\mathbf{\Omega})$ is given by

$$R(\mathbf{\Omega}) = \exp\,[-\mathrm{i}\,\mathbf{J}\cdot\mathbf{\Omega}] \qquad (3)$$

and acting on the state $|\mathbf{p} = 0, m_s\rangle$ one obtains

$$R(\mathbf{\Omega})|\mathbf{p} = 0, m_s\rangle \equiv D^s(\mathbf{\Omega})|\mathbf{p} = 0, m_s\rangle = \sum_{m_s'=-s}^{s} D^s{}_{m_s'm_s}(\mathbf{\Omega})|\mathbf{p} = 0, m_s'\rangle \qquad (4)$$

If we parametrize the rotation by means of the Euler angles (α, β, γ) then we obtain $(j = s)$

$$D^j(\alpha, \beta, \gamma) \equiv \exp\,(-\mathrm{i}\,J^3\alpha)\exp\,(-\mathrm{i}\,J^2\beta)\exp\,(-\mathrm{i}\,J^3\gamma) \qquad (5)$$

and hence

$$D_{m'm}{}^j(\alpha, \beta, \gamma) \equiv \langle j, m'|\exp\,(-\mathrm{i}\,J^3\alpha)\exp\,(-\mathrm{i}\,J^2\beta) \times \times \exp\,(-\mathrm{i}\,J^3\gamma)|j, m\rangle = \exp\,[-\mathrm{i}\,(m'\alpha + m\gamma)]\, d_{m'm}{}^j(\beta) \qquad (6)$$

with

$$d_{m'm}{}^j(\beta) = \langle j, m'|\exp\,(-\mathrm{i}\,J^2\beta)|j, m\rangle \qquad (7)$$

These d-functions can be expressed in elementary functions[1]. The results so far are based on the fact that the state vectors representing the particle at rest form a $(2s + 1)$-dimensional UIR of the rotation group SO(3).

Starting from a state at rest we may now proceed to define various states in motion. We introduce

$$|\mathbf{p}, m_s\rangle \equiv R(\phi, \theta, 0)L_z(v)R^{-1}(\phi, \theta, 0)|\mathbf{p} = 0, m_s\rangle \qquad (8)$$

and

$$|\mathbf{p}, \lambda\rangle \equiv R(\phi, \theta, 0)L_z(v)|\mathbf{p} = 0, \lambda\rangle \qquad (9)$$

where ϕ and θ are the polar angles of $\mathbf{p}$ and $\mathbf{v} = \mathbf{p}/p_0$. We shall refer to the vectors $|\mathbf{p}, \lambda\rangle$ as helicity states since they are eigenvectors of the helicity operator $\mathbf{J} \cdot \mathbf{P}/|\mathbf{P}|$. The Lorentz transformation $L_z(v)$ along the z-axis can be expressed in terms of the infinitesimal generator K^3.

$$L_z(v) = \exp\,[-\mathrm{i}\,K^3 u] \tag{10}$$

and $\tanh u = v$. The normalizations of the vectors (8) and (9) are given by

$$\langle \mathbf{p}', m_s' | \mathbf{p}, m_s \rangle = p_0 \delta^3(\mathbf{p}' - \mathbf{p}) \delta_{m_s' m_s} \tag{11}$$

$$\langle \mathbf{p}', \lambda' | \mathbf{p}, \lambda \rangle = p_0 \delta^3(\mathbf{p}' - \mathbf{p}) \delta_{\lambda' \lambda} \tag{12}$$

To demonstrate the consistency of our procedure we shall compute the x-component of the momentum for a particle represented by $|\mathbf{p}, \lambda\rangle \equiv$ $\equiv |\phi, \theta, p, \lambda\rangle$. Thus

$$\begin{aligned} P^1|\phi, \theta, p, \lambda\rangle &= P^1 R(\phi, \theta, 0) L_z(v) |\mathbf{p} = 0, \lambda\rangle = \\ &= P^1 \exp\,[-\mathrm{i}\,J^3\phi] \exp\,[-\mathrm{i}\,J^2\theta] \exp\,[-\mathrm{i}\,K^3 u] |\mathbf{p} = 0, \lambda\rangle = \\ &= \exp\,[-\mathrm{i}\,J^3\phi] \exp\,[-\mathrm{i}\,J^2\theta] \exp\,[-\mathrm{i}\,K^3 u] \times \{\exp\,[\mathrm{i}\,K^3 u] \cdot \\ &\quad \cdot \exp\,[\mathrm{i}\,J^2\theta] \exp\,[\mathrm{i}\,J^3\phi] P^1 \exp\,[-\mathrm{i}\,J^3\phi] \exp\,[-\mathrm{i}\,J^2\theta] \cdot \\ &\quad \cdot \exp\,[-\mathrm{i}\,K^3 u]\} |\mathbf{p} = 0, \lambda\rangle \end{aligned}$$

We first compute the expression within the braces ({}) and make use of the following relations, which can easily be derived from the commutation rules

$$\begin{aligned} \exp\,[\mathrm{i}\,J^3\phi] P^1 \exp\,[-\mathrm{i}\,J^3\phi] &= P^1 \cos\phi - P^2 \sin\phi \\ \exp\,[\mathrm{i}\,J^2\theta] P^1 \exp\,[-\mathrm{i}\,J^2\theta] &= P^1 \cos\theta + P^3 \sin\theta \\ \exp\,[\mathrm{i}\,J^2\theta] P^2 \exp\,[-\mathrm{i}\,J^2\theta] &= P^2 \\ \exp\,[\mathrm{i}\,K^3 u] P^1 \exp\,[-\mathrm{i}\,K^3 u] &= P^1 \\ \exp\,[\mathrm{i}\,K^3 u] P^2 \exp\,[-\mathrm{i}\,K^3 u] &= P^2 \\ \exp\,[\mathrm{i}\,K^3 u] P^3 \exp\,[-\mathrm{i}\,K^3 u] &= P^3 \cosh u + P^0 \sinh u \end{aligned}$$

Hence, we arrive at the following result

$$\begin{aligned} &\{\exp\,[\mathrm{i}\,K^3 u] \exp\,[\mathrm{i}\,J^2\theta] \exp\,[\mathrm{i}\,J^3\phi] P^1 \exp\,[-\mathrm{i}\,J^3\phi] \exp\,[-\mathrm{i}\,J^2\theta] \\ &\quad \exp\,[-\mathrm{i}\,K^3 u]\} = P^0 \cos\phi \sin\theta \sinh u + P^1 \cos\phi \cos\theta - P^2 \sin\phi + \\ &\quad + P^3 \cos\phi \sin\theta \cosh u \end{aligned}$$

Acting on $|\mathbf{p} = 0, \lambda\rangle$ only the first term contributes

$$P^0 |\mathbf{p} = 0, \lambda\rangle = m |\mathbf{p} = 0, \lambda\rangle$$

and we obtain

$$P^1 |\phi, \theta, p, \lambda\rangle = m \cos\phi \sin\theta \sinh u |\phi, \theta, p, \lambda\rangle$$

We finally take notice of the following identity

$$m \sinh u = m \frac{\tanh u}{\sqrt{(1 - \tanh^2 u)}} = m \cdot \frac{p/E}{\sqrt{\left(\frac{E^2 - p^2}{E^2}\right)}} = p$$

so that

$$P^1|\phi, \theta, p, \lambda\rangle = p \cos \phi \sin \theta |\phi, \theta, p, \lambda\rangle$$

which is the correct result.

The two sets of basis vectors given by (8) and (9) are clearly closely related. The transformation from one set to the other is immediately seen to be

$$|\mathbf{p}, \lambda\rangle = \sum_{m_s=-s}^{+s} D_{m_s}{}^s{}_\lambda(\phi, \theta, 0)|\mathbf{p}, m_s\rangle \tag{13}$$

It should be noted that the angles θ and ϕ in the definitions (8) and (9) range over the intervals

$$0 < \theta < \pi$$
$$-\pi < \phi \leq \pi$$

for which the definitions are unique. For $\theta = 0$ and $\theta = \pi$ this is no longer the case; similarly with half-integer spin representations for $\phi = -\pi$. We define the state vectors corresponding to the momentum parallel or antiparallel with the z-axes in the following way

$$|0, 0, p, \lambda\rangle \equiv L_z(v)|0, 0, 0, \lambda\rangle \tag{14}$$

and

$$|0, \pi, p, \lambda\rangle \equiv L_{-z}(v)|0, 0, 0, -\lambda\rangle = $$
$$= \exp [-\mathrm{i}\, \pi s] R(\pi, \pi, 0) L_z(v)|0, 0, 0, \lambda\rangle \tag{15}$$

To derive the transformation properties of the vectors $|\mathbf{p}, \lambda\rangle$ under a Lorentz transformation we introduce the abbreviated notation (cf. equation 9)

$$|\mathbf{p}, \lambda\rangle = L(\mathbf{p})|\mathbf{p} = 0, \lambda\rangle \tag{16}$$

and thus the operator $L(\mathbf{p})$ takes a state at rest into a state of momentum $\mathbf{p}$. To deduce the action of an arbitrary Lorentz transformation $L(\Lambda)$ on $|\mathbf{p}, \lambda\rangle$ we note the following decomposition of $L(\Lambda)$

$$L(\Lambda) = L(\Lambda\mathbf{p}) R(p, \Lambda) L^{-1}(\mathbf{p}) \tag{17}$$

where $R(p, \Lambda)$ is a pure rotation, the Wigner rotation, and $\Lambda\mathbf{p}$ is defined by

$$p_\mu \rightarrow (\Lambda p)_\mu = \Lambda_\mu{}^\nu p_\nu$$

and $\Lambda_\mu{}^\nu$ is the 4×4 matrix defining the Lorentz transformation. This decomposition yields

$$L(\Lambda)|\mathbf{p}, \lambda\rangle = L(\Lambda\mathbf{p})R(p, \Lambda)|\mathbf{p} = 0, \lambda\rangle = \\ = \sum_{\lambda'=-s}^{+s} D^s{}_{\lambda'\lambda}[R(p, \Lambda)]|\Lambda\mathbf{p}, \lambda'\rangle \qquad (18)$$

and corresponds to first transforming the state to rest and then from there to the state of final momentum $\Lambda\mathbf{p}$. However, in general there is an extra rotation for the 'intermediate' state at rest. The explicit form of the rotation $R(p, \Lambda)$ is given by (17). We shall not derive it but note that the transformation rule (18) for $s \neq 0$ is rather complicated.

We shall finally derive the transformation rule for the vector $|\mathbf{p}, \lambda\rangle$ under a space reflection (cf. equation 48 in Section 3.1). To this end we start from the state $|0, 0, 0, \lambda\rangle$. Since the space reflection operator P commutes with $\mathbf{J}$ we have

$$P|0, 0, 0, \lambda\rangle = \eta_P|0, 0, 0, \lambda\rangle \qquad (19)$$

and further P clearly also commutes with the raising and lowering operators $J_\pm$ so that η_P is independent of λ. The operator P is unitary and hence the intrinsic parity must be a phase factor. Next consider the operator R_{xz} of reflection in the xz-plane

$$R_{xz} = \exp[-\mathrm{i}\,\pi J^2]P = R(0, \pi, 0)P \qquad (20)$$

It commutes with $L_z(v)$ and further

$$R_{xz}|0, 0, 0, \lambda\rangle = \eta_P R(0, \pi, 0)|0, 0, 0, \lambda\rangle = \\ = \eta_P \sum_{\lambda'=-s}^{+s} D^s{}_{\lambda'\lambda}(0, \pi, 0)|0, 0, 0, \lambda'\rangle$$

It can be shown that

$$D^s{}_{\lambda'\lambda}(0, \pi, 0) = (-1)^{s-\lambda}\delta_{\lambda',-\lambda} \qquad (21)$$

so that

$$R_{xz}|0, 0, 0, \lambda\rangle = \eta_P(-1)^{s-\lambda}|0, 0, 0, -\lambda\rangle \qquad (22)$$

and thus

$$P|0, 0, p, \lambda\rangle = R^{-1}(0, \pi, 0)R_{xz}L_z(v)|0, 0, 0, \lambda\rangle = \\ = \eta_P(-1)^{s-\lambda}R^{-1}(0, \pi, 0)|0, 0, p, -\lambda\rangle \qquad (23)$$

Recalling that P commutes with rotations and noting the following relation

$$R(\phi, \theta, 0)R^{-1}(0, \pi, 0) = R(\phi + \pi, \pi - \theta, -\pi) \tag{24}$$

we finally arrive at

$$P|\phi, \theta, p, \lambda\rangle = \eta_p \exp[-\mathrm{i}\,\pi s]|\phi + \pi, \pi - \theta, p, -\lambda\rangle \tag{25}$$

which is the result previously quoted in Section 3.1.

Beyond the vectors $|\mathbf{p}, \lambda\rangle$ and $|\mathbf{p}, m_s\rangle$ we shall also encounter the angular momentum states $|j, m, p, \lambda\rangle$ defined by the partial wave expansion

$$|\mathbf{p}, \lambda\rangle = \sum_{j=0}^{\infty} \sum_{m=-j}^{+j} \left(\frac{2j+1}{4\pi}\right)^{\frac{1}{2}} D^{j}{}_{m\lambda}(\phi, \theta, 0)|j, m, p, \lambda\rangle \tag{26}$$

To invert this relation we recall the following orthogonality relation

$$\frac{2j+1}{4\pi}\int \mathrm{d}\omega\, D^{j^*}{}_{mn}(\phi, \theta, 0) D^{j'}{}_{m'n'}(\phi, \theta, 0) = \delta_{jj'}\delta_{mm'}\delta_{nn'} \tag{27}$$

where $\mathrm{d}\omega = \sin\theta\, \mathrm{d}\theta\, \mathrm{d}\phi$ is the volume element. We then obtain

$$|j, m, p, \lambda\rangle = \left(\frac{(2j+1)}{4\pi}\right)^{\frac{1}{2}} \int \mathrm{d}\omega\, D^{j^*}{}_{m\lambda}(\phi, \theta, 0)|\phi, \theta, p, \lambda\rangle \tag{28}$$

An arbitrary single-particle state vector $|\psi\rangle$ can now always be expressed as an expansion in whichever base is more convenient. For example, we may express it as a plane wave expansion

$$|\psi\rangle = \sum_{\lambda} \int \frac{\mathrm{d}^3 p}{E}\, \psi(\mathbf{p}, \lambda)|\mathbf{p}, \lambda\rangle \tag{29}$$

where the weight function ψ must be square integrable in order that the norm of $|\psi\rangle$ be finite (cf. equation 12)

$$\langle\psi|\psi\rangle = \sum_{\lambda} \int \frac{\mathrm{d}^3 p}{E}\, \psi^*(\mathbf{p}, \lambda)\psi(\mathbf{p}, \lambda) < \infty \tag{30}$$

We shall not pursue the subject.

We finally note that the previous analysis is restricted to the case $m > 0$. Of practical importance is also the case $m = 0$ because there exist massless particles. For $m = 0$, (8) does not make sense since we cannot define a rest system in this case. However, we may start from some other standard state to generate the whole space. As a reference state one usually chooses one in which one has $\tilde{p}^{\mu} = (p, 0, 0, p)$, that is, the particle travels along the positive z-axis. An irreducible representation is characterized by $m = 0$ but not any more by the spin invariant $W^\mu W_\mu$ for the physically important case where $W^\mu W_\mu = 0$. Instead then one finds $W^\mu = \lambda p^\mu$, where λ is the helicity eigenvalue. The helicity is there the second label which characterizes the UIR's, and we write the reference state vector $|(0, \lambda); 0, 0, p\rangle$. An arbitrary state is then defined by the relation

$$|(0, \lambda); \phi, \theta, p\rangle = R(\phi, \theta, 0)L_z(v)|(0, \lambda); 0, 0, \tilde{p}'\rangle$$

with $p^\mu \equiv [R(\phi, \theta, 0)L_z(v)]_\nu{}^\mu \tilde{p}'^\nu$. Since we shall not make explicit use of massless state vectors we shall not pursue the subject any further.

2. *2-particle states*

In most experiments involving 2-particle or many-particle states the experimental arrangements are designed to permit a rather accurate determination of the individual momenta. Therefore, it is from this point of view natural to work with states represented by state vectors of the following type

$$|\mathbf{p}_1, \lambda_1; \mathbf{p}_2, \lambda_2\rangle \equiv |\mathbf{p}_1, \lambda_1\rangle|\mathbf{p}_2, \lambda_2\rangle \tag{31}$$

These state vectors span a space for which the value of $P^\mu P_\mu = (P_1{}^\mu + P_2{}^\mu)(P_{1\mu} + P_{2\mu})$ is no longer constant, hence we are no longer dealing with a UIR of the Poincaré group. If we denote the single-particle Hilbert space $H(m, s)$, then we have

$$H(m_1, s_1) \otimes H(m_2, s_2) = \int_{m_1+m_2}^{\infty} \mathrm{d}M \sum_S \sum_a H(M, S, a) \tag{32}$$

where a is a degeneration index accounting for the possibility that the same irreducible space $H(M, S)$ appears more than once in the decomposition.

For a theoretical analysis it is often advantageous to work with basis vectors different from the ones given by (31), and we shall find it useful to introduce several alternative sets, some of which in fact correspond to irreducible subspaces $H(M, S)$. We shall exclusively consider states in the centre of mass system and thus we shall introduce the total momentum $\mathbf{P} = \mathbf{P}_1 + \mathbf{P}_2$ and the relative momentum $\mathbf{p}$. The restriction to the centre of mass system then corresponds to choosing $\mathbf{P} = 0$. We now define the state $|\mathbf{P} = 0; \phi, \theta, p; \lambda_1, \lambda_2\rangle$ in the following way

$$|\mathbf{P} = 0; 0, 0, p; \lambda_1, \lambda_2\rangle = \sqrt{(p/E)}|0, 0, p, \lambda\rangle|0, 0, -p, \lambda_2\rangle \tag{33}$$

$$|\mathbf{P} = 0; \phi, \theta, p; \lambda_1, \lambda_2\rangle = R(\phi, \theta, 0)|\mathbf{P} = 0; 0, 0, p; \lambda_1, \lambda_2\rangle$$

From the normalization of the single-particle state vectors involved in the definition it follows that

$$\langle \mathbf{P} = 0; \phi', \theta', p'; \lambda_1', \lambda_2' | \mathbf{P} = 0; \phi, \theta, p; \lambda_1, \lambda_2 > \\ = \delta^4(P_\mu' - P_\mu)\ \theta^2(\omega' - \omega)\ \delta_{\lambda_1'\lambda_1}\ \delta_{\lambda_2'\lambda_2} \quad (34)$$

where $\omega = (\theta, \phi)$ is the solid angle so that

$$\int d^2\omega \delta^2(\omega - \omega') = 1; \quad d^2\omega = d(\cos\theta) d\phi$$

if integrated over the whole solid angle.

By the definition (33) we have established the phase relations between single-particle and 2-particle state vectors, and we may now proceed from there and on to define new and suitable sets of 2-particle state vectors. We shall leave out the detailed computations and just quote the definitions essentially.

In some cases we shall use angular momentum vectors of the following type:

$$|\mathbf{P} = 0; J, M, p; \lambda_1, \lambda_2\rangle \equiv \\ \equiv \frac{1}{2\pi}\left(\frac{2J+1}{4\pi}\right)^{\frac{1}{2}} \int dR D^{J^*}{}_{M\lambda}(R) R |\mathbf{P} = 0; 0, 0, p; \lambda_1, \lambda_2\rangle \quad (35)$$

where $\lambda = \lambda_1 - \lambda_2$, R stands for an arbitrary rotation and dR is the corresponding volume element in the parameter space. For example, if R is defined by the Euler angles, then $R = R(\alpha, \beta, \gamma)$ and $dR = d\alpha\, d(\cos\beta)\, d\gamma$. It is easily shown that the vectors $|\mathbf{P} = 0; J, M, p; \lambda_1, \lambda_2\rangle$ for fixed J span a UIR of the rotation group SO(3) corresponding to the angular momentum J. Integrating (35) over γ one finds

$$|\mathbf{P} = 0; J, M, p; \lambda_1, \lambda_2\rangle = \left(\frac{2J+1}{4\pi}\right)^{\frac{1}{2}} \int d^2\omega D^{J^*}{}_{M\lambda}(\phi, \theta, 0) \times \\ \times |\mathbf{P} = 0; \phi, \theta, p; \lambda_1, \lambda_2\rangle \quad (36)$$

If we make use of (27) we may invert (36) to obtain the following *partial wave expansion:*

$$|\mathbf{P} = 0; \phi, \theta, p; \lambda_1, \lambda_2\rangle = \\ = \sum_J \sum_{M=-J}^{+J} \left(\frac{2J+1}{4\pi}\right)^{\frac{1}{2}} D^J{}_{m\lambda}(\phi, \theta, 0) |\mathbf{P} = 0; J, M, p; \lambda_1, \lambda_2 > \quad (37)$$

and as before $\lambda = \lambda_1 - \lambda_2$.

For discussions of selection rules such as space parity it is often convenient to use a somewhat different angular momentum basis in which one first couples the two individual spins $\mathbf{s}_1$ and $\mathbf{s}_2$ to a resultant spin $\boldsymbol{\sigma}$. Then $\boldsymbol{\sigma}$ is coupled to the relative orbital angular momentum $\mathbf{l}$ so that $\mathbf{J} = \mathbf{l} + \boldsymbol{\sigma}$ in the centre of mass. With this in mind we define the vectors $|\mathbf{P} = 0; J, M, p; l, \sigma\rangle$ by the following relation:

$$|\mathbf{P} = 0; J, M, p; l, \sigma\rangle =$$

$$= \sum_{\lambda_1, \lambda_2} \left(\frac{2l+1}{2J+1}\right)^{\frac{1}{2}} \langle l\sigma 0\lambda | J\lambda\rangle \langle s_1 s_2, \lambda_1, -\lambda_2 | \sigma, \lambda\rangle \times$$

$$\times \, |\mathbf{P} = 0; J, M, p; \lambda_1, \lambda_2\rangle \tag{38}$$

where $\langle j_1, j_2, m_1, m_2 | j, m\rangle$ are Clebsch-Gordan coefficients with phase conventions as given for example by Werle[1].

3. 3- *and many-particle states*

Just as in the case of 2-particle states we may form multi-particle states as outer products of single-particle states, that is

$$|\mathbf{p}_1, \lambda_1; \mathbf{p}_2, \lambda_2; \ldots; \mathbf{p}_n, \lambda_n\rangle \equiv |\mathbf{p}_1, \lambda_1\rangle |\mathbf{p}_2, \lambda_2\rangle \ldots |\mathbf{p}_n, \lambda_n\rangle \tag{39}$$

and an arbitrary state $|\psi_m\rangle$ can then always be written as

$$|\psi_n\rangle = \sum_{\lambda_1, \ldots \lambda_n} \int \frac{d^3 p_1}{E_1} \cdots \frac{d^3 p_n}{E_n} \psi(\mathbf{p}_1, \lambda_1; \ldots; \mathbf{p}_n, \lambda_n)$$

$$|\mathbf{p}_1, \lambda_1; \ldots; \mathbf{p}_n, \lambda_n\rangle \tag{40}$$

where $\psi(\mathbf{p}_1, \lambda_1; \ldots; \mathbf{p}_n, \lambda_n)$ must be square integrable in order that $|\psi_n\rangle$ be normalizable.

For theoretical discussions, the states (39) are often complicated since they do not correspond directly to UIR for the Poincaré group. Of course we have, just as in the case of 2-particle states, the following reduction of the Hilbert space with regard to the Poincaré group:

$$H(m_1, s_1) \otimes H(m_2, s_2) \otimes \ldots \otimes H(m_n, s_n) = \int dM \sum_S \sum_a H(M, S; a) \tag{41}$$

For an n-particle state there are in general $(4n - 6)$ degeneracy labels a, some of which range over a continuous spectrum so that the summation is replaced by an appropriate integration. The $(4n - 6)$ additional quantum numbers which are necessary to remove the degeneracy can of course be chosen in many different ways. We shall not discuss this rather complicated problem which is treated by several authors[1]. We just note that the technique used for the 2-particle states can be extended in the sense that we first consider, for example, the particles 1 and 2 and reduce the corresponding product space. We then proceed to consider each of the components in this reduction coupled to particle 3 and reduce these products and so on. Physically we may view this as a successive coupling of n particles into states of definite relative angular momenta and invariant masses. Thus, first we consider the particles 1 and 2 in their centre of mass and couple them to states of definite relative angular momentum $j_{1,2}$ and invariant mass $m_{1,2}$. Then we perform a suitable Lorentz transformation to the centre of mass of the systems (1, 2) and 3 and couple these subsystems to a state of definite relative angular momentum $j_{1,2,3}$ and invariant mass $m_{1,2,3}$ and so on. Of course, this prescription will quickly become very complicated, but in most cases of practical importance $n = 3$ and it is not too elaborate a task to carry out the analysis. We shall not do it but refer the interested reader to the standard literature, for example the excellent book by Werle[1].

References

1. Werle, J., *Relativistic Theory of Reactions*, North-Holland Publishing Company, Amsterdam, 1966; Gasiorowicz, S., *Elementary Particle Physics*, J. Wiley and Sons, Inc., New York, 1966, Chap. 4.
2. For a derivation see for example: Källén, G., *Elementary Particle Physics*, Addison-Wesley Publishing Company, 1964.
3. Yang, C. N., and Tiomno, J., *Phys. Rev.*, **79**, 498 (1950).
4. Dalitz, R. H., *Phil. Mag.*, **44**, 1068 (1953); Dalitz, R. H., *Phil. Mag.*, **94**, 1046 (1954).
5. Wu, C. S., and others, *Phys. Rev.*, **105**, 1413 (1957).
6. We use the same metric and the same set of γ-matrices as Schweber, S., *Relativistic Quantum Field Theory*, Harper and Row, New York, 1962.
7. Wilkinson, D. H., *Phys. Rev.*, **109**, 1603 (1958).
8. Boehm, F., and Kankeleit, E., *Phys. Rev. Letters*, **14**, 312 (1965).
9. Bock, P., and Schopper, H., *Phys. Rev. Letters*, **16**, 284 (1965); Lobashor and others, *JETP*, **3**, 76, 268 (1966); Paul, H., and others, *Report to the Berkeley Conf.*, 1966.
10. Haas, R., and others, *Phys. Rev.*, **116**, 1221 (1959).
11. Bernstein, J., Feinberg, G., and Lee, T. D., *Phys. Rev.*, **139**, B1650 (1965).
12. See for example Cabibbo, N., and Veltman, M., 'Weak Interactions,' *CERN* 65–30 (Yellow report).
13. Day, T. B., and others, *Phys. Rev. Letters*, **3**, 61 (1959); Desay, B., *Phys. Rev.*, **119**, 1385 (1960).
14. Baltay, C., and others, *Phys. Rev. Letters*, **15**, 591 (1965).
15. Duclos, J., and others, *Phys. Letters*, **19**, 253 (1965).
16. Baltay, C., and others, *Phys. Rev. Letters*, **16**, 1224 (1966); *CERN, Report to the Berkeley Conf.*, 1966.
17. Cnops, A. M., and others, *Phys. Rev. Letters*, **22**, 546 (1966); Bowen, R. A., and others, *Phys. Letters*, **24B**, 207 (1967); Litchfield, P. J., and others *Phys. Letters*, **24B**, 486 (1967).

18. Baglin, C., and others, *Phys. Letters*, **22,** 219 (1966).
19. Landau, L. D., *Nucl. Phys.*, **3,** 127 (1957).
20. Christenson, J. H., Cronin, J. W., Fitch, V. L., and Turlay, R., *Phys. Rev. Letters*, **13,** 138 (1964).
21. Cronin, J. W., and others, *Phys. Rev. Letters*, **18,** 25 (1967); Bott-Bodenhausen, M., and others, *Phys. Letters*, **24B,** 438 (1967); Rubbia, C., and Steinberger, J., *Phys. Letters*, **23,** 167 (1966), and **24B,** 531 (1967).
22. Lee, T. D., and Wu, C. S., *Ann. Rev. Nucl. Sci.*, **16** (1966).
23. Wick, G. C., Wightman, A. S., and Wigner, E. P., *Phys. Rev.*, **88,** 101 (1952).
24. Marshak, R. E., *Phys. Rev.*, **82,** 313 (1951); Cheston, W. B., *Phys. Rev.*, **83,** 1118 (1951).
25. Fitch, V. L., *Proc. XIIIth Int. Conf. on High Energy Phys.*, University of California Press, Berkeley, 1967.
26. See the discussion after the talk by Fitch, V. L., (reference 25).
27. Abashian, A., and Hafner, E. M., *Phys. Rev. Letters*, **1,** 255 (1958); Rosen, L. and Brolley, J. E., *Phys. Rev. Letters*, **2,** 98 (1959).
28. Glasser, R. G., and others, *Report to the Berkeley Conf.*, 1966; see also Glasser, R. G., and others, *Proc. Int. Conf. on Weak Interactions*, Argonne National Laboratory report ANL-7130, 1965.
29. Burgy, M. T., and others, *Phys. Rev. Letters*, **1,** 324 (1958).
30. Longo, M. J., and others, see Fitch, V. L., *Rapporteur's talk at the Berkeley Conference* (reference 25).
31. See for example Källén, G., *Elementary Particle Physics*, Addison-Wesley Publishing Company, Evanston, 1964, p. 478.
32. Overseth, O., and Roth, R., see Fitch, V. L., *Rapporteur's talk at the Berkeley Conference* (reference 25).
33. Lüders, G., *Kongl. Dansk Medd. Fys.*, **28,** No. 5 (1954) and *Ann. Phys.* (N.Y.) **2,** 1 (1957); Pauli, W., *Niels Bohr and the Development of Physics*, McGraw-Hill Book Co., New York, 1955; Schwinger, J., *Phys. Rev.*, **82,** 914 (1951); Schwinger, J., *Phys. Rev.*, **91,** 714 (1953); Schwinger, J., *Proc. Nat. Acad. Sci. U.S.*, **44,** 223 (1958).
34. Farley, M., and others, *Report to the Berkeley Conf.*, 1966.
35. Lobkowicz, F., and others, *Phys. Rev. Letters*, **17,** 548 (1966); Lobkowicz, F., and others, *Report to the Berkeley Conf.*, 1966.
36. Lee, T. D., and others, *Phys. Rev.*, **106,** 1367 (1957).
37. Okubo, S., *Phys. Rev.*, **109,** 984 (1958).
38. Nilsson, J., and Beskow, A., *Arkiv Fysik*, **34,** 307 (1967); Beskow, A., and Nilsson, J., *Arkiv Fysik*.
39. Lee, T. D., and Wick, G. C., *Phys. Rev.*, **148,** 1385 (1966).

Ten Years of the Universal *V–A* Weak Interaction Theory and Some Remarks on a Universal Theory of Primary Interactions*

E. C. G. SUDARSHAN

1 Beta Decay Prior to 1957

Beta radioactivity has been known ever since the turn of the century but it was only about thirty years ago that some quantitative understanding of the phenomenon began to emerge. In 1932–33, B. W. Sargent[1] gave a compilation of the electron energy distribution for various beta emitters. He classified beta emitters by observing groupings of radioactive nuclides in a plot of the partial decay constants versus the upper energy limits of the beta particles for various components of complex beta spectra. In a log-log plot he identified several groups of nuclei clustering around what came to be called the Sargent curves, I, II, III, etc. We may also recall that W. Pauli had already proposed, in 1930, the hypothesis of the neutrino to account for the continuous energy distribution of electron energies from a quantum nucleus.

These clues were sufficient for E. Fermi[2] to propose a remarkably successful theory of beta decay. He assumed, by analogy with electromagnetism, that the beta interaction was a new interaction-by-contact and that its law could be expressed by considering an interaction energy density proportional to the product of the neutron, proton, electron and neutrino wave functions. Fermi chose the interaction density to be as similar as possible to electric interaction, and in the approximation of treating the nuclear particles to be at rest only the analogue of the electrostatic interaction survives. In view of the fact that the de Broglie wavelength of the

* Supported in part by the U.S. Atomic Energy Commission.

electron and neutrino were large compared with nuclear dimensions, only the overall effect of the nucleus was felt: in other words only the analogue of the total electric charge was effective in causing the radioactive transmutation in this approximation. We now call this the 'allowed Fermi' transition. If we neglect the corrections due to the motion of the nuclear particles (velocity effects) and the variation of the electron and neutrino wave-functions over the nucleus (retardation effects) but include the Coulomb distortion (enhancement) of the electron wave-function, we can calculate the expected electron energy distribution. This (Coulomb-corrected) 'allowed shape' was the first quantitative verification of Fermi's theory. The proper measurement of the low energy end of the spectrum did not come about for another decade or so but it did eventually come. Fermi had attempted a calculation of the expected transition rate taking the nuclear matrix elements to be of the order of unity. The transition rate is then, for 'allowed' decays, the square of the postulated Fermi coupling constant G multiplied by a known function of the upper limit of the beta spectrum. Hence the product of this function and the lifetime must be a constant for allowed decays; any variation among the allowed decays must be attributed to variation of the nuclear matrix elements (the beta decay 'transition charges'). Fermi was thus able, on the basis of this ft-value classification to give a correspondence with Sargent's curves.*

Two important supplementary developments came about in the two succeeding years. H. Yukawa proposed his fundamental theory of meson fields[3], in which he wished to relate the meson field with the source of weak interactions. At that time Yukawa thought in terms of scalar meson fields but later on the Fermi interaction turned out to be vector and axial vector so that the detailed implementation of his ideas required modern developments. But his was the first idea of relating strongly interacting meson fields and weak interactions. The second development was the proposal of G. Gamow and E. Teller[4] that there are beta decay interactions which involved the spin of the nuclear particles. They were motivated by the radioactive thorium series for which one could not consistently assign spin values and understand their Sargent classification in terms of the degree of forbiddenness with Fermi selection rules. These new Gamow-Teller interactions were thus the analogues of the magnetic part of the electromagnetic interactions in their dependence on the spin of the nuclear particles (but were non-vanishing even for zero momentum transfer). It is now well known that pure Gamow-Teller transitions like

* Fermi's paper and many of the more important subsequent papers are reprinted in *The Development of Weak Interaction Theory*, edited by P. K. Kabir and published by Gordon and Breach, New York, 1963.

$Co^{60}(5^+ \to 4^+)$ and $He^6(0^+ \to 1^+)$ as well as pure Fermi transitions like $O^{14}(0^+ \to 0^+)$ and $C^{14}(0^+ \to 0^+)$ exist.

Using both the Fermi type and the Gamow-Teller type interactions one can calculate the electron energy distribution. The 'allowed shape' is equally valid for the Gamow-Teller allowed interactions. For forbidden interaction the shapes are more complicated, but for low energy release beta transitions in heavy nuclei the effect of relativity and Coulomb attraction is to restore the allowed shape to first forbidden (retarded) decays (like Ce^{141}, Au^{198}); for a class of unique Gamow-Teller forbidden transitions, again one can work out the unique forbidden shapes (as examples we have Be^{10}, Y^{91} unique first forbidden, Na^{22} unique second forbidden, K^{40} unique third forbidden). These shape predictions have been carefully verified in the succeeding years, particularly in the late forties and early fifties.

When relativity is taken into account the local non-derivative couplings in beta decay increase to five, called respectively scalar, vector, tensor, axial vector and pseudoscalar, depending upon the transformation property of the combination of the nuclear wave-functions. By studying the non-relativistic limit one classifies the scalar (S) and vector (V) terms as Fermi couplings, the tensor (T) and axial vector (A) as Gamow-Teller couplings and the pseudoscalar may be ignored in the non-relativistic limit. The question therefore arose as to which of these interactions did really enter nuclear beta decay. We already know that both Fermi and Gamow-Teller interactions were about equally important by a study of ft-values for various allowed transitions. The common coupling could be assigned a coupling constant of about $1{\cdot}5 \times 10^{-49}$ erg cm^3, which could equally well be given as 10^{-23} if the electron mass is chosen as the inverse unit of length (or 10^{-13} with the pion Compton wavelength as the unit of length).

The allowed spectrum shapes gave little information on the type of the interaction; at one time it was thought that if both S and V were present we would get a correction factor of the form $[1 + b(m_e/E_e)]$; and similarly for T and A. We know *now* that these so-called Fierz interference terms would vanish with the special kind of maximal parity violation. But before this was known people assumed that the absence of the Fierz terms was evidence of the absence of both S and V or both T and A. It turns out that the conclusion was in fact true though the argument supporting it was false!

More definitive evidence comes from the angular correlation of the electron and neutrino. In the decay of a particle (parent nucleus) into three particles (daughter nucleus, electron, neutrino) the non-trivial part of the decay momentum configuration is specified if we give either the energies of two particles or energy of the one particle and the angle made by another particle with this direction. It turns out that his correlation has a particularly simple form $\left(1 + a\frac{v_e}{c}\cos\theta_{ev}\right)$ where the correlation coefficient $a = -1, +1, +\frac{1}{3}, -\frac{1}{3}$ for *S*, *V*, *T*, *A* interactions respectively. The measurement of the mixed transitions of Ne^{19} and neutron gave a value near zero which suggested a mixture of *V-A* or *S-T*. The first measurement of the pure Gamow-Teller beta emitter He^6 gave a positive value favouring *T* and hence the *S*, *T* combination. But the nearly pure Fermi beta transition in A^{35} gave also a large positive value suggesting *V* and hence the *V-A* combination. Clearly there was some disagreement in the measurement of angular correlations; and *until this was resolved* we could not determine the relevant beta interactions.

2 The Discovery of the *V-A* Interaction

This brings us to the exciting year 1957. It was a fateful year for beta decay and for nuclear physics in general. The possibility of parity violation that was pointed out by T. D. Lee and C. N. Yang[5] was brilliantly confirmed by C. S. Wu and others[6] in their experiments on the correlation of electron emission direction with the spin direction of decaying Co^{60}. A new era was initiated in that many new quantities could be measured in nuclear beta decay, but along with this the possible number of beta decay interactions also increased since there could be parity-conserving or parity-violating interactions. The possibility was also raised that some or all of the various coupling constants could be complex signalling violating of time reversal invariance. A period of intense activity resulted in a number of significant indications on the nature of the beta interactions. Starting in the summer of 1956 I had followed these developments with great interest at the suggestion of my professor, R. E. Marshak. By the end of spring 1957, I was convinced that the available experimental information crystallized into the following items:

(1) The electron-neutrino angular correlation data favoured *V-A* if the A^{35} was correctly and He^6 was wrongly measured and *S-T* if the He^6 was rightly and A^{35} wrongly measured.

(2) The measurement of electron asymmetry for polarized nuclei or the longitudinal polarization of beta electrons from the Gamow-Teller decay of Co^{60} gave a number negative in sign and very near the maximum absolute value. This was consistent with *A* or *T* according as the antineutrino was right or left polarized.

(3) The corresponding measurement for the Fermi decay of Ga^{66} was

consistent with *V* or *S* according as the antineutrino was right or left polarized.

(4) The mixed transitions Sc^{46} and Au^{198} gave results consistent with both the above determinations. If the antineutrino was the same in both Fermi and Gamow-Teller transitions, we would get *V-A* or *S-T* according as the antineutrino was right or left polarized.

(5) By measuring the beta-polarized gamma correlations in the mixed transition Sc^{46} we get a nearly maximum effect which is consistent with approximately equal *V* and *A* or equal *S* and *T* (with approximately real phase factor) and the antineutrinos in both Fermi and Gamow-Teller transitions had to have the same handedness.

(6) If we thought of pion decay as the decay of a nucleus with atomic weight zero this would give additional information about nuclear beta decay. This could take place only via *A* or *P* interaction and any evidence for it would imply *A* or *P* should be present. The indications were that the electron mode was absent.

On the basis of these data, one had to choose either the *V-A* or the *S-T* combination. By appealing to the principle of universal Fermi interaction I was led to the choice of *V-A* and thus rejected the He^6 experiment and the experiment on pion decay into electron and neutrino. This result, that is the postulation of a maximal parity-violating *V-A* beta interaction, was first presented in a paper by E. C. G. Sudarshan and R. E. Marshak to the Padua-Venice Conference in early September of 1957. It is important to stress that *this was the deduction of a universal law from analysis of empirical evidence*; the use of symmetry principles came as an afterthought! Since that time increasing evidence has accumulated, adding remarkable *confirmation* to the conclusion that was made in this paper on the basis of my analysis of the experimental data. In particular, Goldhaber, Grodzins and Sunyar[7] made a direct determination of the neutrino helicity in Gamow-Teller transitions by measuring the helicity of gamma rays emerging in a direction opposite to that of the neutrino in the proceeding beta transition. This gave the helicity of the neutrino in the Gamow-Teller electron capture process in Eu^{152*} to be negative. This, together with the previous information that beta electrons from Gamow-Teller transitions are left-handed, tells us that the Gamow-Teller interaction must be *A* and hence beta decay must be via the *V-A* combination. The relative sign was determined to be negative from Sc^{46} and neutron decays. Since that time both the He^6 experiment and the electron-neutrino mode of the pion have been redone with new results which confirm the *V-A* prediction.

The discovery of the *V-A* interaction for nuclear beta decay was very satisfying from another point. As early as 1948, O. Klein[8] and G. Puppi[9] had noticed the marked similarity between the beta decay, muon decay and muon capture processes and proposed a universal Fermi interaction based on the approximately equal strength for these couplings. The discovery of parity violation gave new direction to this work and one could conclude, from the negative correlation between the direction of emission of the electrons from stopped muons of pion decay with the original muon momentum direction, that the muon decay coupling should be of the *V-A* type with maximal parity violation. The charged pion decays almost all the time into a two-body state involving the muon but an electron-neutrino mode with a branching ratio $1{\cdot}2 \times 10^{-4}$ was subsequently found experimentally. An *A* interaction with equal coupling of muon-neutrino and electron-neutrino would give precisely this ratio if the weak interaction involving muons or electrons was universal. These facts, together with other fragmentary data, again suggest *V-A* as a universal interaction. Sudarshan and Marshak therefore deduced that if the weak interactions were universal they had to be *V-A*; this proposal was also contained in their original paper.

Once the coupling was determined to be V-A many new elegant symmetry features could be discerned for this form of coupling. The best determination showed that the *A* interaction was 1·2 times the *V* interaction in nuclear beta decay but they were equally strong in muon decay. Assuming this feature was a renormalization effect due to strong interactions, we may think of vector and axial vector couplings of equal strength in their primary form. Such an interaction may be written in the form

$$g\bar{A}\gamma_\lambda(1 + \gamma_5)B, \quad \bar{C}\gamma^\lambda(1 + \gamma_5)D$$

where *A*, *B*, *C*, *D* are the four spinor fields which are coupled together. It then appears that only the positive 'chiral' projections

$$A' = \tfrac{1}{2}(1 + \gamma_5)A, \quad B' = \tfrac{1}{2}(1 + \gamma_5)B$$

etc., appear in the coupling; moreover the coupling above, which may be rewritten in the form

$$4g\bar{A}'\gamma_\lambda B', \quad \bar{C}'\gamma^\lambda D'$$

is the only invariant coupling. We therefore referred to this property of the coupling as 'chirality invariance'.

Previous attempts at using general invariance properties to deduce the general form for the beta interaction have had little success. The difference here is that here we 'admire' the structure of the interaction *after* it has been deduced from an analysis of the experimental data.

Other people found different ways of displaying the remarkable symmetry properties of the universal chiral coupling. R. P. Feynman and M. Gell-Mann, in a paper which followed almost immediately[10], showed that by a suitable form of restriction to non-derivative coupling of the chiral *V-A* interaction could be deduced. J. J. Sakurai[11] showed how chirality invariance could be related to the mass reversal transformation. Several people discovered that the *V-A* form was invariant under recoupling of the spinor fields. Altogether the *V-A* chiral coupling is a most remarkable interaction. But these symmetry properties were incidental. As was stated in the original paper of Sudarshan and Marshak, 'The analysis of the decay of pions and kaons thus seems to point unequivocally to a dominant *A* interaction among those operative in the decay process. The muon decay data is consistent with *V-T*, *A-T* or *A-V*. Among these three possible assignments the only one involving the *A* interaction is *A-V*. Since this is the only assignment consistent with beta decay data . . . the only possibility for a universal Fermi interaction is to choose a vector and axial vector coupling . . .'

3 The Present Status of the *V-A* Interaction

In the ten years since the original *V-A* interaction was proposed, considerable refinement of the experimental results has taken place, and it is of interest to examine the experimental confirmation of our four-fermion interaction structure. In the case of the purely leptonic process of muon decay, no strong interaction effects enter the discussion and we could hope to make a direct comparison of the experimental results with the theoretical prediction. The general four-fermion interaction (which is invariant under *CP*) yields the following differential spectrum:

$$P(\varepsilon, \cos\theta)\, d\varepsilon\, d\Omega = \frac{m_\mu{}^5}{3\pi^4 2^9}\, \varepsilon^2 \left\{ \left[1 + 4\left(\frac{m_e}{m_\mu}\right)\eta \right]^{-1} \left[3(1-\varepsilon) + 2\rho\left(\tfrac{4}{3}\varepsilon - 1\right) \right.\right.$$
$$\left.\left. + 6\left(\frac{m_e}{m_\mu}\right)\frac{1-\varepsilon}{\varepsilon}\eta \right] - \xi\cos\theta[1 - \varepsilon + 2\delta(\tfrac{4}{3}\varepsilon - 1)] \right\} d\varepsilon\, d\Omega$$

for the *V-A* interaction, where ε is the ratio of the electron energy to its maximum value $\frac{1}{2}m_\mu$, $\cos\theta$ the cosine of the angle between the muon spin and electron momentum, ξ the asymmetry parameter, δ the energy dependence of the asymmetry parameter, ρ the spectrum shape (Michel) parameter, and η a low energy correction to the spectrum shape. For the *V-A* interaction the predicted values of ρ and δ are both 0·750 while the experimental values are 0·747 ± 0·005 and 0·78 ± 0·05 respectively, and

ξ should be 1·00 while experiment yields 0·978 $\pm$ 0·030. These quantities are in principle to be corrected for electromagnetic effects. The latter turn out to be finite in lowest order. We could compute the muon transition rate in the form:

$$\tau_\mu^{-1} = \frac{G^2 m_\mu^5}{192\pi^3}\left[1 - \frac{\alpha}{\pi}\left(\pi^2 - \frac{25}{4}\right)\right]$$

where α is the fine structure constant. For the lifetime the radiative correction is less than a percent. From the observed muon lifetime of $(2{\cdot}198 \pm {\cdot}001) \times 10^{-6}$s we can deduce the numerical value for the weak coupling constant

$$G = (1{\cdot}435 \pm 0{\cdot}001) \times 10^{-49} \text{ erg cm}^3$$

Most of the weak processes involve strongly interacting particles; hence the comparison of the experimental results with the predictions of the *V-A* theory becomes entangled with the difficulties of making reliable strong interaction calculations. The one exception is the comparison of processes involving muon and its neutrino and the corresponding processes involving electron and its neutrino. We have already remarked about the significance of the $(\pi \to e\nu)/(\pi \to \mu\nu)$ branching ratio for the *V-A* theory. For the strange particle decays we have to make sure that the comparison be made at the same total lepton momentum. Such a comparison is given in the accompanying table. We can see that the electron and muon are coupled in the same manner in all weak interaction processes.

TABLE 1

Test of electron-muon universality[a]

Decay Process	Theory	Experiment
$\dfrac{\Gamma(\pi^+ \to e^+ + \nu_e)}{\Gamma(\pi^+ \to \mu^+ + \nu_\mu)}$	$1{\cdot}23 \times 10^{-4}$	$(1{\cdot}24 \pm 0{\cdot}03) \times 10^{-4}$
$\dfrac{\Gamma(K^+ \to e^+ + \nu_e)}{\Gamma(K^+ \to \mu^+ + \nu_\mu)}$	$2{\cdot}47 \times 10^{-5}$	$(3 \pm 1{\cdot}89) \times 10^{-5}$
$\dfrac{\Gamma(K^+ \to \pi^0 + \mu^+ + \nu_\mu)}{\Gamma(K^+ \to \pi^0 + e^+ + \nu_e)}$	0·69	0·703 $\pm$ 0·056
$\dfrac{\Gamma(\Lambda \to p + e^- + \bar{\nu}_e)}{\Gamma(\Lambda \to p + \mu^- + \bar{\nu}_\mu)}$	5·88	5·87 $\pm$ 0·75
$\dfrac{\Gamma(\Sigma^- \to n + \mu^- + \bar{\nu}_\mu)}{\Gamma(\Sigma^- \to n + e^- + \nu_e)}$	0·45	0·496 $\pm$ 0·26

[a] R. E. Marshak: University of Rochester Report UR–875–187

The general order of magnitude of the coupling strength, the violation of parity and charge conjugation in these processes and the electron-muon universality demonstrated above all point to the leptonic weak interactions of strange particles to be due to the same interaction. We could then ask whether the baryonic currents which are responsible for strangeness conserving and strangeness violating decays are the 'same'. To give a sense to this question we must have a method of considering an object which has both strangeness conserving and strangeness violating components. We shall see that such a scheme for assignment of the low lying baryons and mesons exist and we shall discuss this question within this context. Before doing this, however, let us return to the nuclear beta decay phenomenon and review it from the point of view of strong interactions.

The *V-A* structure of the nuclear beta decay interaction, if it is to be understood in terms of the primary neutron and proton fields, does not automatically predict the same $(G/\sqrt{2})\gamma_\lambda(1 + \gamma_5)$ structure for the effective nuclear beta decay matrix element. This is just as well, since it is more properly of the form $(G/\sqrt{2})\gamma_\lambda(1 + g_A\gamma_5)$, with $g_A \simeq 1{\cdot}2$. We have already remarked how we could understand the numerical equality of the vector coupling coefficients of the neutron beta decay and the muon beta decay in terms of the conservation of the vector current of nucleons. But there is still the embarrassment of the axial vector coupling constant being different by about 20%. The problem here is not so much to understand why there is a change of the effective coupling constant but to be able to compute this correction in a quantitative manner. Such a calculation has been made only very recently and the success of this calculation *removes the last significant obstacle to the form of the four-fermion interaction that we had proposed for nuclear beta decay*. The principle of this calculation is related to Yukawa's suggestion that the *weak and strong* interactions are related. Quantitatively we may take Yukawa's hypothesis as stating that the meson field is proportional to the nuclear source of beta decay. At the time of Yukawa's hypothesis of the meson field and the connection between the strong and weak interactions, neither vector mesons nor pseudoscalar mesons were known and no precise analytic formulation of the hypothesis could be made. But at the present time we may take the axial vector current A_λ and the pseudoscalar field ϕ to be proportional. We may implement this either by taking A_λ to be proportional to $\partial_\lambda\phi$ or by taking $\partial^\lambda A_\lambda$ to be proportional to ϕ. The first alternative is incompatible with the existence of Gamow-Teller interactions and we may therefore put:

$$\partial^\lambda A_\lambda = c\phi$$

Using this partial conservation of axial vector current we could relate g_A to the ratio of the pion and muon lifetimes; this relation is in reasonable agreement with experiment.

The partial conservation law related weak axial vector coupling to pseudoscalar pion coupling. To use this to deduce the axial vector renormalization constant g_A we have to make use of a non-linear relation involving the axial vector current A_λ. If the axial vector currents have the form

$$A_\lambda{}^\alpha(x) = \bar{\psi}(x)\gamma_\lambda\gamma_5\tau^\alpha\psi(x)$$

where $\psi(x)$ is the spinor-isospinor nucleon field which satisfies standard equal-time anticommutation relations, they would satisfy the equal-time commutation relations

$$[A_0{}^\alpha(\mathbf{x}, t), A_0{}^\beta(\mathbf{y}, t)] = 2\mathrm{i}\, \varepsilon^{\alpha\beta\gamma}\delta(\mathbf{x} - \mathbf{y})V_0{}^\gamma(\mathbf{x}, t)$$

where

$$V_\lambda{}^\gamma(x) = \bar{\psi}(x)\gamma_\lambda\tau^\gamma\psi(x)$$

is the corresponding vector density. This non-linear relation is clearly for the primary ('unrenormalized') fermion fields and can be taken to include the contribution of pions, etc. Using this equal-time commutation relation along with the partial conservation law we can show that the axial vector renormalization constant g_A satisfies the sum rule

$$g_A{}^{-2} = 1 - \frac{c^2}{2\pi m_\pi{}^4}\int \frac{\mathrm{d}\nu}{\nu}\,[\sigma^+(\nu) - \sigma^-(\nu)]$$

where $\sigma^\pm(\nu)$ is the scattering cross-section for pions on nucleons (extrapolated to pions with zero momenta). This was derived by W. I. Weisberger[12] and S. L. Adler[13]. The technique of derivation is to compute the scattering amplitude for zero 4-momentum pions on nucleons which could be expressed as the contribution from the equal-time commutator plus the contribution from the double divergence of the axial vector-nucleon scattering amplitude, and then expressing this zero energy scattering amplitude as an integral over the nucleon-pion scattering over all energies. The numerical evaluation making a careful estimate of the extrapolation of the physical amplitudes gives a remarkably close value for the quantity g_A. The general theory of such relations have been developed by K. Raman and E. C. G. Sudarshan[14]. Y. Tomozawa[15] and several others showed that instead of this we could equally well relate the axial vector coupling constant in terms of the S-wave pion-nucleon scattering length which again gives a remarkably good prediction for g_A.

We have thus ample reason to justify our hypothesis that the beta decay interaction is chiral *V-A* and has the same strength as in muon decay and muon capture.

If the basic beta decay interaction is the four-fermion coupling in terms of the primary nucleon fields, the vector and axial vector currents of beta decay are both the charge changing components of isotopic vectors. We have

$$\frac{G}{\sqrt{2}} j_\lambda{}^+(x) = \frac{G}{\sqrt{2}} \bar{\psi}\gamma_\lambda\tau^+\psi = \frac{G}{\sqrt{2}} \bar{P}\gamma_\lambda N$$

$$\frac{G}{\sqrt{2}} j_\lambda{}^{5+}(x) = \frac{G}{\sqrt{2}} \bar{\psi}\gamma_\lambda\gamma_5\tau^+\psi = \frac{G}{\sqrt{2}} \bar{P}\gamma_\lambda\gamma_5 N$$

On the other hand the corresponding electric current is a pure vector of the form

$$\frac{e}{2}[j_\lambda{}^0(x) + j_\lambda{}^{00}(x)] = \frac{e}{2}(\bar{\psi}\gamma_\lambda\tau_3\psi + \bar{\psi}\gamma_\lambda\psi)$$
$$= e\bar{P}\gamma_\lambda P$$

Consequently, apart from a scale factor, the isotopic vector part of the electric current and the vector part of the beta decay current are components of the same isotopic vector operator. Hence the matrix elements of both these currents must have the same momentum dependence (apart from purely electromagnetic corrections). In particular, there should be a term in vector beta decay matrix element that is the analogue of the anomalous magnetic moment. What is more, we can compute this weak magnetism term in terms of the difference of the proton and neutron anomalous magnetic moments. The effective Fermi matrix element can be written

$$\frac{G_v}{\sqrt{2}} \bar{u}_p \left[\gamma_\lambda F_1(t) + \frac{\kappa}{2m_n} \sigma_{\lambda\nu} q^\nu F_2(t) \right] u_n$$

where $F_1(t)$ and $F_2(t)$ are the isovector charge and magnetic form factors, t is the invariant momentum transfer squared and

$$\kappa = \mu_p - \mu_n$$

From experiments on the shape of the electron and position spectra of B^{12} and N^{12} this weak magnetism term has been quantitatively verified.

In summary then, the universal chiral *V-A* four-fermion interaction for non-strange particles is completely confirmed by the continuing experimental work of the last ten years. The weak magnetism and the departure of the axial vector effective coupling constant are now quantitatively understood. The chirality invariant theory that we had formulated a decade ago has been accepted as *the* theory of universal Fermi interaction.

4 Further Developments in Weak Interactions

Several new experimental developments, however, suggest that it is profitable to reexamine the fundamental hypothesis of a four-fermion interaction. Within the realm of weak interactions it has been found that the leptonic decays of kaons and hyperons are slower by a factor of about ten as compared with the predictions of the universal four-fermion interaction extended to include the strangeness violating decays of hyperons. In the decay of the neutral long lived kaon there is unmistakable evidence for a small violation of *CP*. Among strong interactions a whole collection of vector mesons have been identified and the coupling of the rho meson seems to be essential for an understanding of low energy pion-nucleon scattering and nucleon electromagnetic properties. It is very suggestive to consider that the conserved vector currents of electromagnetism, Fermi-type beta decay and strong interactions are all related. In such a theory the four-fermion coupling in beta decay is no longer primary; we are led to resurrect Yukawa's hypothesis of a beta interaction mediated by meson fields.

Before outlining such a theory let us examine the transformation property of the leptonic weak interaction current in strangeness changing decays. Since the kaon decays into leptons and the Λ hyperon decays into proton there must be an $I = 1/2$, $\Delta S = \pm 1$ component to the current with $\Delta Q = \Delta S$. It is tempting to assume that this must hold generally. This $I = 1/2$ current rule is in good agreement with experiment both for vector and for axial vector currents. But one could go beyond this and ask if the various strangeness violating transitions in hyperon decay can themselves be related. For this purpose we have to consider a higher symmetry group which contains both strange and non-strange particles. Following Sakata, the simplest such group is the SU(3) group, which has eight-dimensional representations to accommodate the eight baryons or eight pseudoscalar mesons. It then appears that the baryon currents of leptonic weak interaction can be taken to transform like the pseudoscalar mesons, namely as octets. The $\Delta Q = \Delta S$ property is then an automatic consequence. Since we could have two different ways of coupling octets to octets via an octet operator (the D and F type couplings) we must specify the D/F ratio also. The generators of SU(3) themselves transform as pure F type; and the present view is that the pseudoscalar couplings have $D/F = 3/2$. The choices of pure F type for vector and a mixture of D and F types with $D/F = 3/2$ for the axial vector seem to be in agreement with experiment. But on comparing the strangeness conserving and strangeness violating decays we find that the strangeness violating decay amplitudes are suppressed by factors of tan θ_V and tan θ_A where

$$\theta_V = \tan^{-1} 0{\cdot}22$$

$$\theta_A = \tan^{-1} 0{\cdot}28$$

are called the Cabibbo angles. It is quite interesting to note that these suppression factors are very close to the ratio m_π/m_K of the pseudoscalar masses. If only we could include the inverse pseudoscalar meson mass as a factor in the interaction we could restore universality of weak interactions. *Within a purely four-fermion theory such a factor cannot enter*; *but if we could take a theory in which the interactions are mediated by meson exchange such a possibility may obtain.* We shall see that the theory of primary interactions does precisely this.

There are also non-leptonic weak decays. In this case the strong interaction effects make it difficult to identify the current-current structure, but if we assume that these decays result from a four-fermion coupling of the primary baryon fields through the same octet type current-current coupling we would expect an octet and 27-type contribution to strangeness violating decays. These imply contributions with the isospin transformation properties of $\Delta I = 1/2$ and $\Delta I = 3/2$, but not $\Delta I = 5/2$. For kaon decay into two pions these imply a sum rule which is satisfied reasonably well. Similar relations also obtain for the SU(3) transformation property with only the 8-type and 27-type contributions. To a fair accuracy it appears also than an 8-type contribution alone can account for all the non-leptonic transition amplitudes.

5 Universal Primary Interactions

We now propose a theory of primary interactions which retains much of the successes of our chiral *V-A* interaction but extends it in a universal fashion to strange particle decays and related strong, electromagnetic and weak interactions. The basic idea of the theory is that weak interactions and electromagnetism of strongly interacting particles are not primary interactions but are induced by the direct coupling of vector and axial vector fields. The primary interactions are the direct couplings of the vector and axial vector meson fields with leptons and the photon.

The electrons and muons are directly coupled to the Maxwell field a^λ according to the standard interaction

$$-e(\bar{\mu}\gamma_\lambda\mu + \bar{e}\gamma_\lambda e)a^\lambda$$

This is a primary interaction; so is the electric interaction

$$-e'\left(\frac{m_\rho{}^2}{g}\rho_\lambda + \frac{m_\omega{}^2}{g}\omega_\lambda\right)a^\lambda$$

where g is the (strong) coupling constant for the vector mesons. The fact that electric charge must be absolutely conserved in the beta decay of neutrons demands the equality of the two coupling parameters e and e'. The vector fields ρ_λ and ω_λ are both the divergence-free neutral vector meson fields. As a consequence the electric current

$$ej_\lambda = -e\left(\bar{\mu}\gamma_\lambda\mu + \bar{e}\gamma_\lambda e + \frac{m_\rho}{g}\rho_\lambda + \frac{m_\omega{}^2}{g}\omega_\lambda\right)$$

is conserved.

As an immediate consequence of this form of the coupling it would follow that the electric form-factor of the nucleon (or any other strong interaction) is given by the form-factor of the vector meson vertex multiplied by an additional pole term to take account of the vector meson propagator. This is in qualitative accord with the two-pole structure observed in the nucleon electromagnetic form factors. In addition to this, the nucleon would exhibit anomalous magnetic properties which would be a direct reflection of the effective magnetic coupling of the vector meson with the nucleon. If we denote the vector meson-nucleon coupling by

$$\tfrac{1}{2}g\bar{N}[\gamma^\lambda\boldsymbol{\tau}\boldsymbol{\rho}_\lambda + (g'/g)\sigma^{\lambda\nu}\tfrac{1}{2}(\boldsymbol{\tau}\boldsymbol{\rho}_{\lambda\nu})]N$$

the nucleon isovector magnetic moment is given by

$$\mu_1 = \frac{2m_N}{m_\rho}\cdot\frac{g'}{g}$$

If we take $g'/g = 5/3$ as suggested by the SU(4) symmetry scheme we predict an isovector magnetic moment of

$$\mu_1 = 4{\cdot}1$$

which is to be compared with the experimental value

$$\mu_1 = 3{\cdot}7$$

An extension of this method leads to a calculation of the ratio of proton and neutron total magnetic moments of

$$-\mu_p/\mu_n = 1 + (g/g')(m_\rho/m_N) = 1{\cdot}49$$

which is to be compared with the experimental value 1·46. It is now appropriate to see how we could use a similar scheme for weak interactions.

The primary weak interactions of the leptons is of the chiral *V-A* form:

$$\frac{G}{\sqrt{2}}\,[\bar{\mu}\gamma_\lambda(1+\gamma_5)\nu_\mu]^\dagger[\bar{e}\gamma^\lambda(1+\gamma_5)\nu_e]$$

with a *possible self-coupling* of the electron covariant or muon covariant. But as far as the nucleons are concerned they have no primary weak interactions; but the charged vector and axial vector mesons couple according to

$$-G\left(\frac{m_\rho{}^2}{g}\rho^\lambda + \frac{m_A{}^2}{g}A^\lambda\right)[\bar{\mu}\gamma_\lambda(1+\gamma_5)\nu_\mu + \bar{e}\gamma_\lambda(1+\gamma_5)\nu_e]$$

By virtue of the strong interaction of the ρ meson we get the effective Fermi interaction:

$$\frac{G}{\sqrt{2}}(\bar{P}\gamma_\lambda N)[\bar{e}\gamma^\lambda(1+\gamma_5)\nu_e]$$

The effective Gamow-Teller interaction is of the form

$$\left(\frac{f}{g}\right)\cdot\frac{G}{\sqrt{2}}\cdot(\bar{P}\gamma_\lambda\gamma_5 N)[\bar{e}\gamma^\lambda(1+\gamma_5)\nu_e]$$

with the axial vector field coupling to nucleons being given by:

$$\tfrac{1}{2}f\bar{N}[\gamma^\lambda\gamma_5\boldsymbol{\tau}\mathbf{A}_\lambda + (f'/f)\sigma^{\lambda\nu}\gamma_5\tfrac{1}{2}\boldsymbol{\partial}\mathbf{A}_{\lambda\nu}]\cdot N$$

Comparison with beta decay experiment suggest that

$$f/g = g_A \simeq 1{\cdot}2$$

We also find a small *CP*-violating interaction from the coupling of $A_{\lambda\nu}$. Its contribution to effective nuclear beta decay interaction seems to be beyond the present experimental limit. It is interesting to note that the weak magnetism term is predicted in this theory with the correct numerical magnitude.

The values of (g'/g), (f/g), (f'/g) can all be derived from the following line of reasoning. As emphasized in connection with the symmetries of the beta interaction, the important question is not whether the symmetry is elegant but whether it is in accordance with experiment. We come back to this question later. We note that the vector and axial vector meson interactions in the low energy limit consist of both Fermi and Gamow-Teller type couplings. The coupling of the vector meson through ρ_λ is Fermi type; but the coupling of the vector meson through $\rho_{\lambda\nu}$ and the axial vector meson through A_λ or through $A_{\lambda\nu}$ are Gamow-Teller type. For vector meson couplings by identifying the two types with the respective generators of SU(4) we arrive at the ratio

$$g'/g = \tfrac{5}{3}$$

As far as the axial vector fields are concerned, since both of them are Gamow-Teller type we should require by a similar identification:

$$\sqrt{(f^2 + f'^2)}/g = \tfrac{5}{3}$$

If in addition we take

$$f'/f = 1$$

we get

$$g_A = f/g = f'/g = \frac{5}{3\sqrt{2}} \simeq 1{\cdot}2$$

in reasonable agreement with experiment.

If these arguments are to be trusted, we should be able to explain low-energy meson-nucleon scattering in terms of a single coupling constant for strong interactions. We should of course associate the vector meson field with the observed vector mesons. Since the neutral fields ρ_λ, ω_λ must be divergence-free to assure the conservation of electric current, this implies, in turn, that no neutral currents are expected to be present; none are, of course, found. On the other hand, as far as the axial vector meson fields are concerned no such conservation law is required; we could then associate the pseudoscalar mesons with the divergence of the axial vector field. Such an assignment is closely related to the partial conservation law for the axial current that we had discussed earlier. We write:

$$\partial^\lambda V_\lambda = 0$$
$$\partial^\lambda B_\lambda = 0; \; A_\lambda = B_\lambda + (\xi/m_\pi)\partial_\lambda \phi_\pi$$

Here V_λ stands for the divergence-free vector fields and B_λ for the divergence-free part of the axial vector field. We have introduced a dimensionless parameter ξ to indicate the relative strength of the pion field. We now get a pseudovector coupling of the pion to the nucleon:

$$(f_1/m_\pi)\bar{N}\gamma^\lambda\gamma_5\boldsymbol{\tau}\cdot\partial_\lambda\phi_\pi N; \;\; f_1 = \tfrac{1}{2}g g_A \xi$$

We may determine f_1 from purely strong interaction data and consider f_1 and g to be the relevant strong interaction parameters.

Using the values $f_1 = 0{\cdot}85$, $g = 9{\cdot}0$ we can get excellent agreement for the low energy pion-nucleon scattering lengths for all *S*-waves and *P*-waves (with the exception of the resonant channel). We have for the *S*-wave scattering lengths:

$$a_1 = +0{\cdot}20 \quad (+0{\cdot}183)$$

$$a_3 = -0{\cdot}10 \quad (-0{\cdot}109)$$

and for *P*-waves:

$$a_{11} = -0{\cdot}091 \quad (-0{\cdot}101)$$

$$a_{13} = -0{\cdot}022 \quad (-0{\cdot}029)$$

$$a_{31} = -0{\cdot}022 \quad (-0{\cdot}039)$$

$$a_{33} = +0{\cdot}133 \quad (+0{\cdot}215)$$

(where the figures in parentheses are the experimental values).

This scheme gives qualitative accord with the main features of the nuclear force. In particular, the pion strength parameter

$$\xi = 2f_1/f = 0{\cdot}16$$

together with the absence of unacceptable singularities in the two-nucleon force imply the relation:

$$(f\xi/m_\pi)^2 - (g/m_\rho)^2 = 0$$

This leads to the prediction

$$(m_\pi/m_\rho) = 0{\cdot}188$$

which is to be compared with the observed value

$$(m_\pi/m_\rho) = 0{\cdot}182$$

We see that the theory of primary interactions is able to relate strong, electromagnetic and weak interaction phenomena involving the non-strange particles.

In the extension to strange particles we have to consider strange vector and axial vector mesons. We would like to incorporate the absence of strange scalar mesons by demanding that

$$\partial^\lambda V_\lambda' = 0$$

This implies that there can be no coupling of strange vector mesons to baryons through V_λ' because of the inequality of the strange and non-strange baryon masses. Instead the entire coupling should proceed through the coupling via $V_{\lambda\nu}'$. This would tend to suppress the vector decays. For the axial vector fields we write:

$$\partial^\lambda B_\lambda' = 0;\ A_\lambda' = B_\lambda' + (\xi/m_K)\partial_\lambda \phi_K$$

We choose the *same* value of the dimensionless parameter ξ because we can now demand that the coupling of the strange and non-strange axial vector fields are *universal*. We can immediately deduce for the ratio of the π and K decay widths:

$$\frac{\Gamma(K \to \mu\nu)}{\Gamma(\pi \to \mu\nu)} = \frac{m_\pi}{m_K}\left[\frac{1 - (m_\mu/m_K)^2}{1 - (m_\mu/m_\pi)^2}\right]^2$$

which yields an excellent value for the lifetime of the kaon. For the axial vector decays of the hyperons we could show that making use of the partial conservation law we could deduce an effective Cabibbo angle

$$\theta_B = \tan^{-1}\left\{\frac{m_\pi}{m_K}\frac{M + m}{2m}\right\}$$

which is in excellent agreement with experiment. Thus the choice of a universal value for the pseudoscalar meson strength parameter enables us to restore universality of the weak interactions. Needless to say, as long as the mesons form an octet coupled with the baryons, the octet property of the leptonic decays is assumed.

For the non-leptonic decays we could consider weak coupling of the vector and/or axial vector mesons with baryons and mesons. Independent of the details of this coupling we get the transformation properties of a current-current interaction with an octet current coupled to a suitable baryonic current. It will take us too far afield to discuss the question of non-leptonic decays in detail.

We have outlined here a theory of universal primary interactions of particles including strong, electromagnetic and weak interactions. The most important idea is that the primary interactions of the baryon consist of the strong coupling to vector and axial vector fields only. Both electromagnetic and weak interactions are acquired characteristics. Thus our theory may be viewed as the logical completion of the hypothesis that the beta decay of the nucleon arose only by virtue of its coupling to the meson. The theory has the ability to correlate such diverse aspects as nucleon magnetic moments, weak magnetism, ratio of Gamow-Teller and Fermi coupling constants, absence of neutral lepton currents, apparent suppression of strange particle leptonic decays and pion-nucleon scattering.

The hypothesis of chiral V-A interaction may thus be considered to be fully verified within the realm of non-strange particle decays for which it was formulated. The continuing work of the last decade has established this scheme. But it appears that its extension to strange particles and the desire to relate strong and weak interactions suggest that we modify the theory. The theory of primary interactions shows that the theoretical scheme acquires a much simpler form if we postulate that electromagnetism and weak interactions are not primary interactions of the nucleons but are acquired by virtue of their interaction with vector and axial vector fields. The next decade will tell as to what extent this theory is a step towards understanding nature.

References

1. Sargent, B. W., *Proc. Cambridge Phil. Soc.*, **28,** 538 (1932); Sargent, B. W., *Proc. Roy. Soc.* (*London*), **A139,** 659 (1933).
2. Fermi, E., *Z. Physik*, **88,** 161 (1934).
3. Yukawa, H., *Proc. Phys. Math. Soc.* (*Japan*), **17,** 48 (1935).
4. Gamow, G., and Teller, E., *Phys. Rev.*, **49,** 895 (1936).
5. Lee, T. D., and Yang, C. N., *Phys. Rev.*, **104,** 254 (1956).
6. Wu, C. S., Ambler, Hayward, Hoppes and Hudson, *Phys. Rev.*, **105,** 1413 (1957).
7. Goldhaber, Grodzins and Sunyar, *Phys. Rev.*, **109,** 1015 (1958).
8. Klein, O., *Nature*, **161,** 897 (1948).
9. Puppi, G., *Nuovo Cimento*, **5,** 587 (1948).
10. Feynman, R. P., and Gell-Mann, M., *Phys. Rev.*, **109,** 193 (1958).
11. Sakurai, J., *Nuovo Cimento*, **7,** 649 (1958).
12. Weisberger, W. I., *Phys. Rev. Letters*, **14,** 1047 (1965).
13. Adler, S. L., *Phys. Rev. Letters*, **14,** 1051 (1965).
14. Raman, K., and Sudarshan, E. C. G., *Phys. Letters*, **21,** 450 (1967).
15. Tomozawa, Y., *Nuovo Cimento*, **46A,** 707 (1966).

Compositeness Criteria in Field Theory

C. R. HAGEN

1 Introduction

The vast increase in the number of known particles and resonances which we have witnessed in recent years has done much to increase our knowledge of the underlying symmetries which govern strong interaction physics. On the other hand this proliferation has also been a source of some embarrassment to high energy physicists who were previously only slightly uncomfortable in referring to nucleons, photons and electrons as 'elementary' particles and deuterons as 'composite' particles. Since the prospect of classifying all the recently discovered particles and low lying resonant states as elementary seems to preclude the possibility of having a basically 'simple' theory of strong interactions, a number of attempts have been made to reexamine the elementary particle concept with a view to establishing some meaningful criteria whereby one can determine whether a given particle should be considered as elementary or composite.

Since some of the criteria which have been proposed in this context appear to be highly arbitrary, one might well be led to the extreme view that either all (or none) of the known particles are composite structures. Although we have dismissed (on purely philosophical grounds) the desirability of taking all particles as elementary, Chew's suggestion[1] of 'nuclear particle democracy' in which all particles are to be treated as composites of all others, at first sight appears to present a convenient way in which one can completely avoid the problem of deciding whether or not a given particle is elementary. Unfortunately (from the viewpoint to be presented here) Chew's formulation is closely tied in with the bootstrap idea and as such cannot be formulated in the language of field theory. Although this point in itself need not be considered a fault, our basic premise here shall be that field theory is essentially correct and that its well-known divergence difficulties are merely the consequences of an inadequate perturbation theory. We shall therefore attempt to formulate the desired compositeness criteria entirely within the framework of Lagrangian field theory, avoiding wherever possible the problem of ultra-violet divergences.

The approach we develop here is motivated by a result obtained several years ago by Howard and Jouvet[2] and also by Vaughn, Aaron and Amado[3]. They observed in the case of the Lee model that in the limit, where the wave-function renormalization constant of the V particle tends to zero, the theory becomes equivalent to a direct interaction theory in which the V particle is described as a bound state. This result has been elaborated upon by many authors in a variety of soluble models in which one invariably finds this equivalence between vanishing wave-function renormalization and the compositeness of the associated particle. In the review presented here this result will be reduced to its bare essentials so that it can be seen to follow as an almost trivial consequence of the equations of motion of a given theory. This approach will have the advantage of emphasizing the absence of any real dynamical content in the so-called $Z = 0$ rule. Although we shall demonstrate here the equivalence of vanishing wave-function renormalization and compositeness, it will be seen that this result is little more than a definition and as such can provide no insight into the details of the dynamics.

2 A Simple Example

In order to fix ideas more firmly it will be useful at this point to illustrate more precisely what we shall mean by the $Z = 0$ condition, by specific reference to the case of the Lee model[4]. The Hamiltonian is taken to be of the usual form

$$H = m_V V^*V + m_N N^*N + \int \mathrm{d}^3k\,\omega\theta^*(\mathbf{k})\theta(\mathbf{k}) - g_0\int \mathrm{d}^3k\,\alpha(|\mathbf{k}|)[V^*N\theta(\mathbf{k}) + N^*V\theta^*(\mathbf{k})]$$

where

$$\omega = (\mathbf{k}^2 + \mu^2)^{\frac{1}{2}}$$

and

$$\alpha(|\mathbf{k}|) = \frac{u(|\mathbf{k}|)}{(2\pi)^{\frac{3}{2}}}\frac{1}{(2\omega)^{\frac{1}{2}}}$$

$u(|\mathbf{k}|)$ being a form-factor inserted to guarantee the convergence of the theory. The only non-vanishing commutation relations are taken to be

$$(V, V^*) = (N, N^*) = 1$$
$$[\theta(\mathbf{k}), \theta^*(\mathbf{k}')] = \delta(\mathbf{k} - \mathbf{k}')$$

The model admits only the single virtual transition

$$V \rightleftharpoons N + \theta \tag{1}$$

which implies the existence of the two conserved operators

$$Q_1 = V^*V + N^*N$$
$$Q_2 = V^*V + \int d^3k\theta^*(\mathbf{k})\theta(\mathbf{k})$$

Although the first of these two quantities is familiar as the fermion number, it is the existence of the second which is peculiar to the Lee model and is ultimately responsible for the solubility of the theory in the usual representation

$$V|0\rangle = N|0\rangle = 0$$
$$\theta(\mathbf{k})|0\rangle = 0$$

of the commutation relations.

Despite the fact that the single N and θ states are not affected by the interaction, i.e.

$$H\theta^*(\mathbf{k})|0\rangle = \omega\theta^*(\mathbf{k})|0\rangle$$
$$HN^*|0\rangle = m_N N^*|0\rangle$$

it is well known that the V state has a non-trivial structure as a consequence of the virtual transition (1). One thus writes the 'dressed' V state as

$$|V\rangle = N[V^*|0\rangle + \int d^3kN^*\theta^*(\mathbf{k})\phi(\mathbf{k})|0\rangle]$$

where N is a normalization constant defined by

$$\langle V|V\rangle = 1 \tag{2}$$

and $\phi(\mathbf{k})$ is to be determined by the condition that V be an eigenstate of H, i.e.

$$H|V\rangle = E_V|V\rangle$$

It follows by straightforward application of the commutation relations that

$$\phi(\mathbf{k}) = \frac{\alpha(|\mathbf{k}|)}{m_N + \omega - E_V}$$

$$E_V = m_V - g_0^2\int d^3k\alpha(\mathbf{k})\phi(\mathbf{k})$$

thus implying the eigenvalue condition

$$E_V = m_V - \frac{g_0^2}{4\pi}\frac{1}{\pi}\int_\mu^\infty \frac{u^2(|\mathbf{k}|)k\,\mathrm{d}\omega}{m_N + \omega - E_V} \tag{3}$$

It may be remarked here that if the condition (3) does not have a solution for $E_V \langle m_N + \mu$, there exists no stable V particle in the theory. Although we shall subsequently consider this latter possibility in more detail, for the present it will be assumed that the V particle exists and that there is consequently a solution of (3) for some real E_V. (It is trivial to show that there can be no more than one such solution.)

Application of (2) now yields the result

$$N^2 = \left[1 + \frac{g_0^2}{4\pi}\frac{1}{\pi}\int_\mu^\infty \frac{u^2(|\mathbf{k}|)k\,\mathrm{d}\omega}{(m_N + \omega - E_V)^2}\right]^{-1}$$

so that the definition of the wave-function renormalization constant Z, by the condition

$$Z^{\frac{1}{2}} = \langle 0|V|V\rangle$$

immediately implies

$$\begin{aligned} Z &= N^2 \\ &= \left[1 + \frac{g_0^2}{4\pi}\frac{1}{\pi}\int_\mu^\infty \frac{u^2(|\mathbf{k}|)k\,\mathrm{d}\omega}{(m_N + \omega - E_V)^2}\right]^{-1} \end{aligned} \tag{4}$$

In order to complete the discussion of the V particle sector we turn our attention to the $N\theta$ scattering states. In this case it is clear that one should attempt to construct a scattering state of energy $E = m_N + \omega_\mathbf{k}$ in the form

$$|N\theta_\mathbf{k}\rangle = N^*\theta^*(\mathbf{k})|0\rangle + \int \mathrm{d}^3k'\phi(\mathbf{k},\mathbf{k}')N^*\theta^*(\mathbf{k}')|0\rangle + \psi(\mathbf{k})V^*|0\rangle$$

As before, one can evaluate the unknown functions [in this case $\phi(\mathbf{k},\mathbf{k}')$ and $\psi(\mathbf{k})$] by straightforward application of the commutation relations, obtaining

$$\phi(\mathbf{k},\mathbf{k}') = \frac{1}{\omega' - \omega \mp \mathrm{i}\,\varepsilon}\psi(\mathbf{k})\alpha(|\mathbf{k}'|)$$

$$\psi(\mathbf{k}) = \frac{\alpha(|\mathbf{k}|)}{m_V - m_N - \omega - g_0^2\int \mathrm{d}^3k' \dfrac{\alpha^2(|\mathbf{k}'|)}{\omega' - \omega \mp \mathrm{i}\,\varepsilon}}$$

where $\omega + i\varepsilon$ and $\omega - i\varepsilon$ correspond respectively to outgoing and incoming spherical waves. Defining the *S*-matrix by

$$\langle N\theta_{k'}{}^{\text{in}}|S|N\theta_k{}^{\text{in}}\rangle = \langle N\theta_{k'}{}^{\text{in}}|N\theta_k{}^{\text{out}}\rangle$$
$$= \delta(\mathbf{k}-\mathbf{k}') + 2\pi \mathrm{i}\,\delta(\omega-\omega')g_0{}^2\alpha^2(|\mathbf{k}|) \times$$
$$\times \left[m_V - m_N - \omega - g_0{}^2\!\int \mathrm{d}^3k''\,\frac{\alpha^2(|\mathbf{k}''|)}{\omega''-\omega-\mathrm{i}\,\varepsilon}\right]^{-1}$$

one easily extracts the (*S*-wave) phase shift

$$\frac{1}{k}\,\mathrm{e}^{\mathrm{i}\delta}\sin\delta = \frac{g_0{}^2}{4\pi}\,\frac{u^2(|\mathbf{k}|)}{m_V - m_N - \omega - \dfrac{g_0{}^2}{4\pi}\dfrac{1}{\pi}\displaystyle\int_\mu^\infty \frac{k'\,\mathrm{d}\omega' u^2(|\mathbf{k}'|)}{\omega'-\omega-\mathrm{i}\,\varepsilon}} \tag{5}$$

Although the above result will be sufficient for the purposes of this discussion, it is customary to express the scattering amplitude in terms of renormalized (i.e. observable) quantities. To this end one can use (3) to write (5) in the once subtracted form

$$\frac{1}{k}\,\mathrm{e}^{\mathrm{i}\delta}\sin\delta =$$
$$= \frac{g_0{}^2}{4\pi}\,\frac{u^2(|\mathbf{k}|)}{(E_V - m_N - \omega)\left[1 - \dfrac{g_0{}^2}{4\pi}\dfrac{1}{\pi}\displaystyle\int_\mu^\infty \frac{k'\,\mathrm{d}\omega' u^2(|\mathbf{k}'|)}{(\omega'-\omega-\mathrm{i}\,\varepsilon)(E_V - m_N - \omega')}\right]} \tag{6}$$

which explicitly displays the *V*-state contribution. In addition one can define the renormalized coupling constant g^2 by the condition*

$$u^2(|\mathbf{k}|)(E_V - E)\,\frac{1}{k}\,\mathrm{e}^{\mathrm{i}\delta}\sin\delta\Big|_{E=E_V} = \frac{g^2}{4\pi}$$

Upon comparison with (4) this yields the identification

$$g^2 = Z g_0{}^2$$

enabling one to write (4) in the alternative form

$$Z = 1 - \frac{g_0{}^2}{4\pi}\,\frac{1}{\pi}\int_\mu^\infty \mathrm{d}\omega\,\frac{u^2(|\mathbf{k}|)k}{(m_N+\omega-E_V)^2}$$

* We assume here that the form-factor $u(|\mathbf{k}|)$ is such that there is no difficulty in performing a continuation below the elastic threshold.

The scattering amplitude can consequently be expressed entirely in terms of renormalized quantities as

$$\frac{1}{k}\,\mathrm{e}^{\mathrm{i}\delta}\sin\delta = \frac{g^2}{4\pi}\frac{u^2(|\mathbf{k}|)}{(E_V - m_N - \omega)} \times$$
$$\times\left[1 - (E_V - m_N - \omega)\frac{g^2}{4\pi}\frac{1}{\pi}\int_\mu^\infty \frac{k'\,\mathrm{d}\omega' u^2(|\mathbf{k}'|)}{(\omega' - \omega - \mathrm{i}\,\varepsilon)(E_V - m_N - \omega')^2}\right]^{-1} \tag{7}$$

thereby completing the discussion of the Nθ sector.

To this point our discussion has quite clearly been devoted entirely to a theory in which the V particle is elementary. Therefore in order to allow any contact with a discussion of compositeness one must find a theory which has the same physical content save for the single feature that the V particle be capable of description as a bound state. In particular this alternative description must yield the same $N\theta$ scattering amplitude at all finite ω and consequently have a stable V particle at the same position as in the Lee model. We shall 'guess' that the correct Hamiltonian is

$$H = m_N N^*N + \int\mathrm{d}^3k\,\omega\theta^*(\mathbf{k})\theta(\mathbf{k}) - \lambda N^*N\int\mathrm{d}^3k\,\alpha(|\mathbf{k}|)\theta^*(\mathbf{k})\int\mathrm{d}^3k'\,\alpha(|\mathbf{k}'|)\theta(\mathbf{k}') \tag{8}$$

where $\alpha(|\mathbf{k}|)$ and ω are defined as before. By identifying the N and θ masses which appear in (8) with the corresponding mass parameters of the Lee model, the N and θ sectors will clearly be identical in the two theories. The Hamiltonian (8) implies the conservation of

$$Q_1' = N^*N$$

and

$$Q_2' = \int\mathrm{d}^3k\,\theta^*(\mathbf{k})\theta(\mathbf{k})$$

corresponding to the two charge operators of the Lee model.

Despite the absence of an elementary V particle in the present case, one can look for a stable particle with the quantum numbers associated with the $N\theta$ sector. Such a particle (if it actually exists) is necessarily composite. We call such a state $|V\rangle$ and assume a structure of the form

$$|V\rangle = \int\mathrm{d}^3k\,\phi'(\mathbf{k})N^*\theta^*(\mathbf{k})|0\rangle$$

The eigenvalue condition

$$H|V\rangle = E_V|V\rangle$$

yields the integral equation

$$(m_N + \omega - E_V)\phi'(\mathbf{k}) = \lambda\alpha(|\mathbf{k}|)\int\phi'(\mathbf{k}')\alpha(|\mathbf{k}'|)\mathrm{d}^3k'$$

This equation is readily solved to yield the result

$$\phi'(\mathbf{k}) = C\frac{\alpha(|\mathbf{k}|)}{m_N + \omega - E_V}$$

where C is a constant, together with the condition

$$1 = \lambda \int \frac{\alpha^2(|\mathbf{k}|)\mathrm{d}^3k}{m_N + \omega - E_V} \tag{9}$$

It may be convenient to view this latter equation as determining λ for a preassigned value of E_V (corresponding to the V particle state of the Lee model) in view of our desire to ensure that both theories have the same physical content.

Proceeding in analogy to the case of the Lee model, one can attempt to determine the $N\theta$ scattering states by assuming the form

$$|N\theta_k\rangle = N^*\theta^*(\mathbf{k})|0\rangle + \int \mathrm{d}^3k'\phi'(\mathbf{k}, \mathbf{k}')N^*\theta^*(\mathbf{k}')|0\rangle$$

Straightforward calculation yields for $\phi'(\mathbf{k}, \mathbf{k}')$ the result

$$\phi'(\mathbf{k}, \mathbf{k}') = \frac{\lambda\alpha(|\mathbf{k}|)\alpha(|\mathbf{k}'|)}{m_N + \omega' - E \mp \mathrm{i}\,\varepsilon}\left[1 - \lambda\int\frac{\alpha^2(|\mathbf{k}''|)\,\mathrm{d}^3k''}{m_N + \omega'' - E \mp \mathrm{i}\,\varepsilon}\right]^{-1}$$

where $E = m_N + \omega$ and the $\pm\,\mathrm{i}\varepsilon$ notation has the same meaning as before. This immediately yields the S-matrix as

$$\langle N\theta_{k'}{}^{\mathrm{in}}|S|N\theta_k{}^{\mathrm{in}}\rangle = \delta(\mathbf{k} - \mathbf{k}') + 2\pi\,\mathrm{i}\delta(\omega - \omega')\lambda\alpha^2(|\mathbf{k}|)\,\times \\ \times\left[1 - \lambda\int\frac{\alpha^2(|\mathbf{k}''|)\,\mathrm{d}^3k'}{m_N + \omega'' - E - \mathrm{i}\,\varepsilon}\right]^{-1}$$

from which one deduces the form of the phase shift

$$\frac{1}{k}\,\mathrm{e}^{\mathrm{i}\delta}\sin\delta = \lambda\frac{u^2(|\mathbf{k}|)}{4\pi}\left[1 - \frac{\lambda}{4\pi}\frac{1}{\pi}\int_\mu^\infty\frac{k'\,\mathrm{d}\omega' u^2(|\mathbf{k}'|)}{(m_N + \omega' - E)}\right]^{-1} \tag{10}$$

With this result one is now able to compare the predictions (with respect to $N\theta$ scattering processes) of the two theories under consideration. To this end we use the eigenvalue condition to rewrite (10) as

$$\frac{1}{k}\,\mathrm{e}^{\mathrm{i}\delta}\sin\delta = \frac{\pi u^2(|\mathbf{k}|)}{E_V - E}\left[\int_\mu^\infty\frac{k'\,\mathrm{d}\omega' u^2(|\mathbf{k}'|)}{(m_N + \omega' - E_V)(m_N + \omega' - E)}\right]^{-1} \tag{11}$$

By a further subtraction (11) assumes the form

$$\frac{1}{k}\,e^{i\delta}\sin\delta = \frac{\pi u^2(|\mathbf{k}|)}{E_V - E}\left[\int_\mu^\infty \frac{k'\,d\omega' u^2(|\mathbf{k}'|)}{(m_N + \omega' - E_V)^2} + \right.$$
$$\left. + (E - E_V)\int_\mu^\infty \frac{k'\,d\omega' u^2(|\mathbf{k}'|)}{(m_N + \omega' - E_V)^2(m_N + \omega' - E)}\right]^{-1}$$

which is readily seen to be identical to (7) for the case that the coupling constant is given by

$$\left(\frac{g_0{}^2}{4\pi}\right)^{-1} = \frac{1}{\pi}\int_\mu^\infty \frac{u^2(|\mathbf{k}'|)k'\,d\omega'}{(m_N + \omega' - E_V)^2}$$

This, however, is precisely the condition that $Z = 0$ in the Lee model and thus establishes the asserted equivalence of the vanishing of the wave-function renormalization constant of the V field to the compositeness of the V particle.

Having completed the proof of equivalence by direct calculation of the scattering amplitudes in these two models, it will be useful from the viewpoint of subsequent sections to indicate a somewhat more direct approach to the problem. In particular we note that the eigenvalue condition (3) may be written in terms of the renormalized coupling constant as

$$m_V = E_V + \frac{1}{Z}\frac{g^2}{4\pi}\frac{1}{\pi}\int_\mu^\infty \frac{u^2(|\mathbf{k}|)k\,d\omega}{m_N + \omega - E_V}$$

It is clear from this expression that the $Z = 0$ limit implies the divergence of the bare mass m_V as Z^{-1}. Therefore one might well anticipate that all the above results can be obtained more simply merely by application of this almost trivial observation without actually carrying out the renormalization explicitly. In particular we note that the expression

$$\frac{1}{k}\,e^{i\delta}\sin\delta = \frac{g_0{}^2}{4\pi}\,\frac{u^2(|\mathbf{k}|)}{m_V - m_N - \omega - \dfrac{g_0{}^2}{4\pi}\dfrac{1}{\pi}\displaystyle\int_\mu^\infty \frac{k'\,d\omega' u^2(|\mathbf{k}'|)}{\omega' - \omega - i\,\varepsilon}}$$

for the Lee model in the $Z = 0$ limit becomes

$$\frac{u^2(|\mathbf{k}|)}{\left(\dfrac{g_0{}^2}{4\pi m_V}\right)^{-1} - \dfrac{1}{\pi}\displaystyle\int_u^\infty \frac{k'\,d\omega' u^2(|\mathbf{k}'|)}{\omega' - \omega - i\,\varepsilon}}$$

for all finite ω. Upon making the identification

$$\lambda = \frac{g_0^2}{m_V} \tag{12}$$

the equivalence of the two expressions for the $N\theta$ scattering amplitude follows immediately. The replacement (12) can also be inferred from the equations of motion of the Lee model

$$[V(t), H] = \mathrm{i}\frac{\partial}{\partial t}V(t)$$
$$= m_V V - g_0 \int \mathrm{d}^3k\,\alpha(|\mathbf{k}|)N\theta(\mathbf{k})$$

In the limit $m_V \to \infty$ this equation degenerates from a true equation of motion to the constraint

$$V = \frac{g_0}{m_V}\int \mathrm{d}^3k\,\alpha(|\mathbf{k}|)N\theta(\mathbf{k})$$

so that the remaining equations of motion

$$\left(\mathrm{i}\frac{\partial}{\partial t} - \omega\right)\theta(\mathbf{k}) = -g_0\alpha(|\mathbf{k}|)N^*V$$
$$\left(\mathrm{i}\frac{\partial}{\partial t} - m_N\right)N = -g_0\int \mathrm{d}^3k\,\alpha(|\mathbf{k}|)V\theta^*(\mathbf{k})$$

become respectively

$$\left(i\frac{\partial}{\partial t} - \omega\right)\theta(\mathbf{k}) = -\frac{g_0^2}{m_V}N^*N\alpha(|\mathbf{k}|)\int \mathrm{d}^3k'\,\alpha(|\mathbf{k}'|)\theta(\mathbf{k}')$$
$$\left(i\frac{\partial}{\partial t} - m_N\right)N = -\frac{g_0^2}{m_V}N\int \mathrm{d}^3k\,\alpha(|\mathbf{k}|)\theta^*(\mathbf{k})\int \mathrm{d}^3k'\,\alpha(|\mathbf{k}'|)\theta(\mathbf{k}')$$

Upon comparison with the equations of motion of the theory described by (8), one again establishes the equivalence of the two models provided only one makes the identification (12).

3 Extension to the Relativistic Case

In the example of the preceding section one had the distinct advantage of being able to explicitly display the $Z = 0$ limit in terms of the solutions of the models under consideration. In the domain of relativistic field theory, however, there exists only an insignificant number of examples of fully soluble models. In consequence of this fact it is necessary to adopt either a relatively heuristic but quite general formulation which completely avoids the usual divergence difficulties or alternatively to develop an approach based upon perturbation theory in which one adopts some specific prescription for dealing with divergent integrals. Since this latter approach does not significantly add to the rigour of the argument and furthermore tends to considerably obscure its basic simplicity, we shall choose the former alternative and thereby avoid basing the results on a specific type of perturbation theory. It is not suggested here that the assumptions we shall make will prove adequate to deal with all situations. On the other hand the motivation for this work[5] is primarily to account for all the presently known successes of the $Z = 0$ rule within a very general framework. Once this is accomplished we shall surmise that further applications of the $Z = 0$ condition can similarly be understood by relatively minor extensions of the method.

We shall begin by considering a rather general class of interactions of a real spin zero field such as described by the Lagrangian*

$$L = \phi^{\mu}\partial_{\mu}\phi + \tfrac{1}{2}\phi^{\mu}\phi_{\mu} - \tfrac{1}{2}\mu_0{}^2\phi^2 + g_0\phi j + L' \tag{13}$$

where $j(x)$ does not depend upon $\phi(x)$ or its derivatives and L' includes all terms in the Lagrangian which do not explicitly include the field $\phi(x)$. In the case of the Lee model these conditions on the form of the interaction are equivalent to the requirement that the coupling terms depend only linearly upon the V field. The Lagrangian (13) yields the equation of motion

$$(-\partial^2 + \mu_0{}^2)\phi(x) = g_0 j(x) \tag{14}$$

and implies the equal-time commutation relation

$$[\partial_0\phi(x), \phi(x')] = -\mathrm{i}\delta(\mathbf{x} - \mathbf{x}') \tag{15}$$

For the present we shall assume that the dynamics is such that there is a single discrete state associated with the field $\phi(x)$ so that the unrenormalized 2-point function

$$\mathscr{G}(x) = \mathrm{i}\langle 0|(\phi(x)\phi(0))_+|0\rangle - \mathrm{i}\langle 0|\phi(0)|0\rangle^2$$

has the usual (unsubtracted) Lehmann representation[6]

$$\mathscr{G}(p) = \frac{Z_3}{p^2 + \mu^2} + \int_{M^2}^{\infty} \frac{B(\kappa)\,\mathrm{d}\kappa^2}{p^2 + \kappa^2 - \mathrm{i}\,\varepsilon}$$

* We take $\hbar = c = l$ and the metric (1, 1, 1, −1).

where M is some mass parameter greater than μ. Our assumptions concerning the structure of $j(x)$ imply the usual sum rules

$$1 = Z_3 + \int_{M^2}^{\infty} B(\kappa)\,\mathrm{d}\kappa^2 \qquad (16)$$

$$\mu_0^2 = Z_3\mu^2 + \int_{M^2}^{\infty} \kappa^2 B(\kappa)\,\mathrm{d}\kappa^2 \qquad (17)$$

The positive definite metric of the associated Hilbert space requires that Z_3 and $B(\kappa)$ be positive definite so that (16) implies that the wave-function renormalization constant Z_3 satisfies the condition

$$0 \leq Z_3 \leq 1$$

Although it is well known that the integrals (16) and (17) are respectively logarithmically and quadratically divergent in perturbation theory for a Yukawa coupling, it is in the spirit of the approach being presented here to assume that they are in fact finite. With these optimistic assumptions it follows that $\mathscr{G}^{-1}(p)$ has the structure

$$\mathscr{G}^{-1}(p) = p^2 + \mu_0^2 - \int_{M^2}^{\infty} \frac{\sigma(\kappa)\,\mathrm{d}\kappa^2}{p^2 + \kappa^2 - i\,\varepsilon} \qquad (18)$$

where $\sigma(\kappa)$ is the positive definite function

$$\sigma(\kappa) = B(\kappa)|\mathscr{G}(p^2 = -\kappa^2)|^2$$

In writing (18) we have excluded the possibility of having a pole of $G^{-1}(p)$ on the real axis between μ^2 and M^2. Since this assumption does not affect any of our subsequent results, we shall not further remark upon this point.

In terms of the representation (18) the condition that there exists a single-particle state in the spectrum of the field $\phi(x)$ is expressed by the condition

$$\mu_0^2 = \mu^2 + \int_{M^2}^{\infty} \frac{\sigma(\kappa)\,\mathrm{d}\kappa^2}{\kappa^2 - \mu^2}$$

or in terms of the renormalized spectral function $\bar{\sigma}(\kappa)$

$$\mu_0^2 = \mu^2 + \frac{1}{Z_3}\int_{M^2}^{\infty} \frac{\bar{\sigma}(\kappa)}{\kappa^2 - \mu^2}\,\mathrm{d}\kappa^2$$

where

$$\bar{\sigma}(\kappa) = Z_3\sigma(\kappa)$$

and

$$Z_3 = 1 - \int_{M^2}^{\infty} \frac{\bar{\sigma}(\kappa)\,\mathrm{d}\kappa^2}{(\kappa^2 - \mu^2)^2}$$

We now explicitly recognize the fact that it is the function $\bar{\sigma}(\kappa)$ which one customarily computes in perturbation theory, so that (at least in a renormalizable theory) the renormalized spectral function $\bar{\sigma}(\kappa)$ is expected to exist as a well-defined function of the physical masses and coupling constants. One thus obtains, in direct analogy to the situation encountered in the Lee model, the divergence of the bare mass as

$$\mu_0^2 \sim \frac{1}{Z_3} \int_{M^2}^{\infty} \frac{\bar{\sigma}(\kappa)\, d\kappa^2}{\kappa^2 - \mu^2} \tag{19}$$

Since there is in the literature a number of apparently contradictory statements concerning the behaviour of the bare mass parameter in the $Z = 0$ limit, it is appropriate to briefly comment upon a frequent source of confusion concerning this point. Clearly the only way in which one can circumvent the conclusions we have reached is for $\bar{\sigma}(\kappa)$ to vanish in the $Z_3 = 0$ limit. However, this would imply that the field is a free field and that Z_3 must consequently be equal to unity. However, in the proposed counter-examples to the result expressed by (19), different definitions of Z_3 are used (other than its field theoretical definition) such that the field $\phi(x)$ is decoupled in what is sometimes incorrectly described as the $Z = 0$ limit. We shall not, however, elaborate upon these results as they do not in any way require a modification of the above arguments.

The divergence of the bare mass in the limit of vanishing wave function renormalization means that (14) has degenerated from an equation of motion to an equation of constraint. The neglect of the kinematical term reduces the field equation to

$$\phi(x) = \frac{g_0}{\mu_0^2} j(x)$$

and allows the reduction of the coupling terms in the Lagrangian to

$$\frac{g^2}{2\mu_0^2} j(x)j(x) = \frac{g^2}{2Z_3\mu_0^2} j(x)j(x)$$

where we have introduced the partially* renormalized coupling constant g. This interaction term is more conveniently written as

$$\frac{\lambda}{2} j(x)j(x)$$

* We assume that all renormalizations except the mass and wave-function renormalizations of the field are finite so that the presence of Z_1 and Z_2 factors in g^2 is of no relevance to this discussion.

where

$$\lambda \equiv g^2 \left[\int_{M^2}^{\infty} \frac{\bar{\sigma}(\kappa)\, d\kappa^2}{\kappa^2 - \mu^2}\right]^{-1} \tag{20}$$

thus demonstrating the equivalence between the direct interaction theory with coupling constant given by (20) and the linearly coupled scalar boson in the $Z_3 = 0$ limit. Although the equivalence of the Yukawa and direct interaction theories for $\mu_0{}^2 \to \infty$ has long been known, the connection of this result with the $Z = 0$ condition has only been made relatively recently.

Despite the scarcity of soluble relativistic models, there is one important (though somewhat trivial) example of a theory in which one can verify by direct calculation the validity of the above remarks. This is the Zachariasen model[7] which together with its numerous variations comprises the only known case of a soluble theory which admits the $Z = 0$ limit. Since a significant fraction of the literature on the $Z = 0$ condition concerns itself primarily with a discussion of this model, we shall briefly discuss this important example. Although the theory was originally formulated by Zachariasen in terms of the language of dispersion theory, it is more appropriate from the viewpoint presented here to use the field theoretical approach first discovered by Thirring[8].

It was observed by Thirring that the Zachariasen model can be formulated in terms of the bilinear coupling between a field $\phi(x)$ and a field $\phi(x, s)$ with a continuous mass parameter s. The Lagrangian is thus of the form

$$\begin{aligned} L = {} & \phi^{\mu}\partial_{\mu}\phi + \tfrac{1}{2}\phi^{\mu}\phi_{\mu} - \tfrac{1}{2}\mu_0{}^2\phi^2 \\ & + \tfrac{1}{2}\int_{4M_A{}^2}^{\infty} ds\, [\phi^{\mu}(s)\partial_{\mu}\phi(s) + \tfrac{1}{2}\phi^{\mu}(s)\phi_{\mu}(s) - s\phi^2(s)] \\ & + \phi \int_{4M_A{}^2}^{\infty} ds\, f(s)\phi(s) \end{aligned} \tag{21}$$

At this point the function $f(s)$ is an arbitrary real function. However, in order to identify the theory described by (21) with the usual form of the Zachariasen model describing the interaction of a particle (called B) with $A\bar{A}$ pairs it is necessary to identify $f^2(s)$ with the spectral weight of the $A\bar{A}$ bubble diagram as calculated in lowest order of perturbation theory, i.e.

$$f^2(s) = \frac{1}{16\pi^2}\,\theta_+(s - 4M_A{}^2)\left(\frac{s - 4M_A{}^2}{s}\right)^{1/2} \tag{22}$$

where $\theta_+(x) \equiv \frac{1}{2}\left(1 + \frac{x}{|x|}\right)$ is the unit step function. Though not all the integrals which appear in the solution will exist for the $f(s)$ of (22), one can readily modify the high energy behaviour of this function so as to avoid such convergence problems.

The equations of motion implied by (21)

$$(-\partial^2 + \mu_0{}^2)\phi(x) = g_0 \int_{4M_A{}^2}^{\infty} \mathrm{d}s\, f(s)\phi(x, s)$$

$$(-\partial^2 + s)\phi(x, s) = g_0 f(s)\phi(x)$$

together with the only non-vanishing equal-time commutation relations

$$[\partial_0\phi(x), \phi(x')] = -\mathrm{i}\,\delta(\mathbf{x} - \mathbf{x}')$$

$$[\partial_0\phi(x, s), \phi(x', s')] = -\mathrm{i}\,\delta(\mathbf{x} - \mathbf{x}')\delta(s - s')$$

yield a coupled set of integral equations for the Green's functions

$$\mathcal{G}(x) = \mathrm{i}\langle 0|(\phi(x)\phi(0))_+|0\rangle$$

$$\mathcal{G}(x; s, s') = \mathrm{i}\langle 0|(\phi(x, s)\phi(0, s'))_+|0\rangle$$

These integral equations can be solved by standard elementary techniques to yield

$$\mathcal{G}(p) = \left[p^2 + \mu_0{}^2 - g_0{}^2 \int_{4M_A{}^2}^{\infty} \mathrm{d}s\, \frac{f^2(s)}{p^2 + s}\right]^{-1} \tag{23}$$

$$\mathcal{G}(p; s, s') = \frac{\delta(s - s')}{p^2 + s} + g_0{}^2 \frac{f(s)}{p^2 + s} \mathcal{G}(p) \frac{f(s')}{p^2 + s'} \tag{24}$$

Since we have already remarked that one can make $f(s)$ fall off arbitrarily rapidly at large s without altering any of the essential features of the model, the general arguments of this section apply without modification and one can fully anticipate that in the $Z = 0$ limit the terms

$$-\tfrac{1}{2}\mu_0{}^2\phi^2 + g_0\phi \int_{4M_A{}^2}^{\infty} \mathrm{d}s\, f(s)\phi(s)$$

will become equivalent to

$$\frac{\lambda}{2}\left[\int_{4M_A{}^2}^{\infty} \mathrm{d}s\, f(s)\phi(x, s)\right]^2 \tag{25}$$

This can be verified directly by calculating the Green's function of the field $\phi(x, s)$ with the coupling (25). One trivially finds in this case

$$\mathscr{G}_\lambda(p; s, s') = \frac{\delta(s - s')}{p^2 + s} + \frac{f(s)}{p^2 + s} \frac{\lambda}{1 - \lambda \int_{4M_A^2}^{\infty} ds'' \frac{f^2(s'')}{p^2 + s''}} \frac{f(s')}{p^2 + s'}$$

which form is identical to that of (24) if one can make the identification

$$g_0^{-2}(p^2 + \mu_0^2) = \lambda^{-1}$$

In the $Z_3 = 0$ limit the kinematical term can be neglected and equivalence is established with

$$\lambda = g_0^2/\mu_0^2$$

in accord with our general arguments.

One can generalize the above discussion to the spin one-half case in a fairly straightforward way. Since there is, however, at least one significant point of contrast with the scalar field situation, it will be worthwhile to enter into a brief discussion of this extension of the above arguments. To this end we consider a Lagrangian of the form

$$L = \frac{i}{2} \psi\beta\gamma^\mu\partial_\mu\psi - \frac{m_0}{2} \psi\beta\psi + g_0\psi\beta\eta + L' \tag{26}$$

where for simplicity we take $\psi(x)$ to be hermitian and as before $\eta(x)$ and $L'(x)$ do not explicitly contain $\psi(x)$. In particular, for a Yukawa type coupling, $\eta(x)$ could have the form $\psi'(x)\phi(x)$ where $\psi'(x)$ and $\phi(x)$ are respectively spin one-half and spin zero fields.

The two-point function

$$G(x) = i\varepsilon(x)\langle 0|(\psi(x)\psi(0)\beta)_+|0\rangle$$

in this case has the Lehmann representation

$$G(p) = \int_{-\infty}^{\infty} d\kappa \frac{A(\kappa)}{\gamma p + \kappa}$$

where in general $A(\kappa)$ will have the form

$$A(\kappa) = Z_2\delta(\kappa - m) + \theta_+(\kappa^2 - M^2)A'(\kappa)$$

corresponding to the existence of a stable particle of mass m with a continuum threshold at $\kappa^2 = M^2 > m^2$. The equation of motion implied by (26), together with the equal-time commutation relation

$$[\psi(x), \psi(x')] = \delta(\mathbf{x} - \mathbf{x}')$$

yields the sum rules

$$1 = Z_2 + \int_{-\infty}^{\infty} A'(\kappa)\theta_+(\kappa^2 - M^2)\, d\kappa$$

$$m_0 = Z_2 m + \int_{-\infty}^{\infty} \kappa A'(\kappa)\theta_+(\kappa^2 - M^2)\, d\kappa$$

If one assumes the existence of these integrals, the inverse of $G(p)$ has the unsubtracted representation

$$G^{-1}(p) = \gamma p + m_0 - \int_M^{\infty} d\kappa \left[\frac{\rho_+(\kappa)}{\gamma p + \kappa} + \frac{\rho_-(\kappa)}{\gamma p - \kappa}\right]$$

Although the positive definite metric of the Hilbert space ensures that both $A(\kappa)$ and $\rho(\kappa)$ will be non-negative functions of κ, the fact that the equation of motion is a first order differential equation precludes the possibility of an immediate generalization of the results obtained for spin zero.

The condition for a bound state of mass m is

$$m_0 = m + \frac{1}{Z_2}\int_M^{\infty} d\kappa \left[\frac{\bar\rho_+(\kappa)}{\kappa - m} - \frac{\bar\rho_-(\kappa)}{\kappa + m}\right] \tag{27}$$

where we have introduced the renormalized spectral function $\bar\rho(\kappa) = Z_2\rho(\kappa)$ and

$$Z_2 = 1 - \int_M^{\infty} d\kappa \left[\frac{\bar\rho_+(\kappa)}{(\kappa - m)^2} + \frac{\bar\rho_-(\kappa)}{(\kappa + m)^2}\right]$$

For the case of Z_2 going to zero it is clear that m_0 will diverge unless the integral appearing in (27) vanishes in that limit. Barring the existence of an additional invariance property which could lead to such a result, $m_0 Z_2$ can be expected to approach a finite value. The consequent degeneration of the equation of motion leads to an effective interaction term in the Lagrangian of the form

$$\frac{g_0^2}{2m_0}\,\eta(x)\beta\eta(x) = \tfrac{1}{2}g^2 \left\{\int_M^{\infty} d\kappa \left[\frac{\bar\rho_+(\kappa)}{\kappa - m} - \frac{\bar\rho_-(\kappa)}{\kappa + m}\right]\right\}^{-1} \eta(x)\beta\eta(x)$$

where we have defined $g^2 \equiv Z_2 g_0^2$.

As in the spin zero case there exists a simple illustration of these remarks in terms of a field theory of the Zachariasen type. In this example the Lagrangian containing an elementary particle is

$$L = \frac{i}{2}\,\psi\beta\gamma^\mu\partial_\mu\psi - \frac{m_0}{2}\,\psi\beta\psi + \int_{-\infty}^{\infty} d\kappa\, \theta_+(\kappa^2 - (M+\mu)^2)$$

$$\tfrac{1}{2}[i\psi(\kappa)\beta\gamma^\mu\partial_\mu\psi(\kappa) - \kappa\psi(\kappa)\beta\psi(\kappa)] + g_0\psi\beta \int_{-\infty}^{\infty} d\kappa\, f(\kappa)\psi(\kappa) \tag{28}$$

where in analogy to the spin zero case we consider the field $\psi(x)$ to describe an elementary particle of bare mass m_0 coupled to a field $\psi(x, \kappa)$ with a continuous mass parameter κ. The model is equivalent to summing the bubble diagrams containing a fermion of mass M and a boson of mass μ provided one makes the identification

$$-\mathrm{i}\int \frac{\mathrm{d}\,k}{(2\pi)^4}\,\frac{1}{k^2+\mu^2}\,\frac{1}{\gamma(p-k)+M} = \int_{-\infty}^{\infty} \mathrm{d}\kappa\,\frac{f^2(\kappa)}{\gamma p+\kappa}$$

i.e.

$$f^2(\kappa) = \frac{1}{16\pi^2}\,\theta_+(\kappa^2-(M+\mu)^2)\,\frac{(\kappa+M)^2-\mu^2}{2\kappa^2}\,q(\kappa)$$

where

$$q^2(\kappa) = \frac{[\kappa^2-(M+\mu)^2][\kappa^2-(M-\mu)^2]}{4\kappa^2}$$

The Lagrangian (28) implies the equation of motion

$$\left(\gamma^{\mu}\frac{1}{\mathrm{i}}\,\partial_{\mu}+m_0\right)\psi(x) = g_0\int_{-\infty}^{\infty} f(\kappa)\psi(x,\kappa)\,\mathrm{d}\kappa$$

$$\left(\gamma^{\mu}\frac{1}{\mathrm{i}}\,\partial_{\mu}+\kappa\right)\psi(x,\kappa) = g_0 f(\kappa)\psi(x)$$

and the equal-time commutation relations

$$[\psi(x), \psi(x')] = \delta(\mathbf{x}-\mathbf{x}')$$
$$[\psi(x,\kappa), \psi(x',\kappa')] = \delta(\mathbf{x}-\mathbf{x}')\delta(\kappa-\kappa')$$
$$[\psi(x), \psi(x,\kappa)] = 0$$

The equations for the Green's functions

$$G(x) = \mathrm{i}\;\varepsilon(x)\langle 0|(\psi(x)\psi(0)\beta)_+|0\rangle$$
$$G(x;\kappa,\kappa') = \mathrm{i}\;\varepsilon(x)\langle 0|(\psi(x,\kappa)\psi(0,\kappa')\beta)_+|0\rangle$$

are readily solved to yield

$$G(p) = \left[\gamma p+m_0-g_0{}^2\int_{-\infty}^{\infty}\mathrm{d}\kappa\,\frac{f^2(\kappa)}{\gamma p+\kappa}\right]^{-1} \tag{29}$$

$$G(p;\kappa,\kappa') = \frac{\delta(\kappa-\kappa')}{\gamma p+\kappa}+g_0{}^2\,\frac{f(\kappa)}{\gamma p+\kappa}\,G(p)\,\frac{f(\kappa')}{\gamma p+\kappa'} \tag{30}$$

As anticipated, the condition for a particle of mass m

$$m_0 = m + \frac{g^2}{Z_2}\int_{-\infty}^{\infty} \mathrm{d}\kappa \, \frac{f^2(\kappa)}{\kappa - m}$$

does not necessarily require the divergence of the bare mass in the limit in which Z_2 goes to zero. On the other hand it is clear that this will indeed occur except for the case where $m_0 = 0$ and $f^2(\kappa) = f^2(-\kappa)$ or where the deviations from this full symmetry somehow cancel out. We are thus led to expect equivalence, except in these special cases, to the direct interaction theory with the coupling term

$$\tfrac{1}{2}\lambda \int_{-\infty}^{\infty} f(\kappa)\, \mathrm{d}\kappa\, \psi(\kappa)\beta \int_{-\infty}^{\infty} f(\kappa)\psi(\kappa)\, \mathrm{d}\kappa \tag{31}$$

where

$$\lambda = \lim_{Z_2 \to 0} g_0{}^2/m_0$$

The Green's function appropriate to the interaction (31) is readily found to be

$$G_\lambda(p;\ \kappa, \kappa') = \frac{\delta(\kappa - \kappa')}{\gamma p + \kappa} + \frac{f(\kappa)}{\gamma p + \kappa} \, \frac{1}{\lambda^{-1} - \displaystyle\int_{-\infty}^{\infty} \mathrm{d}\kappa'' \, \frac{f^2(\kappa'')}{\gamma p + \kappa''}} \, \frac{f(\kappa')}{\gamma p + \kappa'}$$

which is clearly identical to (30) for the case in which m_0 diverges as $Z_2{}^{-1}$.

4 The Multi-particle Case

To this point it has been convenient to simplify our discussion as much as possible by considering a situation in which there is associated only a single stable particle with a given channel. It is therefore appropriate to briefly consider here the possible relevance of this idealization to the conclusions obtained in the preceding section. There are, of course, two ways in which we can increase the number of particles in a given channel. One possibility is to keep fixed the number of elementary fields and merely change the coupling constants associated with the system in such a way as to induce an additional $N - 1$ bound states. This, of course, implies the existence of N pole terms in the Lehmann representation as well as the existence of at least $N - 1$ poles in the inverse propagator. A second approach is to increase the number of stable particles merely by increasing the number of elementary fields by $N - 1$. These two possibilities are thus distinguished by the fact that as the couplings are 'turned off' they reduce respectively to a single free field and a set of N free fields. The former situation we shall not elaborate upon as one can readily convince oneself that the results of the preceding section can be immediately generalized to this case merely by looking at the eigenvalue equation for the lowest lying state which is implied by the representation of the inverse propagator. The latter possibility is somewhat more interesting as it is associated with a point which has recently given rise to some conflicting statements concerning the $Z = 0$ rule in the case of several stable particles.

We shall try to avoid inessential complications by considering the case $N = 2$ so that one is dealing with a situation in which there exist two stable particles and two spin zero fields $\phi_1(x)$ and $\phi_2(x)$. The Green's function

$$\mathscr{G}_{ij}(x) = \mathrm{i}\langle 0|(\phi_i(x)\phi_j(0))]_+|0\rangle$$

is thus a 2×2 matrix in the 2-dimensional internal space associated with the doubling of the number of degrees of freedom of the system. (Such a statement should not, of course, be construed to imply the existence of any symmetry, either 'broken' or 'unbroken', within this internal space.) One clearly has the representation

$$\mathscr{G}_{ij}(p) = \sum_{\alpha=1}^{2} \frac{Z_{ij}(\alpha)}{p^2 + \mu_\alpha{}^2} + \int_{s_{ij}}^{\infty} \frac{B_{ij}(\kappa)\,\mathrm{d}\kappa^2}{p^2 + \kappa^2 - \mathrm{i}\varepsilon} \tag{32}$$

where the masses μ_1 and μ_2 refer to the two discrete particle states in the theory. In writing (32) we have particularly avoided making an identification of $\phi_1(x)$ with μ_1 and $\phi_2(x)$ with μ_2 (or vice versa). This point is in fact the source of much of the confusion which has been associated with the multi-particle case.

Since the wave-function renormalization constants $Z_{ij}(\alpha)$ are related to the matrix elements of $\phi_i(x)$ between the vacuum and single-particle states, i.e.

$$Z_{ij}(\alpha) \sim \langle 0|\phi_i(0)|p^2 = -\mu_\alpha{}^2\rangle\langle p^2 = -\mu_\alpha{}^2|\phi_j(0)|0\rangle$$

one immediately obtains the result that the matrix $Z(\alpha)$ has vanishing determinant and (by appropriate choice of an arbitrary phase factor in the definition of the physical states) can be taken to be symmetric, i.e.

$$Z_{12}(\alpha) = Z_{21}(\alpha) \quad \det Z(\alpha) = 0 \quad \alpha = 1, 2$$

Of the 8 renormalization constants $Z_{ij}(\alpha)$ we thus see that only 4 are independent. By imposing the equal-time commutation relations

$$[\partial_0\phi_i(x), \phi_j(x')] = -\mathrm{i}\delta(\mathbf{x} - \mathbf{x}')\delta_{ij} \tag{33}$$

one has the additional constraints

$$\sum_{\alpha=1}^{2} Z_{ij}(\alpha) + \int_{s_{ij}}^{\infty} B_{ij}(\kappa)\, \mathrm{d}\kappa^2 = \delta_{ij}$$

We now observe that one has here considerably more freedom in choosing the fundamental fields $\phi_i(x)$ than in the case of a single elementary field. Since $\phi_1(x)$ and $\phi_2(x)$ have the same quantum numbers it is clear that a linear combination of $\phi_1(x)$ and $\phi_2(x)$ can equally well serve in place of $\phi_1(x)$ or $\phi_2(x)$. Thus one has available many possible ways in which to further specify a complete set of fields, each of which leads to different numerical values for the $Z_{ij}(\alpha)$. In particular one could impose the conditions[9]

$$Z_{11}(2) = Z_{22}(1) = 0 \tag{34}$$

thereby requiring that the functions $\mathscr{G}_{11}(p)$ and $\mathscr{G}_{22}(p)$ only have poles associated with μ_1, and μ_2 respectively. This ambiguity, which is associated with the invariance of (33) under the orthogonal transformation

$$\phi_i'(x) = \sum A_{ij}\phi_j(x)$$

where

$$\sum A_{ik}A_{jk} = \delta_{ij}$$

we shall resolve[10] by making explicit reference to the field equations of the theory. The equations of motion can be taken to be of the form

$$(-\partial^2 + \mu_{0i}{}^2)\phi_i(x) = j_i(x)$$

where $j_i(x)$ has no terms containing the fields $\phi_i(x)$. This is quite clearly equivalent to the assumption that the Lagrangian has the structure

$$L = \sum_{i=1}^{2}\left[\phi_i{}^{\mu}\partial_{\mu}\phi_i - \frac{\mu_{0i}{}^2}{2}\phi_i{}^2 + \tfrac{1}{2}\phi_i{}^{\mu}\phi_{\mu}{}^{i} + \phi_i j_i\right]$$

the essential point being the specification of our particular choice of canonical fields $\phi_i(x)$ as that which brings the most general mass term

$$\sum_{i,j}\mu_{0ij}{}^2\phi_i\phi_j$$

to the diagonal form

$$\sum_{i}\mu_{0i}{}^2\phi_i{}^2$$

On the basis of these remarks one can now generalize the approach of the preceding section. Since the diagonalization of the mass matrix reduces each equation of motion to the form

$$(-\partial^2 + \mu_{0i}^2)\phi_i(x) = j_i(x)$$

with $j_i(x)$ independent of $\phi_i(x)$, the conditions required for the application of the previous results to the multi-particle case are quite clearly fulfilled. It thus follows that a given field becomes composite when the corresponding wave-function renormalization constant tends to zero. It should be noted that since we shall assume that none of the particles decouple in that limit, $Z_{ii}(\alpha)/Z_{ii}(\beta)$ must remain finite so that the vanishing of $Z_{ii}(\alpha)$ for any α implies $Z_{ii}(\alpha) = 0$ for all α. Thus all the wave-function renormalization constants $Z_{11}(\alpha)$ associated with the field $\phi_1(x)$ tend to zero when $\phi_1(x)$ becomes composite while at the same time no further conditions are imposed on the set $Z_{22}(\alpha)$.

Although the multi-particle case as treated within the framework of the preceding section thus requires no essential modification of our approach to the $Z = 0$ condition, it can readily be shown that the use of a set of canonical fields such as that described by (34) requires that all the renormalization constants vanish simultaneously in the limit in which one of the basic fields becomes composite. To demonstrate this result we transform to a new set of fields $\phi_i'(x)$ for which (34) holds. One thus writes

$$\phi_1' = a\phi_1 + b\phi_2$$

$$\phi_2' = c\phi_1 + d\phi_2$$

where the coefficients of the transformation are real and normalized by the condition

$$a^2 + b^2 = c^2 + d^2 = 1$$

The single-pole structure of the Green's function of the field $\phi_1'(x)$ has the form

$$\mathscr{G}_{11}' \sim \frac{a^2 Z_{11}(1) + b^2 Z_{22}(1) + 2ab[Z_{11}(1)Z_{22}(1)]^{1/2}}{p^2 + \mu_1^2} + \frac{a^2 Z_{11}(2) + b^2 Z_{22}(2) + 2ab[Z_{11}(2)Z_{22}(2)]^{1/2}}{p^2 + \mu_2^2}$$

On taking

$$b = -a\left[\frac{Z_{11}(2)}{Z_{22}(2)}\right]^{1/2}$$

$$c = -d\left[\frac{Z_{22}(1)}{Z_{11}(1)}\right]^{1/2}$$

one readily obtains

$$\mathscr{G}_{ii}' \sim \frac{Z_i'}{p^2 + \mu_2{}^2}$$

where

$$Z_1' = Z_{11}(1)\xi^2 \left[1 + \frac{Z_{11}(2)}{Z_{22}(2)}\right]^{-1}$$

$$Z_2' = Z_{22}(2)\xi^2 \left[1 + \frac{Z_{22}(1)}{Z_{11}(1)}\right]^{-1}$$

ξ being given by

$$\xi = 1 - \left[\frac{Z_{11}(2)Z_{22}(1)}{Z_{11}(1)Z_{22}(2)}\right]^{1/2}$$

which is assumed to be non-zero. Since the $Z_{ii}(\alpha)$ are positive definite it now clearly follows that the ratio

$$\frac{Z_1'}{Z_2'} = \frac{Z_{11}(1) + Z_{22}(1)}{Z_{22}(2) + Z_{11}(2)}$$

remains finite for $Z_{22}(2) \to 0$ provided only that $Z_{11}(2)$ does not vanish. Since this latter situation corresponds to the relatively uninteresting case in which the two particles decouple, it follows that Z_1' and Z_2' are both zero or both finite as long as the two particles have a non-trivial interaction. Although this has elsewhere[9] been interpreted to mean that the compositeness criteria break down in the multi-particle case, it is clear from our discussion that the above result can be readily understood in terms of the general approach which has been presented here. We have thus seen that no essential modification of the $Z = 0$ rule should be required in the multi-particle case, provided one starts from a field theoretical formulation which allows one to take full advantage of the ambiguity associated with the choice of a complete set of canonical fields appropriate to a theory containing more than one elementary particle.

5 Composite Virtual States

In order to discuss more precisely the circumstances under which the $Z = 0$ rule can be expected to apply we now seek to relax the condition that there be a stable particle associated with the field under consideration. Since the discussion of the preceding sections has tended to associate the concept of compositeness more with the field operator than with the particle, one might well anticipate that it is possible to drop the assumption that the field $\phi(x)$ excites a stable particle. In fact, the essential requirement appears to be only the existence of a pole on some reasonably accessible sheet of the Riemann surface associated with the analytic structure of the 2-point function. Since the definition of Z_3 as the residue of the Green's function at the single-particle pole can be readily generalized merely by identifying it with the residue of a pole on an unphysical sheet, one has only to effect the analytic continuation to such an unphysical sheet in order to examine the $Z_3 = 0$ limit.

In order to carry out this analytic continuation we shall assume that the continuum contribution to the spectral weight of the Lehmann representation

$$\mathscr{G}(-p^2) = \int_{s_0}^{\infty} \frac{B(s)\,\mathrm{d}s}{p^2 + s}$$

(which we have written in a form intended to emphasize the absence of a single-particle state) refers to a 2-particle threshold, i.e. $s_0 = (m_a + m_b)^2$ where m_a and m_b are the masses of the least massive particles to which $\phi(x)$ is coupled. Then the contribution of the (a, b) intermediate state to $B(s)$ (which we call $B_{\mathrm{el}}(s)$) is of the form

$$B_{\mathrm{el}}(s) = \frac{1}{2\pi \mathrm{i}} [\mathscr{G}(s + \mathrm{i}\,\varepsilon) - \mathscr{G}(s - \mathrm{i}\,\varepsilon)]$$

$$= Z_2{}^2 \frac{1}{8\pi^2} g_0{}^2 \rho(s) |\Gamma(s)\mathscr{G}(s)|^2 \tag{35}$$

where

$$\rho(s) = \frac{1}{2s} \{[s - (m_a + m_b)^2][s - (m_a - m_b)^2]\}^{1/2}$$

and $\Gamma(s)$ is the unrenormalized vertex function. The quantity g_0 is the bare coupling constant describing the interaction of $\phi(x)$ with the fields associated with a and b and $Z_2{}^2$ is the product of the wave-function renormalization constants of these two fields. Making the usual assumption concerning the analyticity of $\Gamma(s)$ in the cut plane, one can deduce the contribution of the 2-particle state to the discontinuity of $\Lambda(s) \equiv \Gamma(s)\mathscr{G}(s)$ in the interval $s_0 < s < s_1$ (s_1 being the next highest threshold)

$$\Lambda(s + \mathrm{i}\,\varepsilon) - \Lambda(s - \mathrm{i}\,\varepsilon) = 2\mathrm{i}\,\rho(s)\Lambda(s)f^*(s) \tag{36}$$

where $f(s)$ is the usual S-wave amplitude for (a, b) scattering

$$f(s) = \frac{1}{\rho}\, \mathrm{e}^{\mathrm{i}\delta} \sin \delta$$

Since $\mathscr{G}(s)$, $\Lambda(s)$ and $f(s)$ are all real analytic functions of s, one can use (35) and (36) and the unitarity relation

$$f(s + \mathrm{i}\,\varepsilon) - f(s - \mathrm{i}\,\varepsilon) = 2\mathrm{i}\,\rho(s) f(s + \mathrm{i}\,\varepsilon) f(s - \mathrm{i}\,\varepsilon)$$

to rewrite the expressions for the discontinuities of $\mathscr{G}(s)$, $\Lambda(s)$ and $f(s)$ in the more suggestive form

$$\mathscr{G}_{\mathrm{I}}(s) - \mathscr{G}_{\mathrm{II}}(s) = 2\mathrm{i}\,\rho(s) g_0{}^2 \frac{1}{8\pi^2} Z_2{}^2 \Lambda_{\mathrm{I}}(s) \Lambda_{\mathrm{II}}(s)$$

$$\Lambda_{\mathrm{I}}(s) - \Lambda_{\mathrm{II}}(s) = 2\mathrm{i}\,\rho(s) \Lambda_{\mathrm{I}}(s) f_{\mathrm{II}}(s)$$

$$f_{\mathrm{I}}(s) - f_{\mathrm{II}}(s) = 2\mathrm{i}\,\rho(s) f_{\mathrm{I}}(s) f_{\mathrm{II}}(s)$$

where we have introduced a subscript notation to refer to functions defined on the first and second sheets (the second sheet being that obtained by continuation through the 2-particle unitarity cut). It now requires only simple algebraic manipulation to deduce the results[11]

$$\mathscr{G}_{\mathrm{II}}(s) = \mathscr{G}_{\mathrm{I}}(s) \tilde{S} S^{-1} \tag{37}$$

$$\Lambda_{\mathrm{II}}(s) = \Lambda_{\mathrm{I}}(s) S^{-1} \tag{38}$$

$$f_{\mathrm{II}}(s) = f_{\mathrm{I}}(s) S^{-1} \tag{39}$$

where

$$S = 1 + 2\mathrm{i}\,\rho(s) f_{\mathrm{I}}(s)$$

$$\tilde{S} = 1 + 2\mathrm{i}\,\rho(s) \tilde{f}_{\mathrm{I}}(s)$$

and

$$\tilde{f}_{\mathrm{I}}(s) = f_{\mathrm{I}}(s) - Z_2{}^2 g_0{}^2 \frac{1}{8\pi} \Gamma_{\mathrm{I}}(s) \mathscr{G}_{\mathrm{I}}(s) \Gamma_{\mathrm{I}}(s)$$

It is to be noted that (37) and (38) imply

$$\Gamma_{\mathrm{II}}(s) = \Gamma_{\mathrm{I}}(s) \tilde{S}^{-1} \tag{40}$$

which together with (39) yield

$$\tilde{f}_{\mathrm{II}}(s) = \tilde{f}_{\mathrm{I}}(s) \tilde{S}^{-1} \tag{41}$$

thereby demonstrating that $\tilde{f}(s)$ is also a unitary amplitude, i.e.

$$\tilde{f}(s) - \tilde{f}^*(s) = 2\mathrm{i}\,\rho(s) \tilde{f}(s) \tilde{f}(s)^*$$

Although equations (37) to (41) thus far refer only to the elastic interval $s_0 < s < s_1$, because of the fact that only analytic functions occur therein they are valid in the entire cut place and represent the desired analytic continuation of the functions $\mathcal{G}(s)$, $\Gamma(s)$, $f(s)$ and $\tilde{f}(s)$ into the second sheet.

There is one further function which is essential to our discussion, namely the inverse of the propagator, the analytic continuation of which follows trivially from the preceding results. As before, we write an unsubtracted dispersion relation for $\mathcal{G}^{-1}(-p^2)$ in the form

$$\mathcal{G}^{-1}(-p^2) = p^2 + \mu_0{}^2 - \int_{s_0}^{\infty} \frac{\sigma(s)\,\mathrm{d}s}{p^2 + s} \tag{42}$$

where

$$\sigma(s) = B(s)|\mathcal{G}(s)|^{-2}$$

and we have made use of the fact that in the case in which there is no stable particle $\mathcal{G}^{-1}(-p^2)$ has no poles. In the elastic interval one has from (35)

$$\sigma_{\mathrm{el}}(s) = Z_2{}^2 \frac{1}{8\pi^2} g_0{}^2 \rho(s) |\Gamma(s)|^2$$

which enables one to express the discontinuity of the inverse propagator entirely in terms of analytic functions

$$\mathcal{G}_{\mathrm{II}}{}^{-1} - \mathcal{G}_{\mathrm{I}}{}^{-1} = 2\mathrm{i}\, Z_2{}^2 \frac{1}{8\pi} g_0{}^2 \rho(s) \Gamma_{\mathrm{I}}{}^2(s) \tilde{S}^{-1}$$

This result is thus a prescription for the analytic continuation of $\mathcal{G}^{-1}(-p^2)$ into the second sheet.

It should be remarked here that although $f(s)$ and $\tilde{f}(s)$ may have rather complicated analytic properties on the physical sheet (i.e. there may exist cuts in the complex plane other than the usual left- and right-hand cuts), it is sufficient for our purposes merely to effect the continuation through the lowest portion of the unitarity cut. We shall in fact require only that the functions as defined here on the second sheet be valid in a non-vanishing region including the elastic threshold.

Since we have not yet defined what we shall call a virtual state, it is well to briefly discuss here the significance of virtual states in ordinary potential theory. It is well known that the bound states in non-relativistic quantum mechanics are associated with the zeros of the Jost function in the upper half k-plane, where k is the usual complex wave number. These zeros which occur on the positive imaginary axis are mapped onto the negative real axis on the first sheet of the complex energy surface and can, of course, disappear from the discrete spectrum by moving to the lower half k plane. Now because of the reality property

$$f(\mathrm{i}\,k) = f^*(-\mathrm{i}\,k^*)$$

it follows that the zeros of $f(\mathrm{i}\,k)$ are symmetrically placed with respect to the imaginary k-axis. Thus a variation of the parameters of a given potential can cause a bound state to move off the negative imaginary k-axis only when there is a confluence of two or more such zeros. For P and higher partial waves it is well known that such confluence invariably occurs at the origin with the consequent result that a bound state passing through threshold immediately appears as a resonance, i.e. a positive energy 'state' on the second sheet of the associated energy surface. In the case of S-waves, however, the bound state zero must generally move a finite distance into the lower half k-plane before encountering another zero and subsequently moving off the imaginary k-axis. In the interval between the passage of this zero through threshold and its departure from the imaginary k-axis one customarily refers to this zero as describing a virtual state. It thus appears as a negative energy 'state' much as a bound state with however, the important distinction that it lies on an unphysical sheet.

In order to arrive at the corresponding field theoretical description, we have recourse to the useful representation (42) of the inverse propagator. In the usual case in which there occurs a single-particle state of mass μ once one has

$$\mu_0{}^2 = \mu^2 + \int_{s_0}^{\infty} \frac{\sigma(s)\,\mathrm{d}s}{s - \mu^2}$$

from the vanishing of $\mathscr{G}^{-1}(-p^2)$ at $p^2 = -\mu^2$. This leads to the familiar once subtracted form

$$\mathscr{G}^{-1}(-p^2) = (p^2 + \mu^2)\left[1 + \int_{s_0}^{\infty} \frac{\sigma(s)\,\mathrm{d}s}{(s - \mu^2)(p^2 + s)}\right]$$

and the definition of Z_3

$$Z_3{}^{-1} = 1 + \int_{s_0}^{\infty} \frac{\sigma(s)\,\mathrm{d}s}{(s - \mu^2)^2} \tag{43}$$

as the residue of $\mathscr{G}(-p^2)$ at the single-particle pole. Upon defining the renormalized spectral function

$$\bar{\sigma}(s) = \frac{1}{Z_3}\,\sigma(s)$$

(43) becomes

$$Z_3 = 1 - \int_{s_0}^{\infty} \frac{\bar{\sigma}(s)\,\mathrm{d}s}{(s - \mu^2)^2}$$

It is worth noting that in the stable-particle case one can write

$$\mathcal{G}_{\mathrm{II}}^{-1}(-p^2) = p^2 + \mu_0^2 - \int_{s_0}^{\infty} \frac{\sigma(s)\,\mathrm{d}s}{p^2 + s} + \\ + 2\mathrm{i}\, Z_2^2 \frac{1}{8\pi} g_0^2 \rho(s) \Gamma_{\mathrm{I}}^2(-p^2) \tilde{S}^{-1}(-p^2)$$

(where for simplicity we assume no first sheet zero of $\mathcal{G}$) to argue the existence of a 'partner' zero on the second sheet*. Since $\mathcal{G}_{\mathrm{II}}^{-1}$ is negative at threshold and is asymptotically dominated by the p^2 term for large p^2, it follows that $\mathcal{G}_{\mathrm{II}}^{-1}$ must have a zero on the real axis of the second sheet. Although this conclusion is largely independent of the properties of $\tilde{S}(s)$, it is clear that additional zeros of $\mathcal{G}^{-1}$ can occur on the second sheet if, for example, $\tilde{S}(s)$ has one or more zeros in the region below the elastic threshold.

Let us now consider the development of a virtual state from the more familiar stable-particle case. In particular we assume a situation in which there exists a single pole in $\mathcal{G}_{\mathrm{I}}(-p^2)$ with possible additional poles and zeros in $\mathcal{G}_{\mathrm{II}}(-p^2)$. As the parameters of the theory are varied, the stable-particle pole can pass through s_0 and emerge on the second sheet. We may continue to use this pole as a convenient subtraction point, obtaining from

$$\mu_0^2 = \mu^2 + \int_{s_0}^{\infty} \frac{\sigma(s)\,\mathrm{d}s}{s - \mu^2} - 2\mathrm{i}\, Z_2^2 \frac{1}{8\pi} g_0^2 \rho(\mu^2) \Gamma_{\mathrm{I}}^2(\mu^2) \tilde{S}^{-1}(\mu^2)$$

the representation

$$\mathcal{G}_{\mathrm{II}}^{-1} = (p^2 + \mu^2) \left[1 + \int_{s_0}^{\infty} \frac{\sigma(s)\,\mathrm{d}s}{(s - \mu^2)(p^2 + s)} \right] + \\ + 2\mathrm{i}\, Z_2^2 \frac{1}{8\pi} g_0^2 \left[\rho(s) \Gamma_{\mathrm{I}}^2(s) \tilde{S}^{-1}(s) \right]_{\mu^2}^{-p^2}$$

$$\mathcal{G}_{\mathrm{I}}^{-1} = (p^2 + \mu^2) \left[1 + \int_{s_0}^{\infty} \frac{\sigma(s)\,\mathrm{d}s}{(s - \mu^2)(p^2 + s)} \right] - \\ - 2\mathrm{i}\, Z_2^2 \frac{1}{8\pi} g_0^2 \rho(\mu^2) \Gamma_{\mathrm{I}}^2(\mu^2) \tilde{S}^{-1}(\mu^2)$$

where we have introduced the notation

$$\left[f(x) \right]_a^b = f(b) - f(a)$$

* It must be remarked that in the subsequent argument we neglect the effect of cuts along the real axis other than the usual right-hand cut. More specifically we assume that these asymptotic considerations can be applied in the interval between the left- and right-hand cuts. The alternative possibility is that one is dealing with a situation in which the second sheet zero has migrated through the left-hand cut and is not immediately accessible within the present context.

It should be noted that (43), together with the usual expression for $\sigma_{el}(s)$, implies the vanishing of Z_3 as the pole passes through threshold. Furthermore, the term $2i\rho\Gamma_I^2\tilde{S}^{-1}$ is negative just below s_0 ($\tilde{S}$ is clearly required by (40) to be negative at threshold) so that in the absence of a stable particle the two conditions

$$\mathscr{G}(s_0) > 0$$

$$\frac{d}{ds}\mathscr{G}(s)|_{s=s_0} > 0$$

essential for the occurrence of a zero of $\mathscr{G}_{II}^{-1}$ near threshold are quite clearly satisfied.

Before going on to discuss the wave-function renormalization, we briefly comment on the possible effect of isolated zeros of $\tilde{S}(s)$ on the structure of $\mathscr{G}_{II}^{-1}$. Because of the absence of a single-particle state both S and $\tilde{S}$ will have no poles on the physical sheet. This means that $\tilde{S}$ changes sign only by passing through zero with each such zero generally introducing one more second sheet pole of $\mathscr{G}$. These zeros of $\tilde{S}$ as well as the poles which they induce in $\mathscr{G}_{II}$ can subsequently be made to coalesce and move off the real axis as complex conjugate points in the p^2 plane by suitable variation of coupling constants. Ultimately, all such zeros of $\tilde{S}$ can presumably be caused to move off the real axis in this way and one then need consider only the case of two zeros of $\mathscr{G}_{II}^{-1}$.

The wave-function renormalization constant is conveniently defined to be the residue of $\mathscr{G}_{II}$ at the pole, thereby leading to the expression

$$[Z_3^{II}]^{-1} = 1 + \int_{s_0}^{\infty} \frac{\sigma(s)\,ds}{(s-\mu^2)^2} - 2i\,Z_2^2\,\frac{1}{8\pi}\,g_0^2\left[\frac{d}{ds}(\rho\Gamma_I^2\tilde{S}^{-1}\right]_{s=\mu^2} \tag{44}$$

where we have introduced the notation Z_3^{II} for the wave-function renormalization of the virtual state. The crucial point here is the sign of $(Z_3^{II})^{-1}$ as the pole passes through threshold and emerges onto the second sheet. Since this can be readily seen from (44) to diverge as $(s_0-\mu)^{-\frac{1}{2}}$, its sign is best determined by writing a dispension relation for $\mathscr{G}_{II}^{-1}$ keeping only σ_{el} and the discontinuity associated with the right-hand cut. One finds for $\mu^2 \sim s_0$

$$\mathscr{G}_{II}^{-1} \sim (p^2+\mu^2)\left[1 - \frac{Z_2^2 g_0^2}{8\pi^2}\int_{s_0}^{\infty}\frac{\rho|\Gamma|^2\,ds}{(p^2+s)(s-\mu^2)}\right]$$

so that

$$[Z_3^{II}]^{-1} \sim 1 - \frac{Z_2^2 g_0^2}{8\pi^2}\int_{s_0}^{\infty}\frac{\rho|\Gamma|^2\,ds}{(s-\mu^2)^2}$$

clearly implying that Z_3^{II} is negative for μ^2 near s_0. However, it follows more generally from the definition

$$Z_3^{-1} = -\frac{\mathrm{d}}{\mathrm{d}s}\mathscr{G}^{-1}(s)|_{s=\mu^2}$$

that Z_3^{II} is necessarily negative (for the virtual state nearest threshold) a situation which is in marked contrast with the stable-particle case. One can now complete the renormalization of $\mathscr{G}_{\mathrm{II}}^{-1}$ by defining the renormalized vertex function

$$\bar{\Gamma}(s) = Z_1\Gamma(s)$$

$\bar{\Gamma}_{\mathrm{II}}(s)$ being normalized such that

$$\bar{\Gamma}_{\mathrm{II}}(s = \mu^2) = 1$$

and the renormalized (positive) coupling constant

$$g^2 = -Z_3\frac{Z_2{}^2}{Z_1{}^2}g_0{}^2$$

The propagator thus assumes the form

$$\bar{\mathscr{G}}_{\mathrm{II}}^{-1} = (p^2+\mu^2)\left\{-1-(p^2+\mu^2)\left[\int_{s_0}^{\infty}\frac{\bar{\sigma}(s)\,\mathrm{d}s}{(s-\mu^2)^2(p^2+s)} - \right.\right.$$

$$\left.\left. -\frac{\mathrm{i}\,g^2}{4\pi}\frac{\left[\rho\bar{\Gamma}_{\mathrm{I}}{}^2\tilde{S}^{-1}\right]_{\mu^2}^{-p^2} - (p^2+\mu^2)\left[\frac{\mathrm{d}}{\mathrm{d}p^2}(\rho\bar{\Gamma}_{\mathrm{I}}{}^2\tilde{S}^{-1})\right]_{\mu^2}}{(p^2+\mu^2)^2}\right]\right\}$$

$$\bar{\mathscr{G}}_{\mathrm{I}}^{-1} = (p^2+\mu^2)\left\{-1-(p^2+\mu^2)\int_{s_0}^{\infty}\frac{\bar{\sigma}(s)\,\mathrm{d}s}{(s-\mu^2)^2(p^2+s)}\right\} -$$

$$-\frac{\mathrm{i}\,g^2}{4\pi}\left\{\left[\rho\bar{\Gamma}_{\mathrm{I}}{}^2\tilde{S}^{-1}\right]_{s=\mu^2} + (p^2+\mu^2)\left[\frac{\mathrm{d}}{\mathrm{d}p^2}(\rho\bar{\Gamma}_{\mathrm{I}}{}^2\tilde{S}^{-1})\right]_{\mu^2}\right\}$$

where the renormalized spectral function $\bar{\sigma}(s)$ and renormalized propagator are given by

$$\bar{\sigma}(s) = -Z_3^{\mathrm{II}}\sigma(s)$$

$$\bar{\mathscr{G}}(-p^2) = -[Z_3^{\mathrm{II}}]^{-1}\mathscr{G}(-p^2)$$

Similarly the wave-function renormalization constant can be written as

$$Z_3^{\mathrm{II}} = 1 + \int_{s_0}^{\infty}\frac{\bar{\sigma}(s)\,\mathrm{d}s}{(s-\mu^2)^2} - \mathrm{i}\frac{g^2}{4\pi}\left[\frac{\mathrm{d}}{\mathrm{d}(-p^2)}\rho|\bar{\Gamma}|^2\right]_{p^2=-\mu^2}$$

which is a real quantity despite the explicit appearance of the imaginary unit.

With this general discussion of virtual states as a background, the extension of the $Z = 0$ condition can be readily carried out. In this case we start from the eigenvalue equation

$$\mu_0^2 = \mu^2 + \frac{1}{|Z_3^{\mathrm{II}}|}\left\{\int_{s_0}^{\infty} \frac{\bar{\sigma}(s)\,\mathrm{d}s}{s - \mu^2} + \left|\frac{g^2}{4\pi}\rho(\mu^2)\Gamma_{\mathrm{I}}^2(\mu^2)\right| \tilde{S}^{-1}(\mu^2)\right\} \qquad (45)$$

Since $\tilde{S}(s)$ is positive below s_0 and can in fact only change sign at a value of s below the position of the virtual state, it follows that each term on the right-hand side of (45) is positive definite. This has the immediate consequence that as Z_3^{II} goes to zero from below, μ_0^2 diverges and the usual equation of motion

$$(-\partial^2 + \mu_0^2)\phi = g_0 j(x)$$

degenerates as in the stable-particle case to the constraint

$$\bar{\phi}(x) = \tilde{g}j(x)$$

where

$$\bar{\phi}(x) = [-Z_3^{\mathrm{II}}]^{-\frac{1}{2}}\phi(x)$$

and

$$\tilde{g} = [-Z_3^{\mathrm{II}}]^{\frac{1}{2}} g_0 \left[\int_{s_0}^{\infty} \frac{\bar{\sigma}(s)\,\mathrm{d}s}{s - \mu^2} + \left|\frac{g^2}{4\pi}\rho\Gamma_{\mathrm{I}}^2\tilde{S}^{-1}\right|\right]^{-1}$$

Since the rest of the equivalence argument proceeds exactly as before, we shall not pursue this point in greater detail. It may be well to remark, however, that the Zachariasen model (in the case in which there is a virtual particle) provides a specific example of a theory to which the general remarks of this section can be readily applied. Since the calculations involved are entirely straightforward, the reader is referred elsewhere[12] for a detailed discussion.

6 Concluding Remarks

In the preceding sections it has been shown that all the successes thus far achieved by the $Z = 0$ rule can be interpreted simply in terms of the degeneration of an equation of motion as a consequence of the divergence of the bare mass term in the $Z = 0$ limit. By carrying out the generalization of this result to the multi-particle and virtual particle cases, it has been possible to gain further insight into the question as to what is the minimal set of assumptions required for the application of the $Z = 0$ rule. Although the preceding section has only attempted to generalize to the case of poles in the vicinity of the threshold on the second sheet, it does not require much imagination to conjecture that the usual requirement of a particle-like excitation is quite irrelevant. In fact the only essential condition appears to be the vanishing of the residue at a pole of the unrenormalized propagator on some arbitrary sheet of the Riemann surface, provided that the analytic continuation of the scattering amplitude to that pole yields a non-vanishing coupling constant. This generalization furthermore serves to emphasize that the so-called compositeness criteria should properly refer to fields rather than particles, and that in fact the only reasonably compelling reason for associating the concept of compositeness with the particles themselves appears to be the mathematical simplicity which ensues if one has only to deal with the physical sheet. From this point of view one might well question the motivation of a number of the dispersion theoretical approaches to the $Z = 0$ rule. Since the practitioners of dispersion theory have long stressed the possible advantages of dealing only with renormalized expressions (i.e. particles and coupling constants), it would seem (aside even from the fact that Z is intrinsically a field theoretical quantity) that dispersion theory is hardly likely to be a useful tool for purposes of discussing the question of the compositeness or elementarity of field operators.

Perhaps the most useful application of the field theoretical viewpoint advanced here arises in connection with Chew's view of elementary particle physics in terms of what he calls 'nuclear particle democracy'. According to this idea all particles are to be considered as composites of all others. However, from the field theoretical approach presented here, we have seen that compositeness is to be viewed in terms of the degeneration of an equation of motion in such a way that an elementary field can be eliminated in favour of a more complicated operator constructed from the remaining fields of the system. One can thus imagine a situation in which one constructs a theory from a certain set of fundamental fields, in which one or more of the wave-function renormalization constants are allowed to vanish. Upon eliminating the corresponding 'elementary' fields by means of the techniques we have described here, all fields with vanishing Z can be eliminated. However, this process must clearly terminate at some point. If all the Z's were to vanish the equations of motion would all become constraints and no field theory would be possible. This then immediately implies a hierarchy rather than a 'democracy' of elementary particles. Clearly those particles which are created by a single elementary field operator acting on the vacuum must be considered as being 'more elementary' than those which require a product of two such operators.

Similarly, those which require a product of three or more field operators are still 'less elementary'. Outside of the remote possibility that all particles are connected to the vacuum by the same number of elementary field operators, it seems that 'nuclear particle democracy' is indeed conceptually difficult in a field theoretical formalism. At our present level of understanding it would appear to be totally accidental and in no way related to the $Z = 0$ rule.

It may be useful to remark that in field theories which possess Galilean invariance[13], the above remarks can be made somewhat more rigorous. In this latter situation one has a set of eleven generators **P**, **J**, **K**, H and M which satisfy the commutation relations

$$[J_i, J_j] = \mathrm{i}\varepsilon_{ijk}J_k, \quad [J_i, K_j] = \mathrm{i}\varepsilon_{ijk}K_k, \quad [J_i, P_j] = \mathrm{i}\varepsilon_{ijk}P_k, \quad [J_i, H] = 0$$

$$[K_i, K_j] = [P_i, P_j] = [P_i, H] = 0, \quad [K_i, H] = \mathrm{i}\, P_i, \quad [K_i, P_j] = \mathrm{i}\,\delta_{ij}M$$

$$[J_i, M] = [K_i, M] = [P_i, M] = [H, M] = 0$$

The operators **P**, **J**, and H are familiar as the total momentum, angular momentum, and energy of the system. The features peculiar to the Galilean invariance consist in the existence of the operator **K**, the generator of pure Galilean transformations, and M, the mass operator of the theory. Since M commutes with all the generators of the group, it implies the existence of a super-selection rule in the theory, i.e. states referring to different eigenvalues of M belong to mutually incoherent subspaces of the total Hilbert space. This has the consequence that a particle A of mass m can couple to particles A_1 and A_2 if $m = m_1 + m_2$ (or more generally to $A_1, A_2, \ldots A_n$ if $m = \sum_{i=1}^{n} m_i$) so that A can in principle be a composite formed by the interaction of A_1 and A_2. However, the existence of the super-selection rule clearly denies the possibility of considering A_1 to be a composite of A and A_2. In general, although Galilean invariance allows for the possibility of all heavy particles being composites of all the lighter particles, the reverse cannot be true and Chew's hypothesis must fail for such systems. Although the argument is not so impeccable in the fully relativistic case, it would be somewhat surprising if the generalization of Galilean invariance to Lorentz invariance were to save this proposal. Although these remarks are not to be construed in such a way as to deny the utility of the bootstrap idea as an approximation technique in strong interaction dynamics, the concept of 'nuclear particle democracy' appears at the present time to be totally incompatible with any reasonable field theoretical approach.

References

1. Chew, G. F., *Phys. Today*, **17,** No. 4, 30 (1964).
2. Houard, J., and Jouvet, B., *Nuovo Cimento*, **18,** 466 (1960).
3. Vaughn, M. T., Aaron, R., and Amado, R. D., *Phys. Rev.*, **124,** 1258 (1961).
4. Lee, T. D., *Phys. Rev.*, **95,** 1329 (1954).
5. Hagen, C. R., *Ann. Phys.* (*N.Y.*), **31,** 185 (1965).
6. Lehmann, H., *Nuovo Cimento*, **11,** 342 (1954).
7. Zachariasen, F., *Phys. Rev.*, **121,** 1851 (1961).
8. Thirring, W., *Phys. Rev.*, **126,** 1209 (1962).
9. Srivastava, P. K., and Choudhury, S. R., *Nuovo Cimento*, **43A,** 239 (1966).
10. Hagen, C. R., *Nuovo Cimento*, **45,** 505 (1966).
11. Jin, Y. S., and MacDowell, S. W., *Phys. Rev.*, **137,** B688 (1965).
12. Hagen, C. R., *Nuovo Cimento*, **50,** 545 (1967).
13. For an example of a field theory possessing Galilean invariance see Levy-Leblond, J. M., *Commun. math. Phys.*, **4,** 157 (1967).

Topics in Bound State Theory

D. LURIE

1 Field Operators for Composite Particles

1.1 *Introduction*

The fact that the observed baryons and mesons are probably composite structures of some kind makes the question of assigning field operators to composite particles a particularly timely one[1,2]. The basic work on this problem was carried out in 1958 by Haag[3], Nishijima[4] and Zimmermann[5], who noticed that there exists a fundamental arbitrariness in the choice of the interpolating field which tends asymptotically to a given in-field or out-field[6]. It is this arbitrariness which allows field operators to be assigned to composite particles and places them on the same footing as elementary particles as far as the asymptotic condition and its consequences (i.e. reduction formulae) are concerned.

We begin by recalling the framework. This is the non-perturbative approach to quantum field theory developed twelve years ago by Lehmann, Symanzik and Zimmermann[7]. Although everything we will say is model-independent, it is convenient to have a certain model in mind for purposes of illustration. We shall take as our model neutral pseudoscalar meson theory.

1.2 *In- and out-fields*

The equations of motion for neutral pseudoscalar meson theory are

$$(\Box - m_{0\pi}{}^2)\phi = -\mathrm{i}\, G_0\bar{\psi}\gamma_5\psi \tag{1}$$

$$(\gamma\,.\,\partial + m_{0n})\psi = \mathrm{i}\, G_0\gamma_5\psi\phi \tag{2}$$

$$\bar{\psi}(\gamma\,.\,\overleftarrow{\partial} - m_{0n}) = -\mathrm{i}\, G_0\bar{\psi}\gamma_5\phi \tag{3}$$

and the equal-time commutation relations

$$[\psi_\alpha(x),\, \psi_\beta{}^\dagger(x')]_{t=t'} = \delta_{\alpha\beta}\delta^{(3)}(\mathbf{x} - \mathbf{x}') \tag{4}$$

$$[\phi(x),\, \dot{\phi}(x')]_{t=t'} = \mathrm{i}\, \delta^{(3)}(\mathbf{x} - \mathbf{x}') \tag{5}$$

all other fermion field anticommutators and boson field commutators being equal to zero. In this model, the nucleon and the π^0 are regarded as 'elementary'. Let us also assume that as a result of the π^0-nucleon interaction there appears a spinless 'deuteron'-like 2-nucleon bound state. We would like to be able to associate a field operator to this composite particle.

Note that the existence of a deuteron-like bound state in this model is an explicit assumption. Since the field equations are insoluble, we are obliged to *postulate* the existence of physically reasonable eigenstates of the Hamiltonian. Thus we assume that in addition to the vacuum state $|0\rangle$ satisfying

$$P_\mu|0\rangle = 0$$

where $P_\mu = (\mathbf{P}, \mathrm{i}\, H)$ is the 4-momentum operator, there exist discrete 1-particle states

$$\begin{aligned} &|\mathbf{k}\rangle \qquad (k^2 = \mathbf{k}^2 - k_0{}^2 = -m_\pi{}^2) \\ &|\mathbf{p}, \sigma\rangle \quad (p^2 = \mathbf{p}^2 - p_0{}^2 = -m_n{}^2) \\ &|\mathbf{q}\rangle \qquad (q^2 = \mathbf{q}^2 - q_0{}^2 = -m_d{}^2) \end{aligned}$$

representing single π^0, nucleon and 'deuteron' states respectively. In order to complete the description of the Hilbert space, we further assume that there exist *continuum* states with 2, 3, . . ., n, . . . incoming particles (elementary or bound), as well as the corresponding out-states. (For the vacuum and single-particle states there is no distinction between in-and out-states in the absence of external fields.) The set of all such in-states (or out-states) is assumed to form a complete set of states spanning the Hilbert space. The S-matrix for a transition from an initial state with momenta $\mathbf{k}_1, \ldots \mathbf{k}_n$ to a final state with momenta $\mathbf{k}_1', \ldots, \mathbf{k}_m'$ is given by the scalar product

$$\langle \mathbf{k}_1', \ldots \mathbf{k}_m' \,\mathrm{out}|\mathbf{k}_1 \ldots \mathbf{k}_n \,\mathrm{in}\rangle$$

Thus, in the non-perturbative approach to field theory, the structure of the Hilbert space is *postulated*. Once this is done, we can trivially define in-and out-fields by specifying their matrix elements in the in- and out-basis respectively. To construct the in-field for the π^0, for example, we construct destruction and creation operators $a_\mathbf{k}{}^{\mathrm{in}}$ and $a_\mathbf{k}{}^{\mathrm{in}\dagger}$ which have the property of, respectively, destroying and creating single mesons of momentum $\mathbf{k}$, for example

$$\begin{aligned} a_\mathbf{k}{}^{\mathrm{in}\dagger}|0\rangle &= |\mathbf{k}\rangle \\ a_\mathbf{k}{}^{\mathrm{in}\dagger}|\mathbf{p}\sigma\rangle &= |\mathbf{k}, \mathbf{p}\sigma \,\mathrm{in}\rangle \\ &\ldots \text{etc.} \end{aligned}$$

Specification of the matrix elements of $a_{\mathbf{k}}^{\text{in}}$ and $a_{\mathbf{k}}^{\text{in}\dagger}$ in the given basis serves to define these operators completely. From their definition, $a_{\mathbf{k}}^{\text{in}}$ and $a_{\mathbf{k}}^{\text{in}\dagger}$ must of course satisfy the standard commutation relations

$$[a_{\mathbf{k}}^{\text{in}}, a_{\mathbf{k}'}^{\text{in}\dagger}] = \delta_{\mathbf{k}\mathbf{k}'}$$
$$[a_{\mathbf{k}}^{\text{in}}, a_{\mathbf{k}'}^{\text{in}}] = [a_{\mathbf{k}}^{\text{in}\dagger}, a_{\mathbf{k}'}^{\text{in}\dagger}] = 0 \tag{6}$$

and the in-field $\phi^{\text{in}}(x)$, constructed according to

$$\phi^{\text{in}}(x) = \frac{1}{\sqrt{V}} \sum_{\mathbf{k},k_0=\omega_{\mathbf{k}}} \frac{1}{\sqrt{(2\omega_{\mathbf{k}})}} (a_{\mathbf{k}}^{\text{in}}\, \mathrm{e}^{\mathrm{i}k\cdot x} + a_{\mathbf{k}}^{\text{in}\dagger}\, \mathrm{e}^{-\mathrm{i}k\cdot x}) \tag{7}$$

with $\omega_{\mathbf{k}} = (\mathbf{k}^2 + m_\pi{}^2)^{\frac{1}{2}}$, satisfies the usual free-field equation

$$(\Box - m_\pi{}^2)\phi^{\text{in}}(x) = 0 \tag{8}$$

and equal-time commutation rules

$$[\phi^{\text{in}}(x), \dot{\phi}^{\text{in}}(x')]_{t=t'} = \mathrm{i}\, \delta^{(3)}(\mathbf{x} - \mathbf{x}') \tag{9}$$

The out-field $\phi^{\text{out}}(x)$ is defined in exactly the same way in terms of its matrix elements in the out-basis. Similarly, we can define in-and out-fields $\psi^{\text{in}}(x)$ and $\psi^{\text{out}}(x)$ for the nucleon and in-and out-fields $B^{\text{in}}(x)$ and $B^{\text{out}}(x)$ for the bound state. In the latter case we define creation and destruction operators $b_{\mathbf{q}}^{\text{in}\dagger}$ and $b_{\mathbf{q}}^{\text{in}}$ with the property that they respectively create and destroy single 'deuterons' when acting on an arbitrary basis vector, e.g.

$$b_{\mathbf{q}}^{\text{in}\dagger}|0\rangle = |\mathbf{q}\rangle \qquad (q^2 = -m_d{}^2)$$
$$b_{\mathbf{q}}^{\text{in}\dagger}|\mathbf{k}\rangle = |\mathbf{k}, \mathbf{q}\ \text{in}\rangle$$
$$\ldots \text{etc.}$$

$B^{\text{in}}(x)$ is then given by

$$B^{\text{in}}(x) = \frac{1}{\sqrt{V}} \sum_{\mathbf{q},q_0=E_{\mathbf{q}}} \frac{1}{\sqrt{(2E_{\mathbf{q}})}} (b_{\mathbf{q}}^{\text{in}}\, \mathrm{e}^{\mathrm{i}q\cdot x} + b_{\mathbf{q}}^{\text{in}\dagger}\, \mathrm{e}^{-\mathrm{i}q\cdot x}) \tag{10}$$

with $E_{\mathbf{q}} = (\mathbf{q}^2 + m_d{}^2)$.

The Asymptotic Condition. For the time being, we restrict our attention to the elementary in- and out-fields, i.e. to ϕ^{in}, ϕ^{out}, ψ^{in} and ψ^{out}. The connection between these fields and the field operators $\phi(x)$ and $\psi(x)$ is expressed by the *asymptotic condition* of Lehmann, Symanzik and

Zimmermann[7]. For the pion field, for example, this condition states, roughly, that

$$\lim_{t\to-\infty} \langle m|\phi(x)|n\rangle = \lim_{t\to-\infty} Z_3^{\frac{1}{2}}\langle m|\phi^{\text{in}}(x)|n\rangle \tag{11a}$$

$$\lim_{t\to+\infty} \langle m|\phi(x)|n\rangle = \lim_{t\to+\infty} Z_3^{\frac{1}{2}}\langle m|\phi^{\text{out}}(x)|n\rangle \tag{11b}$$

for any two normalizable state vectors $|m\rangle$ and $|n\rangle$. In fact, the statement (11a, b) is not very precise, since the right-hand side involves oscillating exponentials whose limit as $t \to \pm\infty$ is undefined. The correct statement is obtained by expanding $\phi(x)$ according to

$$\phi(x) = \frac{1}{\sqrt{V}} \sum_{\mathbf{k},k_0=\omega_\mathbf{k}} \frac{1}{\sqrt{(2\omega_\mathbf{k})}} [a_\mathbf{k}(t)\,\mathrm{e}^{\mathrm{i}k\cdot x} + a_\mathbf{k}^\dagger(t)\,\mathrm{e}^{-\mathrm{i}k\cdot x}] \tag{12}$$

where the expansion coefficients, given by

$$a_\mathbf{k}(t) = \mathrm{i}\int \mathrm{d}^3x \frac{1}{\sqrt{V}} \frac{1}{\sqrt{(2\omega_\mathbf{k})}} \mathrm{e}^{-\mathrm{i}k\cdot x}(\partial_t - \overleftarrow{\partial_t})\phi(x) \tag{13}$$

are now functions of t, and expressing the asymptotic conditions as

$$\lim_{t\to-\infty} \langle m|a_\mathbf{k}(t)|n\rangle = Z_3^{\frac{1}{2}}\langle m|a_\mathbf{k}^{\text{in}}|n\rangle \tag{14a}$$

$$\lim_{t\to+\infty} \langle m|a_\mathbf{k}(t)|n\rangle = Z_3^{\frac{1}{2}}\langle m|a_\mathbf{k}^{\text{out}}|n\rangle \tag{14b}$$

Thus, we shall always understand (11a, b) in the sense (14a, b). Since $\phi(x)$ tends (in the above sense) to $Z_3^{\frac{1}{2}}\phi^{\text{in}}(x)$ for $t \to -\infty$ and to $Z_3^{\frac{1}{2}}\phi^{\text{out}}(x)$ for $t \to +\infty$, it is known as the *interpolating* field. The factor $Z_3^{\frac{1}{2}}$ is required because of the difference between the vacuum to 1-particle matrix elements of $\phi^{\text{in}}_{\text{out}}$ and ϕ. For ϕ^{in} (or ϕ^{out}) we have, by virtue of (7)

$$\langle 0|\phi^{\text{in}}(x)|\mathbf{k}\rangle = \frac{1}{\sqrt{V}} \frac{1}{\sqrt{(2\omega_\mathbf{k})}} \mathrm{e}^{\mathrm{i}k\cdot x} \tag{15a}$$

On the other hand, for the matrix elements of $\phi(x)$ we have, on grounds of space-time translation invariance

$$\begin{aligned}\langle 0|\phi(x)|\mathbf{k}\rangle &= \langle 0|\phi(0)|\mathbf{k}\rangle\,\mathrm{e}^{\mathrm{i}k\cdot x} \\ &= Z_3^{\frac{1}{2}} \frac{1}{\sqrt{V}} \frac{1}{\sqrt{(2\omega_\mathbf{k})}} \mathrm{e}^{\mathrm{i}k\cdot x}\end{aligned} \tag{15b}$$

where the last line follows from Lorentz invariance. Z_3—the meson wave-function renormalization constant—can be shown to be necessarily

$\neq 1$ in the case of interacting fields. The necessity for the $Z_3^{\frac{1}{2}}$ factor in the asymptotic condition is now obvious if we apply (11a, b) for $|m\rangle = |0\rangle$ and $|n\rangle = |\mathbf{k}\rangle$ and compare with (15a, b).

We also remark that it is essential to take matrix elements when writing down the asymptotic condition. This is known as *weak convergence*. The stronger statement, that $\phi \to Z_3^{\frac{1}{2}}\phi^{\text{in}}$, runs into trouble, since it would mean that, for example

$$\lim_{t\to-\infty} [\phi(\mathbf{x}, t), \dot{\phi}(\mathbf{x}', t)] = \lim_{t\to-\infty} Z_3[\phi^{\text{in}}(\mathbf{x}, t), \dot{\phi}^{\text{in}}(\mathbf{x}', t)]$$

or, by (5) and (9),

$$Z_3 = 1$$

The asymptotic condition for the nucleon field follows the same pattern as for ϕ. We have

$$\lim_{t\to\mp\infty} \langle m|\psi(x)|n\rangle = \lim_{t\to\mp\infty} Z_2^{\frac{1}{2}}\langle m|\psi^{\text{in}}_{\text{out}}(x)|n\rangle \tag{16}$$

where Z_2 is the nucleon wave-function renormalization constant. Again, (16) must be interpreted as a condition relating the expansion coefficients of $\psi(x)$ and $\psi^{\text{in}}_{\text{out}}(x)$.

The most important consequence of the asymptotic conditions are the so-called reduction formulae for S-matrix elements[7]. For $\pi^0 - N$ scattering for example, the asymptotic conditions lead to the formula

$$\begin{aligned}\langle \mathbf{k}', \mathbf{p}'\sigma' \text{ out}|\mathbf{k}, \mathbf{p}\sigma \text{ in}\rangle &= \delta_{\mathbf{k}'\mathbf{k}}\delta_{\mathbf{p}'\mathbf{p}}\delta_{\sigma'\sigma} + \\ &+ (Z_3^{-\frac{1}{2}}Z_2^{-\frac{1}{2}})^2 \int \mathrm{d}^4x'\, \mathrm{d}^4y'\, \mathrm{d}^4x\, \mathrm{d}^4y\, f_{\mathbf{k}'}^*(x')f_{\mathbf{k}}(x)(\Box_{x'} - m_\pi^2)(\Box_x - m_\pi^2) \times \\ &\times \bar{u}_{\mathbf{p}'\sigma'}(y')(\gamma\,.\,\partial_{y'} + m_n)\langle 0|T\psi(y')\bar{\psi}(y)\phi(x')\phi(x)|0\rangle(\gamma\,.\,\overleftarrow{\partial}_y - m_n)u_{\mathbf{p}\sigma}(y)\end{aligned} \tag{17}$$

where T denotes the time-ordered product and $f_{\mathbf{k}}$ and $u_{\mathbf{p}\sigma}$ are the initial meson and nucleon wave-functions

$$f_{\mathbf{k}}(x) = \frac{1}{\sqrt{V}}\frac{1}{\sqrt{(2\omega_{\mathbf{k}})}}\,\mathrm{e}^{\mathrm{i}\mathbf{k}\cdot\mathbf{x}-\mathrm{i}\omega_{\mathbf{k}}t} \tag{18}$$

and

$$u_{\mathbf{p}\sigma}(y) = \frac{1}{\sqrt{V}}\left(\frac{m}{E_{\mathbf{p}}}\right)^{\frac{1}{2}} u_{\mathbf{p}\sigma}\,\mathrm{e}^{\mathrm{i}\mathbf{p}\cdot\mathbf{x}-\mathrm{i}E_{\mathbf{p}}t} \tag{19}$$

with $\omega_{\mathbf{k}} = (\mathbf{k}^2 + m_\pi^2)^{\frac{1}{2}}$ and $E_{\mathbf{p}} = (\mathbf{p}^2 + m_n^2)^{\frac{1}{2}}$. For the derivation of reduction formulae from the asymptotic conditions, see any standard text on field theory[8]. The important point to note about these reduction formulae is that they relate S-matrix elements to the *mass-shell values* of corresponding Green's functions. Equation (17), for example, says that to get the S-matrix element for meson-nucleon scattering, you take the meson-nucleon Green's function

$$\langle 0|T\psi(y')\bar{\psi}(y)\phi(x')\phi(x)|0\rangle \tag{20}$$

and remove all external propagator factors, placing the external particles on their respective mass shells. For instance, the factor $-\mathrm{i}\,(\Box - m_\pi{}^2)$ in (17) yields $\mathrm{i}\,(k^2 + m_\pi{}^2)$ in momentum space. On the mass shell, this cancels the external propagator factor $-\mathrm{i}\,(k^2 + m_\pi{}^2)^{-1}$ in the expansion of the Green's function.

Thus the function of the reduction formula is to single out the mass shell values of the Green's function, to yield the corresponding S-matrix element. The reduction formulae provide the basic connection between the interpolating field operators and the particle properties, as expressed by the S-matrix elements. Now the only information about the interpolating field $\phi(x)$ which is needed for the S-matrix is the fact that $\phi \to Z_3{}^{\frac{1}{2}}\phi^{\mathrm{in}}$ for $t \to -\infty$ and $\phi \to Z_3{}^{\frac{1}{2}}\phi^{\mathrm{out}}$ for $t \to +\infty$, in the sense of weak convergence. Our next task is to show that this asymptotic connection is a very weak restriction on ϕ, in the sense that a number of field operators other than ϕ can equally well be cast as the interpolating field[3,4,5], and can therefore be regarded as legitimate field operators for the meson. To understand this, we must examine the *proof* of the asymptotic condition.

1.3 *Zimmermann's proof*

The most convenient derivation of the asymptotic condition for our purposes is that given by Zimmermann[5]. Let us *define* Heisenberg fields ϕ^{in} and ϕ^{out} in terms of the interpolating field ϕ by means of the equations

$$Z_3{}^{\frac{1}{2}}\phi^{\mathrm{in}}(x) = \phi(x) + \int \Delta_R(x-y)(\Box_y - m_\pi{}^2)\phi(y)\,\mathrm{d}^4y \tag{21a}$$

$$Z_3{}^{\frac{1}{2}}\phi^{\mathrm{out}}(x) = \phi(x) + \int \Delta_A(x-y)(\Box_y - m_\pi{}^2)\phi(y)\,\mathrm{d}^4y \tag{21b}$$

where we assume Z_3 to be non-zero and where (21a, b) are to be understood in the sense of weak convergence, that is, as being meaningful only when sandwiched between normalizable states. The functions Δ_R and Δ_A are the retarded and advanced Green's functions satisfying

$$(\Box - m_\pi{}^2)\Delta_R(x) = -\delta^{(4)}(x), \quad \Delta_R(x) = 0 \text{ for } x_0 < 0 \tag{22a}$$
$$(\Box - m_\pi{}^2)\Delta_A(x) = -\delta^{(4)}(x), \quad \Delta_A(x) = 0 \text{ for } x_0 > 0 \tag{22b}$$

From (21a, b) it follows that

$$\lim_{t \to -\infty} \langle m|a_{\mathbf{k}}(t)|n\rangle = Z_3^{\frac{1}{2}}\langle m|a_{\mathbf{k}}^{\text{in}}|n\rangle \tag{23a}$$

$$\lim_{t \to +\infty} \langle m|a_{\mathbf{k}}(t)|n\rangle = Z_3^{\frac{1}{2}}\langle m|a_{\mathbf{k}}^{\text{out}}|n\rangle \tag{23b}$$

for $a_{\mathbf{k}}(t)$ constructed according to (13). Indeed, the integral terms on the right-hand side of (21a) and (21b) vanish for $x_0 \to -\infty$ and $x_0 \to +\infty$ respectively, the regions of integration shrinking to zero by virtue of the retarded and advanced character of Δ_R and Δ_A. To show that (23a) and (23b) are really the asymptotic conditions however, we must show that ϕ^{in} and ϕ^{out} *are*, in fact, the in and out meson fields which we have previously introduced. This entails showing that

(*a*) $\phi^{\text{in}}(x)$ and $\phi^{\text{out}}(x)$ obey the free-field equations for a particle of mass m_π;

(*b*) $\phi^{\text{in}}(x)$ and $\phi^{\text{out}}(x)$ obey the free-field commutation rules

$$[\phi^{\text{in}}(x), \phi^{\text{in}}(y)] = \mathrm{i}\,\Delta(x-y) \tag{24a}$$

$$[\phi^{\text{out}}(x), \phi^{\text{out}}(y)] = \mathrm{i}\,\Delta(x-y) \tag{24b}$$

The proof of (*a*) is immediate and follows from the fact that Δ_R and Δ_A satisfy the inhomogeneous Klein-Gordon equations (22a) and (22b) for mass m_π. We have for $\phi^{\text{in}}(x)$, for example,

$$\begin{aligned}(\Box - m_\pi{}^2)Z_3{}^{\frac{1}{2}}\phi^{\text{in}}(x) &= (\Box - m_\pi{}^2)\phi(x) - \int \delta^{(4)}(x-y)(\Box - m_\pi{}^2)\phi(y)\,\mathrm{d}^4y \\ &= 0\end{aligned} \tag{25}$$

and similarly for ϕ^{out}.

As a preliminary to establishing (*b*) let us show that ϕ^{in} and ϕ^{out} have the correct property that, when operating on the vacuum state they create properly normalized 1-meson states only. Consider an arbitrary state $|a\rangle$ with energy-momentum eigenvalue $k_\mu{}^a$

$$P_\mu|a\rangle = k_\mu{}^a|a\rangle \tag{26}$$

By invariance under space-time displacements we have

$$\langle a|\phi^{\text{in}}(x)|0\rangle = \mathrm{e}^{-\mathrm{i}k^a\cdot x}\langle a|\phi^{\text{in}}(0)|0\rangle$$

and hence

$$\mathrm{i}\partial_\mu\langle a|\phi^{\text{in}}(x)|0\rangle = k_\mu{}^a\langle a|\phi^{\text{in}}(0)|0\rangle$$

and

$$(-\partial_\mu\partial_\mu + m_\pi{}^2)\langle a|\phi^{\text{in}}(x)|0\rangle = (k_\mu{}^a k_\mu{}^a + m_\pi{}^2)\langle a|\phi^{\text{in}}(x)|0\rangle$$

The left-hand side is zero by virtue of (25), so that

$$\langle a|\phi^{\text{in}}(x)|0\rangle = 0 \text{ unless } k_a{}^2 = -m_\pi{}^2 \tag{27}$$

Thus $\phi^{\text{in}}(x)$ can only connect 1-meson states to the vacuum. Setting $\langle a| = \langle \mathbf{k}|$ we have, from (21a, b),

$$Z_3^{\frac{1}{2}}\langle \mathbf{k}|\phi^{\text{in}}(x)|0\rangle = \langle \mathbf{k}|\phi(x)|0\rangle + \int \Delta_R(x-y)(\Box_y - m_\pi{}^2)\,\langle \mathbf{k}|\phi(y)|0\rangle\, \mathrm{d}^4y \tag{28}$$

Since, by (15b),

$$\langle \mathbf{k}|\phi(y)|0\rangle = \frac{1}{\sqrt{V}}\frac{1}{\sqrt{(2\omega_{\mathbf{k}})}}\,\mathrm{e}^{ik\cdot x}$$

for a neutral scalar field, the second term on the right-hand side of (28) is zero and (28) reduces to

$$\langle \mathbf{k}|\phi^{\text{in}}(x)|0\rangle = \frac{1}{\sqrt{V}}\frac{1}{\sqrt{(2\omega_{\mathbf{k}})}}\,\mathrm{e}^{ik\cdot x} \tag{29}$$

in agreement with the usual normalization (see 15a).

To show that $\phi^{\text{in}}(x)$ and $\phi^{\text{out}}(x)$ obey the correct free-field commutation relations (24a, b) we proceed in three steps. First we show that

$$\langle 0|[\phi^{\text{in}}(x), \phi^{\text{in}}(y)]|0\rangle = \mathrm{i}\Delta(x-y) \tag{30}$$

Secondly we show that

$$[\phi^{\text{in}}(x), \phi^{\text{in}}(y)] = [\phi^{\text{out}}(x), \phi^{\text{out}}(y)] \tag{31}$$

and finally that

$$[\phi^{\text{in}}(x), \phi^{\text{in}}(y)] = c\text{-number} \tag{32}$$

Equations (30), (31) and (32) are obviously sufficient for (24a, b). To prove (30) we use (27) and (29) to compute

$$\begin{aligned}\langle 0|\phi^{\text{in}}(x)\phi^{\text{in}}(y)|0\rangle &= \sum_{\mathbf{k}}\langle 0|\phi^{\text{in}}(x)|\mathbf{k}\rangle\langle \mathbf{k}|\phi^{\text{in}}(y)|0\rangle \\ &= \frac{1}{V}\sum_{\mathbf{k}}\frac{1}{2\omega_{\mathbf{k}}}\mathrm{e}^{\,ik\cdot(x-y)} \\ &= \frac{1}{(2\pi)^3}\int\frac{\mathrm{d}^3k}{2\omega_{\mathbf{k}}}\,\mathrm{e}^{ik\cdot(x-y)} \\ &= \mathrm{i}\,\Delta^{(+)}(x-y)\end{aligned} \tag{33}$$

Similarly we find that

$$\langle 0|\phi^{\text{in}}(y)\phi^{\text{in}}(x)|0\rangle = -\mathrm{i}\,\Delta^{(-)}(x-y) \tag{34}$$

Combining (33) and (34) we recover (30), since $\Delta = \Delta^{(+)} + \Delta^{(-)}$.

To establish (31) we start from the identity

$$\int d^4x\, d^4y f_{\mathbf{q}}^*(x) f_{\mathbf{k}}(y)(\Box_x - m_\pi^2)(\Box_y - m_\pi^2) T\phi(x)\phi(y) =$$
$$= \int d^4y\, d^4x f_{\mathbf{q}}^*(x) f_{\mathbf{k}}(y)(\Box_x - m_\pi^2)(\Box_y - m_\pi^2) T\phi(x)\phi(y) \quad (35)$$

where $f_{\mathbf{k}}$ is given by (18). This is a simple interchange in the order of integration, but it actually requires justification. In particular, one must invoke the microcausality requirement that $\phi(x)$ is a local field which commutes with itself at space-like separations[5]. Also, (35) must be understood in the sense of weak operator convergence, i.e. as valid only when sandwiched between two normalizable state vectors. Now for an arbitrary function $F(y)$ we have, by integration by parts

$$\int d^4y f_{\mathbf{k}}(y)(\Box_y - m_\pi^2) F(y) = \int d^3y \int_{-\infty}^{\infty} dy_0 f_{\mathbf{k}}(y)\left(\overleftarrow{\nabla}^2 - \frac{\partial}{\partial y_0{}^2} - m_\pi^2\right) F(y)$$
$$= \int d^3y \int_{-\infty}^{\infty} dy_0 f_{\mathbf{k}}(y)\left(\frac{\overleftarrow{\partial^2}}{\partial y_0{}^2} - \frac{\partial^2}{\partial y_0{}^2}\right) F(y)$$
$$= \int d^3y \int_{-\infty}^{\infty} dy_0 \frac{\partial}{\partial y_0}\left[F(y)\left(\frac{\partial}{\partial y_0} - \frac{\overleftarrow{\partial}}{\partial y_0}\right) f_{\mathbf{k}}(y)\right] \quad (36)$$

where in deriving the second line we have used the Klein-Gordon equation for $f_{\mathbf{k}}(y)$. Applying (36) to the left-hand side of (35) and recalling (13) we get

$$-i\int d^4y f_{\mathbf{k}}(y)(\Box_y - m_\pi^2) T\phi(x)\phi(y)$$
$$= -i\int d^3y \int_{-\infty}^{\infty} dy_0 \frac{\partial}{\partial y_0}\left[T\phi(x)\phi(y)\left(\frac{\partial}{\partial y_0} - \frac{\overleftarrow{\partial}}{\partial y_0}\right) f_{\mathbf{k}}(y)\right]$$
$$= -\int_{-\infty}^{\infty} dy_0 \frac{\partial}{\partial y_0} T\phi(x) a_{\mathbf{k}}^\dagger(y_0)$$
$$= \phi(x) a_{\mathbf{k}}^{\mathrm{in}\dagger} - a_{\mathbf{k}}^{\mathrm{out}\dagger}\phi(x) \quad (37)$$

and dealing with the x-integration in the same way, we get

$$\int d^4x \int d^4y f_{\mathbf{q}}^*(x) f_{\mathbf{k}}(y)(\Box_x - m_\pi^2)(\Box_y - m_\pi^2) T\phi(x)\phi(y) =$$
$$= a_{\mathbf{k}}^{\mathrm{out}\dagger} a_{\mathbf{q}}^{\mathrm{out}} - a_{\mathbf{k}}^{\mathrm{out}\dagger} a_{\mathbf{q}}^{\mathrm{in}} - a_{\mathbf{q}}^{\mathrm{out}} a_{\mathbf{k}}^{\mathrm{in}\dagger} + a_{\mathbf{q}}^{\mathrm{in}} a_{\mathbf{k}}^{\mathrm{in}\dagger} \quad (38)$$

Similarly, the integral on the right-hand side of (35) can be written as

$$\int d^4y \int d^4x f_{\mathbf{q}}{}^*(x) f_{\mathbf{k}}(y)(\Box_x - m_\pi{}^2)(\Box_y - m_\pi{}^2) T\phi(x)\phi(y) = \\ = a_{\mathbf{q}}{}^{\text{out}} a_{\mathbf{k}}{}^{\text{out}\dagger} - a_{\mathbf{q}}{}^{\text{out}} a_{\mathbf{k}}{}^{\text{in}\dagger} - a_{\mathbf{k}}{}^{\text{out}\dagger} a_{\mathbf{q}}{}^{\text{in}} + a_{\mathbf{k}}{}^{\text{in}\dagger} a_{\mathbf{q}}{}^{\text{in}} \quad (39)$$

Equating (37) with (39), we obtain the result

$$[a_{\mathbf{q}}{}^{\text{out}}, a_{\mathbf{k}}{}^{\text{out}\dagger}] = [a_{\mathbf{q}}{}^{\text{in}}, a_{\mathbf{k}}{}^{\text{in}\dagger}] \quad (40)$$

Proceeding in the same way, we can show that

$$[a_{\mathbf{q}}{}^{\text{out}\dagger}, a_{\mathbf{k}}{}^{\text{out}\dagger}] = [a_{\mathbf{q}}{}^{\text{in}\dagger}, a_{\mathbf{k}}{}^{\text{in}\dagger}] \quad (41)$$

$$[a_{\mathbf{q}}{}^{\text{out}}, a_{\mathbf{k}}{}^{\text{out}}] = [a_{\mathbf{q}}{}^{\text{in}}, a_{\mathbf{k}}{}^{\text{in}}] \quad (42)$$

by replacing $f_{\mathbf{q}}{}^* f_{\mathbf{k}}$ in equality (35) by $f_{\mathbf{q}} f_{\mathbf{k}}$ and $f_{\mathbf{q}}{}^* f_{\mathbf{k}}{}^*$ respectively. This completes the proof of (31). It remains to show that the commutator of $\phi^{\text{in}}(x)$ with $\phi^{\text{in}}(y)$ is a *c*-number. To do this we use the identity

$$\int d^4x \int d^4y f_{\mathbf{q}}{}^*(x) f_{\mathbf{k}}(y)(\Box_x - m_\pi{}^2)(\Box_y - m_\pi{}^2) T\phi(x)\phi(y)\phi(z) = \\ = \int d^4y \int d^4x f_{\mathbf{q}}{}^*(x) f_{\mathbf{k}}(y)(\Box_x - m_\pi{}^2)(\Box_y - m_\pi{}^2) T\phi(x)\phi(y)\phi(z)$$

and manipulate each side by the same technique as in deriving (40). This yields the result

$$\phi(z)[a_{\mathbf{q}}{}^{\text{in}}, a_{\mathbf{k}}{}^{\text{in}\dagger}] = [a_{\mathbf{q}}{}^{\text{out}}, a_{\mathbf{k}}{}^{\text{out}\dagger}]\phi(z) \quad (43)$$

or, by virtue of our previous result (40)

$$\{[a_{\mathbf{q}}{}^{\text{in}}, a_{\mathbf{k}}{}^{\text{in}\dagger}], \phi(z)\} = 0 \quad (44)$$

We have therefore shown that $[a_{\mathbf{q}}{}^{\text{in}}, a_{\mathbf{k}}{}^{\text{in}\dagger}]$ commutes with the basic meson field $\phi(z)$ and we can show by exactly the same argument that $[a_{\mathbf{q}}{}^{\text{in}}, a_{\mathbf{k}}{}^{\text{in}\dagger}]$ commutes with the basic nucleon field operator $\psi(x)$. Since $\phi(x)$ and $\psi(x)$ are the basic fields in terms of which we have formulated the theory, they must form an irreducible ring of operators and it follows that $[a_{\mathbf{q}}{}^{\text{in}}, a_{\mathbf{k}}{}^{\text{in}\dagger}]$ must be a *c*-number. Applying the same reasoning to $[a_{\mathbf{q}}{}^{\text{in}\dagger}, a_{\mathbf{k}}{}^{\text{in}\dagger}]$ and $[a_{\mathbf{q}}{}^{\text{in}}, a_{\mathbf{k}}{}^{\text{in}}]$ we deduce that $[\phi^{\text{in}}(x), \phi^{\text{in}}(y)]$ must be a *c*-number, which completes the proof.

1.4 *Arbitrariness in the choice of interpolating field*

So far it has been understood that the interpolating field ϕ is identical with the basic meson field operator featured in the Lagrangian and equations of motion. In fact this need not be so. If we examine Zimmermann's proof of the asymptotic condition, we observe that the only properties of ϕ which were actually used were:

(*a*) the normalization condition

$$\langle 0|\phi(x)|\mathbf{k}\rangle = Z_3^{\frac{1}{2}} \frac{1}{\sqrt{V}} \frac{1}{\sqrt{(2\omega_{\mathbf{k}})}} \mathrm{e}^{ik\cdot x} \tag{45}$$

with Z_3 assumed to be non-zero, and

(*b*) locality, or microscopic causality. This is used to justify the interchange in the order of integration in (35)[5].

It follows that *any* local field $\varphi(x)$ satisfying (45) will tend asymptotically to $\phi^{\mathrm{in}}(x)$ and $\phi^{\mathrm{out}}(x)$ and is as good a candidate for the role of interpolating field as $\phi(x)$. For any such field φ we can define, following Zimmermann,

$$Z_3^{\frac{1}{2}}\phi^{\mathrm{in}}(x) = \varphi(x) + \int \Delta_R(x-y)(\Box_y - m_\pi^2)\varphi(y)\,\mathrm{d}^4y \tag{46}$$

with a corresponding equation for ϕ^{out}, and proceed to establish that ϕ^{in} and ϕ^{out} are the correct in- and out-fields for the meson.

An immediate corollary is that φ can be used in place of ϕ in reduction formulae for S-matrix elements.

To construct fields $\varphi(x)$ satisfying (45) we must of course ensure that $\varphi(x)$ has the same Lorentz transformation properties as $\phi(x)$. An obvious choice for $\varphi(x)$ is the pseudoscalar density $\bar{\psi}(x)\gamma_5\psi(x)$ divided by the matrix element $\langle 0|\bar{\psi}(0)\gamma_5\psi(0)|\mathbf{k}\rangle$ and multiplied by $Z_3^{\frac{1}{2}}(2\omega_{\mathbf{k}}V)^{-\frac{1}{2}}$ to ensure the normalization condition (45). Explicitly,

$$\varphi(x) = Z_3^{\frac{1}{2}} \frac{1}{\sqrt{V}} \frac{1}{\sqrt{(2\omega_{\mathbf{k}})}} \frac{1}{\langle 0|\bar{\psi}(0)\gamma_5\psi(0)|\mathbf{k}\rangle} \bar{\psi}(x)\gamma_5\psi(x) \tag{47}$$

This satisfies the correct normalization (45)

$$\langle 0|\varphi(x)|\mathbf{k}\rangle = Z_3^{\frac{1}{2}} \frac{1}{\sqrt{V}} \frac{1}{\sqrt{(2\omega_{\mathbf{k}})}} \mathrm{e}^{ik\cdot x} \tag{48}$$

since, on grounds of space-time displacement invariance

$$\langle 0|\bar{\psi}(x)\gamma_5\psi(x)|\mathbf{k}\rangle = \langle 0|\bar{\psi}(0)\gamma_5\psi(0)|\mathbf{k}\rangle\, \mathrm{e}^{ik\cdot x}$$

Naturally, a crucial condition is that

$$\langle 0|\bar{\psi}(0)\gamma_5\psi(0)|\mathbf{k}\rangle \neq 0$$

Actually, care must be taken to define the limit of the operator product $\bar{\psi}(x')\gamma_5\psi(x)$ for $x' \to x$. Haag, Nishijima and Zimmermann adopt the prescription

$$\varphi(x) = \lim_{a\to 0} Z_3^{\frac{1}{2}} \frac{1}{\sqrt{V}} \frac{1}{\sqrt{(2\omega_{\mathbf{k}})}} \frac{1}{\langle 0|T\bar{\psi}(a)\gamma_5\psi(-a)|\mathbf{k}\rangle} T\bar{\psi}(x+a)\gamma_5\psi(x-a) \tag{49}$$

where T denotes the time-ordered product

$$T\bar{\psi}_\alpha(x)\psi_\beta(y) = \begin{cases} \bar{\psi}_\alpha(x)\psi_\beta(y) & x_0 > y_0 \\ -\psi_\beta(y)\bar{\psi}_\alpha(x) & y_0 > x_0 \end{cases} \tag{50}$$

and where the limit $a \to 0$ is taken along a space-like direction.

An alternative choice for $\varphi(x)$, assuming that $\langle 0|\bar{\psi}(0)\gamma_5\gamma_\mu\psi(0)|\mathbf{k}\rangle \neq 0$, is to take $\varphi(x)$ proportional to the divergence of the axial vector current $\bar{\psi}(x)\gamma_5\gamma_\mu\psi(x)$:

$$\varphi(x) = c\partial_\mu\bar{\psi}(x)\gamma_5\gamma_\mu\psi(x) \tag{51}$$

This recalls the partially conserved axial vector current hypothesis (PCAC) of particle physics[10]. However, it should be clearly understood that the arbitrariness in the choice of field applies only to the *mass shell* values of Green's functions constructed from the fields. That is, the possibility of replacing the meson field operator ϕ by the divergence of an axial vector current (or by a pseudoscalar density as in 47) is limited only to the asymptotic condition and the reduction formulae. On the other hand, the PCAC hypothesis as applied to particle physics states that the meson field may be identified (to within a multiplicative constant) with the divergence of the axial vector current as an *operator equality*. This is a much stronger statement than the result of the Haag-Nishijima-Zimmermann analysis.

1.5 *Composite particles*

We have concluded that the interpolating field for the meson can be taken to be any local field $\varphi(x)$ satisfying $\langle 0|\varphi(x)|\mathbf{k}\rangle \neq 0$ and properly normalized so as to guarantee the condition (45). From this point of view the fact that $\phi(x)$ itself is available as a possible candidate for the interpolating field becomes irrelevant and we are in a position to treat composite and elementary particles on an equal footing. Thus, in our model, if we seek an interpolating field $B(x)$ which tends asymptotically to $B^{\text{in}}(x)$, as given by (10), the Haag-Nishijima-Zimmermann analysis tells us that *any* local field having non-vanishing matrix elements between $|0\rangle$ and the 2-nucleon bound state $|\mathbf{q}\rangle(q^2 = -m_d{}^2)$ is a good candidate for $B(x)$. Since

$$\langle 0|T\psi_A(x_1)\psi_B(x_2)|\mathbf{q}\rangle \neq 0$$

the most obvious choice is the product

$$B(x) = \frac{1}{\sqrt{V}}\frac{1}{\sqrt{(2E_\mathbf{q})}}\frac{1}{\langle 0|T\psi_A(0)C^{-1}\psi_B(0)|\mathbf{q}\rangle}T\psi_A(x)C^{-1}\psi_B(x) \tag{52}$$

where a limiting process similar to (49) is understood, and where we have taken the combination $\psi C^{-1}\psi$ in order to form a Lorentz scalar, C being the charge conjugation matrix. The normalization of $B(x)$ is

$$\langle 0|B(x)|\mathbf{q}\rangle = \frac{1}{\sqrt{V}}\frac{1}{\sqrt{(2E_{\mathbf{q}})}}\,\mathrm{e}^{\mathrm{i}q\cdot x} \tag{53}$$

owing to our choice of factors in (52). The normalization is, of course, arbitrary, but once it is fixed it determines the proportionality coefficient appearing in the asymptotic condition. With our choice (53), we have, by Zimmermann's argument,

$$\lim_{t\to\mp\infty} \langle m|b_{\mathbf{k}}(t)|n\rangle = \langle m|b_{\mathbf{k}}^{\,\substack{\mathrm{in}\\ \mathrm{out}}}|n\rangle \tag{54}$$

with a proportionality coefficient equal to 1.

We can conclude that as far as the asymptotic condition and its consequences—i.e. the reduction formulae—are concerned, there is no essential distinction between an elementary and a composite particle. Both can be represented by a variety of local fields. Moreover, it is clear that the *same* basic field featured in the Lagrangian can be employed to represent both elementary and composite particles, assuming the latter has the same quantum numbers as one of the basic fields.

2 Bethe-Salpeter Wave-Functions

2.1 *Non-relativistic theory*

We begin by considering the non-relativistic Schrödinger theory of n identical particles interacting through a general 2-body potential $v(|\mathbf{x}-\mathbf{x}'|)$. It is well known that this theory can be cast in field theoretic form with the Hamiltonian

$$H = \int \mathrm{d}^3x\psi^\dagger(\mathbf{x},t)\left(-\frac{1}{2m}\nabla^2\right)\psi(\mathbf{x},t) + \\ +\frac{1}{2}\int \mathrm{d}^3x\int \mathrm{d}^3x'\psi^\dagger(\mathbf{x}',t)\psi^\dagger(\mathbf{x},t)v(|\mathbf{x}-\mathbf{x}'|)\psi(\mathbf{x},t)\psi(\mathbf{x}',t) \tag{55}$$

The field operator $\psi(x)$ satisfies the equation of motion

$$\mathrm{i}\dot\psi(\mathbf{x},t) = [\psi(\mathbf{x},t), H] \\ = -\frac{1}{2m}\nabla^2\psi(\mathbf{x},t) + \int \mathrm{d}^3x'\psi^\dagger(\mathbf{x}',t)v(|\mathbf{x}-\mathbf{x}'|)\psi(\mathbf{x},t)\psi(\mathbf{x}',t) \tag{56}$$

and the canonical equal-time commutation relations

$$[\psi(\mathbf{x}, t), \psi^\dagger(\mathbf{x}', t)] = \delta^{(3)}(\mathbf{x} - \mathbf{x}') \tag{57a}$$

$$[\psi(\mathbf{x}, t), \psi(\mathbf{x}', t)] = [\psi^\dagger(\mathbf{x}, t), \psi^\dagger(\mathbf{x}', t)] = 0 \tag{57b}$$

in the case of Bose-Einstein statistics. For Fermi-Dirac statistics, the above commutation rules must of course be replaced by anticommutation rules.

Let us define the particle-number operator

$$N = \int \psi^\dagger(\mathbf{x}, t)\psi(\mathbf{x}, t)\, \mathrm{d}^3x \tag{58}$$

Applying (57a, b) we find that

$$[N, H] = 0 \tag{59}$$

so that the number of particles is a constant of the motion for the non-relativistic theory. Moreover, the momentum

$$\mathbf{P} = \int \psi^\dagger(\mathbf{x}, t)(-\mathrm{i}\,\nabla)\psi(\mathbf{x}, t)\, \mathrm{d}^3x \tag{60}$$

also commutes with H

$$[\mathbf{P}, H] = 0 \tag{61}$$

so that H, $\mathbf{P}$ and N are simultaneously diagonalizable. A complete set of states can therefore be formed by taking the set of all states

$$|n, \mathbf{k}, E, \alpha\rangle \tag{62}$$

where n, $\mathbf{k}$, and E denote eigenvalues of N, $\mathbf{P}$ and H respectively, while α represents the eigenvalues of whatever other quantities are required to form a complete set of observables. From the equal-time commutation rules we deduce that

$$[\psi, N] = \psi \tag{63a}$$

$$[\psi^\dagger, N] = -\psi^\dagger \tag{63b}$$

implying that ψ and $\psi^\dagger$ are single-particle destruction and creation operators respectively. For example, (63b) yields

$$N\psi^\dagger|n\rangle = \psi^\dagger N|n\rangle + \psi^\dagger|n\rangle = (n + 1)\psi^\dagger|n\rangle$$

Defining the vacuum state by

$$\psi(\mathbf{x}, t)|0\rangle = 0 \quad (\text{for all } \mathbf{x}, t) \tag{64}$$

we have, by (55), (58) and (60):

$$H|0\rangle = \mathbf{P}|0\rangle = N|0\rangle = 0 \tag{65}$$

We now define a *local* number operator

$$N_v(t) = \int_v \psi^\dagger(\mathbf{x}, t)\psi(\mathbf{x}, t)\, \mathrm{d}^3x \tag{66}$$

whose eigenvalues represent the number of particles contained in the volume v. From the equal-time commutation rules we have

$$[N_v(t), \psi^\dagger(\mathbf{x}, t)] = \begin{cases} \psi^\dagger(\mathbf{x}, t) & \text{for } \mathbf{x} \text{ contained in } v \\ 0 & \text{for } \mathbf{x} \text{ outside } v \end{cases} \tag{67}$$

and if $v(y)$ is an infinitesimally small volume centred at the point y, (67) implies that

$$N_{v(y)}(t)\psi^\dagger(\mathbf{x}, t)|0\rangle = \begin{cases} \psi^\dagger(\mathbf{x}, t)|0\rangle & \text{if } x = y \\ 0 & \text{if } x \neq y \end{cases} \tag{68}$$

Thus we can interpret $\psi^\dagger(\mathbf{x}, t)$ as the creation operator for a single particle localized at $\mathbf{x}$ at time t. The state

$$|\mathbf{x}_1, \mathbf{x}_2, t\rangle = \frac{1}{\sqrt{2!}}\, \psi^\dagger(\mathbf{x}_1, t)\psi^\dagger(\mathbf{x}_2, t)|0\rangle \tag{69}$$

represents a 2-particle state localized at $\mathbf{x}_1$ and $\mathbf{x}_2$ at time t. The possibility of constructing perfectly localized states is a unique feature of the non-relativistic theory; it does not occur in relativistic quantum field theory.

The scalar product

$$\langle \mathbf{x}_1, \mathbf{x}_2, t|2, \mathbf{k}, E, \alpha\rangle = \frac{1}{\sqrt{2!}}\langle 0|\psi(\mathbf{x}_1, t)\psi(\mathbf{x}_2, t)|2, \mathbf{k}, E, \alpha\rangle \tag{70}$$

is the probability amplitude for finding the two particles in the state $|2, \mathbf{k}, E, \alpha\rangle$ at the positions $\mathbf{x}_1$ and $\mathbf{x}_2$ at time t. In fact the amplitude (70) is just the Schrödinger wave-function for the two-boson system. It is symmetric under interchange of $\mathbf{x}_1$ and $\mathbf{x}_2$ by virtue of (57b). The correct normalization

$$\int \mathrm{d}^3x_1\, \mathrm{d}^3x_2 \langle 2, \mathbf{k}', E', \alpha'|\mathbf{x}_1, \mathbf{x}_2, t\rangle\langle \mathbf{x}_1, \mathbf{x}_2, t|2, \mathbf{k}, E, \alpha\rangle = \delta_{\mathbf{k}'\mathbf{k}}\delta_{E'E}\delta_{\alpha'\alpha} \tag{71}$$

follows from the definition (70) by writing

$$\begin{aligned} &\frac{1}{2!}\int \mathrm{d}^3x_1 \int \mathrm{d}^3x_2 \langle 2, \mathbf{k}', E', \alpha'|\psi^\dagger(\mathbf{x}_1, t)\psi^\dagger(\mathbf{x}_2, t)|0\rangle \\ &\qquad\qquad \langle 0|\psi(\mathbf{x}_2, t)\psi(\mathbf{x}_1, t)|2, \mathbf{k}, E, \alpha\rangle \\ &= \frac{1}{2!}\int \mathrm{d}^3x_1 \int \mathrm{d}^3x_2 \langle 2, \mathbf{k}', E', \alpha'|\psi^\dagger(\mathbf{x}_1, t)\psi^\dagger(\mathbf{x}_2, t)\psi(\mathbf{x}_2, t) \\ &\qquad\qquad \psi(\mathbf{x}_1, t)|2, \mathbf{k}, E, \alpha\rangle \end{aligned} \tag{72}$$

where we have used the fact that $\psi(\mathbf{x}_2, t)\psi(\mathbf{x}_1, t)|2, \mathbf{k}, E, \alpha\rangle$ has non-vanishing components only on $|0\rangle$ to replace $|0\rangle\langle 0|$ by a sum over a complete set of states:

$$|0\rangle\langle 0| \to \sum_n |n\rangle\langle n| = 1$$

Using the equal-time commutation relation (57a) and the definition (58), we easily simplify (72) to

$$\frac{1}{2!}\int d^3x_1\langle 2, \mathbf{k}', E', \alpha'|N^2 - N|2, \mathbf{k}, E, \alpha\rangle = \frac{2!}{2!}\,\delta_{\mathbf{k}'\mathbf{k}}\delta_{E'E}\delta_{\alpha'\alpha} \tag{73}$$

assuming the proper orthonormality relations for the states $|2, \mathbf{k}, E, \alpha\rangle$. Finally, the Schrödinger equation for (70) is recovered by taking the time derivative of both sides and applying the equation of motion (56) for ψ. Using the property

$$\langle 0|\psi^\dagger(\mathbf{x}, t) = 0 \tag{74}$$

we find

$$\begin{aligned} i\,\partial_t\langle \mathbf{x}_1, \mathbf{x}_2, t|2, \mathbf{k}, E, \alpha\rangle &= -\frac{1}{2m}(\nabla_1{}^2 + \nabla_2{}^2)\langle \mathbf{x}_1, \mathbf{x}_2, t|2, \mathbf{k}, E, \alpha\rangle + \\ &+ \int d^3x'\,v(|\mathbf{x}' - \mathbf{x}_2|)\langle 0|\psi(\mathbf{x}_1, t)\psi^\dagger(\mathbf{x}', t)\psi(\mathbf{x}', t)\psi(\mathbf{x}_2, t)|2, \mathbf{k}, E, \alpha\rangle \end{aligned}$$

The second term on the right-hand side may be simplified with the aid of (57a) and (74) and we recover the 2-body Schrödinger equation

$$\begin{aligned} i\,\partial_t\langle \mathbf{x}_1, \mathbf{x}_2, t|2, \mathbf{k}, E, \alpha\rangle &= -\frac{1}{2m}(\nabla_1{}^2 + \nabla_2{}^2)\langle \mathbf{x}_1, \mathbf{x}_2, t|2, \mathbf{k}, E, \alpha\rangle + \\ &+ v(|\mathbf{x}_1 - \mathbf{x}_2|)\langle \mathbf{x}_1, \mathbf{x}_2, t|2, \mathbf{k}, E, \alpha\rangle \end{aligned} \tag{75}$$

Having identified the 2-body Schrödinger wave-function in terms of field theoretic quantities, we now generalize (70) by lifting the equal-time restriction for $\psi(\mathbf{x}_1, t)$ and $\psi(\mathbf{x}_2, t)$. Introducing a time-ordered product

$$T\psi(x_1)\psi(x_2) = \begin{cases} \psi(\mathbf{x}_1, t_1)\psi(\mathbf{x}_2, t_2) & t_1 > t_2 \\ \psi(\mathbf{x}_2, t_2)\psi(\mathbf{x}_1, t_1) & t_2 > t_1 \end{cases} \tag{76a}$$

$$= \tfrac{1}{2}[\psi(x_1), \psi(x_2)] + \tfrac{1}{2}\varepsilon(t_1 - t_2)[\psi(x_1), \psi(x_2)] \tag{76b}$$

we define the Bethe-Salpeter amplitude to be

$$\chi_{\mathbf{k},E,\alpha}(\mathbf{x}_1, t_1; \mathbf{x}_2, t_2) = \langle 0|T\psi(\mathbf{x}_1, t_1)\psi(\mathbf{x}_2, t_2)|2, \mathbf{k}, E, \alpha\rangle \tag{77}$$

To derive the equation of motion for $\chi_{\mathbf{k}E\alpha}$, we apply the operator $(\mathrm{i}\,\partial t_1 + \nabla_1^2/2m)(\mathrm{i}\,\partial t_2 + \nabla_2^2/2m)$ to both sides of (67), with the T-product written in the form (76b), and use the equation of motion (56) in the form

$$\left(\mathrm{i}\frac{\partial}{\partial t} + \frac{1}{2m}\nabla^2\right)\psi(\mathbf{x}, t) = J(\mathbf{x}, t) \tag{78}$$

with

$$J(\mathbf{x}, t) = \int \mathrm{d}^3x'\psi^\dagger(\mathbf{x}', t)v(|\mathbf{x} - \mathbf{x}'|)\psi(\mathbf{x}', t)\psi(\mathbf{x}, t) \tag{79}$$

This yields

$$\begin{aligned}\left(\mathrm{i}\frac{\partial}{\partial t_1} + \frac{1}{2m}\nabla_1^2\right)\left(\mathrm{i}\frac{\partial}{\partial t_2} + \frac{1}{2m}\nabla_2^2\right)\chi_{\mathbf{k},E,\alpha}(\mathbf{x}_1, t_1; \mathbf{x}_2, t_2) = \\ = \langle 0|TJ(\mathbf{x}_1, t_1)J(\mathbf{x}_2, t_2)|2, \mathbf{k}, E, \alpha\rangle + \\ + \mathrm{i}\,\delta(t_1 - t_2)\langle 0|[\psi(\mathbf{x}_1, t_1), J(x_2, t_2)]|2, \mathbf{k}, E, \alpha\rangle\end{aligned} \tag{80}$$

where we have recalled the elementary property $\partial_t\varepsilon(t) = \delta(t)$ and used the equal-time commutation rules to dispose of an additional term proportional to $\delta(t_1 - t_2)\langle 0|[\psi(\mathbf{x}_1, t_1)\psi(\mathbf{x}_2, t_2)]|2, \mathbf{k}, E, \alpha\rangle$ on the right-hand side. Since

$$\langle 0|J(\mathbf{x}, t) = 0 \tag{81}$$

the first-term on the right-hand side of (80) vanishes identically and we are left with the equation

$$\begin{aligned}\left(\mathrm{i}\frac{\partial}{\partial t_1} + \frac{1}{2m}\nabla_1^2\right)\left(\mathrm{i}\frac{\partial}{\partial t_2} + \frac{1}{2m}\nabla_2^2\right)\chi_{\mathbf{k},E,\alpha}(x_1, x_2) = \\ = \mathrm{i}\,\delta(t_1 - t_2)v(|\mathbf{x}_1 - \mathbf{x}_2|)\chi_{\mathbf{k},E,\alpha}(x_1, x_2)\end{aligned} \tag{82}$$

where the right-hand side of (82) results from the equal-time commutator

$$[\psi(\mathbf{x}_1, t_2), J(\mathbf{x}_2, t_2)] = v(|\mathbf{x}_1 - \mathbf{x}_2|)\psi(\mathbf{x}_1, t_2)\psi(\mathbf{x}_2, t_2)$$

Equation (82) is the non-relativistic Bethe-Salpeter equation in differential form[11]. It is valid for both scattering (continuum) and bound (discrete) 2-particle states. The difference between bound states and scattering states emerges upon converting (82) to an integral equation. Let us introduce the non-relativistic propagator function

$$S_F(x_1 - x_2) = \langle 0|T\psi(x_1)\psi^\dagger(x_2)|0\rangle \tag{83}$$

By virtue of (74) we have

$$S_F(x_1 - x_2) = 0 \quad \text{for } t_1 < t_2$$

Operating on both sides of (83) with $i\,\partial_{t_1} + \nabla_1^2/2m$ and using (78) and (81), we see that S_F satisfies

$$\left(i\frac{\partial}{\partial t_1} + \frac{1}{2m}\nabla_1^2\right) S_F(x_1 - x_2) = i\,\delta^{(4)}(x_1 - x_2) \tag{84}$$

We may therefore employ S_F as a Green's function to integrate (82):

$$\chi_{\mathbf{k}E\alpha}(x_1, x_2) = \chi^0_{\mathbf{k}E\alpha}(x_1, x_2) - \int d^4x\, d^4x' S_F(x_1 - x)S_F(x_2 - x')\\ i\,\delta(t - t')v(|\mathbf{x} - \mathbf{x}'|)\chi_{\mathbf{k}E\alpha}(x, x') \tag{85}$$

where $\chi^0_{\mathbf{k}E\alpha}$ satisfies

$$\left(i\frac{\partial}{\partial t_1} + \frac{1}{2m}\nabla_1^2\right)\left(i\frac{\partial}{\partial t_2} + \frac{1}{2m}\nabla_2^2\right) \chi^0_{\mathbf{k}E\alpha}(x_1, x_2) = 0 \tag{86}$$

We now show that

$$\chi^0_{\mathbf{k}E\alpha} = 0 \text{ for bound states} \tag{87}$$

First we note that the Bethe-Salpeter wave-function must depend exponentially on the centre of mass coordinates $\mathbf{X} = \frac{1}{2}(\mathbf{x}_1 + \mathbf{x}_2)$, $T = \frac{1}{2}(t_1 + t_2)$ according to

$$\chi_{\mathbf{k},E,\alpha}(\mathbf{x}_1, t_1; \mathbf{x}_2, t_2) \sim e^{i\mathbf{k}\cdot\mathbf{X} - iET} f(\mathbf{x}_1 - \mathbf{x}_2, t_1 - t_2) \tag{88}$$

as a consequence of invariance under space and time translations. This is proved by using the property that $\mathbf{P}$ and H are, respectively, the generators of space and time translations

$$\psi(\mathbf{x} + x_0, t + t_0) = e^{-i\mathbf{P}\cdot\mathbf{x}_0 + iHt_0}\psi(\mathbf{x}, t)\, e^{i\mathbf{P}\cdot\mathbf{x}_0 - iHt_0} \tag{89}$$

Applying (89) and the properties $\mathbf{P}|0\rangle = H|0\rangle = 0$ to the definition (77) of $\chi_{\mathbf{k},E,\alpha}(x_1, x_2)$ and setting $\mathbf{x}_0 = -\mathbf{X}$, $t_0 = -T$ we recover the result (88). Since (88) must also apply to $\chi^0_{\mathbf{k}E\alpha}$ we may write the general solution to (86) in the form of a Fourier integral

$$\chi^0_{\mathbf{k}E\alpha}(x_1, x_2) = \int d^4k_1\, d^4k_2\, e^{ik_1\cdot x_1 + ik_2\cdot x_2}\delta^{(4)}(k_1 + k_2 - k) \times\\ \times \left[f(k_1, k_2)\delta\left(k_{10} - \frac{\mathbf{k}_1^2}{2m}\right)\delta\left(k_{20} - \frac{\mathbf{k}_2^2}{2m}\right)\right.\\ + g(k_1, k_2)\delta\left(k_{10} - \frac{\mathbf{k}_1^2}{2m}\right)\\ \left. + h(k_1, k_2)\delta\left(k_{20} - \frac{\mathbf{k}_2^2}{2m}\right)\right] \tag{90}$$

where the factor $\delta^{(4)}(k_1 + k_2 - k)$ with $k = (\mathbf{k}, \mathrm{i}\, E)$ guarantees that $\chi^0{}_{\mathbf{k},E,\alpha}(x_1, x_2)$ has the correct dependence on $X = \frac{1}{2}(x_1 + x_2)$ in accordance with (88), and where the factors $\delta\left(k_{10} - \dfrac{\mathbf{k}_1{}^2}{2m}\right)$ and $\delta\left(k_{20} - \dfrac{\mathbf{k}_2{}^2}{2m}\right)$ ensure that $\chi^0{}_{\mathbf{k},E,\alpha}$ satisfies (86). Now by applying $(\mathrm{i}\,\partial t_2 + \nabla_2{}^2/2m)$ to both sides of (77) with the T-product written in the form (76b), it is a simple matter to check that

$$
\begin{aligned}
\left(\mathrm{i}\frac{\partial}{\partial t_2} + \frac{1}{2m}\nabla_2{}^2\right)\chi_{\mathbf{k},E,\alpha}(x_1, x_2) &= \langle 0|T\psi(x_1)J(x_2)|\mathbf{k}, E, \alpha\rangle \\
&= \int \mathrm{d}^4x \langle 0|T\psi(x_1)\psi^\dagger(x)|0\rangle\langle 0|\psi(x)\psi(x_2)|\mathbf{k}, E, \alpha\rangle\delta(t_2 - t)v(|\mathbf{x}_2 - \mathbf{x}|) \\
&= \int \mathrm{d}^4x S_F(x_1 - x)\chi_{\mathbf{k},E,\alpha}(x, x_2)\delta(t_2 - t)v(|\mathbf{x}_2 - \mathbf{x}|) \qquad (91)
\end{aligned}
$$

On the other hand, (85) yields

$$
\begin{aligned}
\left(\mathrm{i}\frac{\partial}{\partial t_2} + \frac{1}{2m}\nabla_2{}^2\right)\chi_{\mathbf{k},E,\alpha}(x_1, x_2) &= \left(\mathrm{i}\frac{\partial}{\partial t_2} + \frac{1}{2m}\nabla_2{}^2\right)\chi^0{}_{\mathbf{k},E,\alpha}(x_1, x_2) + \\
&+ \int \mathrm{d}^4x S_F(x_1 - x)\chi_{\mathbf{k},E,\alpha}(x, x_2)\delta(t_2 - t)v(|\mathbf{x}_2 - \mathbf{x}|) \qquad (92)
\end{aligned}
$$

by virtue of (84). Comparing (92) with (91), we infer that

$$\left(\mathrm{i}\frac{\partial}{\partial t_2} + \frac{1}{2m}\nabla_2{}^2\right)\chi^0{}_{\mathbf{k},E,\alpha}(x_1, x_2) = 0 \qquad (93)$$

and hence that

$$g(k_1, k_2) = 0 \qquad (94)$$

in (90). In a similar way, we can show that

$$\left(\mathrm{i}\frac{\partial}{\partial t_1} + \frac{1}{2m}\nabla_1{}^2\right)\chi^0{}_{\mathbf{k},E,\alpha}(x_1, x_2) = 0 \qquad (95)$$

and hence that

$$h(k_1, k_2) = 0 \qquad (96)$$

It follows that

$$
\begin{aligned}
\chi^0{}_{\mathbf{k},E,\alpha}(x_1, x_2) = \int \mathrm{d}^4k_1\, \mathrm{d}^4k_2\, e^{ik_1\cdot x_1 + ik_2\cdot x_2}\delta^{(4)}(k_1 + k_2 - k) \times \\
\times f(k_1, k_2)\delta\left(k_{10} - \frac{\mathbf{k}_1{}^2}{2m}\right)\delta\left(k_{20} - \frac{\mathbf{k}_2{}^2}{2m}\right) \qquad (97)
\end{aligned}
$$

But this integral vanishes identically in the case of a bound state, as the delta function restrictions $k_{10} + k_{20} = E$, $k_{10} = \dfrac{\mathbf{k}_1{}^2}{2m}$, $k_{20} = \dfrac{\mathbf{k}_2{}^2}{2m}$, are incompatible with a negative value $E < 0$ for the bound state energy. Thus the inhomogeneous term in (85) drops out for a bound state and we are left the homogeneous Bethe-Salpeter integral equation

$$\chi_{\mathbf{k}E\alpha}(x_1, x_2) = -\int \mathrm{d}^4x\, \mathrm{d}^4x' S_F(x_1 - x) S_F(x_2 - x') \mathrm{i}\, \delta(t - t') v(|\mathbf{x} - \mathbf{x}'|) \times \\ \times \chi_{\mathbf{k}E\alpha}(x, x') \qquad (98)$$

The integral form of the Bethe-Salpeter equation can also be obtained directly from the integral equation for the *2-body propagator*. This approach will be followed in dealing with the relativistic field theoretic problem, to which we now turn.

2.2 *Relativistic Bethe-Salpeter equation*

As a convenient model, we consider the interaction of two distinguishable spinor particles A and B with a neutral scalar field $\phi(x)$

$$\mathscr{L}_{\mathrm{int}}(x) = G_0 \tfrac{1}{2}[\bar{\psi}_A(x), \psi_A(x)]\phi(x) + (A \to B) \qquad (99)$$

and we assume the existence of a bound state $|k\rangle$ of mass M. For simplicity we assume the bound state to be non-degenerate, so that it is uniquely characterized by its energy-momentum eigenvalue k_μ. We seek to derive the Bethe-Salpeter integral equation for

$$\chi_k(x_1, x_2) = \langle 0|T\psi_A(x_1)\psi_B(x_2)|k\rangle \qquad (100)$$

from the integral equation for the 2-body propagator

$$K(x_1, x_2; x_3, x_4) = -\langle 0|T\psi_A(x_1)\psi_B(x_2)\bar{\psi}_A(x_3)\bar{\psi}_B(x_4)|0\rangle \qquad (101)$$

The integral equation for K, derived from perturbation theory, has the form[12]

$$K(1, 2; 3, 4) = S_F'^A(1, 3) S_F'^B(2, 4) - \int \mathrm{d}^4x_5\, \mathrm{d}^4x_6\, \mathrm{d}^4x_7\, \mathrm{d}^4x_8 \times \\ \times K(1, 2; 5, 6) G(5, 6; 7, 8) S_F'^A(7, 3) S_F'^B(8, 4) \qquad (102)$$

where G is the interaction function corresponding to the sum of all *irreducible* graphs, that is all graphs which cannot be divided into two disjoint parts by cutting only two fermion propagator lines. An example of a *reducible* graph is shown in figure 1. The iterative solution of the

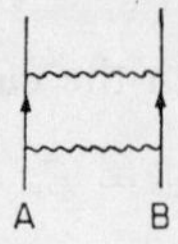

FIGURE 1

inhomogeneous integral equation (102),

$$K(1, 2; 3, 4) = S_F'^A(1, 3)S_F'^B(2, 4) - \int d^4x_5\, d^4x_6\, d^4x_7\, d^4x_8 \times$$
$$\times S_F'^A(1, 5)S_F'^B(2, 6)G(5, 6; 7, 8)S_F'^A(7, 3)S_F'^B(8, 4) + \ldots \quad (103)$$

reproduces the perturbation expansion of the right-hand side of (101), as given by expressing the Heisenberg picture operators and states in terms of the interaction picture and using Wick's theorem[13]. The integral equation for K can also be written in the equivalent form

$$K(1, 2; 3, 4) = S_F'^A(1, 3)S_F'^B(2, 4) - \int d^4x_5\, d^4x_6\, d^4x_7\, d^4x_8 \times$$
$$\times S_F'^A(1, 5)S_F'^B(2, 6)G(5, 6; 7, 8)K(7, 8; 3, 4) \quad (104)$$

In practice one works with the *ladder approximation*, in which $G(5, 6; 7, 8)$ is replaced by its lowest order value

$$G^0(5, 6; 7, 8) = G_0{}^2\delta^{(4)}(x_5 - x_7)\delta^{(4)}(x_6 - x_8)\Delta_F(x_5 - x_6) \quad (105)$$

and the exact propagator S_F' is replaced by S_F. Then (103) reduces to

$$K(1, 2; 3, 4) = S_F{}^A(1, 3)S_F{}^B(2, 4) - \int d^4x_5\, d^4x_6 S_F{}^A(1, 5)S_F{}^B(2, 6) \times$$
$$\times G_0{}^2\Delta_F(5, 6)S_F{}^A(5, 3)S_F{}^B(6, 4) + \ldots \quad (106)$$

represented diagrammatically by the ladder graph expansion of figure 2.

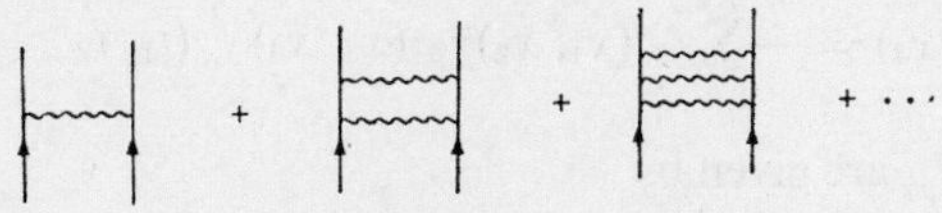

FIGURE 2

Let us transform (102) and (104) to momentum space. It is easy to show, using (89) and the invariance of the vacuum under space-time translations, that $K(x_1, x_2; x_3, x_4)$ and $G(x_1, x_2; x_3, x_4)$ can only depend on differences of coordinates, say

$$x = x_1 - x_2$$
$$x' = x_3 - x_4$$
$$X - X' = \tfrac{1}{2}(x_1 + x_2) - \tfrac{1}{2}(x_3 + x_4)$$

Accordingly, we may define Fourier transforms $K(p, q, k)$ and $G(p, q, k)$ by

$$K(x_1, x_2; x_3, x_4) = \frac{1}{(2\pi)^8}\int d^4p\, d^4q\, d^4k\, e^{ik\cdot(X-X')}\, e^{ip\cdot x}\, e^{-iq\cdot x'} K(p, q, k) \quad (107)$$

and

$$G(x_1, x_2; x_3, x_4) = \frac{1}{(2\pi)^8} \int d^4p\, d^4q\, d^4k\, e^{ik\cdot(X-X')}\, e^{ip\cdot x}\, e^{-iq\cdot x'} G(p, q, k) \quad (108)$$

We also define the Fourier transform $S_F'(p)$ by

$$S_F'(x_1, x_2) = \frac{1}{(2\pi)^4} \int d^4p\, e^{ip\cdot(x_1-x_2)} S_F'(p) \quad (109)$$

and set

$$I(p, q, k) = \delta^{(4)}(p - q)[S_F'^A(\tfrac{1}{2}k + p) S_F'^B(\tfrac{1}{2}k - p)]^{-1} \quad (110)$$

It is then a matter of straightforward calculation to show that (102) and (104) can be written in the form

$$\int d^4q' K(p, q', k)[I(q', q, k) + G(q', q, k)] = \delta^{(4)}(p - q) \quad (111a)$$

and

$$\int d^4p'[I(p, p', k) + G(p, p', k)]K(p', q, k) = \delta^{(4)}(p - q) \quad (111b)$$

respectively.

What is the contribution of the bound state $|k\rangle$ to the 2-body propagator? If we select the time-ordering $t_1, t_2 > t_3, t_4$ in (101) and insert a sum over a complete set of states $|p, \alpha\rangle$ between ψ_B and $\bar{\psi}_A$ we have

$$K(x_1, x_2; x_3, x_4) = -\sum_{p,\alpha} \chi_{p\alpha}(x_1, x_2) \bar{\chi}_{p\alpha}(x_3, x_4) \quad (t_1, t_2 > t_3, t_4) \quad (112)$$

where $\chi_{p\alpha}$ and $\bar{\chi}_{p\alpha}$ are given by

$$\chi_{p\alpha}(x_1, x_2) = \langle 0|T\psi_A(x_1)\psi_B(x_2)|p, \alpha\rangle \quad (113a)$$

$$\bar{\chi}_{p\alpha}(x_1, x_2) = \langle p\alpha|T\bar{\psi}_A(x_1)\bar{\psi}_B(x_2)|0\rangle \quad (113b)$$

The contribution of the bound state to $K(x_1, x_2; x_3, x_4)$ is therefore given by

$$-\int d^4k\, \chi_k(x_1, x_2)\bar{\chi}_k(x_3, x_4)\theta(k_0)\delta(k^2 + M^2)\theta(X_0 - X_0' - \tfrac{1}{2}|x_0| - \tfrac{1}{2}|x_0'|) \quad (114)$$

where we have imposed the restriction $t_1, t_2 > t_3, t_4$ by means of the factor $\theta(X_0 - X_0' - \frac{1}{2}|x_0| - \frac{1}{2}|x_0'|)$ which satisfies

$$\theta(X_0 - X_0' - \tfrac{1}{2}|x_0| - \tfrac{1}{2}|x_0'|) = \begin{cases} 1 & \text{for } t_1, t_2 > t_3, t_4 \\ 0 & \text{for all other time orderings} \end{cases} \quad (115)$$

as is easily verified. Factoring out the centre-of-mass dependence of $\chi_k(x_1, x_2)$ and $\bar{\chi}_k(x_3, x_4)$ as in (88), i.e.

$$\chi_k(x_1, x_2) = \frac{1}{(2\pi)^{3/2}} e^{ik\cdot X}\chi_{k\alpha}(x) \tag{116a}$$

$$\bar{\chi}_k(x_1, x_2) = \frac{1}{(2\pi)^{3/2}} e^{-ik\cdot X}\bar{\chi}_{k\alpha}(x) \tag{116b}$$

and observing that

$$\theta(k_0)\delta(k^2 + M^2) = (2k_0)^{-1}\delta(k_0 - \omega_{\mathbf{k}}) \tag{117}$$

with $\omega_{\mathbf{k}} = (\mathbf{k}^2 + M^2)^{1/\alpha}$ we can write (114) in the form

$$-\frac{1}{(2\pi)^3}\int \frac{d^3k}{2\omega_{\mathbf{k}}}\chi_k(x)\bar{\chi}_k(x')\, e^{i\mathbf{k}\cdot(\mathbf{X}-\mathbf{X}')}\, e^{-i\omega_{\mathbf{k}}(X_0-X_0')}$$
$$\theta(X_0 - X_0' - \tfrac{1}{2}|x_0| - \tfrac{1}{2}|x_0'|) \tag{118}$$

Substituting the formula

$$\theta(y_0) = -\frac{1}{2\pi i}\int dp_0 \frac{1}{p_0 + i\varepsilon}\, e^{-ip_0 y_0} \tag{119}$$

and making the change of variables $p_0 \to k_0 - \omega_{\mathbf{k}}$, we obtain the contribution of the bound state to $K(x_1, x_2; x_3, x_4)$ in the form

$$-\frac{i}{(2\pi)^4}\int d^3k \frac{dk_0}{2\omega_{\mathbf{k}}}\, e^{i\mathbf{k}\cdot(\mathbf{X}-\mathbf{X}')} e^{-ik_0(X_0-X_0')}\frac{1}{k_0 - \omega_{\mathbf{k}} + i\varepsilon}\chi_k'(x)\chi_k'(x') \tag{120}$$

where we have defined new amplitudes

$$\chi_k'(x) = e^{\frac{1}{2}i(k_0-\omega_{\mathbf{k}})|x_0|}\chi_k(x) \tag{121a}$$

$$\bar{\chi}_k'(x) = e^{\frac{1}{2}i(k_0-\omega_{\mathbf{k}})|x_0|}\bar{\chi}_k(x) \tag{121b}$$

Defining Fourier transforms $\chi_k'(p)$ and $\bar{\chi}_k'(q)$ by

$$\chi_k'(x) = \frac{1}{(2\pi)^4}\int d^4p\, e^{ip\cdot x}\chi_k'(p) \tag{122a}$$

$$\bar{\chi}_k'(x) = \frac{1}{(2\pi)^4}\int d^4q\, e^{-iq\cdot x}\bar{\chi}_k'(q) \tag{122b}$$

the bound state contribution (120) becomes

$$-\frac{i}{(2\pi)^{12}}\int d^4p\, d^4q\, d^4k\, e^{ik\cdot(X-X')}\, e^{ip\cdot x}\, e^{-iq\cdot x'}$$
$$\times\frac{1}{2\omega_{\mathbf{k}}}\frac{1}{k_0 - \omega_{\mathbf{k}} + i\varepsilon}\chi_k'(p)\bar{\chi}_k'(q) \tag{123}$$

This means that the Fourier transform $K(p, q, k)$ defined by (107) develops a pole at

$$k_0 = \omega_{\mathbf{k}} = (\mathbf{k}^2 + M^2)^{\frac{1}{2}} \tag{124}$$

due to the presence of the bound state. Since χ_k' and $\bar{\chi}_k'$ reduce to the Bethe-Salpeter amplitudes χ_k and $\bar{\chi}_k$ on the $k_0 = \omega_{\mathbf{k}}$ mass shell, the *residue* of $K(p, q, k)$ at the pole is just

$$\frac{-\mathrm{i}}{(2\pi)^4} \frac{1}{2k_0} \chi_k(p)\bar{\chi}_k(q) \tag{125}$$

and we have

$$K(p, q, k) = \frac{-\mathrm{i}}{(2\pi)^4} \frac{1}{2k_0} \frac{1}{k_0 - (\mathbf{k}^2 + M^2)^{\frac{1}{2}} + \mathrm{i}\varepsilon} \chi_k(p)\bar{\chi}_k(q) + \\ + \text{terms regular at } k_0 = \omega_{\mathbf{k}} \tag{126}$$

From (126) and (111b) we can immediately derive the integral equation for the Bethe-Salpeter amplitude $\chi_k(p)$. Multiplying (111b) by $k_0 - \omega_{\mathbf{k}}$ and taking the limit $k_0 \to \omega_{\mathbf{k}}$, the inhomogeneous term

$$(k_0 - \omega_{\mathbf{k}})\delta^{(4)}(p - q)$$

drops out and we are left with

$$\int \mathrm{d}^4p'[I(p, p', k) + G(p, p', k)]\chi_k(p') = 0 \qquad (k_0 = \omega_{\mathbf{k}}) \tag{127}$$

Thus χ_k satisfies the *homogeneous* integral equation associated with (111b). Transforming (127) to configuration space, we find the equation

$$\chi_k(1, 2) = -\int \mathrm{d}^4x_5\, \mathrm{d}^4x_6\, \mathrm{d}^4x_7\, \mathrm{d}^4x_8 S_F'^A(1, 5) S_F'^B(2, 6) \times \\ \times G(5, 6; 7, 8)\chi_k(7, 8) \tag{128}$$

In the ladder approximation, for which $G(5, 6, 7, 8)$ is given by (105), the Bethe-Salpeter equation reduces to

$$\chi_k(x_1, x_2) = -\int \mathrm{d}^4x_5\, \mathrm{d}^4x_6 S_F^A(x_1 - x_5) S_F^B(x_2 - x_6) \times \\ \times G_0^2 \Delta_F(x_5 - x_6)\chi_k(x_5, x_6) \tag{129}$$

or, in differential form

$$(\gamma^A \cdot \partial + m)(\gamma^B \cdot \partial + m)\chi_k(x_1, x_2) = G_0^2 \Delta_F(x_1 - x_2)\chi_k(x_1, x_2) \tag{130}$$

Equations (129) and (130) are structurally similar to the non-relativistic equations (98) and (82) respectively. The conversion of the relativistic Bethe-Salpeter equation into a differential equation is only possible in the ladder approximation. In general, (130) would be replaced by an integro-differential equation.

2.3 *Normalization condition*

The Bethe-Salpeter equation is, like the Schrödinger equation, a homogeneous equation which leaves the normalization of χ_k undetermined. However, the magnitude of χ_k cannot be arbitrary. Like the Schrödinger amplitude (70), which was shown to satisfy (71), we expect that χ_k, defined by (100), will satisfy a certain normalization condition. To determine this condition we can apply the relation (126) between χ_k, $\bar{\chi}_k$ and K. Since K satisfies an *inhomogeneous* equation, its magnitude is essentially fixed. Thus the magnitude of the residue of K at the $k_0 = \omega_{\mathbf{k}}$ pole must be fixed and our task is to extract this information. Let us define an auxiliary quantity[14]

$$Q(p, q, k) = \int \mathrm{d}^4q'(k_0 - \omega_{\mathbf{k}})K(p, q', k) \frac{\partial}{\partial k_0} [I(q', q, k) + G(q', q, k)] \tag{131}$$

defined for both $k_0 = \omega_{\mathbf{k}}$ and $k_0 \neq \omega_{\mathbf{k}}$, i.e. both on and off the bound state mass shell. The use of $Q(p, q, k)$ enables us to derive the normalization condition for χ_k in a very simple, elegant way. Let us regard p and q as continuous indices and write (131) in the compact operator form

$$Q(k) = (k_0 - \omega_{\mathbf{k}})K(k) \frac{\partial}{\partial k_0} [I(k) + G(k)] \tag{132}$$

In terms of this notation, we can write (111a), (127) and (126) as

$$K(k)[I(k) + G(k)] = 1 \tag{133}$$

$$[I(k) + G(k)]\chi_k = 0 \quad (k_0 = \omega_{\mathbf{k}}) \tag{134}$$

$$\lim_{k_0 \to \omega_{\mathbf{k}}} (k_0 - \omega_{\mathbf{k}})K(k) = \frac{-\mathrm{i}}{(2\pi)^4} \frac{1}{2k_0} \chi_k \bar{\chi}_k \tag{135}$$

respectively. Using (133) we can obtain an alternative form for Q by writing

$$Q(k) = \frac{\partial}{\partial k_0} \left\{(k_0 - \omega_{\mathbf{k}})K(k)[I(k) + G(k)]\right\} - \frac{\partial[(k_0 - \omega_{\mathbf{k}})K(k)]}{\partial k_0} \times [I(k) + G(k)]$$

$$= 1 - \frac{\partial[(k_0 - \omega_{\mathbf{k}})K(k)]}{\partial k_0} [I(k) + G(k)] \tag{136}$$

Operating on χ_k with $Q(k)$ in the form (132) we get, using (135) for $k_0 = \omega_{\mathbf{k}}$

$$Q(k)\chi_k = \frac{-i}{(2\pi)^4}\frac{1}{2k_0}\chi_k\bar{\chi}_k\frac{\partial}{\partial k_0}[I(k) + G(k)]\chi_k \qquad (k_0 = \omega_{\mathbf{k}})$$

On the other hand, (136) gives, using (134)

$$Q(k)\chi_k = \chi_k \qquad (k_0 = \omega_{\mathbf{k}})$$

It follows that

$$-\frac{i}{(2\pi)^4}\bar{\chi}_k\frac{\partial}{\partial k_0}[I(k) + G(k)]\chi_k = 2k_0 \qquad (k_0 = \omega_{\mathbf{k}}) \qquad (137)$$

or, explicitly,

$$-\frac{i}{(2\pi)^4}\int d^4q\, d^4q'\,\bar{\chi}_k(q')\frac{\partial}{\partial k_0}[I(q', q, k) + G(q', q, k)]\chi_k(q) = 2k_0$$
$$(k_0 = \omega_{\mathbf{k}}) \qquad (138)$$

Equation (138) is the normalization condition for Bethe-Salpeter amplitudes[14,15,16]. It generalizes the normalization condition (71) for the non-relativistic Schrödinger wave-function.

In practical calculations with the ladder approximation, (138) is considerably simplified by virtue of the fact that $G(p, q, k)$ is independent of k. Corresponding to (105) we have

$$G(p, q, k) = \frac{G_0^2}{(2\pi)^4}\Delta_F(p - q) \qquad (139)$$

The Bethe-Salpeter equation reads, in this approximation,

$$S_F{}^A(\tfrac{1}{2}k + p)^{-1}S_F{}^B(\tfrac{1}{2}k - p)^{-1}\chi_k(p) = -\frac{G_0^2}{(2\pi)^4}\int d^4p'\,\Delta_F(p - p')\chi_k(p') \qquad (140)$$

and the normalization condition reduces to

$$-\frac{i}{(2\pi)^4}\int d^4q\,\bar{\chi}_k(q)\frac{\partial}{\partial k_0}[S_F{}^A(\tfrac{1}{2}k + p)S_F{}^B(\tfrac{1}{2}k - p)]^{-1}\chi_k(q) = 2k_0$$
$$(k_0 = \omega_{\mathbf{k}}) \qquad (141)$$

Although (137) has a somewhat unfamiliar appearance, it is actually the exact analogue of the ordinary normalization condition for elementary particle amplitudes. For example, we can write the normalization $\bar{u}_{\mathbf{p}}u_{\mathbf{p}} = 1$, or equivalently

$$\bar{u}_{\mathbf{p}}\gamma_4 u_{\mathbf{p}} = \frac{p_0}{m} \qquad (142)$$

for Dirac spinors satisfying

$$(\gamma \cdot p - \mathrm{i}\, m)u_{\mathrm{p}} = 0 \tag{143}$$

in the form

$$\mathrm{i}\, \bar{u}_{\mathrm{p}} \frac{\partial}{\partial p_0} (\gamma \cdot p - \mathrm{i}\, m)u_{\mathrm{p}} = -\frac{p_0}{m} \tag{144}$$

in obvious analogy with (137). In fact we can 'derive' (144) in exactly the same way as (137) by starting from the three equations

$$S_F(p)(\gamma \cdot p - \mathrm{i}\, m) = -1$$
$$(\gamma \cdot p - \mathrm{i}\, m)u_{\mathrm{p}} = 0$$
$$\lim_{p_0 \to E_{\mathrm{p}}} (p_0 - E_{\mathrm{p}})S_F(p) = \mathrm{i}\frac{m}{p_0} \sum_{\text{spins}} u_{\mathrm{p}}\bar{u}_{\mathrm{p}}$$

defining the auxiliary quantity $(p_0 - E_{\mathrm{p}})S_F(p) \dfrac{\partial}{\partial p_0} (\gamma \cdot p - \mathrm{i}\, m)$, and following the same steps as in the Bethe-Salpeter case[14]. The technique is quite general. It allows us to extract the normalization condition for the residue of a single-particle pole in any higher order Green's function, once we have the inhomogeneous equation for the Green's function and the homogeneous equation for the residue[17].

3 $Z_3 = 0$ Condition for Composite Bosons

3.1 *Introduction*

We discuss the assertion that a composite boson may be characterized by the vanishing of its wave-function renormalization constant[18,19] by reference to the following model. We consider a 4-fermion coupling theory with the Lagrangian density

$$L_{\mathrm{ff}} = -\bar{\psi}(\gamma \cdot \partial + m_0)\psi - g_0\bar{\psi}\gamma_5\psi\bar{\psi}\gamma_5\psi \tag{145}$$

for which we assume the existence of a pseudoscalar boson bound state of mass μ. We ask whether the above theory can then be replaced by an equivalent Yukawa coupling theory

$$L_{\mathrm{Y}} = -\bar{\psi}(\gamma \cdot \partial + m_0)\psi - \tfrac{1}{2}\partial_\mu\phi\partial_\mu\phi - \tfrac{1}{2}\mu_0{}^2\phi^2 + \mathrm{i}\, G_0\bar{\psi}\gamma_5\psi\phi \tag{146}$$

in which the pseudoscalar boson is represented by an 'elementary' field. One can show[20] that to all orders of perturbation theory, the necessary and sufficient condition is

$$Z_3 = 0 \tag{147}$$

i.e. that the boson wave-function renormalization constant be equal to zero in the Yukawa theory. Here we shall restrict ourselves to deriving this result in the chain approximation and to commenting on its significance.

3.2 *Chain approximation*

We wish to compare the scattering amplitudes generated by the infinite chains of diagrams shown in figure 3 and figure 4 for the 4-fermion

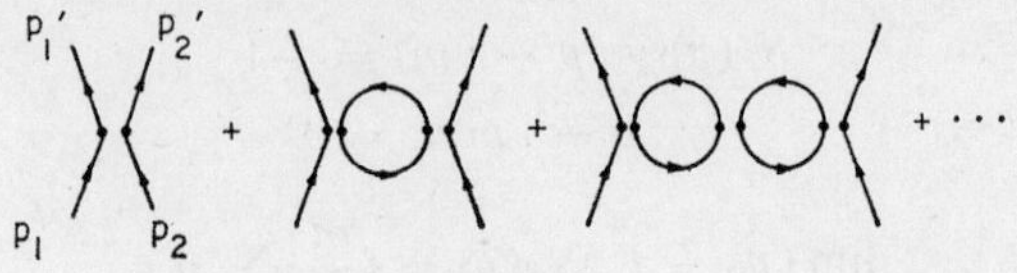

FIGURE 3

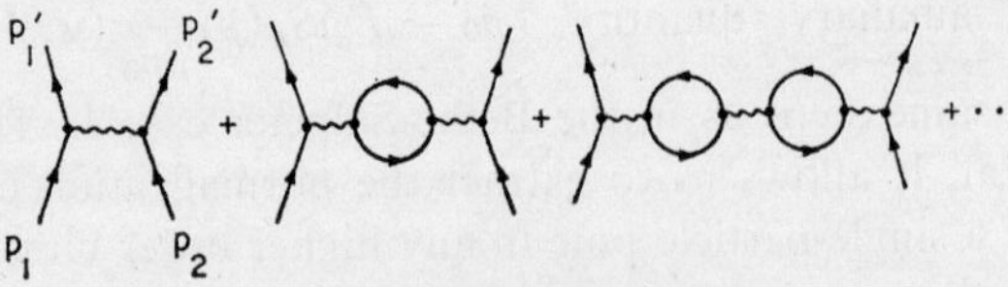

FIGURE 4

and Yukawa theories respectively. Denoting the 4-fermion scattering amplitude by M_f we find, after summing the geometric progression,

$$M_f = 2g_0 \frac{1}{1 - 2g_0 \Pi(q^2)} \tag{148}$$

where $q = p_1' - p_1$ is the momentum transfer. The quantity $\Pi(q^2)$ is given by

$$\Pi(q^2) = \frac{i}{(2\pi)^4} \int d^4 p T_r \left(\gamma_5 \frac{1}{\gamma . p - \gamma . q - i m} \gamma_5 \frac{1}{\gamma . p - i m} \right) \tag{149}$$

and represents the fermion closed loop whose iteration generates the chain approximation. In accordance with our assumption that the 4-fermion theory gives rise to a pseudoscalar bound state of mass μ, we suppose that (148) has a pole at $q^2 = -\mu^2$, i.e. that

$$2g_0 \Pi(-\mu^2) = 1 \tag{150}$$

Expanding $\Pi(q^2)$ around $q^2 = -\mu^2$ we have

$$\Pi(q^2) = \Pi(-\mu^2) + (q^2 + \mu^2)\Pi'(-\mu^2) + \Pi_c(q^2) \tag{151}$$

where Π_c represents the sum of the remaining terms in the expansion. Thus,

$$M_f = \frac{-1}{(q^2 + \mu^2)\Pi'(-\mu^2) + \Pi_c(q^2)} \tag{152}$$

Let us compare (152) with the chain approximation result in the Yukawa theory. Summing the geometric progression corresponding to the infinite set of graphs in figure 4, we get for the Yukawa theory scattering amplitude M_Y

$$M_Y = G_0{}^2 \frac{1}{q^2 + \mu^2 - G_0{}^2\Pi(q^2) - \delta\mu^2} \tag{153}$$

where μ, the physical boson mass, is taken to be identical with the mass of the bound state in the 4-fermion theory, $\delta\mu^2$ represents the contribution of the boson mass renormalization counter-term, while $\Pi(q^2)$ is again given by (149).

Now $\Pi(q^2)$ is given by (149) as a quadratically divergent integral. To isolate and dispose of the quadratically divergent part of $\Pi(q^2)$, renormalization theory instructs us to set

$$\delta\mu^2 = -G_0{}^2\Pi(-\mu^2) \tag{154}$$

so that (153) becomes, using (151),

$$M_Y = G_0{}^2 \frac{1}{(q^2 + \mu^2)[1 - G_0{}^2\Pi'(-\mu^2)] - G_0{}^2\Pi_c(q^2)} \tag{155}$$

Since, as can easily be shown, $\Pi(-\mu^2)$ contains the entire quadratic divergence, M_Y is now only logarithmically divergent. To get rid of the logarithmic divergence, we perform coupling-constant renormalization and define

$$G_R{}^2 = Z_3 G_0{}^2 \tag{156}$$

where

$$Z_3 = 1 + G_R{}^2\Pi'(-\mu^2) \tag{157}$$

Substituting into (155) we find

$$M_Y = G_0{}^2 Z_3 \frac{1}{q^2 + \mu^2 - G_R{}^2\Pi_c(q^2)}$$

$$= G_R{}^2 \frac{1}{q^2 + \mu^2 - G_R{}^2\Pi_c(q^2)} \tag{158}$$

M_Y is now completely divergence-free and expressed in terms of the (finite) renormalized coupling constant G_R and renormalized mass μ.

We may now compare (158) with (152). We immediately see that equality is obtained when

$$G_R^2 = -\frac{1}{\Pi'(-\mu^2)} \tag{159}$$

But by (157) this is just the condition for the boson renormalization constant to vanish:

$$Z_3 = 0 \tag{160}$$

Thus the vanishing of Z_3 appears as the condition for the equivalence of the Yukawa theory with a 4-fermion coupling model in which the pseudoscalar boson appears as a bound state. In this sense, $Z_3 = 0$ may be regarded as a compositeness criterion for the boson. These considerations can be extended to all orders of perturbation theory of the Yukawa and 4-fermion theories[20].

3.3 *General remarks and interpretation*

From the condition $Z_3 = 0$, we deduce from (156) that the *bare* constant G_0 must be *infinite* if the physical constant G_R is finite. From (154) and (150) we get

$$\delta\mu^2 = \mu^2 - \mu_0^2 = -G_0^2\Pi(-\mu^2) = -G_0^2/2g_0 \tag{161}$$

so that μ_0 must also be infinite if μ and $2g_0$ are finite. This suggests another, more intuitive, way of understanding the $Z_3 = 0$ equivalence property. Let us return to the two Lagrangian densities (145) and (146). The equation of motion for the boson field ϕ which follows from (146) is

$$(\Box - \mu_0^2)\phi = -\mathrm{i}\, G_0\bar{\psi}\gamma_5\psi \tag{162}$$

When G_0 and μ_0 are infinite we may write (162) simply as a proportionality relation between ϕ and $\bar{\psi}\gamma_5\psi$

$$\phi = \mathrm{i}\,\frac{G_0}{\mu_0^2}\,\bar{\psi}\gamma_5\psi \tag{163}$$

assuming that we can disregard infinite momentum states. Substituting (163) into the Yukawa Lagrangian (146) we obtain

$$L_y = -\bar{\psi}(\gamma\,.\,\partial + m_0)\psi - \frac{G_0^2}{2\mu_0^2}\,\bar{\psi}\gamma_5\psi\bar{\psi}\gamma_5\psi \tag{164}$$

where we have again neglected the kinetic energy term $-\frac{1}{2}\partial_\mu\phi\partial_\mu\phi$ relative to $-\frac{1}{2}\mu_0^2\phi\phi$. The expression (164) is identical to the 4-fermion coupling Lagrangian (145) if we set

$$g_0 = \frac{G_0^2}{2\mu_0^2} \tag{165}$$

which is identical to the condition (161) for $\mu_0 \to \infty$. We have thus reproduced our earlier results, although in a rougher form.

Strictly speaking, the neglect of $\Box$ relative to μ_0^2 in (162) is justified only if the theory contains a high momentum *cut-off* of some sort. We can also see this directly, in terms of our earlier argument, if we examine the structure of the $Z_3 = 0$ condition (159):

$$G_R^2 = -\frac{1}{\Pi'(-\mu^2)}$$

Since $\Pi'(-\mu^2)$ is logarithmically divergent, we again require a high momentum cut-off to make G_R finite! In this light, the $Z_3 = 0$ condition seems to lie outside the framework of local field theory. It appears as a condition relating the renormalized Yukawa coupling constant, the physical boson and fermion masses and the cut-off.

References

1. Delbourgo, R., Salam, A., and Strathdee, J., 'Multiple field relations', ICTP preprint.
2. Fayyazuddin and Riazuddin, 'Equal-time commutation rules for composite fields and strong interaction sum rules,' EFINS preprint.
3. Haag, R., *Phys. Rev.*, **112,** 669 (1958).
4. Nishijima, K., *Phys. Rev.*, **111,** 995 (1958).
5. Zimmermann, W., *Nuovo Cimento*, **10,** 597 (1958).
6. See also Nishijima, K., in *High Energy Physics and Elementary Particles*, Trieste, 1965, p. 137. This paper contains references to earlier work.
7. Lehmann, H., Symanzik, K., and Zimmermann, W., *Nuovo Cimento*, **1,** 205 (1955).
8. See for example Schweber, S. S., *Introduction to Relativistic Quantum Field Theory*, Row, Peterson & Co., 1961.
9. Ruelle, D., *Helv. Phys. Acta*, **35,** 147 (1962).
10. Gell-Mann, M., and Lévy, M., *Nuovo Cimento*, **16,** 705 (1960).
11. Schweber, S. S., *Ann. Phys.*, **20,** 61 (1962).
12. Salpeter, E. E., and Bethe, H. A., *Phys. Rev.*, **84,** 1232 (1951).
13. Gell-Mann, M., and Low, F. E., *Phys. Rev.*, **84,** 350 (1951).
14. Lurié, D., Macfarlane, A. J., and Takahashi, Y., *Phys. Rev.*, **140B,** 1091 (1965).
15. Allcock, G. R., *Phys. Rev.*, **104,** 1799 (1956).
16. Cutkosky, R. E., and Leon, M., *Phys. Rev.*, **135B,** 1445 (1964).
17. Lurié, D., and Takahashi, Y., *Nuovo Cimento*, **40,** 295 (1965).
18. Jouvet, B., *Nuovo Cimento*, **3,** 1133 (1956); *Nuovo Cimento*, **5,** 1 (1957).
19. Vaughn, M. T., Aaron, R., and Amado, R. D., *Phys. Rev.*, **124,** 1258 (1961).
20. Lurié, D., and Macfarlane, A. J., *Phys. Rev.*, **136B,** 816 (1964).

Some Aspects of the Quark Model for Hadrons

RIAZUDDIN and A. Q. SARKER

1 Introduction

The idea that all strongly interacting particles (hadrons) are composite of a few fundamental objects, was put forward as early as 1956 by Sakata[1]. The shortcomings of the original proposal that the nucleons and Λ^0 are these fundamental units, are well known and will therefore not be elaborated in this review[2]. Later[3], in early 1964, an elegant and simple scheme that the hadrons are composite of a (single) unitary triplet of quarks $q_i(=\mathscr{P},\mathscr{N},\lambda)$, with spin $\frac{1}{2}$, but with fractional charges $Q_i(=\frac{2}{3},-\frac{1}{3},-\frac{1}{3})$, and baryon number $B(=\frac{1}{3},\frac{1}{3},\frac{1}{3})$, was proposed by Gell-Mann[4] and Zweig[5]. The lowest configuration of three quarks, (qqq), then gives the representations **1**, **8** and **10** for baryons that have been observed, while the lowest meson configuration $(q\bar{q})$ gives **1** and **8**. The search for the quarks has not yet been fruitful, and even if they exist, the (free) mass must be greater than 4·5 BeV[6] (there are even theoretical arguments[7] that even if the quarks exist they may not be observable). Or it may be that these fractionally charged fundamental units do not correspond to any physical particles, rather they form a convenient set of symbols in terms of which the symmetries of the strong, electromagnetic and weak interactions are conveniently expressed.

The quark model of hadrons is, in a sense, a dynamical model, but the most attractive feature of it is that it predicts the most of the successful relations, obtained from the SU(3) and SU(6) symmetries (and many more) based on assumptions (sometimes) weaker than the above-mentioned symmetries in a direct way by treating the quarks non-relativistically and leaving the unanswered problems connected with the detailed dynamics of interaction of quarks among themselves and other hadrons. Pertinent remarks in this direction were first made by Beg, Lee and Pais[8] that all their SU(6) results for the baryon magnetic moments can also be obtained by the method of vector addition (assumption of additivity) of magnetic moments by regarding the baryons as composite of quarks with symmetric wave-functions. Following this remark and taking the quark model seriously, Thirring[9], Becchi and Morpurgo[10], Anisovitch, Anselm, Azimov, Damlov and Daytlor[11] and L. D. Soloviev[12] independently calculated the partial decay width of $\omega^0 \rightarrow \pi^0 + \gamma$ which agrees quite well with the experimental value[13].

The relation between the decay coupling constants of $N^* \rightarrow N + \pi$ and $\rho \rightarrow \pi + \pi$, and the pion-nucleon coupling constant can be discussed within the non-relativistic quark model with fair success. Further, the predicted decay rates for the 2^+ mesons from this model agree with the experimental data reasonably well.

Striking predictions have also been obtained for the high energy scattering by Levin and Frankfurt[14], Lipkin and Scheck[15], where the quark model gives for the total cross-sections*

$$\sigma_t(\pi^+ p)/\sigma_t(pp) = \tfrac{2}{3}, \text{ as } s \rightarrow \infty \tag{1}$$

which is to be compared with $\sigma_t(\pi^- p) = 24{\cdot}8$ mb and $\sigma_t(pp) = 38{\cdot}4$ mb at 20 BeV/c[15]. The other predictions that we should mention are the Barger-Rubin[16], and the Johnson-Treiman[17] relations for total cross-sections which agree with experiment reasonably well.

The assumptions that go into the derivation of these high energy cross-section relations are (i) the simple additivity of the quark-quark scattering amplitudes and (ii) the impulse approximation. The assumption of additivity says that the hadron-hadron scattering amplitude is the sum of the 2-body quark-quark scattering amplitudes (multiplied by a form-factor depending upon the momentum transfer only and being unity in the forward direction), in which a single quark of the incident hadron makes encounter with a single quark of the target, the remaining quarks being unaffected.

Double scattering corrections for the meson-baryon scattering have also been evaluated within the quark model of baryons[18]. The cross-section relations which follow are those of the collinear U(3) X U(3) symmetry[19].

The plan of the present review is as follows: in the following two sections the electromagnetic and strong decays of hadrons are discussed and cross-section relations at high energies are derived within the quark model.

* The corresponding prediction from the Sakata model is:

$$\sigma_t(\pi p)/\sigma_t(pp) = 2, \text{ as } s \rightarrow \infty$$

which is in contradiction with the experimental value, 0·64, as quoted.

Only those relations which are at present accessible to experimental verification are dealt with in the main text and the others are given in the Appendix. The double scattering corrections have been evaluated in Section 4.

In the latter half of the paper, we do not consider the quark model as in the first half, but use it simply to abstract the algebraic properties for certain operators which are in principle measurable. It is now well known that the space integrals of the fourth components of vector and axial vector current densities, namely F_i and F_i^5, generate the algebra of chiral SU(3) X SU(3) as suggested by the quark model. This algebra together with the partially conserved axial vector current (PCAC) hypothesis has been used with remarkable success both in strong and weak interactions[20]. This is because the vector and axial vector currents are observables, i.e. their appearance in electromagnetic and weak interactions can be related to physically observable quantities. It was observed by Gell-Mann[21] that if one also includes scalar and pseudoscalar operators S_i and $S_i^5 (i = 0, 1, \ldots, 8)$, which transform as the space integrals of pseudoscalar and scalar densities of quarks, the algebra thus generated leads to SU(6). The question now arises: are scalar and pseudoscalar densities observables? We give them a character of observables by assuming that they give rise to physically observable effects in the form of Gell-Mann-Okubo mass splitting within any SU(3) multiplet and of non-leptonic decays. In particular, except for a scale factor, we identify[22,23,24] S^8 (transforming as eighth component of scalar quark density $\bar{q}\lambda_i q$) with the mass splitting Hamiltonian. Then S_i together with F_i form a non-chiral SU(3) X SU(3) algebra. Using this algebra, we discuss in Section 6 that it is then possible to obtain a uniform picture of Gell-Mann-Okubo mass splittings in any SU(3) multiplet in terms of a more or less uniform mass-splitting parameter for every SU(3) multiplet. One in fact gets more information than simply Gell-Mann-Okubo mass formula for an octet. In particular, one obtains for 1^- and 2^+ nonets, Schwinger[25] formulae in agreement with experiment. We also show that S_3 may play a part in electromagnetic mass differences[26] between members of an isospin multiplet. By including the contribution of S_3 together with the contribution from usual minimal electromagnetic coupling, it is possible to understand all electromagnetic mass differences in terms of a single parameter with the possible exception of $K^0 - K^+$ mass difference where one also obtains the right sign but the magnitude comes out to be somewhat smaller. Possible reasons for this discrepancy are also discussed.

One can also discuss sum rules for meson-baryon and meson-meson coupling constants[27] by assuming that the Hamiltonian, which breaks the

SU(3) symmetry of these vertices, is also proportional to S_8. For this purpose one has to utilize the PCAC hypothesis and a commutation relation between F_i^5 and S_j belonging to the SU(6) algebra. These sum rules come out to be much stronger than those previously obtained on purely symmetry arguments[28].

In Section 7 we discuss non-leptonic decays where we take the Hamiltonian to be of the form[24,29,30]

$$H_W^{NL} = G_s S_7^5 + G_p S_6 \tag{2}$$

With the use of PCAC, techniques of algebra of currents, and some equal-time commutation relations of the SU(6) algebra, it is possible to understand all non-leptonic s-wave or parity-violating (p.v.) hyperon decays in terms of a single parameter, and results come out in close agreement with experiment. However, for p-wave or parity-conserving (p.c.) non-leptonic decays of hyperons, the model is in trouble. The trouble arises because in the model S_6, appearing in H_W^{pc}, and S_8, appearing in the mass-splitting Hamiltonian, belong to the same unitary octet. For such a case it was shown by Coleman and Glashow[31] and B. W. Lee[32] that the H_W^{pc} can be transformed away by a unitary transformation so that no p-wave hyperon decay can occur. In pole model for p-wave decays to which algebra of currents leads[33], it would mean that if strong meson-baryon coupling constants have their SU(3) values, the p-wave amplitudes vanish. Thus in this model the p-wave hyperon decays can occur only through deviations of these coupling constants from their SU(3) values. Since the experimental information about meson-baryon coupling constants is very meagre, one cannot say anything more about the p-wave hyperon decays in this model.

We also discuss K-2π and K-3π decays in the model. By algebra of current techniques, these are related[29,34,35,36,37] and the Hamiltonian (2) predicts decay rate of K-3π relative to K-2π as well as the slope parameters for energy spectra of odd pions in K-3π in agreement with experiment.

2 Electromagnetic and Strong Decays of Hadrons

It has been mentioned in the introduction that the magnetic moments of the baryons can be obtained from that of the quarks on the assumption of simple additivity. One can, of course, take the reverse view that the magnetic moments of quarks are determined in terms of that of protons and test the model by predicting the electromagnetic properties of other hadrons. One of the simplest possibility appeared to be that of calculating the rate, $V \rightarrow P + \gamma$, where V and P represent respectively the vector and pseudoscalar nonets. These transitions are pure $M1$ transitions and depend essentially on the magnetic moments of the quarks. From the assumptions of our model, the magnetic moments of the quarks can be written as

$$\boldsymbol{\mu}_i = \mu_p Q_i \boldsymbol{\sigma}_i \tag{3}$$

where the index $i(= 1, 2, 3)$ specifies the member of the triplet of quarks with charge Q_i and μ_p is the magnetic moment of the proton. The $M1$ transition operator with polarization $\boldsymbol{\epsilon}_a$ of the proton for the ith quark can be written as*

$$M_i^a = \mu_p Q_i \boldsymbol{\sigma}_i \cdot (\mathbf{k} \times \boldsymbol{\epsilon}_a)\, e^{i\mathbf{k}\cdot\mathbf{r}_i} \tag{4}$$

As we are interested in calculating the transition rate for the following processes,

$$\begin{aligned} &\text{(i)} \qquad \omega^0 \rightarrow \pi^0 + \gamma \\ &\text{(ii)} \qquad \rho \rightarrow \pi + \gamma \\ &\text{(iii)} \qquad K^{*0} \rightarrow K^0 + \gamma \end{aligned} \tag{5}$$

from (4) and the wave-functions as given in reference 5 for the mesons, the squares of the transition matrix elements of interest, summed over the polarization of the final proton and averaged over the spin states of the initial vector meson, are easily evaluated as follows:

$$\begin{aligned} |\langle\omega^0|\pi^0\rangle|^2_{\text{av}} &= \tfrac{4}{3}\,\mu_p{}^2k^2 \\ |\langle\rho|\pi\rangle|^2_{\text{av}} &= \tfrac{2}{27}\,\mu_p{}^2k^2 \\ |\langle K^{*0}|K^0\rangle|^2_{\text{av}} &= \tfrac{8}{27}\,\mu_p{}^2k^2 \end{aligned} \tag{6}$$

The decay rates are obtained by multiplying (6) by $(2\pi\rho_f/2k)$ where $\frac{1}{2}k$ is the normalization factor for the photon and $2\pi\rho_f$ is the usual phase-space factor given by

$$2\pi\rho_f = 2\pi\frac{4\pi k^2}{(2\pi)^3} \tag{7}$$

The decay rates are, therefore, given by

$$\Gamma_{iJ} = \left|\langle V|P\rangle\right|^2_{\text{av}} \frac{k}{2\pi} \tag{8}$$

These decay rates are shown in Table 1 and the predicted values are compared with the partial decay rates observed. The agreement in case of

* The orbital magnetic moment term, being proportional to $1/M_q$, is neglected.

$\omega^0 \to \pi^0 + \gamma$, is quite good, and for others no experimental information is yet available.

TABLE 1

The decay rates from expression (8) for the processes $V \to P + \gamma$ in the quark model and their comparison with the experimental data[38]

Process	k(MeV/c)	Decay rate	Partial rate %	Experimental data (%)
$\omega^0 \to \pi^0 + \gamma$	380	1·17	12·0	9·7 ± 0·8
$\rho \to \pi + \gamma$	370	0·12	0·11	—
$K^{*0} \to K^0 + \gamma$	308	0·28	0·56	—

We shall now calculate some of the strong vertices within the quark model; in particular, we obtain the relation between $NN\pi$, $N^*N\pi$ and $\rho \to \pi\pi$ vertices. This has also been discussed by Gursey, Pais and Radicati[40] within the SU(6) symmetry. Their assumptions are that (1) the pseudoscalar octet and the vector nonet belong to the *35* dimensional adjoint representation of SU(6) and (2) the coupling of the vector meson to the isospin current is universal. The following discussion, which follows closely Becchi and Morpurgo[10], depends mainly on the static quark-quark meson interaction, which can be written as

$$H_{q\pi} = \sum_i \sqrt{2}(f_q/\mu)(\tau_i^+\sigma_i)\nabla \cdot \phi(x) + h \cdot c \tag{9}$$

where the summation is over the constituent quarks. τ^+ transforms '$\mathscr{P}$' quarks into a '$\mathscr{N}$' quark and f_q is the $qq\pi$ coupling constant; σ_i is the spin operator and μ is the mass constant taken to be the pion mass. The conventional pion-nucleon interaction Hamiltonian, on the other hand, is given by

$$H_{N\pi} = \sqrt{2}(f/\mu)(\tau^+\sigma)\nabla \cdot \phi(x) + h \cdot c \tag{10}$$

with $f = 0{\cdot}08$.

Taking the matrix elements of (9) and (10) between the neutron and proton states, we obtain

$$f = f_q \langle p\uparrow | \sum_i (\tau_i^+ \sigma_{iz}) | n\uparrow \rangle \tag{11}$$

Using the quark wave-functions for p and n, the right-hand side of the equation is easily seen to be $\frac{5}{3}$. Hence one obtains

$$f_q = \tfrac{3}{5} f \tag{12}$$

the axial vector quark-pion coupling constant in terms of the pion-nucleon coupling constant.

Now the matrix element for the decay, $N^* \to p + \pi^+$ with the Z-components of spins, $\frac{3}{2}$ and $\frac{1}{2}$ respectively for N^* and p, is

$$M_{\uparrow\uparrow} = \langle N^*\uparrow | \sqrt{2}(f_q/\mu) \sum_i (\tau_i{}^+ \sigma_i) | p\uparrow \rangle \nabla \cdot \phi(x)$$

$$= \frac{2f_q}{\mu} \cdot \frac{p_x - \mathrm{i}\, p_y}{\sqrt{(2\omega_p)}} \tag{13}$$

where $\mathbf{p}$ and ω_p are the pion momentum and energy. The decay width is then given by

$$\Gamma(N^* \to p + \pi^+) = \frac{4}{3\pi} \frac{f_q{}^2}{\mu^2} p^3 \frac{m_p}{M_{N^*}}$$

$$= \frac{48}{25} \frac{f^2}{\mu^2} \frac{p^3}{4\pi} \frac{m_p}{M_{N^*}} \tag{14}$$

where m_p and M_{N^*} are the proton and N^* masses. This gives a total width of 80 MeV compared to the experimental value[38] of 120 MeV for the decay $N^* \to p + \pi^+$.

The above procedure can also be applied to the decay $\rho \to \pi + \pi$, in which the triplet $(q\bar{q})$ state makes transition into a singlet $(q\bar{q})$ by spin flip and a pion. The relevant matrix element is

$$M(\rho\uparrow \to \pi + \pi) = \frac{4f_q}{\mu} \frac{p_x - \mathrm{i}\, p_y}{\sqrt{2}} \frac{1}{\sqrt{(2\omega_p)}} \tag{15}$$

This when compared with the relativistic expression

$$M\uparrow^{\mathrm{rel}} = 2g\boldsymbol{\epsilon} \cdot \mathbf{p}/(8M_\rho \omega_p{}^2)^{\frac{1}{2}} \tag{16}$$

gives, with $\omega_p = \frac{1}{2} M_\rho$,

$$g = 2\sqrt{2} f_q M_\rho/\mu \tag{17}$$

Substituting the value of f_q from (12), we obtain

$$\frac{g^2}{4\pi} = \frac{9}{25}\, 8 \left(\frac{M_\rho}{\mu}\right)^2 \frac{f^2}{4\pi} \simeq 7{\cdot}5 \tag{18}$$

with $M_\rho = 765$ MeV. This should be compared with $g^2/4\pi = 2$ obtained from the decay width of ρ and the relativistic expression for the matrix element (16)*.

The above treatment of the $\rho \to \pi + \pi$ decay admits the following criticism:

(i) In the decay $\rho \to \pi + \pi$, the two pions have been treated in an asymmetrical way, one being a quantum of a field and the other a bound state of two quarks $(q\bar{q})$.

(ii) The interaction Hamiltonian (9) is non-relativistic and the resulting matrix element, (15), has then been compared with the relativistic one, (16), introducing certain ambiguities in such comparison.

The strong decay properties of the other well-established hadron multiplet that is described by the quark model are those of the spin 2^+ mesons. Elitzur and others[41] described these 2^+ mesons by wave-functions that are eigenstates of n_λ (number of the strange quark) and that corresponds to the representation (21, 21*) of U(6) X U(6), i.e. the symmetric coupling of both quarks and antiquarks. Since the spins are all parallel, this means also a symmetric coupling in U(3) X U(3), i.e. the representation (6, 6*) of U(3) X U(3). The wave-functions can be written explicitly:

$$
\begin{aligned}
f &= (q_i q_i \bar{q}_i \bar{q}_i)_{I=0,S=2} \\
A_2 &= (q_i q_i \bar{q}_i \bar{q}_i)_{I=1,S=2} \\
K^* &= (q_i q_i \bar{q}_i \bar{\lambda})_{I=\frac{1}{2},S=2} \\
f' &= (q_i \lambda \bar{q}_i \bar{\lambda})_{I=0,S=0}
\end{aligned}
\tag{19}
$$

where i now indicates the non-strange iso-doublet quark. The wave-functions form well-defined linear combinations of the SU(3) representations *1*, *8* and *27*.

The possible 2-body decay rates are (i) two pseudoscalar mesons and (ii) one vector meson and one pseudoscalar meson. Both decays are D-wave to conserve total angular momentum and parity. Since the spin S changes from 2 to zero or 1 in the final state, Elitzur, Rubinstein, Stern and Lipkin first recoupled the total quark spin S from 2 to 0 or 1, then evaluated the decay by simple quark rearrangement from the transformed state. The results and their comparison with the experimental data[38] are shown in Table 2.

* It should be pointed out that the quark model gives a high value for $g^2/4\pi(\simeq 7{\cdot}5)$, as compared with the SU(6) result, $g^2/4\pi \simeq 2{\cdot}9$ and that $(g^2/4\pi \simeq 2)$ from the decay width of ρ. This may be due to the fact that one does not know the effective non-relativistic mass of ρ, as the calculations in the quark model are based upon non-relativistic approximations.

TABLE 2

The predicted decay rates of 2^+ mesons from the quark model and their comparison with the experimental data[38]. The underlined predicted decay rates are taken input. For the SU(3) results for these decay rates the reader is referred to the paper by Glashow and Socolow[63]

Decay modes	Squares of C.G. factor	Quark model prediction with phase space	Experimental data[39]
$A_2 \rightarrow \rho + \pi$	$\frac{1}{4}$	$\underline{100}$	75 ± 15
$K^{**} \rightarrow K^* + \pi$	$\frac{1}{8}$	50·3	33 ± 8
$\rightarrow \rho + K$	$\frac{1}{8}$	14·8	8 ± 5
$\rightarrow \omega + K$	1/24	3·9	$1 \pm 1{\cdot}6$
$f' \rightarrow K^* + \bar{K}$	$\frac{1}{2}$	42·5	<34
$f \rightarrow \pi + \pi$	$\frac{1}{4}$	$\underline{100}$	113 ± 14
$\rightarrow K + \bar{K}$	0	0	3 ± 1
$\rightarrow \eta + \eta$	$\frac{3}{8} \sin^2 \alpha$	$4{\cdot}2 \sin^2 \alpha$	Small
$A_2 \rightarrow \eta + \pi$	$\frac{1}{2} \sin^2 \alpha$	$94 \sin^2 \alpha$	$2{\cdot}4 \pm 2{\cdot}1$
$\rightarrow K + \bar{K}$	0	0	$3{\cdot}1 \pm 1{\cdot}4$
$K^{**} \rightarrow K + \pi$	$\frac{1}{8}$	42	48 ± 8
$\rightarrow K + \eta$	$\frac{3}{8} \sin^2 \alpha$	$38{\cdot}4 \sin^2 \alpha$	$2 \pm 2{\cdot}8$
$f' \rightarrow \pi + \pi$	0	0	<12
$\rightarrow K + \bar{K}$	$\frac{1}{4}$	53·0	$>51 \pm 14$
$\rightarrow \eta + \eta$	$\frac{1}{2} \cos^2 \alpha \sin^2 \alpha$	$(67) \cos^2 \alpha \sin^2 \alpha$	Small

3 Cross-Section Relations from Additivity and Symmetry Assumptions

We now consider the hadron-hadron scattering process

$$A + B \rightarrow A' + B' \tag{20}$$

within the quark model, assuming the simple additivity of 2-body quark scattering amplitudes. The 4-momentum transfer is denoted by

$$\Delta p_\mu = p_\mu{}^{A'} - p_\mu{}^{A} = p_\mu{}^{B} - p_\mu{}^{B'} \tag{21}$$

The invariant scattering amplitude for the process (20) can be written as

$$\begin{aligned} \langle A'B'|T|AB\rangle &= \delta_4(p^{A'} + p^{B'} - p^A - p^B) \times \\ &\times \sum_{ij} \langle q_i{}^{A'} q_j{}^{B'}|t_{ij}|q_i{}^A q_j{}^B\rangle \cdot f_i{}^{AA'}(\Delta p_\mu) \times \\ &\times f_j{}^{BB'}(-\Delta p_\mu) \end{aligned} \tag{22}$$

where $\langle q_i{}^{A'} q_j{}^{B'}|t_{ij}|q_i{}^A q_j{}^B\rangle$ describes the scattering of the ith quark, q_i belonging to A with the jth quark, q_j, of B to produce the ith and jth quark states, $q_i^{A'}$ and $q_j^{B'}$, of A' and B', respectively, the other quarks within A and B remaining unaffected. The additivity assumption says that the amplitude for (20) is the sum of these 2-body quark scattering amplitudes. The form-factors $f_i^{AA'}(\Delta p_\mu)$ and $f_j^{BB'}(-\Delta p_\mu)$ represent the overlap of the quark wave-functions of A and A', and B and B', respectively. The summation (ij) is over all the quark states of A and B.

The dependence of the various factors in (22) on the incident energy s has been discussed by Van Hove and Kokkedee[42]. The form-factors $f_i^{A'A}$, $f_i^{B'B}$, which are unity for the forward scattering, are taken to be independent of s. The same has been assumed also for $\langle q_i^A q_j^B | t_{ij} | q_i^A q_j^B \rangle$ in the case for the elastic scattering at very high energy, provided the cross-sections become constant. This is not so, however, for inelastic processes with decreasing cross-sections.

Since the main assumption of the quark model is the additivity, it is then assumed that the properties carried by the hadrons are the sum of those carried by the constituent quarks, and that this holds not only for baryon number, isospin, etc., but also for the momentum and energy. This assumption cannot be true for a free quark of large mass, $M_q \gg M_B$. However, if one takes an effective mass M_q^{eff} for the bound quark within the hadron, then it can easily be shown that M_q^{eff} is of the order of

$$M_q^{\text{eff}} \approx \tfrac{1}{3} M_B \approx \tfrac{1}{2} \mu_M \tag{23}$$

where M_B and μ_M are the average values of the baryons and mesons respectively. Such a low effective mass of the bound quark is also suggested by the value of the baryon magnetic moments in the quark model, as mentioned in the last section.

3.1 *Total cross-sections*

The quark model for the hadrons along with the assumption of additivity and SU(2) and SU(3) symmetries predicts a large number of relations for the total cross-sections for meson-baryon and baryon-baryon scattering. We shall mention here only those which can be verified with the available experimental data, others are given in the Appendix.

We shall examine the consequences of the simple additivity assumption first and introduce the internal symmetries in steps later.

For the forward elastic scattering amplitude,

$$A' = A,\ B' = B, \text{ and } f_i^{AA}(0) = f_j^{BB}(0) = 1 \tag{24}$$

Now, using the quark wave-functions for the hadrons, as given in reference 5, one easily obtains the following relations between the hadron scattering amplitudes and the 2-body quark scattering amplitude (short hand notation MB, etc., has been used to denote, for example, the meson-baryon scattering amplitude, etc.):

$$(pp) = 4(\mathscr{P}\mathscr{P}) + 4(\mathscr{P}\mathscr{N}) + (\mathscr{N}\mathscr{N}) \tag{25}$$

$$(pn) = 2(\mathscr{P}\mathscr{P}) + 5(\mathscr{P}\mathscr{N}) + 2(\mathscr{N}\mathscr{N}) \tag{26}$$

$$(\bar{p}p) = 4(\bar{\mathscr{P}}\mathscr{P}) + 2(\bar{\mathscr{P}}\mathscr{N}) + 2(\bar{\mathscr{N}}\mathscr{P}) + (\bar{\mathscr{N}}\mathscr{N}) \tag{27}$$

$$(\bar{p}n) = 2(\bar{\mathscr{P}}\mathscr{P}) + (\bar{\mathscr{N}}\mathscr{P}) + 4(\bar{\mathscr{P}}\mathscr{N}) + 2(\bar{\mathscr{N}}\mathscr{N}) \tag{28}$$

$$(\pi^+p) = 2(\mathscr{P}\mathscr{P}) + (\mathscr{P}\mathscr{N}) + 2(\bar{\mathscr{N}}\mathscr{P}) + (\bar{\mathscr{N}}\mathscr{N}) \tag{29}$$

$$(\pi^-p) = 2(\bar{\mathscr{P}}\mathscr{P}) + (\bar{\mathscr{P}}\mathscr{N}) + 2(\mathscr{P}\mathscr{N}) + (\mathscr{N}\mathscr{N}) \tag{30}$$

$$(K^+p) = 2(\mathscr{P}\mathscr{P}) + (\mathscr{P}\mathscr{N}) + 2(\bar{\lambda}\mathscr{P}) + (\bar{\lambda}\mathscr{N}) \tag{31}$$

$$(K^-p) = 2(\bar{\mathscr{P}}\mathscr{P}) + 2(\bar{\mathscr{P}}\mathscr{N}) + 2(\lambda\mathscr{P}) + (\lambda\mathscr{N}) \tag{32}$$

$$(K^+n) = (\mathscr{P}\mathscr{P}) + 2(\mathscr{P}\mathscr{N}) + (\bar{\lambda}\mathscr{P}) + 2(\bar{\lambda}\mathscr{N}) \tag{33}$$

$$(K^-n) = (\bar{\mathscr{P}}\mathscr{P}) + 2(\bar{\mathscr{P}}\mathscr{N}) + (\lambda\mathscr{P}) + 2(\lambda\mathscr{N}) \tag{34}$$

From (25) to (34) the following two sum rules emerge:

$$\sigma_t(\pi^+p) + \sigma_t(\pi^-p) = \tfrac{1}{3}[\sigma_t(pp) + \sigma_t(pn) + \sigma_t(\bar{p}p) + \sigma_t(\bar{p}n)] \tag{35}$$

$$\sigma_t(\pi^+p) - \sigma_t(\pi^-p) = [\sigma_t(pp) - \sigma_t(pn) + \sigma_t(\bar{p}n) - \sigma_t(\bar{p}p)] \tag{36}$$

Eliminating $\bar{p}n$ from (35) and (36) one obtains:

$$\sigma_t(\pi^+p) + 2\sigma_t(\pi^-p) = \sigma_t(pn) + \sigma_t(\bar{p}p) \tag{37}$$

The relation (35) was first obtained by Lipkin and Scheck[15], and (36) and (37) by Chan[43]. It is obvious none of these relations can be obtained from any higher symmetries in which the mesons and baryons are put in different multiplets.

It must be stressed that the relations (35) to (37) are the consequence of the assumption of simple additivity alone. Therefore, the experimental verification of these relations tests the validity of the additivity assumption. The available data at high energies as reported by Galbraith and others[44] have been shown in Table 3. Although the relation (36) agrees quite well with experiment (figure 1), the discrepancies in case of relations (35) and (37) are of the order of 20% at the highest available energy.

TABLE 3

The experimental data[44] for the reactions (35) to (37)

Column 1: momentum in BeV/c
Column 2: $\sigma_t(\pi^+p) + \sigma_t(\pi^-p)$
Column 3: $\frac{1}{3}[(\sigma_t(pp) + \sigma_t(pn) + \sigma_t(\bar{p}p) + \sigma_t(\bar{p}n)]$
Column 4: $\sigma_t(\pi^+p) + \sigma_t(pn) + \sigma_t(\bar{p}p)$
Column 5: $\sigma_t(\pi^-p) + \sigma_t(pp) + \sigma_t(\bar{p}n)$
Column 6: $\sigma_t(\pi^+p) + 2\sigma_t(\pi^-p)$
Column 7: $\sigma_t(pn) + \sigma_t(\bar{p}p)$

1 BeV/c	2 (mb)	3 (mb)	4 (mb)	5 (mb)	6 (mb)	7 (mb)
6	54·7 ± 0·5	67·3 ± 2·5	128·1 ± 3·0	128·6 ± 4·9	82·2 ± 0·8	101·9 ± 2·8
8	52·6 ± 0·5	65·2 ± 2·3	123·3 ± 2·7	124·8 ± 4·8	80·1 ± 0·8	98·2 ± 2·5
10	51·3 ± 0·5	—	—	—	77·8 ± 0·8	—
12	50·1 ± 0·5	61·8 ± 2·3	116·3 ± 2·9	119·1 ± 4·6	76·0 ± 0·8	92·1 ± 2·5
14	49·3 ± 0·5	61·1 ± 2·3	114·8 ± 2·8	117·9 ± 4·6	74·7 ± 0·8	90·9 ± 2·6
16	48·5 ± 0·5	60·3 ± 2·3	112·8 ± 2·7	116·5 ± 4·6	73·6 ± 0·8	89·4 ± 2·5
18	48·5 ± 0·5	57·5 ± 5·0	113·0 ± 5·5	108·1 ± 9·9	73·5 ± 0·8	89·5 ± 5·3
20	48·2 ± 0·5	—	—	—	73·0 ± 0·8	—

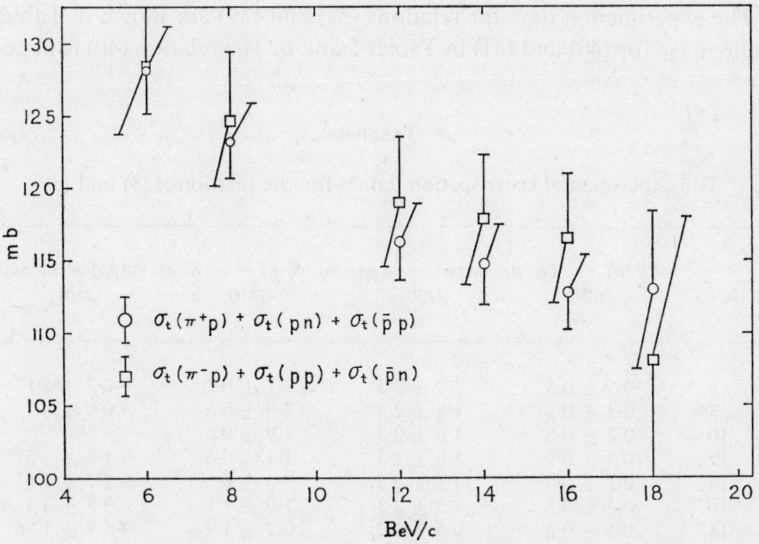

FIGURE 1 The comparison of the relation (36) with the experimental data[45]

If we now assume the SU(2) symmetry, then the following relations hold:

$$(\mathscr{P}\mathscr{P}) = (\mathscr{N}\mathscr{N}), (\mathscr{P}\lambda) = (\mathscr{N}\lambda)$$
$$(\mathscr{P}\bar{\mathscr{P}}) = (\mathscr{N}\bar{\mathscr{N}}), (\mathscr{P}\bar{\lambda}) = (\mathscr{N}\bar{\lambda})$$

Then from (25) to (34), we obtain

$$\sigma_t(K^+p) - \sigma_t(K^+n) = \sigma_t(pp) - \sigma_t(pn) \tag{38}$$

$$\sigma_t(K^-p) - \sigma_t(K^-n) = \sigma_t(\bar{p}p) - \sigma_t(\bar{p}n) \tag{39}$$

$$\sigma_t(K^+p) - \sigma_t(K^-p) = \sigma_t(\pi^+p) - \sigma_t(\pi^-p) + \sigma_t(K^+n) - \sigma_t(K^-n) \tag{40}$$

$$\sigma_t(pp) + \sigma_t(\bar{p}p) = \tfrac{3}{2}[\sigma_t(\pi^+p) + \sigma_t(\pi^-p)] + \tfrac{1}{2}[\sigma_t(K^+p) + \sigma_t(K^-p)] - \tfrac{1}{2}[\sigma_t(K^+n) + \sigma_t(K^-n)] \tag{41}$$

The relation (40) was first obtained by Barger and Rubin[16] from SU(3) with octet dominance in the *t*-channel. It can also be obtained from quark model for only mesons, no model for nucleons.

The experimental data for relations (38) and (39) are shown in Table 4, while those for (40) and (41) in Tables 5 and 6. The relation (40) is in good

TABLE 4

The experimental cross-section data[44] for the relations (38) and (39)

BeV/c	$\sigma_t(K^+n) - \sigma_t(K^+p)$ (mb)	$\sigma_t(pn) - \sigma_t(pp)$ (mb)	$\sigma_t(K^-p) - \sigma_t(K^-n)$ (mb)	$\sigma_t(\bar{p}p) - \sigma_t(\bar{p}n)$ (mb)
6	0·5 ± 0·5	2·0 ± 2·3	2·1 ± 0·7	−0·2 ± 5·1
8	0·3 ± 0·5	1·8 ± 2·3	3·9 ± 0·6	−0·9 ± 4·7
10	0·2 ± 0·5	1·6 ± 2·3	1·9 ± 0·6	—
12	0·3 ± 0·5	1·0 ± 2·3	1·4 ± 0·6	−1·9 ± 4·5
14	0·1 ± 0·5	1·1 ± 2·3	1·4 ± 0·6	−2·7 ± 4·6
16	0·4 ± 0·5	1·5 ± 2·3	1·0 ± 1·1	−3·3 ± 4·6
18	0·5 ± 0·5	0·5 ± 2·3	0·7 ± 1·9	+5·9 ± 12·6
20	0·2 ± 0·5	0·3 ± 2·3	—	—

TABLE 5

The experimental data[44] for the relations (40) and (42)

$$\Delta_{\pi p} = \sigma_t(\pi^- p) - \sigma_t(\pi^+ p)$$

$$\Delta_{Kn} = \sigma_t(K^- n) - \sigma_t(K^+ n)$$

$$\Delta_{Kp} = \sigma_t(K^- p) - \sigma_t(K^+ p)$$

BeV/c	$2\Delta_{\pi p}$ (mb)	$2\Delta_{Kn}$ (mb)	Δ_{Kp} (mb)	$\Delta_{\pi p} + \Delta_{Kn}$ (mb)
6	3·5 ± 0·2	9·6 ± 1·2	6·6 ± 0·7	6·4 ± 0·3
8	4·1 ± 0·3	7·2 ± 0·8	5·7 ± 0·5	5·9 ± 0·2
10	3·2 ± 0·1	5·2 ± 0·8	4·2 ± 0·5	5·3 ± 0·2
12	3·2 ± 0·1	4·8 ± 1·0	4·0 ± 0·6	4·0 ± 0·2
14	2·9 ± 0·1	4·6 ± 1·0	3·7 ± 0·6	3·9 ± 0·2
16	2·9 ± 0·1	6·1 ± 1·4	4·5 ± 0·8	4·0 ± 0·3
18	3·0 ± 1·0	4·4 ± 2·8	3·7 ± 1·9	3·6 ± 0·8
20	2·3 ± 0·4	—	—	3·8 ± 2·5

TABLE 6

The comparison of the relations (41) and (44) with the experimental cross-section data[44]

Column 1: momentum in Bev/c

Column 2: $\sigma_t(pp) + \sigma_t(\bar{p}p)$

Column 3: $\frac{3}{2}[\sigma_t(\pi^+p) + \sigma_t(\pi^-p)] + \frac{1}{2}[\sigma_t(K^+p) + \sigma_t(K^-p)] - \frac{1}{2}[\sigma_t(K^+n) + \sigma_t(K^-n)]$

Column 4: $2[\sigma_t(\pi^+p) + \sigma_t(\pi^-p] - \frac{1}{2}[\sigma_t(K^+p) + \sigma_t(K^-p)]$

1 Bev/c	2 (mb)	3 (mb)	4 (mb)
6	99·9 ± 1·7	82·9 ± 1·4	88·9 ± 1·2
8	96·4 ± 1·4	80·7 ± 1·2	84·7 ± 1·1
10	—	77·8 ± 1·2	82·7 ± 1·1
12	91·1 ± 1·4	75·7 ± 1·2	80·2 ± 1·1
14	89·8 ± 1·5	74·6 ± 1·2	78·7 ± 1·1
16	87·7 ± 2·2	73·6 ± 1·6	77·8 ± 1·2
18	89·0 ± 4·2	72·9 ± 2·0	78·5 ± 1·5
20			76·5 ± 3·3

agreement with the data (figure 2), while for (41) the discrepancies are about 16% (figure 3). Due to the isospin symmetry, either side of (38) and (39) should vanish; therefore, these two relations are not, in fact, any test for the quark model. Experimentally, both sides of (38) are practically zero, while the right-hand side of (39) turns out to be of opposite sign to the left-hand side. Since there are large experimental errors, in particular for $\bar{p}n$ cross-section data, no definite conclusion can yet be made about the validity of the relation (39).

The isospin symmetry and the Pomeranchuk theorem at high energy predicts the asymptotic limit,

$$\sigma_t(\pi^+p)/\sigma_t(pp) = \tfrac{2}{3}, \text{ as } s \to \infty$$

mentioned in the introduction, from both the relations (35) and (41). At 20 BeV/c this ratio is about 0·64 from the available data[38].

(iii) Assuming SU(3) symmetry we have

$$(\mathscr{P}\mathscr{P}) = (\mathscr{N}\mathscr{N}) = (\lambda\lambda)$$
$$(\bar{\mathscr{P}}\mathscr{P}) = (\bar{\mathscr{N}}\mathscr{N}) = (\bar{\lambda}\lambda)$$
$$(\mathscr{P}\mathscr{M}) = (\mathscr{N}\lambda) = (\mathscr{P}\lambda)$$
$$(\mathscr{P}\bar{\mathscr{M}}) = (\mathscr{N}\bar{\lambda}) = (\lambda\bar{\mathscr{P}})$$

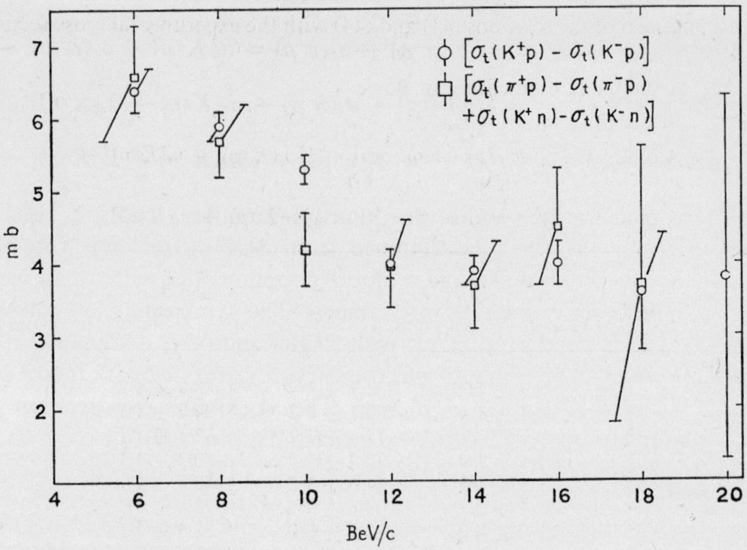

FIGURE 2 The comparison of the relation (40) with the experimental data[45]

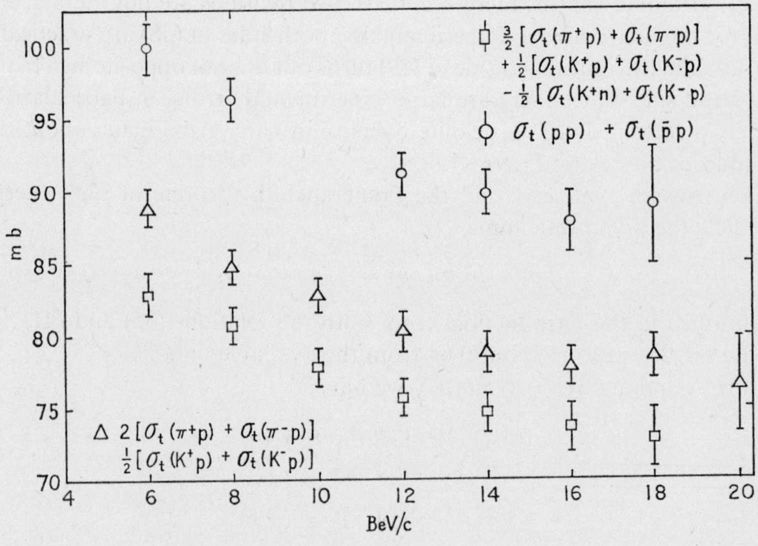

FIGURE 3 The relations (41) and (44) are compared with the experimental data[45]

which result in the following set of sum rules for the total cross-sections:

$$\tfrac{1}{2}[\sigma_t(K^+p) - \sigma_t(K^-p)] = \sigma_t(\pi^+p) - \sigma_t(\pi^-p) = \sigma_t(K^+n) - \sigma_t(K^-n) \quad (42)$$

$$\sigma_t(K^+p) + \sigma_t(K^-p) = \tfrac{1}{2}[\sigma_t(\pi^+p) + \sigma_t(\pi^-p) + \sigma_t(K^+n) + \sigma_t(K^-n)] \quad (43)$$

$$\sigma_t(pp) + \sigma_t(\bar{p}p) = 2[\sigma_t(\pi^+p) + \sigma_t(\pi^-p)] - \tfrac{1}{2}[\sigma_t(K^+p) + \sigma_t(K^-p)] \quad (44)$$

Relation (42) is the well-known Johnson–Treiman[17] relation obtained from SU(6). It can also be obtained from SU(3) symmetry with octet dominance in the *t*-channel and a pure *F* coupling[44a] or with quark model only for nucleons, no model for mesons. The symmetric sum rule (43) can also be obtained from SU(3) with singlet and octet dominance in the *t*-channel and a pure *F* coupling at the baryon vertex or quark model only for the nucleon, no model for mesons. The SU(6) also gives relation (43) with singlet and *35* dominance in the *t*-channel (no. *405*).

The experimental data for relations (42) and (43) are shown in Tables 5 and 7. It is noticed that the latter half of (42) (figure 4) is not so well

TABLE 7

The experimental data[44] for the cross-section relation (43)

BeV/c	$\sigma_t(K^+p) + \sigma_t(K^-p)$ (mb)	$\tfrac{1}{2}[\sigma_t(\pi^+p) + \sigma_t(\pi^-p) + \sigma_t(K^+n) + \sigma_t(K^-n)]$ (mb)
6	41·0 ± 0·4	47·1 ± 0·65
8	40·9 ± 0·3	45·0 ± 0·65
10	39·8 ± 0·3	44·7 ± 0·65
12	38·9 ± 0·3	44·0 + 0·65
14	38·9 ± 0·3	43·5 ± 0·65
16	38·3 + 0·5	43·1 ± 0·75
18	38·1 ± 0·9	43·2 ± 1·0
20	39·9 ± 4·6	—

satisfied compared to the first half of the sum rule. This may be due to the $\pi - K$ mass differences, as the agreement with the data is observed to be better in higher energies than the lower energy range considered. The agreement of (43) with the experimental data is reasonably good; the discrepancies are about 12%. The experimental data for relation (44) are

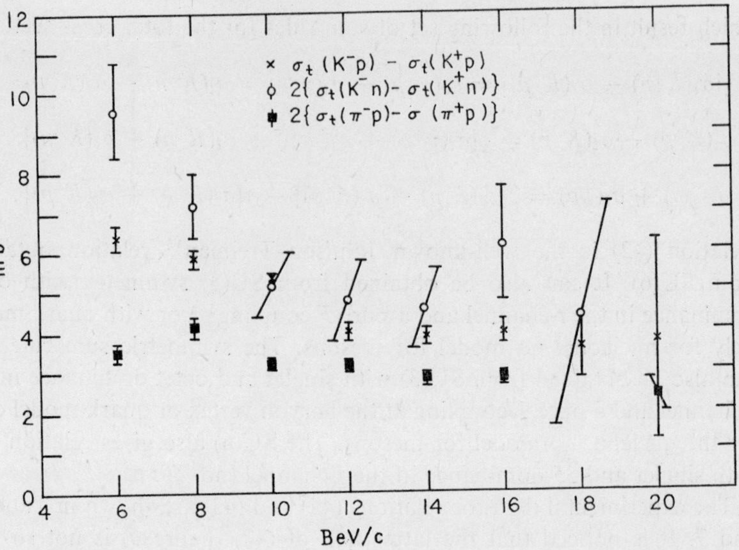

FIGURE 4 The experimental verification of the Johnson–Treiman relation (42)

also shown in Table 6 and figure 3. Relation (44) is found to be in slightly better agreement with the data than relation (41).

3.2 *Production processes*

We shall mention a few examples for inelastic relations, which follow from the quark model with the assumption of additivity and SU(2) symmetry. They have been compared with the experimental data by Lipkin, Scheck and Stern[45]. They are:

$$\bar{\sigma}(\pi^- p \to K^0 \Lambda^0) = \tfrac{9}{4}\bar{\sigma}(\bar{p}p \to \overline{\Sigma}{}^0 \Lambda^0) - \tfrac{1}{12}\bar{\sigma}(\bar{p}p \to \overline{\Lambda}\Lambda) \tag{45}$$

$$\bar{\sigma}(\bar{p}p \to \overline{\Lambda}\Lambda) = 3\bar{\sigma}(\bar{p}p \to \overline{\Sigma}{}^0\Lambda) + 3\bar{\sigma}(\bar{p}p \to \bar{Y}_1{}^{*0}\Lambda) \tag{46}$$

$$\bar{\sigma}(\bar{p}p \to \overline{\Sigma}{}^-\Sigma^+) = \bar{\sigma}(\pi^+ p \to K^+\Sigma^+) + \tfrac{1}{9}\bar{\sigma}(\pi^+ p \to K^{*+}\Sigma^+) \tag{47}$$

where $\bar{\sigma}$ means the cross-section averaged over all initial and summed over all final polarization states. Since the cross-section for $\bar{p}p \to \bar{Y}_1{}^{*0}\Lambda$ is known to be small, then from (46) and (45) one obtains:

$$\bar{\sigma}(\bar{p}p \to \overline{\Lambda}\Lambda) \simeq 3\bar{\sigma}(\bar{p}p \to \overline{\Sigma}{}^0\Lambda) \tag{48}$$

$$\bar{\sigma}(\pi^- p \to K^0\Lambda) \simeq 2\bar{\sigma}(\bar{p}p \to \overline{\Sigma}{}^0\Lambda) \tag{49}$$

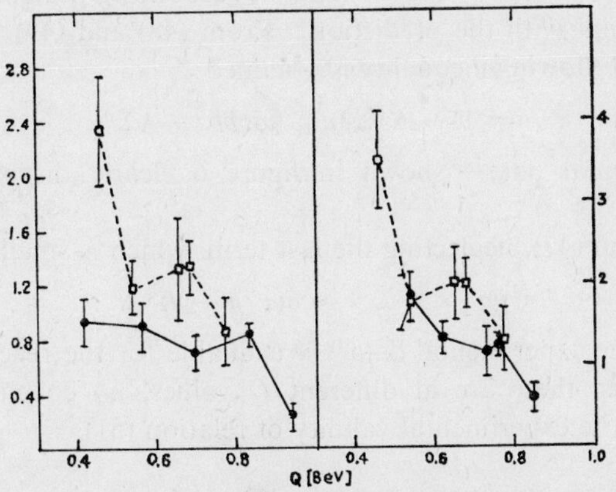

FIGURE 5 (a) The experimental test of the relation (49): □ = $2\bar{\sigma}(\bar{p}p \to \Sigma^0\Lambda)$, ● = $\bar{\sigma}(\pi^- p \to K^0\Lambda)$; (b) Test of the relation (48): □ = $3\bar{\sigma}(\bar{p}p \to \Sigma^0\Lambda)$, ● = $\bar{\sigma}(\bar{p}p \to \bar{\Lambda}\Lambda)$

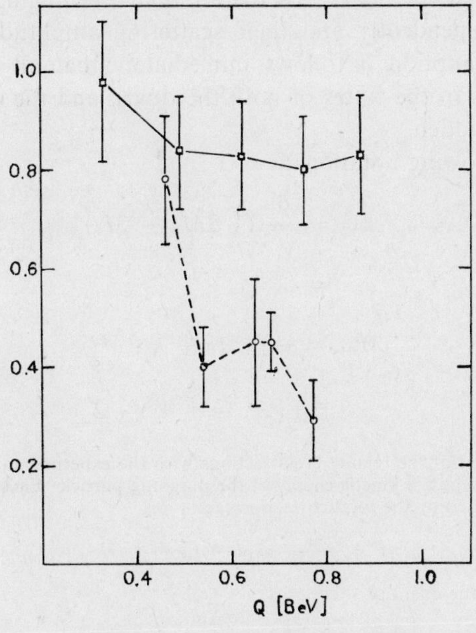

FIGURE 6 The experimental test for the inequality (50): □ = $\bar{\sigma}(\pi^- p \to K^0\Sigma^0)$, o = $\frac{2}{3}\bar{\sigma}(\bar{p}p \to \bar{\Lambda}\Sigma^0)$

The experimental data[46,47] for (48) and (49) are shown in figure 5, showing rough agreement with the prediction*. From (48) and (49), by replacing Λ by Σ^0, the following inequality is obtained:

$$\bar{\sigma}(\pi^- p \to K^0\Sigma^0) < \tfrac{2}{3}\bar{\sigma}(\bar{p}p \to \bar{\Lambda}\Sigma^0) \tag{50}$$

The experimental data[48] shown in figure 6 clearly disagree with the relation (50).

In expression (47), neglecting the last term, which is small, we obtain:

$$\bar{\sigma}(\bar{p}p \to \bar{\Sigma}^-\Sigma^+) \simeq \bar{\sigma}(\pi^+ p \to K^+\Sigma^+) \tag{51}$$

There are some experimental data[48,49] available for the reactions in (51). However, since these are at different Q values, no conclusion can be drawn about the experimental validity of relation (51).

3.3 *Neutral meson production and mixing angles*

Alexander and others[50] have pointed out the possibility of determining the singlet-octet mixing angle for the mesons from the production data. For a reaction of the type

$$\text{meson} + B \to \text{meson}' + B'$$

it is assumed that the quark and the antiquark belonging to the incident meson act independently and their scattering amplitudes are additive. From this assumption it follows immediately that all reactions which require changes in the states of both the quark and the antiquark in the meson are forbidden.

With the following notations:

$$\begin{aligned} M_\lambda &= \frac{1}{\sqrt{3}}(\sqrt{2}M_8 - M_1) \\ &= \lambda\bar{\lambda} \end{aligned} \tag{52}$$

$$\begin{aligned} M_{no} &= \frac{1}{\sqrt{3}}(M_8 + \sqrt{2}M_1) \\ &= (\mathscr{P}\bar{\mathscr{P}} - \mathscr{N}\bar{\mathscr{N}})/\sqrt{2} \end{aligned} \tag{53}$$

* The comparison of the related cross-sections with the experimental data is done at the same Q value, Q being the kinetic energy of the outgoing particle. Each experimental cross-section is then divided by the total correction factor

$$F = \frac{p_{\text{out}}{}^{\text{cm}}}{S p_{\text{in}}{}^{\text{cm}}} \exp[-(A_B + A_M)\Delta^2]$$

in order to obtain the quantity

$$\bar{\sigma} = \sigma_{\text{exp}}/F$$

which appeared in the text. Here Δ is the momentum transfer, and the A's are parameters determined from the elastic cross-section data. Some typical values are: $(A_\pi + A_p) \simeq (A_K + A_p) \simeq 8(\text{BeV/c})^{-2}$ and $(A_{\bar{p}} + A_p) \simeq 13(\text{BeV/c})^{-2}$.

and defining in the standard way the physical ω and ϕ as

$$\omega = M_8 \cos\theta_v - M_1 \sin\theta_v \tag{54}$$

$$\phi = M_8 \sin\theta_v + M_1 \cos\theta_v \tag{55}$$

we obtain the following predictions

$$\begin{aligned}\langle\pi^- p|M_\lambda n\rangle &= \langle\pi^+ n|M_\lambda p\rangle = \langle\pi^+ p|M_\lambda N^{*++}\rangle \\ &= \langle\pi^- p|M_\lambda N^{*0}\rangle = 0\end{aligned} \tag{56}$$

$$\begin{aligned}\frac{\sigma(\pi^- p \to \phi n)}{\sigma(\pi^- p \to \omega n)} &= \frac{\sigma(\pi^+ p \to \phi N^{*++})}{\sigma(\pi^+ p \to \omega N^{*++})} \\ &= \left(\frac{\cos\theta_v - \sqrt{2}\sin\theta_v}{\sin\theta_v + \sqrt{2}\cos\theta_v}\right)^2\end{aligned} \tag{57}$$

and an expression similar to (57) for X^0 and η^0 with the mixing angle α.

The right-hand side of (57) is identically zero for $\theta_v = 35{\cdot}3°$ and is very close to zero for $\theta_v = 40°$. This is in excellent agreement with experimental data[51]. For example, $\sigma(\pi^+ p \to \phi N^{*++})$ is less than 10μb at $p_\pi = 3{\cdot}65$ BeV/c while $\sigma(\pi^+ p \to \omega N^{*++})$ is $700 \pm 80\ \mu$b at 3·65 BeV/c and 300 μb at 4·0 BeV/c. No data are reported for the $\pi^- p$ reactions; however, for $\pi^- n$, appreciable ω production has been reported[52], while no corresponding ϕ production has been observed.

Another set of relations of this kind can be obtained also for reactions induced by the K-meson:

$$\sigma(K^- p \to M_\lambda \Lambda) = \sigma(\pi^- p \to K^{*0}\Lambda) \tag{58}$$

$$\sigma(K^- p \to M_{no}\Lambda) = \sigma(\pi^- p \to \rho\Lambda) \tag{59}$$

Since the physical ϕ and ω can to a good approximation be replaced by M_λ and M_{no} respectively, relations (58) and (59) can be compared directly with experiment. Relation (58), when compared at the same Q value, is in good agreement with the data[53]:

$$\begin{aligned}&\sigma(K^- p \to \phi\Lambda) = 40 \pm 8\ \mu\text{b at } p_K = 3{\cdot}0 \text{ BeV/c} \\ &\sigma(\pi^- p \to K^*\Lambda) = 53 \pm 8\ \mu\text{b at } p_\pi = 2{\cdot}7 \text{ BeV/c} \\ &\quad\text{at about } Q = 0{\cdot}46 \text{ BeV}\end{aligned}$$

From (59) we obtain:

$$\sigma(K^- p \to \omega\Lambda) \simeq \sigma(K^- p \to \rho^0\Lambda) \tag{60}$$

The data for relation (60), as quoted by London and others[54], are in contradiction with the prediction. At 2·24 BeV London and others gave a $\Lambda\omega$ production cross-section much larger than that for $\rho^0\Lambda$; moreover, the $\Lambda\omega$ cross-section is roughly isotropic, whereas the $\rho^0\Lambda$ cross-section is peripheral in character. However, recent data[55,56] of Davis and others at 4·1 and 5·5 BeV/c give quite good agreement with the prediction (60). Although the total cross-sections for the production of $\Lambda\omega$ and $\Lambda\rho^0$ are not yet well known, the preliminary estimate at 4·1 BeV/c are $\sigma(\Lambda\omega) \simeq 41 \pm 15\ \mu b$ and $\sigma(\rho^0\Lambda) \simeq 41 \pm 17\ \mu b$, which are compatible with (60). Further, the ω and ρ^0 distributions agree in showing a strong forward peak together with a secondary backward peak.

3.4 $\bar{p}p$ *Annihilation at rest*

Recently Rubinstein and Stern[57], and Harte and others[58], have discussed the proton-antiproton annihilation within the quark model. The principle assumption is that the three quarks in the proton and the three quarks in the antiproton rearrange themselves into three quark-antiquark pairs, without exchange of spin, isospin or hypercharge. The final state, therefore, contains three non-strange mesons. Moreover, the quark-antiquark pairs are assumed to be $L = 0$ states, so the final state mesons are restricted to be π, η, χ, ρ and ω. The decay rates are then obtained by multiplying the 3-body phase space with the respective rearrangement co-efficient, which is the overlap of the initial $\bar{p}p$ spin wave-function onto the final

TABLE 8

Predicted distribution of final states for $p\bar{p}$ annihilation at rest from the quark model

$\pi^+\pi^-\pi^0$	7·4	$\pi^+\pi^-\rho^0$	25·6	$\pi^0\rho^+\rho^-$	2·9
$\pi^0\pi^0\pi^0$	11·2	$\pi^0\pi^0\rho^0$	4·7	$\pi^+\rho^0\rho^-$	1·4
$\pi^+\pi^-\eta^0$	1·4	$\pi^+\pi^0\rho^-$	1·0	$\pi^-\rho^0\rho^+$	1·4
$\pi^0\pi^0\eta^0$	0·7	$\pi^-\pi^0\rho^+$	1·0	$\pi^+\rho^-\omega^0$	1·0
$\pi^+\pi^-X^0$	1·0	$\pi^+\pi^-\omega^0$	24·0	$\pi^-\rho^+\omega^0$	1·0
$\pi^0\pi^0X^0$	0·5	$\pi^0\pi^0\omega^0$	12·2	$\pi^0\omega^0\omega^0$	0·5

state 3-meson spin state. The model has a number of obvious predictions. The annihilation into a pair of strange mesons, into ϕ mesons (assumed to be composite of $\bar{\lambda}\lambda$) as well as two and four pions is forbidden. Experimentally these decay rates are also found to be small. The other predictions for $\bar{p}p$ annihilation have been shown in Table 8, and their comparison with the experimental data[59] has been made in Table 9. There are discrepancies between the predictions and the observation which are sometimes large. It must, of course, be mentioned that the comparison, as it was made in Table 9, is not free from ambiguities, because many of the channels, which involve several neutral particles, cannot be separated

TABLE 9

Predicted and observed distribution of final states among eight topological categories, for $p\bar{p}$ annihilation at rest. The kaonic annihilation is about 4·6%

Category	1	2	3	4	5	6	7	8
	0 prong	$\pi^+\pi^-$	$\pi^+\pi^-\pi^0$	$\pi^+\pi^-X^0$	$2\pi^+2\pi^-$	$2\pi^+2\pi^-\pi^0$	$2\pi^+2\pi^-X^0$	6 prong
Predicted (%)	12·9	0	7·4	22·3	25·6	28·5	3·1	0·1
Observed (%)	3·2 ± 0·5	0·5 ± 0·1	7·3 ± 0·9	34·8 ± 1·2	5·8 ± 0·3	18·7 ± 0·9	21·3 ± 1·1	3·8 ± 0·2

experimentally, and the identifications of the various resonance states are not well defined. Calculations also have been done for the $\bar{p}p$ annihilation in flight and found to disagree with the experimental data[58,60].

Thus most of the predictions from the quark model about the cross-section relations are in reasonably good agreement with the experimental data, except the relation (50), which predicts

$$\bar{\sigma}(\pi^- p \to K^0\Sigma^0) < \tfrac{2}{3}\bar{\sigma}(\bar{p}p \to \overline{\Lambda}\Sigma^0)$$

while experimentally the left-hand side of (50) is found to be larger than its right side. More data at higher energies for this relation are, therefore, required. There are also large experimental errors for the $(\bar{p}n)$ total cross-sections. Better accuracy in these cross-sections would be very helpful to check the various predictions from the quark model. Availability of Λp and Σp total cross-sections would make some of the cross-section relations given in the Appendix available also for experimental verification.

4 Double Scattering Corrections

We have discussed the high energy scattering of hadrons in the quark model based upon the assumption of simple additivity for the 2-body quark scattering amplitude. We shall now include the effects of the double scattering corrections. Thus prediction can be obtained from the quark model also for those processes for which there are exchanges of two units of charge and (or) hypercharge.

To keep the algebra manageable, we consider only the meson-baryon scattering processes in the quark model for baryons, while the mesons form an SU(3) octet. Further assumptions are that (i) the incident meson does not make more than two encounters with the quarks within the target baryon and (ii) the first scattering is mainly in the forward direction. Then the invariant amplitude for the meson-baryon scattering process

$$M + B \to M' + B' \tag{61}$$

can be written as

$$\langle B'M'_{\beta}{}^{\alpha}|T|\beta M_{\nu}{}^{\mu}\rangle = \delta^4(p^{B'} + p^{M'} - p^B - p^M)$$
$$\sum_{ij}[\langle q^l M'_{\beta}{}^{\alpha}|t_{ij}{}^l|q^i M_{\nu}{}^{\mu}\rangle + \langle q^l q^m M'_{\beta}{}^{\alpha}|t_{ij}{}^{lm}|q^i q^j M_{\nu}{}^{\mu}\rangle] \tag{62}$$

where the summation in the last term (the double scattering term) is over the distinct pairs of quarks in the wave-function of B. The first term on the right-hand side of (62) can be expressed in terms of three independent parameters:

$$\langle q^l M'_{\beta}{}^{\alpha}|t_i{}^l|q^i M_{\nu}{}^{\mu}\rangle = f(\bar{q}_i q^i)(\bar{M}_{\beta}{}^{\alpha} M_{\alpha}{}^{\beta}) + $$
$$+ g(\bar{q}_l q^i)(\bar{M}_{\beta}{}^{\alpha} M_{\mu}{}^{\beta}) + h(\bar{q}_l q^i)(M_{\beta}{}^{\alpha} \bar{M}_{\mu}{}^{\beta})$$

The second term can then be expressed in terms of the single scattering amplitude $t_j{}^l$ as follows:

$$\langle q^l q^m M'_{\beta}{}^{\alpha}|t_{ij}{}^{lm}|q_i{}^i q^j M_{\nu}{}^{\mu}\rangle =$$
$$= \sum_{\rho\lambda}[\langle q^m M'_{\beta}{}^{\alpha}|t_j{}^m|q^j M_{\lambda}{}^{\rho}\rangle\langle q^l M_{\lambda}{}^{\rho}|t_i{}^l|q^i M_{\nu}{}^{\mu}\rangle +$$
$$+ \langle q^l M'_{\beta}{}^{\alpha}|t_j{}^l|q^j M_{\lambda}{}^{\rho}\rangle\langle q^m M_{\lambda}{}^{\rho}|t_i{}^m|q^i M_{\nu}{}^{\mu}\rangle +$$
$$+ \langle q^m M'_{\beta}{}^{\alpha}|t_i{}^m|q^i M_{\lambda}{}^{\rho}\rangle\langle q^l M_{\lambda}{}^{\rho}|t_j{}^l|q^j M_{\nu}{}^{\mu}\rangle +$$
$$+ \langle q^l M'_{\beta}{}^{\alpha}|t_i{}^l|q^i M_{\lambda}{}^{\rho}\rangle\langle q^m M_{\lambda}{}^{\rho}|t_j{}^m|q^j M_{\nu}{}^{\mu}\rangle] \tag{64}$$

where the summation is over all possible SU(3) indices ρ and λ. If there are two identical quarks in the final state, the corresponding scattering terms must be divided by two.

Using the unitary part of the baryon wave-functions from references we obtain the following results for the meson-baryon scattering amplitudes (written MB for short):

$$(\pi^+p) = 3f + 2h + g + 6f^2 + 2h^2 + 4hg + 8fh + 4fg -$$
$$- [\tfrac{4}{3}h^2 + \tfrac{4}{3}g^2 - \tfrac{4}{3}gh] \tag{64a}$$
$$(\pi^-p) = 3f + 2g + h + 6f^2 + 2h^2 + 4gh + 8fg + 4fg -$$
$$- [\tfrac{4}{3}h^2 + \tfrac{4}{3}g^2 - \tfrac{4}{3}gh] \tag{64b}$$
$$(K^+p) = 3f + 2h + 6f^2 + 2h^2 + 8fh - [2h^2] \tag{64c}$$
$$(K^-p) = 3f + 2g + 6f^2 + 2g^2 + 8fg - [2g^2] \tag{64d}$$
$$(K^+n) = 3f + h + 6f^2 + 4fh - [2h^2] \tag{64e}$$
$$(K^-n) = 3f + g + 6f^2 + 4fg - [2g^2] \tag{64f}$$

In the following amplitudes there are no single scattering terms:

$$\tfrac{1}{2}\langle K^+\Xi^-|K^-p\rangle = \langle K^0\Xi^0|K^-p\rangle = \langle \pi^+\Sigma^-|K^-p\rangle$$

$$= \langle K^0\Xi^-|K^-n\rangle = \langle K^+\Sigma^-|\pi^-p\rangle \tag{65}$$

$$= \tfrac{1}{3}h^2 + \tfrac{1}{3}g^2 - \tfrac{4}{3}gh \tag{66}$$

From (64) we obtain the Barger–Rubin relation (40) without any further assumption. The relations (40) and (65) are also obtained from the U(3) X U(3) for the collinear group[56]. The Johnson–Treiman relation (42) is also obtained from the present model if we take into account the ordinary spin part of the baryon wave-function and make the further assumption that (iii) there is always an exchange of ordinary spin along with the exchange of unitary spin in the meson-quark scattering processes. This gives us a change of sign in the last brackets of all the expressions (64). It should be emphasized that the present derivation of the Johnson–Treiman relation is independent of any particular symmetry of the baryon wave-function (i.e. independent of the Fermi or para-statistics of the quarks).

5 SU(6) Algebra of Gell-Mann Suggested by Quark Model

In this section we discuss the algebra generated by V, A, S and P currents. To derive this algebra, we consider the quark model in which the V, A, S and P current densities are respectively given by:

$$\begin{aligned} \mathscr{F}_{i\mu} &= \mathrm{i}\,\bar{q}\frac{\lambda_i}{2}\gamma_\mu q \\ \mathscr{F}_{i\mu}{}^5 &= \mathrm{i}\,\bar{q}(\lambda_i/2)\gamma_\mu\gamma_5 q \\ \mathscr{S}_i &= \bar{q}(\lambda_i/2)q \\ \mathscr{S}_i{}^5 &= \mathrm{i}\,\bar{q}(\lambda_i/2)\gamma_5 q \end{aligned} \tag{67}$$

The space-integrals of the fourth components of $\mathscr{F}_{i\mu}$ and $\mathscr{F}_{i\mu}{}^5$ and those of $\mathscr{S}_i$ and $\mathscr{S}_i{}^5$ are respectively given by:

$$\begin{aligned} F_i(t) &= -\mathrm{i}\int \mathrm{d}^3x\,\mathscr{F}_{i4}(\mathbf{x}, t) \\ F_i{}^5(t) &= -\mathrm{i}\int \mathrm{d}^3x\,\mathscr{F}_{i4}{}^5(\mathbf{x}, t) \\ S_i(t) &= \int \mathrm{d}^3x\,\mathscr{S}_i(\mathbf{x}, t) \\ S_i{}^5(t) &= \int \mathrm{d}^3x\,\mathscr{S}_i{}^5(\mathbf{x}, t) \end{aligned} \tag{68}$$

To generate the algebra of these currents, we need to know how to find the equal-time commutator of two field operator products. Using the equal-time commutator relation for the quark field

$$\left[q_i(x), q_j^+(x')\right]_{+\,t'=t} = \delta_{ij}\delta^3(\mathbf{x} - \mathbf{x}')$$

one would formally obtain:

$$[q^+(x)\mathrm{O}q(x), q^+(x')\mathrm{O}'q(x')]_{t'=t} = \delta^3(\mathbf{x} - \mathbf{x}')q^+(x)[\mathrm{O}, \mathrm{O}']q(x) \quad (69)$$

where O and O′ are of the form $\gamma_A\lambda_i$, $\gamma_B\lambda_j$, γ's being Dirac matrices. Although the foregoing consideration is sufficient for our purpose, one may point out that (69) is correct only to the extent that a spatial derivative of the δ-function, which gives rise to so-called Schwinger terms[20], can be neglected. Since we shall always use equal-time commutation relations of two operators in which at least one of them is in the form of a space integral, the spatial derivative of the δ-function will not contribute. Hence we shall not consider it any longer. Taking now $\mathrm{O} = \gamma_A\lambda_i$, $\mathrm{O}' = \gamma_B\lambda_j$ and noting:

$$[\gamma_A\lambda_i, \gamma_B\lambda_j] = \tfrac{1}{2}[\gamma_A, \gamma_B]_+[\lambda_i, \lambda_j] + \tfrac{1}{2}[\gamma_A, \gamma_B][\lambda_i, \lambda_j]_+ \quad (70)$$

we see from (69) that the 35 operators listed in (68) satisfy the equal-time commutation relations[21]

$$[F_i(t), F_j(t)] = \mathrm{i}\, f_{ijk}F_k(t) \quad (71a)$$

$$[F_i(t), F_j^5(t)] = \mathrm{i}\, f_{ijk}F_k^5(t) \quad (71b)$$

$$[F_i^5(t), F_j^5(t)] = \mathrm{i}\, f_{ijk}F_k(t) \quad (71c)$$

$$[S_i(t), S_j(t)] = \mathrm{i}\, f_{ijk}F_k(t) \quad (71d)$$

$$[F_i(t), S_j(t)] = \mathrm{i}\, f_{ijk}S_k(t) \quad (71e)$$

$$[S_i^5(t), S_j^5(t)] = \mathrm{i}\, f_{ijk}F_k(t) \quad (71f)$$

$$[F_i(t), S_j^5(t)] = \mathrm{i}\, f_{ijk}S_k^5(t) \quad (71g)$$

$$[F_i^5(t), Sj^5(t)] = -\mathrm{i}\, d_{ijk}S_k(t) \quad (71h)$$

$$[F_i^5(t), S_j(t)] = \mathrm{i}\, d_{ijk}S_k^5(t) \quad (71i)$$

$$[S_i(t), S_j^5(t)] = \mathrm{i}\, d_{ijk}F_k^5(t) \quad (71j)$$

These commutation relations, which form an algebra of SU(6), are derived by utilizing explicitly the quark model. However, we do not have to take the model seriously and we may abstract the algebraic properties of currents and discard the model. We thus assume that the above commutation relations hold generally for V, A, S and P hadron currents. The commutation relations (71) are assumed to be exact even though SU(6) symmetry is badly broken.

The sub-algebra given by the commutation relations (71a, b, c) form the chiral SU(3) X SU(3) which together with PCAC has been used to obtain some remarkable results[20] both in strong and weak interactions. This sub-algebra will not be considered here. The sub-algebra given by (71a, d, e) has been used[22,23,24] as will be discussed in the next section, to obtain a uniform picture of mass splitting within an SU(3) multiplet by assuming that mass-breaking Hamiltonian is proportional to S_8. We note that the commutation relations (71a, d, e) form a non-chiral SU(3) X SU(3) algebra

$$[G_i^{+,-}(t), G_j^{+,-}(t)] = \mathrm{i}\, f_{ijk} G_k^{+,-}(t)$$
$$[G_i^{+}(t), G_j^{-}(t)] = 0 \tag{72}$$

where

$$G_i^{\pm} = \tfrac{1}{2}[F_i \pm S_i]$$

so that in quark model

$$G_i^{+}(t) = \tfrac{1}{2}\int \mathrm{d}^3x q^{+} \frac{1+\gamma_4}{2} \lambda_i q$$
$$G_i^{-}(t) = \tfrac{1}{2}\int \mathrm{d}^3x q^{+} \frac{1-\gamma_4}{2} \lambda_i q$$

The sub-algebra given by (71a, f, g) may be useful in strong interactions although so far no definite use has been made of this sub-algebra. The commutation relation (71) has been used to derive[27] sum rules for meson-baryon and meson-meson coupling constants if one assumes that the Hamiltonian which breaks the vertex SU(3) symmetry is also proportional to S_8. It has been suggested[24,29,30] that the commutation relations (71h, i) might be useful in the non-leptonic decays of hadrons when the non-leptonic Hamiltonian is used in the form (2). These will be used in Section 7 in conjunction with the PCAC hypothesis to derive[24,30] some interesting consequences in non-leptonic hadron decays.

6 Mass Splitting

6.1 *Mass splitting within an* SU(3) *multiplet*

One takes the attitude[22,23,24] that Gell-Mann–Okubo mass-splitting for any SU(3) multiplet comes from the eighth component of the scalar operator S_i, where S_i belong to the non-chiral SU(3) X SU(3) algebra (71a, d, e). Thus we write the Hamiltonian for strong interaction as:

$$H = H_0 + \delta m S_8 \tag{73}$$

where H_0 is the SU(3) invariant part and δm is a universal parameter for medium strong mass-splittings, although one should keep in mind the possibility that it may be somewhat different for baryon and meson families of multiplets. Such a Hamiltonian is suggested by the quark model if the isospin singlet quark has a different mass from the members of iso-doublets of quarks.

In the presence of a symmetry-breaking Hamiltonian the SU(3) generators F_i satisfy

$$\begin{aligned}\frac{\mathrm{d}F_i}{\mathrm{d}t} &= \mathrm{i}\,\delta m[S_8, F_i] \\ &= -\delta m f_{8il} S_l \end{aligned} \tag{74}$$

Taking the matrix elements of (74) between states j and k of a multiplet at rest, one obtains

$$\mathrm{i}\,(m_k - m_j)\langle k|F_i|j\rangle = \delta m f_{i8l}\langle k|S_l|j\rangle \tag{75}$$

Since we are calculating the mass differences to order δm, $\langle k|S_l|j\rangle$ and $\langle k|F_i|j\rangle$ need to be evaluated in the SU(3) limit. In order to evaluate the matrix elements $\langle k|S_l|j\rangle$, one can make use of the commutation relation (71d), namely:

$$[S_i, S_j] = \mathrm{i}\,f_{ijk}F_k \tag{76}$$

and try to saturate it by a few low-lying states. It is important to note that when the matrix elements of (76) are taken between 1-particle states of H_0 at rest, then the scale on the right-hand side is fixed since F_k is a generator of the SU(3) and is a conserved operator in that limit; for example $\langle p|F_i|p\rangle$ will give just $\sqrt{3}/2$ times the hypercharge of the proton. Below we discuss specific examples of this procedure for various SU(3) multiplets. For this purpose, it is convenient to rewrite relations (75) and (76) in spherical base having the phase convention of de Swart[61]:

$$(m_\alpha - m_\beta)\langle\alpha|F^\nu|\beta\rangle = -\sqrt{3}\delta m \begin{pmatrix} 8 & 8 & 8a \\ \text{o} & \nu & \lambda \end{pmatrix} \langle\alpha|S^\lambda|\beta\rangle \tag{77}$$

$$[S^\lambda, S^\mu] = -\sqrt{3} \begin{pmatrix} 8 & 8 & 8a \\ \lambda & \mu & \nu \end{pmatrix} F^\nu \tag{78}$$

(i) *Baryon $\frac{1}{2}^+$ octet and baryon $\frac{3}{2}^+$ decuplet.* Let us first consider the decuplet of $\frac{3}{2}^+$ baryons. We take the matrix elements of (78) between zero momentum states of the $\frac{3}{2}^+$ baryon decuplet and introduce a complete set of intermediate states:

$$\sum[\langle\alpha|S^\lambda|n\rangle\langle n|S^\mu|\beta\rangle - (\mu \leftrightarrow \lambda)] = -\sqrt{3}\begin{pmatrix} 8 & 8 & 8a \\ \lambda & \mu & \nu \end{pmatrix}\langle\alpha|F^\nu|\beta\rangle \quad (79)$$

We retain only baryon $\frac{1}{2}^+$ octet and baryon $\frac{3}{2}^+$ decuplet states. Since the octet $\frac{1}{2}^+$ cannot contribute, we obtain:

$$\sum_\gamma[\langle\alpha_{10}|S^\lambda|\gamma_{10}\rangle\langle\gamma_{10}|S^\mu|\beta_{10}\rangle - (\mu \leftrightarrow \lambda)] =$$

$$= -\sqrt{3}\begin{pmatrix} 8 & 8 & 8a \\ \lambda & \mu & \nu \end{pmatrix}\langle\alpha_{10}|F^\nu|\beta_{10}\rangle \quad (80)$$

For zero momentum states, we have

$$\langle\alpha_{10}|F^\nu|\beta_{10}\rangle = \sqrt{6}\begin{pmatrix} 10 & 8 & 10 \\ \beta & \nu & \alpha \end{pmatrix} \quad (81)$$

the normalization here being fixed from the fact that $\langle\alpha|F^8|\alpha\rangle$ should be $\sqrt{3}/2$ times the hypercharge of α. We define

$$\langle\alpha_{10}|S^\lambda|\gamma_{10}\rangle = G_{10}\begin{pmatrix} 10 & 8 & 10 \\ \gamma & \lambda & \alpha \end{pmatrix} \quad (82)$$

Using now the method of B. W. Lee[62], we obtain from (80) in a straightforward way*

$$G_{10} = -\sqrt{6} \quad (83)$$

Equation (78) then gives, on using (83),

$$(m_\alpha - m_\beta)\sqrt{6}\begin{pmatrix} 10 & 8 & 10 \\ \beta & \nu & \alpha \end{pmatrix} = \sqrt{3}\delta m\begin{pmatrix} 8 & 8 & 8a \\ 0 & \nu & \lambda \end{pmatrix}\sqrt{6}\begin{pmatrix} 10 & 8 & 10 \\ \beta & \lambda & \alpha \end{pmatrix} \quad (84)$$

leading to

$$Y_1{}^* - N^* = \Xi^* - Y_1{}^* = \Omega - \Xi^* = -\frac{\sqrt{3}}{2}\,\delta m = 145 \text{ MeV} \quad (85)$$

This gives the equal spacing rule with

$$\delta m = -165 \text{ MeV} \quad (86)$$

It is interesting to note that this value of δm also follows, as will be discussed in Section 7, from a consistency condition that in the current

* Actually, the sign of G_{10} is not determined; one gets $G_{10}{}^2 = 6$. We have arbitrarily selected the minus sign in (83).

algebra approach the p-wave on p.c. non-leptonic hyperon decay amplitudes should vanish due to the Coleman–Glashow theorem[31] if SU(3) invariant meson-baryon vertices are used and if S_6 (proportional to H_W^{pc}) and S_8 (proportional to the mass-breaking Hamiltonian) belong to the same unitary octet. In Section 7 we show that the consistency condition gives:

$$\frac{\sqrt{3}}{2}\,\delta m = -f_\pi \tag{87}$$

where f_π is the decay constant for pion. Experimentally $f_\pi \simeq m_\pi$ and (87) thus gives $\delta m = -160$ MeV in remarkable agreement with (86).

For $\frac{1}{2}^+$ baryon octet, (75) gives, in addition to Gell-Mann–Okubo mass formula, the following relations:

$$\Sigma - \Lambda = -\frac{2}{\sqrt{3}}\,\delta m s_d$$
$$\Xi - N = -\sqrt{3}\delta m s_f \tag{88}$$

where s_d and s_f are reduced matrix elements appearing in the matrix elements

$$\langle k, \mathbf{p} = 0|S_i|j, \mathbf{p} = 0\rangle = \mathrm{i}\, f_{ijk}s_f + d_{ijk}s_d \tag{89}$$

or in the spherical base

$$\langle\alpha|S^\nu|\beta\rangle = \sqrt{3}\begin{pmatrix} 8 & 8 & 8a \\ \beta & \nu & \alpha \end{pmatrix} s_f + \left(\frac{5}{3}\right)^{\frac{1}{2}}\begin{pmatrix} 8 & 8 & 8s \\ \beta & \nu & \alpha \end{pmatrix} s_d \tag{90}$$

Equation (88) gives

$$\frac{s_d}{s_f} = -\frac{3}{2}\frac{\Sigma - \Lambda}{\Xi - N} \tag{91}$$
$$= -0{\cdot}31 \pm 0{\cdot}02 \tag{92}$$

and

$$\frac{\sqrt{3}}{2}\,\delta m = -\frac{\Xi - N}{2s_f} \tag{93}$$

If now one follows the same procedure as illustrated for the decuplet and retains only $\frac{1}{2}^+$ octet and $\frac{3}{2}^+$ decuplet in the intermediate states, then, since $\frac{3}{2}^+$ decuplet states do not contribute in this case, one obtains

$$s_f = 1, \quad s_d = 0 \tag{94}$$

which is approximately the case, since empirically from (91) we have $s_d/s_f = -\frac{1}{3}$. This makes Λ and Σ masses degenerate giving $\Xi - \Lambda = \Lambda - N$ and δm to about -180 MeV, a value somewhat higher than that obtained for the decuplet case. This would imply that the assumption of saturating (78) only by $\frac{1}{2}^+$ octet and $\frac{3}{2}^+$ decuplet is not sufficient in this case and that the scattering states in this case may change (94) to a more realistic one which will give $s_d/s_f = -\frac{1}{3}$ and at the same make δm come closer to -165 MeV. Such a possibility is made quite plausible in references 22 and 23. We would then conclude that universality for the parameter δm for the baryon family of $\frac{1}{2}^+$ octet and $\frac{3}{2}^+$ decuplet is a reasonable possibility.

(ii) *Mass-splitting for* 0^-, 1^- *and* 2^+ *nonets*. The Hamiltonian consists, as before, of an SU(3) invariant part H_0 plus S_8

$$H = H_0 + \delta m' S_8 \tag{95}$$

where the mass-splitting parameter is denoted here by $\delta m'$ to incorporate the possibility that $\delta m'$ for the meson family of multiplets may be slightly different from that of the baryon family.

For simplicity we illustrate the procedure for mass splitting within the framework of non-chiral SU(3) X SU(3) algebra for spin zero mesons. The treatment for spin 1^- and spin 2^+ mesons is exactly similar. Now if we denote the $I = Y = 0$ member of the unitary 0^- octet by η_8 and the unitary singlet by η_1, then the physical particles η and χ are given by:

$$\begin{aligned} \eta &= p\eta_8 + q\eta_1 \\ \chi &= q\eta_8 - p\eta_1 \end{aligned} \tag{96}$$

We now define the following matrix elements in the SU(3) limit:

$$\langle k(p)|F_i|j(k)\rangle = \frac{1}{\sqrt{(4k_0p_0)}}\,\mathrm{i}\,f_{ijk}(k_0 + p_0) \tag{97}$$

$$\langle k(p)|S_i|j(k)\rangle = \frac{1}{\sqrt{(4k_0p_0)}}\,d_{ijk}s_d' \tag{98a}$$

$$\langle 0(p)|S_i|j(k)\rangle = \frac{1}{\sqrt{(4k_0p_0)}}\,d_{0ij}s_0' \tag{98b}$$

Note that only D-type symmetric coupling is possible in (98a) and (98b) due to charge conjugation invariance. In the above equations $\mathbf{p} = \mathbf{k}$. In (98b), 0 denotes the unitary singlet while in (97) the factor $(p_0 + k_0)$ appears since $F_i = -\mathrm{i}\int \mathrm{d}^3x\mathscr{F}_{i4}(\mathbf{x}, t)$ involves the fourth component of the vector $\mathscr{F}_{i\mu}$. For states at rest, $p_0 = m_k$ and $k_0 = m_j$. With the above definitions, (75) gives immediately*

$$K^2 - \pi^2 = -\frac{\sqrt{3}}{2}\,\delta m'\mathscr{S}_d' \tag{99a}$$

$$\eta_8{}^2 - \pi^2 = -\frac{2\delta m'}{\sqrt{3}}\,\mathscr{S}_d' \tag{99b}$$

$$\chi^2 - \eta_8{}^2 = -\sqrt{\left(\frac{2}{3}\right)}\,\delta m'\left(\frac{p}{q}\right)s_0' \tag{99c}$$

$$\eta^2 - \eta_8{}^2 = -\sqrt{\left(\frac{2}{3}\right)}\,\delta m'\left(\frac{q}{p}\right)s_0' \tag{99d}$$

The first two equations give the Gell-Mann–Okubo mass formula.

To get more information we need to know the reduced matrix elements s_d' and s_f'. For that purpose we make use of the commutation relations (78) and take the matrix elements of (78) between zero momentum states of the 0^- nonet to obtain:

$$\langle\alpha|[S^\lambda, S^\mu]|\beta\rangle = -\sqrt{3}\begin{pmatrix}8 & 8 & 8a\\ \lambda & \mu & \nu\end{pmatrix}\langle\alpha|F^\nu|\beta\rangle \tag{100}$$

We now define the following matrix elements in the spherical base:

$$\langle\alpha(k')|F^\nu|\beta(k)\rangle = (4k_0k_0')^{-\frac{1}{2}}(k_0 + k_0')\sqrt{3}\begin{pmatrix}8 & 8 & 8a\\ \beta & \nu & \alpha\end{pmatrix}$$

$$\langle\gamma_8(p)|S^\mu|\beta(k)\rangle = (4p_0k_0)^{-\frac{1}{2}}\begin{pmatrix}8 & 8 & 8s\\ \beta & \mu & \gamma\end{pmatrix}H_8$$

$$\langle\gamma_0(p)|S^\mu|\beta(k)\rangle = (4p_0k_0)^{-\frac{1}{2}}\begin{pmatrix}8 & 8 & 1\\ \beta & \mu & 0\end{pmatrix}H_0 \tag{101}$$

where

$$H_8 = -\sqrt{(\tfrac{5}{3})}s_d'$$

$$H_0 = \frac{4}{\sqrt{3}}s_0' \tag{102}$$

* There is some slight ambiguity here. For states at rest we have used the matrix element $\langle k|F_i|j\rangle$ in the form

$$\langle k|F_i|j\rangle = (4p_0k_0)^{-\frac{1}{2}}if_{ijk}(m_k + m_j)$$

where m_k and m_j are the physical masses of k and j states. It is because of this fact that we get quadratic mass relations for mesons. If we had put $m_k = m_j = m_0'$ where m_0 is the mean mass of the multiplet, in the above matrix element, we would have obtained linear mass relations. Clearly, the difference between using the above matrix element to be proportional to $(m_k + m_j)$ or to $2m_0'$ is a second order effect which we are neglecting. We have preferred the first point of view which leads to the quadratic mass relations.

Since we deal with states at rest and since we are interested in solving (100) in the SU(3) limit, we may put $k_0 = k_0' = p_0 = m_0'$ where m_0' is the mean mass of the meson multiplet. Putting in now a nonet of 0^- mesons as intermediate states, (100) gives

$$\sum_{\gamma}[\langle\alpha|S^{\lambda}|\gamma_8\rangle\langle\gamma_8|S^{\mu}|\beta\rangle + \langle\alpha|S^{\lambda}|\gamma_0\rangle\langle\gamma_0|S^{\mu}|\beta\rangle - (\lambda\leftrightarrow\mu)]$$
$$= -3\begin{pmatrix}8 & 8 & 8a\\ \lambda & \mu & \nu\end{pmatrix}\begin{pmatrix}8 & 8 & 8a\\ \beta & \nu & \alpha\end{pmatrix} \tag{103}$$

From this, using the method of B. W. Lee[62], we obtain in a straightforward way the following two independent equations:

$$8H_d^2 + 2H_0^2 = 96\,m_0'^2$$
$$-4H_d^2 + \tfrac{5}{4}H_0^2 = 0$$

These give the solution:

$$H_d^2 = \tfrac{20}{3}m_0'^2,\; H_0^2 = \tfrac{16}{5}H_d^2$$

or

$$s_d'^2 = 4m_0'^2, \quad \left(\frac{s_0'}{s_d}\right)^2 = 1 \tag{104}$$

With this solution, (99a), (99c) and (99d) give the well-known Schwinger formula[25]

$$(\chi^2 - \eta_8^2)(\eta^2 - \eta_8^2) = -\tfrac{8}{9}(K^2 - \pi^2)^2$$

or

$$(\chi^2 - \pi^2)(\eta^2 - \pi^2) = \tfrac{4}{3}(K^2 - \pi^2)(\chi^2 + \eta^2 - 2K^2) \tag{105}$$

We emphasize that this formula is independent of the mass-splitting parameter $\delta m'$ and is solely a consequence of the relation (104).

The solution (104) and the relations (99) also hold, for 1^- and 2^+ nonets, where m_0' would be the mean mass of the corresponding multiplet, so that for these cases we obtain, independent of the value of $\delta m'$ for each nonet, the Schwinger formula[25]

$$(\phi^2 - \rho^2)(\omega^2 - \rho^2) = \tfrac{4}{3}(K^{*2} - \rho^2)(\phi^2 + \omega^2 - 2K^{*2}) \tag{106}$$

for the 1^- nonet[25], and

$$(f'^2 - A_2^2)(f^2 - A_2^2) = \tfrac{4}{3}(K^{*2} - A_2^2)(f'^2 + f^2 - 2K^{*2}) \tag{107}$$

for the 2^+ nonet[64], where $K^*(1420)$, $A_2(1310)$, $f(1250)$ and $f'(1500)$ form the 2^+ nonet. If the same $\delta m'$ is used for every nonet, one obtains:

$$\frac{K^2 - \pi^2}{2(m_0')_0} = \frac{K'^{*2} - \rho^2}{2(m_0')_1} = \frac{K^{*2} - A_2{}^2}{2(m_0')_2} = -\frac{\sqrt{3}}{2}\,\delta m' \tag{108}$$

where $(m_0') \simeq 500$ MeV, $(m_0')_1 \simeq 853$ MeV and $(m_0')_2 \simeq 1370$ MeV. From (108) one obtains $\delta m'$ to be nearly the same and equal to about -140 MeV for both 1^- and 2^+ nonets; the Schwinger formula is also very well satisfied for these nonets. However, for 0^- mesons, (108) gives $\delta m' \simeq -260$ MeV; and also we know that the Schwinger formula (105) is not satisfied by the 0^- nonet when x is identified with X_0 (960). We therefore take the attitude that $\delta m'$ in this case should be the same as for the other nonets and the approximation which leads to (104) is bad for this particular case, whereas the same approximation is good for 1^- and 2^+ nonets. We believe that if one takes scattering states in addition to single particle states for saturating the relation (100), one will get a different structure for (104) in the case of 0^- which will give $\delta m' = -140$ MeV. We also note that $\delta m' = -140$ MeV obtained for the meson family of multiplets is nearly the same within a few percent as $\delta m \simeq -165$ MeV obtained for the baryon family of multiplets.

To sum up, with the possible exception of the 0^- nonet, one gets a uniform picture of mass splitting within any SU(3) multiplet in terms of a more or less universal splitting parameter if the mass-splitting Hamiltonian is taken proportional to S_8 where S_i belong to the non-chiral SU(3) X SU(3) algebra (71a, d, e).

We may mention in passing that if one assumes the SU(3) symmetry for a 3-point vertex is also broken by a Hamiltonian proportional to S_8, then using PCAC and the commutation relation (71), one can obtain sum rules for coupling constants. For pseudoscalar meson-baryon coupling constants sum rules, the reader is referred to a paper by Bose and Hara[27]. The sum rules for vector meson-baryon coupling constants and for meson-meson coupling constants are discussed respectively by Lai[27] and Biswas and Patil[27]. These sum rules are generally stronger than those obtained on purely symmetry arguments[28]. The general conclusion for pseudoscalar meson-baryon coupling constants sum rules is as follows[27]: The renormalization effects due to symmetry breaking are entirely absent in those sum rules which involve only pion-baryon couplings or only kaon-baryon couplings or only η couplings. Renormalization effects appear only when pion-baryon couplings are compared to kaon-baryon (or η-baryon) couplings. Thus, for example, the decay widths for

$N^* \to N + \pi$, $Y_1^* \to \Sigma + \pi$, $Y_1^* \to \Lambda + \pi$ and $\Xi^* \to \Xi + \pi$ are unrenormalized and are given by their SU(3) values. This conclusion is in agreement with experiment. The other sum rules obtained by Bose and Hara[27] are as yet difficult to test experimentally. Similar conclusions hold for vector meson-baryon and meson-meson coupling constants. For details the reader is referred to papers of Lai[27], Biswas and Patil[27] and Rockmore[27].

6.2 *Electromagnetic mass differences*

In the previous sub-section we have taken the attitude that the medium strong mass splittings come from a scalar S_8 so that the Hamiltonian for strong interaction was written as in (73). It was pointed out that such a Hamiltonian is suggested by the quark model if the isospin singlet quark has a different mass from the members of iso-doublet of quarks. If now two members of an iso-doublet quark have slightly different masses due to electromagnetism, then[26]

$$H = H_0 + \delta m S_8 + \delta m_e S_3 \tag{109}$$

where $\delta m_e/\delta m \ll 1$. Thus for charged isospin currents, we have[26]

$$\begin{aligned} \partial_\mu \mathscr{F}_\mu{}^\pm &= \mathrm{i}\, \delta m_e [S_3, \mathscr{F}_\mu{}^\pm] \\ &= \pm \mathrm{i}\, \delta m_e \mathscr{S}^\pm \end{aligned} \tag{110}$$

whereas for the neutral member, we have

$$\partial_\mu \mathscr{F}_\mu{}^3 = 0 \tag{111}$$

If one takes the matrix elements of the left-hand side of (100) between charged and neutral members of an isospin multiplet, it will give electromagnetic mass difference at zero momentum transfer of that isospin multiplet. But clearly (110) derived from (109), which has the merit of treating medium strong mass splittings as well as electromagnetic mass splittings on similar footing, cannot be a complete story, since it would not give $\pi^\pm - \pi^0$ mass difference. This is because $\mathscr{S}^\pm$ can give only $\Delta I = 1$ transitions whereas $\pi^\pm - \pi^0$ mass difference is a $\Delta I = 2$ transition. Therefore, (110) has to be modified. One can take the point of view that in (110) the minimal electromagnetic coupling has not been included and that it includes only a 'feed-back' contribution from the mass difference of members of iso-doublet quarks. The minimal electromagnetic coupling is now introduced in the usual way; i.e. we replace ∂_μ in (110) by $\partial_\mu \pm \mathrm{i}\, eA_\mu$. Thus instead of (110), we obtain[26]

$$\partial_\mu \mathscr{F}_\mu{}^\pm = \pm \mathrm{i}\, \delta m_e \mathscr{S}^\pm \mp \mathrm{i}\, eA_\mu \mathscr{F}_\mu{}^\pm \tag{112}$$

We now take the matrix elements of (112) between charged and neutral members of an isospin multiplet:

$$\langle\alpha, I, I_z|\partial_\mu \mathscr{F}_\mu{}^+|\alpha, I, I_{z-1}\rangle = \mathrm{i}\,\delta m_e\langle\alpha, I, I_z|\mathscr{S}^+|\alpha, I, I_{z-1}\rangle - \\ - \mathrm{i}\,e\langle\alpha, I, I_z|A_\mu \mathscr{F}_\mu{}^+|\alpha, I, I_{z-1}\rangle \qquad (113)$$

where α denotes a particular isospin multiplet. We observe that the first term on the right-hand side of (113) gives a tadpole type[31] of contribution, obtained here without any postulate of scalar mesons for which there is no experimental evidence and from the point of view of regarding S_i to be an observable whose third component plays a part in electromagnetic mass splittings, just as S_8 is responsible for the medium strong mass splittings. To proceed with the second term on the right-hand side of (113), we observe that it is to be evaluated to order e so that we can use CVC for $\mathscr{F}_\mu{}^+$ and can regard $|\alpha, I, I_{z-1}\rangle$ as an eigenstate of I. Now

$$[I^+, A_\mu j_\mu] = A_\mu[I^+, j_\mu] + [I^+, A_\mu]j_\mu$$

where $j_\mu = \mathscr{F}_{\mu 3} + \dfrac{1}{\sqrt{3}}\mathscr{F}_{\mu 8}$ is the electromagnetic current. Assuming that A_μ commutes with I^+ at equal times, we obtain:

$$[I^+, A_\mu j_\mu] = -A_\mu \mathscr{F}_\mu{}^+ \qquad (114)$$

Using this relation we obtain:

$$\mathrm{i}\,e\langle\alpha, I, I_z|A_\mu \mathscr{F}_\mu{}^+|\alpha, I, I_{z-1}\rangle = -\mathrm{i}\,e[(I - I_z + 1)(I + I_z)]^{\frac{1}{2}} \times \\ \times [\langle\alpha, I, I_{z-1}|A_\mu j_\mu|\alpha, I, I_{z-1}\rangle - \langle\alpha, I, I_z|A_\mu j_\mu|\alpha, I, I_z\rangle] \qquad (115)$$

To simplify notation, we shall write $|\alpha, I, I_z\rangle = |\alpha\rangle$ and $|\alpha, I, I_{z-1}\rangle = |\alpha - 1\rangle$. Using the Yang–Feldman technique[64], $A_\mu(x)$ can be expressed as:

$$A_\mu(x) = A_\mu{}^{\mathrm{in}}(x) + e\int_{-\infty}^{\infty} \mathrm{d}^4y\,\delta_{\mu\nu} D_F(y) j_\nu(y) \qquad (116)$$

where

$$D_F(y) = -\frac{2\mathrm{i}}{(2\pi)^4}\int \mathrm{d}^4q\,\frac{\mathrm{e}^{iq.y}}{q^2 - \mathrm{i}\,\varepsilon} \qquad (117)$$

With the help of these equations, since $A_\mu{}^{\mathrm{in}}$ does not contribute, we can write

$$(-2\pi)^3\langle\alpha|A_\mu j_\mu|A\rangle = \frac{e}{2}\frac{\mathrm{i}}{(2\pi)^4}\int \mathrm{d}^4q\,\frac{\delta_{\mu\nu}\phi_{\mu\nu}{}^\alpha(q^0, \mathbf{q})}{q^2 - \mathrm{i}\,\varepsilon}, \qquad (118)$$

where

$$\phi_{\mu\nu}{}^{\alpha}(q^0, \mathbf{q}) = \mathrm{i}\,(2\pi)^3 \int \mathrm{d}^3z \langle\alpha|T[j_\mu(z/2)j_\nu(-z/2)]|\alpha\rangle \tag{119}$$

Since the particle α is at rest, $\phi_{\mu\nu}{}^{\alpha}$ is a function of q^0 and $\mathbf{q}$ only.

One can easily recognize that $\varepsilon_\mu\varepsilon_\nu\phi_{\mu\nu}{}^{\alpha}$ appearing in (118) is closely related to the Compton scattering amplitude of virtual photon on particle α. It differs slightly from the Compton amplitude $\varepsilon_\mu\varepsilon_\nu T_{\mu\nu}{}^{\alpha}$ in the sense that $T_{\mu\nu}{}^{\alpha}$ contains an equal-time commutator which usually appears in the reduction of a scattering amplitude. It is important to note that in the dispersion theoretic treatment of electromagnetic self-energy in references[65,66], it is for the Compton amplitude $\varepsilon_\mu\varepsilon_\nu T_{\mu\nu}{}^{\alpha}$ that an unsubtracted dispersion relation is written. On the other hand, in reference[26], which we follow, $\varepsilon_\mu\varepsilon_\nu\phi_{\mu\nu}{}^{\alpha}$ is assumed to satisfy an unsubtracted dispersion relation. The 1-particle intermediate states contribution to dispersion integrals are the same in both the cases and thus we may take these contributions in terms of elastic form-factors from references[65,66,67]. We call such contributions conventional type. Hence we may put

$$\begin{aligned} -(2\pi)^3\langle\alpha|A_\mu j_\mu|\alpha\rangle &= e\Delta m_\alpha \bar{u}u, \text{ for baryons} \\ &= e\Delta m_\alpha{}^2, \text{ for pseudoscalar mesons} \end{aligned} \tag{120}$$

where Δm_α denotes the conventional type of contribution to the electromagnetic self-mass of the particle.

From the above discussion it is clear that if one writes a dispersion relation for the Compton amplitude $T(\nu, q^2) = \varepsilon_\mu\varepsilon_\nu T_{\mu\nu}{}^{\alpha}$ and with ν defined as $-q \cdot p/m_\alpha$, p being the momentum of the particle, as

$$T(\nu, q^2) = A(q^2) + \int \rho(\nu', q^2)\left(\frac{1}{\nu' - \nu} + \frac{1}{\nu' + \nu}\right)\mathrm{d}\nu' \tag{121}$$

the approach followed here will be consistent with $A(q^2)$ not being zero. $A(q^2)$ was taken to be zero in references[65,66]. In fact the contribution from $(\delta m_e)S_3$ term in this approach may be interpreted as arising from the subtraction term $A(q^2)$ in the dispersion relation for the Compton amplitude. In other words one can say that the contribution Δm_α, which arises from the minimal electromagnetic coupling and elastic form-factors, emphasizes low energy contribution to the self-energy, whereas the tadpole-type contribution arising from S_3 gives an estimate of the high energy part. Similar conclusions were also drawn by other people[68] from somewhat different considerations. One may note that since S_3 belongs to an octet, it follows that only that part of the Compton amplitude which contributes to $\Delta I = 1$ mass differences needs a subtraction term in its dispersion relation.

Now at zero momentum transfer, the matrix element $\langle\alpha|\partial_\mu \mathscr{F}_\mu^+|\alpha-1\rangle$ is given by

$$\langle\alpha|\partial_\mu \mathscr{F}_\mu^+|\alpha-1\rangle = -\mathrm{i}\frac{C}{(2\pi)^3}(m_{\alpha-1}-m_\alpha)\bar{u}u \tag{122}$$

where $C = 1$ or $\sqrt{2}$ according to α belongs to iso-doublet or iso-vector. Hence from (122), (113), (115) and (120) we have

$$\begin{aligned} C(m_{\alpha-1}-m_\alpha)\bar{u}u = &-(2\pi)^3\delta m_e\langle\alpha|\mathscr{S}^+|\alpha-1\rangle \\ &+ e^2[(I-I_z+1)(I+I_z)]^{\frac{1}{2}}[\Delta m_{\alpha-1}-\Delta m_\alpha]\bar{u}u \end{aligned} \tag{123}$$

which gives:

$$\begin{aligned} n-p &= -\delta m_e(s_f+s_d)+e^2(\Delta m_n-\Delta m_p) \\ \Sigma^- - \Sigma^0 &= -\delta m_e s_f + e^2(\Delta m_{\Sigma^-}-\Delta m_{\Sigma^0}) \\ \Sigma^0 - \Sigma^+ &= -\delta m_e s_f + e^2(\Delta m_{\Sigma^0}-\Delta m_{\Sigma^+}) \\ \Xi^- - \Xi^0 &= -\delta m_e(s_f-s_d)+e^2(\Delta m_{\Xi^-}-\Delta m_{\Xi^0}) \end{aligned} \tag{124}$$

Similarly for mesons, we obtain

$$\begin{aligned} \pi^{+2}-\pi^{02} &= e^2(\Delta m_{\pi^+}{}^2-\Delta m_{\pi^0}{}^2) \\ K^{02}-K^{+2} &= -\delta m_e s'_d + e^2(\Delta m_{K^0}{}^2-\Delta m_{K^+}{}^2) \end{aligned} \tag{125}$$

In the above equations, the first term gives the same result in practice as the tadpole model[31], while the second term gives the conventional contribution to the electromagnetic self-mass. Since the reduced matrix elements s_f, s_d and s_d' have already appeared in sub-section 6a, they are known in terms of medium strong mass difference. Introducing

$$X = \frac{\delta m_e}{2\sqrt{3}\delta m}(2y_1-3y_2) \tag{126}$$

where

$$y_1 = (\Xi - N), \quad y_2 = (\Sigma - \Lambda) \tag{127}$$

we summarize the electromagnetic mass differences in Table 10. We observe that these mass differences depend on the single parameter which in turn depends on $\delta m_e/2\sqrt{3}\delta m$, which we have assumed to be universal, i.e. to be the same both for baryons and pseudoscalar mesons. Taking the conventional contributions from reference[67], we summarize the numerical comparison, which we take from Socolow[67] and Nieh[69], in Table 10. We see from Table 10 that one can explain all electromagnetic

TABLE 10

Electromagnetic mass differences for baryons and mesons

$$y_1 = \Xi - N = 379{\cdot}0 \text{ MeV}, \; y_2 = \Sigma - \Lambda = 77{\cdot}58 \text{ MeV}$$

$$X = \frac{\delta m_e}{2\sqrt{3}\delta m}[2y_1 - 3y_2] = 2{\cdot}8 \text{ MeV}$$

Mass Difference	Contribution from S_3	Conventional contribution	Total	Experimental value (MeV)
$n - p$	$X = 2{\cdot}8$	$-1{\cdot}1$	1·7	1·3
$\Sigma^- - \Sigma^0$	$\frac{2y_1}{2y_1 - 3y_2} X = 1{\cdot}45X = 4{\cdot}06$	1·4	5·46	4·88 ± ·06
$\Sigma^0 - \Sigma^+$	4·06	$-0{\cdot}7$	3·36	3·09 ± ·11
$\Xi^- - \Xi^0$	$\frac{2y_1 + 3y_2}{2y_1 - 3y_2} X = 1{\cdot}9X = 5{\cdot}32$	1·2	6·52	6·5 ± ·2
$\pi^+ - \pi^0$	0	4·9	4·9	4·6
$K^0 - K^+$	$\frac{4(K^2 - \pi^2)}{2y_1 - 3y_2} X = 1{\cdot}75X = 4{\cdot}90$	$-2{\cdot}8$	2·1	4·05 ± ·12

mass differences in terms of a single parameter X, except for kaon mass differences. Even for kaon mass differences one gets the right sign, only the magnitude is about half the experimental value. This may be due to the reason that higher states in the conventional contribution are important. It is possible but difficult for higher mass intermediate states to conspire to yield the experimental kaon mass difference without upsetting the pion mass difference, or instead $\delta m_e/\delta m$ is not universal and is different for baryons and mesons. We saw in sub-section 6a that the medium strong mass-splitting parameter was slightly different for baryons and mesons, the ratio $\delta m/\delta m'$ being about 1·18. A similar difference for $\delta m_e/\delta m$ will enhance the contribution to kaon mass difference, in Table 10.

7 Non-Leptonic Decays Based on Quark Model

It is now well known[20] that the current-current form of $H_W{}^{NL}$ together with the algebra of currents and the PCAC hypothesis have given good predictions for non-leptonic decays except for p-wave hyperon decays. One must point out that there is another model, where $H_W{}^{NL}$ is taken to be a member of an octet transforming like scalar plus pseudoscalar quark densities, which has given equally well predictions for non-leptonic decays. When used in conjunction with the SU(6) algebra and the PCAC hypothesis, the model has even more predictive power, at least for *s*-wave decays, than that of the current-current form of $H_W{}^{NL}$.

In the case of semi-leptonic decays of hadrons one assumes that the Hamiltonian transforms like $F_1 + i\,F_2$ or $F_4 + i\,F_5$. It turns out, using Cabbibo picture and the current commutation relations of the chiral SU(3) X SU(3), that all semi-leptonic decays of hadrons[20] can be expressed in terms of a single universal Fermi constant G and the Cabbibo angle θ. We now take the $H_W{}^{NL}$ to be of the form[24,29,30]

$$H_W{}^{NL} = H_W{}^{pv} + H_W{}^{pc} = G_s \mathscr{S}_7{}^5 + G_p \mathscr{S}_6 \tag{128}$$

where $\mathscr{S}_7{}^5$ transforms like the seventh component of the pseudoscalar quark density, $i\,\bar{q}(\lambda_i/2)\gamma_5 q$, while $\mathscr{S}_6$ transforms like the sixth component of the scalar quark density, $\bar{q}(\lambda_i/2)q$. That $H_W{}^{pv}$ should transform as seventh component of the pseudoscalar octet and $H_W{}^{pc}$ as the sixth component of the scalar octet follow from the *CP* invariance and the fact that $\mathscr{C} = +1$ for quark scalar and pseudoscalar densities. A consequence of the above transformation property is that $K_1{}^0 \rightarrow 2\pi$ is not forbidden in the limit of exact SU(3) in contrast to what is found in the current-current form for $H_W{}^{NL}$ in which case $H_W{}^{pv}$ transforms as λ_6. In this approach one takes the attitude that just as semi-leptonic decays are described by vector and axial vector currents, non-leptonic decays are described by scalar and pseudoscalar densities. Just as in semi-leptonic decays one exploits the commutation relations of the chiral SU(3) X SU(3), in non-leptonic decays one exploits the following commutation relations:

$$[F_i{}^5(t), \mathscr{S}_j{}^5(x)] = -i\, d_{ijk} \mathscr{S}_k(x) \tag{129a}$$

$$[F_i{}^5(t), \mathscr{S}_j(x)] = i\, d_{ijk} \mathscr{S}_k{}^5(x) \tag{129b}$$

The Hamiltonian (128), when used in conjunction with the commutation relations (129) and PCAC, leads to the Lee–Sugawara triangular relation[7] for *s*-wave hyperon non-leptonic amplitudes and to $A(\Sigma_+{}^+) = 0$. In addition all *s*-wave amplitudes[24,30] can be described in terms of a single coupling constant, G_s. The model, however, runs into trouble as far as *p*-wave non-leptonic hyperon decays are concerned. They vanish if SU(3) invariance is used for strong meson-baryon vertices.They can arise only due to deviations of strong meson-baryon coupling constants from their SU(3) values. Since experimental information about these coupling constants is very meagre, one cannot say anything more about *p*-wave hyperon amplitudes in this model. As regards the relationship of $K \rightarrow 3\pi$ between $K \rightarrow 2\pi$ decays the model gives identical predictions, as were obtained[20] for the current-current form of $H_W{}^{NL}$ except that $K_1{}^0 \rightarrow 2\pi$ does not vanish in the SU(3) limit, which is the case for the current-current picture. As $H_W{}^{NL}$ (128) is a member of an octet, the $\Delta I = \frac{1}{2}$ rule follows. In this model $K^+ \rightarrow \pi^+\pi^0$ would probably arise due to electromagnetic corrections to the Hamiltonian (128), although no calculation has yet been done to demonstrate this fact.

7.1 *Hyperon decays*

The matrix element for a non-leptonic hyperon decay process

$$B(p') \rightarrow B(p) + \pi(q)$$

is given by

$$\begin{aligned} S &= (2\pi)^4\delta(p' - p - q)\langle B_k(p)\pi_i(q)| -\mathrm{i}\, H_W{}^{NL}(0)|B_j(p')\rangle \\ &= (2\pi)^4\delta(p' - p - q)\,\frac{(-\mathrm{i})\mathscr{R}}{\sqrt{(2q_0)}} \end{aligned}$$

where i, j, k, etc., denote the SU(3) indices, $B(p)$ denotes a baryon of momentum p and $\mathscr{R}$ is given by:

$$\begin{aligned} \mathscr{R} &= (2q_0)^{\frac{1}{2}}\langle B_k(p)\pi_i(q)|H_W{}^{NL}(0)|B_j(p')\rangle \\ &= \mathrm{i}\,(q^2 + m^2)\int \mathrm{d}^4x\, \mathrm{e}^{-\mathrm{i}q\cdot x}\theta(x_0)\langle B_k(p)|[\pi_i(x), H_W{}^{NL}(0)]|B_j(p')\rangle \\ &= \mathrm{i}\sqrt{\frac{(M_jM_k)}{p_0p_0{}'}}\,\bar{u}(p)[A_{ijk} + B_{ijk}\gamma_5]u(p') \end{aligned} \tag{131}$$

Now using a result of the algebra of currents techniques[20,33], we obtain:

$$-\frac{f_\pi}{\sqrt{2}}\lim_{q_\mu\rightarrow 0}\mathscr{R}^{ijk}(q) = \mathrm{i}\,\langle B_k(p)|[F_i{}^5, H_W{}^{NL}(0)]|B(p')\rangle - \lim_{q_\mu\rightarrow 0}\mathrm{i}\,q_\mu M_\mu{}^{ijk} \tag{132}$$

where

$$M_\mu{}^{ijk} = \mathrm{i}\int \mathrm{d}^4x\, \mathrm{e}^{-\mathrm{i}q\cdot x}\theta(x_0)\langle B_k(p)|[\partial_\mu \mathscr{F}_{i\mu}{}^5(x) H_W{}^{NL}(0)]|B_j(p')\rangle \tag{133}$$

and PCAC gives

$$\begin{aligned} \partial_\mu \mathscr{F}_{i\mu}{}^5 &= \frac{f_\pi}{\sqrt{2}}\pi_i(x) \\ \frac{f_\pi}{\sqrt{2}} &= g_A\,\frac{m_N}{g_\gamma K(0)} \end{aligned} \tag{134}$$

The term $\lim_{q_\mu \to 0} i\, q_\mu M_\mu$ gets contribution only from single-particle intermediate states degenerate in mass with either B_j or B_k. In fact it is $\left[\frac{\sqrt{2}}{f_\pi} i\, q_\mu M_\mu + \mathscr{R}_{\text{Born}}(q)\right]$, where $\mathscr{R}_{\text{Born}}(q)$ denotes the baryon pole contributions to $\mathscr{R}(q)$, which has a well-defined limit when $q_\mu \to 0$, the limit of individual terms $i\, q_\mu M_\mu$ and $\mathscr{R}_{\text{Born}}(q)$ is ill defined. Writing therefore

$$\mathscr{R}(q) = \mathscr{R}_{\text{Born}}(q) + \tilde{\mathscr{R}}(q) \tag{135}$$

where $\tilde{\mathscr{R}}(q)$ is the decay amplitude without the baryon pole terms, one obtains from (132)

$$\tilde{\mathscr{R}}(q = 0) = -\frac{\sqrt{2}}{f_\pi} i \langle B_k | [F_i{}^5, H_{\text{W}}{}^{\text{NL}}(0)] | B_j \rangle - \tilde{\mathscr{R}}_c(0) \tag{136}$$

where

$$\tilde{\mathscr{R}}_c(0) = -\lim_{q_\mu \to 0} \left[\frac{\sqrt{2}}{f_\pi} i\, q_\mu M_\mu - \mathscr{R}_{\text{Born}}(q)\right] \tag{137}$$

Assuming now that $\tilde{\mathscr{R}}(q)$ is a slowly-varying function of q, so that its physical value may be replaced by $\mathscr{R}(0)$, one obtains

$$\mathscr{R}(q^2 = -m_\pi{}^2) = -\frac{\sqrt{2}}{f_\pi} i \langle B_k | [F_i{}^5, H_{\text{W}}{}^{\text{NL}}] | B_j \rangle + \\ + \mathscr{R}_{\text{Born}}(q, q^2 = -m_\pi{}^2) - \tilde{\mathscr{R}}_c(0) \tag{138}$$

where[20,33]

$$\mathscr{R}_{\text{Born}}(q, q^2 = -m_\pi{}^2) = \mathscr{R}_{\text{Born}}{}^{\text{pv}}(q, q^2 = -m_\pi{}^2) + \mathscr{R}_{\text{Born}}{}^{\text{pc}}(q, q^2 = -m_\pi{}^2)$$

$$\mathscr{R}_{\text{Born}}{}^{\text{pv}}(q, q^2 = -m_\pi{}^2) = i \sqrt{\left(\frac{M_j M_k}{p_0 p_0{}'}\right)}\, \bar{u}(p) \left[\sum_l g_{ilk} \, b_{jl} \frac{1}{M_j + M_l} + \sum_{l'} g_{ijl'} b_{l'k} \frac{1}{M_k + M_{l'}}\right] u(p')$$

$$\mathscr{R}_{\text{Born}}{}^{\text{pc}}(q, q^2 = -m_\pi{}^2) = -i \sqrt{\left(\frac{M_j M_k}{p_0 p_0{}'}\right)}\, \bar{u}(p) \gamma_5 \left[\sum_l g_{ilk} \, a_{jl} \frac{1}{M_j - M_l} - \sum_{l'} g_{ijl'} a_{l'k} \frac{1}{M_{l'} - M_k}\right] u(p') \tag{139}$$

$$\tilde{\mathscr{R}}_c(0) = i \sqrt{\left(\frac{M_j M_k}{p_0 p_0{}'}\right)}\, \bar{u}(p) \left[\sum_l g_{ilj} \gamma_5 \frac{1}{M_k + M_l} \times \\ \times (a_{jl} + \gamma_5 b_{jl}) + \sum_{l'} (a_{l'k} + \gamma_5 b_{l'k}) \frac{1}{M_j + M_{l'}} \gamma_5 g_{ijl'}\right] u(p') \tag{140}$$

In the above equations, g_{ijk} are strong meson-baryon coupling constants defined by:

$$\langle B_k|J_i|B_j\rangle = \mathrm{i}\, g_{ijk}\bar{u}(p)\gamma_5 u(p')K(q^2) \tag{141}$$

where $K(q^2 = -m_\pi{}^2) = 1$, while a_{jk} and b_{jk} are the weak vertices defined by

$$\begin{aligned}\langle B_k|H_\mathrm{W}{}^\mathrm{NL}|B_j\rangle &= \langle B_k|H_\mathrm{W}{}^\mathrm{pc} + H_\mathrm{W}{}^\mathrm{pv}|B_j\rangle \\ &= \sqrt{\left(\frac{M_j M_k}{p_0 p_0{}'}\right)}\, \bar{u}(p)[a_{jk} + \gamma_5 b_{jk}]u(p')\end{aligned} \tag{142}$$

If one assumes SU(3) invariance for the meson-baryon vertex, then

$$g_{ijk} = 2g_{\pi NN}[\mathrm{i}\, f_{ijk}f + d_{ijk}(1-f)] \tag{143}$$

where $f/(1-f)$ is the f/d ratio.

(i) *s-wave or p.v. amplitudes.* With

$$H_\mathrm{W}{}^\mathrm{pv} = G_s \mathscr{S}_7{}^5$$

our formula (138) gives

$$\begin{aligned}\mathscr{R}^\mathrm{pv}(q^2 = -m_\pi{}^2) = -\, G_s \frac{\mathrm{i}\sqrt{2}}{f_\pi} \langle B_k|[F_i{}^5, \mathscr{S}_7{}^5]|B_j\rangle\, + \\ + \mathscr{R}_\mathrm{Born}{}^\mathrm{pv}(q^2 = -m_\pi{}^2) - \tilde{\mathscr{R}}_c{}^\mathrm{pv}(0)\end{aligned} \tag{144}$$

First, we notice from (139) and (140) that

$$\{\mathscr{R}_\mathrm{Born}{}^\mathrm{pv} - \tilde{\mathscr{R}}_c{}^\mathrm{pv}\} \simeq G_s \left[\frac{1}{M_k + M_l} - \frac{1}{M_j + M_l}\right]$$

i.e. it is of order $G_s(\Delta M)/2M$ compared to the first term on the right-hand side of (144). For this reason we shall neglect it. Thus

$$\mathscr{R}^\mathrm{pv} \simeq -\, G_s \frac{\mathrm{i}\sqrt{2}}{f_\pi} \langle B_k|[F_i{}^5, \mathscr{S}_7{}^5]|B_j\rangle \tag{145}$$

On making use of the commutation relations (129), one then obtains for the s-wave amplitudes

$$\begin{aligned}A_{ijk} &= -\, G_s \frac{\sqrt{2}}{f_\pi} \sqrt{\left(\frac{p_0 p_0{}'}{M_j M_k}\right)} \langle B_k|[F_i{}^5, \mathscr{S}_7{}^5]|B_j\rangle \\ &= -\, G_s \frac{\sqrt{2}}{f_\pi} \sqrt{\left(\frac{p_0 p_0{}'}{M_j M_k}\right)} (-\mathrm{i}\, d_{i7l})\langle B_k|\mathscr{S}_l|B_j\rangle\end{aligned} \tag{146}$$

Defining the matrix elements

$$\langle B_k|s_l|B_j\rangle = \sqrt{\left(\frac{M_j M_k}{p_0 p_0'}\right)}\, \bar{u}(p)[\mathrm{i} f_{ljk} s_f + d_{ljk} s_d] u(p') \tag{147}$$

we obtain[24,30]

$$\begin{aligned}
A(\Lambda_-{}^0) &= -\, G_s \frac{\sqrt{2}}{f_\pi} \frac{1}{4\sqrt{3}} (s_d + 3s_f) \\
A(\Xi_-{}^-) &= -\, G_s \frac{\sqrt{2}}{f_\pi} \frac{1}{4\sqrt{3}} (s_d - 3s_f) \\
A(\Sigma_-{}^-) &= G_s \frac{\sqrt{2}}{f_\pi} \frac{1}{2\sqrt{2}} (s_d - s_f) \\
A(\Sigma_0{}^+) &= -\, G_s \frac{\sqrt{2}}{f_\pi} \frac{1}{4} (s_d - s_f) \\
A(\Sigma_+{}^+) &= 0
\end{aligned} \tag{148}$$

The other amplitudes are related by the $\Delta I = \frac{1}{2}$ rule. From the above relations, we immediately obtain the Lee–Sugawara relation[70]

$$A(\Lambda_-{}^0) + 2A(\Xi_-{}^-) = \sqrt{3} A(\Sigma_0{}^+) \tag{149}$$

Additional information is obtained by noting the fact that the reduced matrix elements s_d, s_f appearing in (148) also occurred in Section 6 where medium strong mass splittings were discussed. Thus, making use of equations (88) and the Gell-Mann–Okubo mass formula, we can write equations (148) as

$$\begin{aligned}
A(\Lambda_-{}^0) &= C\sqrt{\left(\frac{3}{2}\right)}\,(\Lambda - N) \\
A(\Xi_-{}^-) &= -C\sqrt{\left(\frac{3}{2}\right)}\,(\Xi - \Lambda) \\
A(\Sigma_-{}^-) &= C(\Sigma - N) \\
A(\Sigma_0{}^+) &= -C\frac{1}{\sqrt{2}}(\Sigma - N) \\
A(\Sigma_+{}^+) &= 0
\end{aligned} \tag{150}$$

where $C = -\dfrac{G_s}{f_\pi}\dfrac{s_f}{\Xi - N} = \dfrac{G_s}{f_\pi}\dfrac{1}{\sqrt{3}\delta m}$. We note that the s-amplitudes are determined by a single parameter C and hence we can predict their relative strengths, e.g.

$$A(\Sigma_0^+)/A(\Lambda_-^0) = -\frac{1}{\sqrt{3}}\frac{\Sigma - N}{\Lambda - N} = -0{\cdot}84$$

$$A(\Xi_-^-)/A(\Lambda_-^0) = -\frac{\Xi - \Lambda}{\Lambda - N} = -1{\cdot}23 \tag{151}$$

in very good agreement with the experimental values[72]. Thus we obtain more predictions for the s-wave amplitudes than was possible from the current-current form[20] of H_W^{NL}.

(ii) *p-wave or p.c. amplitudes.* Since in the model the p.c. non-leptonic Hamiltonian transforms like S_6 and the SU(3) mass-breaking Hamiltonian transforms like S_8, where S_6 and S_8 belong to the same octet, in general the p.c. non-leptonic Hamiltonian can be eliminated by a unitary transformation[31,32], thus leading to no p-wave decays, as was mentioned in the introduction. Within the framework of algebra of currents, this can also be seen as follows. We have from (138)

$$\mathscr{R}^{pc} = -\frac{\sqrt{2}}{f_\pi} G_p\, \mathrm{i}\, \langle B_k|[F_i^5, \mathscr{S}_6]|Bj\rangle + \mathscr{R}_{Born}{}^{pc} - \tilde{\mathscr{R}}_c{}^{pc}(0) \tag{152}$$

By using the commutation relation (129b), the first term on the right becomes:

$$-G_p \frac{\mathrm{i}\sqrt{2}}{f_\pi}\, \mathrm{i}\, d_{i6l}\langle B_k|\mathscr{S}_l{}^5|B_j\rangle$$

We now identify $\mathscr{S}_l{}^5$ with the source density J_l of pseudoscalar meson octet. This is justified because, as will be shown below, with this identification the p-wave amplitudes will vanish in the limit when SU(3) invariance is used for the meson-baryon vertices in accordance with the theorem just mentioned. Thus we have:

$$\mathscr{R}^{pc} = -G_p \frac{\mathrm{i}\sqrt{2}}{f_\pi}\, \mathrm{i}\, d_{i6l}\langle B_k|J_l|B_j\rangle + \mathscr{R}_{Born}{}^{pc} - \tilde{\mathscr{R}}_c{}^{pc}(0) \tag{153}$$

Now in $\mathscr{R}_{Born}{}^{pc}$, the weak vertices $a_{BB'}$ being proportional to $\langle B|\mathscr{S}_6|B'\rangle$ are determined in terms of s_f and s_d, which have already appeared in the s-wave decays and the mass relations (88). Noting that $\langle B_k|J_l|B_j\rangle$ are just the strong meson-baryon coupling constants, we obtain by using formulae (139) and (140), the following expressions for the p-wave amplitudes

$$B(\Lambda_-^0) = -\frac{G_p}{\sqrt{2}f_\pi}\left\{-g_{K\Lambda N} + \frac{2f_\pi}{\sqrt{3}\delta m}\frac{M_\Lambda + M_N}{2M_N} \times \right.$$

$$\left. \times \left[\sqrt{3}g_{\pi NN} - \frac{2M_N}{M_\Lambda + M_\Sigma} g_{\pi\Lambda\Sigma}\right]\right\}$$

$$B(\Xi_-^-) = -\frac{G_p}{\sqrt{2}f_\pi}\left\{-g_{K\Lambda\Xi} - \frac{2f_\pi}{\sqrt{3}\delta m}\frac{M_\Lambda + M_\Xi}{2M_\Lambda}\left[\sqrt{3}g_{\pi\Xi\Xi} + \frac{M_\Lambda}{M_\Xi} g_{\pi\Lambda\Sigma}\right]\right\}$$

$$B(\Sigma_+^+) = \frac{G_p}{\sqrt{3}\delta m}\frac{M_\Sigma + M_N}{2M_N}\left\{-2g_{\pi NN} + \frac{M_N}{M_\Sigma} g_{\pi\Sigma\Sigma} + \frac{2M_N}{M_\Lambda + M_\Sigma}\sqrt{3}g_{\pi\Lambda\Sigma}\right\}$$

$$B(\Sigma_-^-) = -\frac{G_p}{\sqrt{2}f_\pi}\left\{-\sqrt{2}g_{KN\Sigma} + \frac{2f_\pi}{\sqrt{3}\delta m}\frac{M_\Sigma + M_N}{2M_N} \times \right.$$

$$\left. \times \left[-\sqrt{(3/2)}\frac{2M_N}{M_\Lambda + M_\Sigma} g_{\pi\Lambda\Sigma} + \frac{M_N}{M_\Sigma} g_{\pi\Sigma\Sigma}\right]\right\}$$

$$B(\Sigma_0^+) = -\frac{G_p}{\sqrt{2}f_\pi}\left\{g_{KN\Sigma} + \frac{2f_\pi}{\sqrt{3}\delta m}\frac{M_\Sigma + M_N}{2M_N}\left[g_{\pi NN} - \frac{M_N}{M_\Sigma} g_{\pi\Sigma\Sigma}\right]\right\} \quad (154)$$

The mass factors in the above equations appear due to the presence of $\mathscr{R}_c^{\text{pc}}$. We observe that these amplitudes vanish if the meson-baryon coupling constants and mass factors appearing in (154) have their SU(3) values, provided that

$$\frac{\sqrt{3}\delta m}{2} = -f_\pi \quad (155)$$

This relation is interesting because it gives $\delta m = -160$ MeV, the same value obtained in Section 6 from a uniform treatment of Gell-Mann–Okubo mass splitting based on non-chiral algebra formed by S_i and F_i. We thus see that the *p*-wave hyperon decays in the model can arise only due to deviations of the strong meson-baryon coupling constants from their SU(3) values. Since experimental information about the coupling constants is meagre, one cannot say anything more about the *p*-wave decays in this model. One may, of course, also take the attitude that the p.c. non-leptonic Hamiltonian does not transform like the member of the same octet to which the mass-breaking Hamiltonian belongs. But in this case one would not be able to relate the ratios of *p*-wave amplitudes to those of *s*-waves. For this reason this possibility is not attractive and we shall not consider it.

(iii) *Relationship between K-3π and K-2π decays.* The current algebra techniques together with the PCAC hypothesis can relate K-3π and K-2π decays[29,35,37] in the soft pion limit. It is known that a good description of K-3π decays is provided by a linear expansion of the decay amplitude in the form[73]

$$A = A_0 \left[1 - \frac{a}{m_\pi^2}(s_3 - s_0) \right] \tag{156}$$

where $s_i = -(p - q_i)^2$, $s_0 = \frac{1}{3}(s_1 + s_2 + s_3)$ and s_3 refers to the unlike or 'odd' pion in K-3π decay. Application of current algebra techniques not only relates the K-3π and the K-2π rates, but also one determines the value of slope parameter in the expansion (156). The slope comes out to be in good agreement with experiment, supporting the point of view that the energy spectra in K-3π decays arise from the structure of the weak interaction itself and that the final state interaction among the pions may be neglected.

We assume that $H_W{}^{NL}$ is of the form (128). Then in the limit when $q \to 0$, one obtains[29,35]

$$\begin{aligned} \lim_{q_{3\mu}\to 0} (2q_{30})^{\frac{1}{2}} G_p \langle \pi_1\pi_2\pi_3 | \mathscr{S}_6 | K \rangle &= -\frac{\mathrm{i}\sqrt{2}}{f_\pi} G_p \pi_{3k} \langle \pi_1\pi_2 | [F_k{}^5, \mathscr{S}_6] | K \rangle \\ &= (\sqrt{2}/f_\pi) G_p \pi_{3k} d_{k6l} \langle \pi_1\pi_2 | \mathscr{S}_l{}^5 | K \rangle \end{aligned} \tag{157}$$

on making use of the commutation relation (129b). In (157) π_{3k} denotes the isospin vector for the pion π_3. Now the matrix element for K-2π decay is given by

$$\begin{aligned} \langle \pi_1\pi_2 | H_W{}^{pv} | K \rangle &= G_s \langle \pi_1\pi_2 | \mathscr{S}_7{}^5 | K \rangle \\ &= \frac{A_{2\pi}}{\sqrt{(8p_0q_{10}q_{20})}} \end{aligned} \tag{158}$$

Since $\mathscr{S}_7{}^5$ behaves as i $K_1{}^0$, we can write

$$\begin{aligned} A_{2\pi} &= \mathrm{i}\, b(\boldsymbol{\pi}_1 \cdot \boldsymbol{\pi}_2) K_1{}^0 \\ &= \frac{\mathrm{i}}{\sqrt{2}} b(\boldsymbol{\pi}_1 \cdot \boldsymbol{\pi}_2) \bar{\beta} K \end{aligned} \tag{159}$$

where β is the spurion $\binom{0}{1}$. Now it is easy to see that one can write in isospin space $\pi_{3k} d_{k6l} \mathscr{S}_l{}^5$ as

$$\pi_{3k} d_{k6l} \mathscr{S}_l{}^5 = \frac{1}{2\sqrt{2}} \bar{\beta} (\boldsymbol{\pi}_3 \,.\, \boldsymbol{\tau}) \mathscr{S}_5{}^K \tag{160}$$

where $\mathscr{S}_5{}^K$ is the pseudoscalar density which behaves as K. Thus we can write:

$$\begin{aligned} \pi_{3k} d_{k6l} \langle \pi_1\pi_2 | \mathscr{S}_l{}^5 | K \rangle &= \frac{1}{2\sqrt{2}} \langle \pi_1\pi_2 | \bar{\beta} (\boldsymbol{\pi}_3 \cdot \boldsymbol{\tau}) \mathscr{S}_5{}^K | K \rangle \\ &= \frac{1}{2\sqrt{2}} (\boldsymbol{\pi}_1 \cdot \boldsymbol{\pi}_2)\, \mathrm{i}\, b \bar{\beta} (\boldsymbol{\pi}_3 \cdot \boldsymbol{\tau}) K \end{aligned} \tag{161}$$

Now since $\mathscr{S}_6$ behaves as $K_2{}^0$ as far as internal quantum numbers are concerned, so the effective Hamiltonian for $K \to 3\pi$ decays in isospin space behaves[74] as

$$\tilde{A}^+K^+ - \tilde{A}^0K_2{}^0 + \tilde{A}^-K^-$$

where $\tilde{\mathbf{A}} = (\tilde{A}^+, \tilde{A}^0, \tilde{A}^-)$ is an isovector. Thus we can write the matrix elements for K-3π decay in isospin space as

$$(2q_{30})^{\frac{1}{2}} G_p \langle \pi_1\pi_2\pi_3 | \mathscr{S}_6 | K \rangle = \frac{i}{\sqrt{(8p_0q_{10}q_{20}}} \frac{1}{\sqrt{2}} \bar{\beta}(\tilde{\mathbf{A}} \cdot \boldsymbol{\tau})K \tag{162}$$

where

$$\begin{aligned} \tilde{\mathbf{A}} &= f(s_1, s_2, s_3)(\boldsymbol{\pi}_1 \cdot \boldsymbol{\pi}_2)\boldsymbol{\pi}_3 + f(s_3, s_1, s_2) \times \\ &\quad \times (\boldsymbol{\pi}_3 \cdot \boldsymbol{\pi}_1)\boldsymbol{\pi}_2 + f(s_2, s_3, s_1)(\boldsymbol{\pi}_2 \cdot \boldsymbol{\pi}_3)\boldsymbol{\pi}_1 \\ f(s_1, s_2, s_3) &= A_0 \left[1 - \frac{a}{m_\pi{}^2}(s_3 - s_0)\right] \\ &= A_0 \left[1 + \frac{a}{3m_\pi{}^2}(s_1 + s_2 - 2s_3)\right] \end{aligned} \tag{163}$$

Equations (157), (158), (159), (161) and (162) thus give

$$\lim_{q_{3\mu} \to 0} \bar{\beta}(\tilde{\mathbf{A}} \cdot \boldsymbol{\tau})K = \frac{G_p}{G_s} \frac{1}{\sqrt{2} f_\pi} b(\boldsymbol{\pi}_1 \cdot \boldsymbol{\pi}_2)\bar{\beta}(\boldsymbol{\pi}_3 \cdot \boldsymbol{\tau})K \tag{164}$$

Since $q_{3\mu} \to 0$ we should put $s_3 = m_K{}^2$, $s_1 = s_2 = m_\pi{}^2$, $s_0 = \frac{1}{3}m_K{}^2 + \frac{2}{3}m_\pi{}^2$ on the left-hand side of (164). With these values of s_1, s_2 and s_3, the relation (164), on using (163), gives

$$\begin{aligned} A_0 \Bigg\{ &(\boldsymbol{\pi}_1 \cdot \boldsymbol{\pi}_2)\bar{\beta}(\boldsymbol{\pi}_3 \cdot \tau)K + (\boldsymbol{\pi}_2 \cdot \boldsymbol{\pi}_3)\bar{\beta}(\boldsymbol{\pi}_1 \cdot \boldsymbol{\tau})K + \\ &+ (\boldsymbol{\pi}_3 \cdot \boldsymbol{\pi}_1)\bar{\beta}(\boldsymbol{\pi}_2 \cdot \boldsymbol{\tau})K - \frac{a(m_K{}^2 - m_\pi{}^2)}{3m_\pi{}^2} \Big[2(\boldsymbol{\pi}_1 \cdot \boldsymbol{\pi}_2) \\ &\bar{\beta}(\boldsymbol{\pi}_3 \cdot \boldsymbol{\tau})K - (\boldsymbol{\pi}_2 \cdot \boldsymbol{\pi}_3)\bar{\beta}(\boldsymbol{\pi}_1 \cdot \boldsymbol{\tau})K - (\boldsymbol{\pi}_3 \cdot \boldsymbol{\pi}_1)\, \bar{\beta}(\boldsymbol{\pi}_2 \cdot \boldsymbol{\tau})K \Big] \Bigg\} \\ &= \frac{G_p}{G_s} \frac{1}{\sqrt{2} f_\pi} b(\boldsymbol{\pi}_1 \cdot \boldsymbol{\pi}_2)\bar{\beta}(\boldsymbol{\pi}_3 \cdot \boldsymbol{\tau})K \end{aligned}$$

leading to

$$A_0 = \frac{G_p}{G_s} \frac{1}{3\sqrt{2} f_\pi} b \tag{165a}$$

$$a = -\frac{3m_\pi{}^2}{m_K{}^2 - m_\pi{}^2} = -0{\cdot}26 \tag{165b}$$

For an alternative derivation of these results in a probably more transparent way, the reader is referred to a paper by Abarhanel[37].

Now according to the $|\Delta \mathbf{I}| = \frac{1}{2}$ rule[74,75] which we are assuming:

$$A_\tau(++-) = \frac{2}{\sqrt{(2!)}} A_0 \left[1 + \frac{a}{2m_\pi^2}(s_3 - s_0)\right]$$

$$A_{\tau'}(0\,0\,+) = \frac{1}{\sqrt{(2!)}} A_0 \left[1 - \frac{a}{m_\pi^2}(s_3 - s_0)\right]$$

$$A_2(+-0) = -A_0 \left[1 - \frac{a}{m_\pi^2}(s_3 - s_0)\right]$$

$$A_2' = -\frac{3}{\sqrt{(3!)}} A_0 \tag{166}$$

where the factors $\sqrt{(2!)}$ and $\sqrt{(3!)}$ appear because a state with n identical pions contains $(n!)^{-\frac{1}{2}}$ in its normalization. Thus finally we obtain:

$$\begin{aligned} -A_{02}(+-0) &= \frac{1}{\sqrt{2}} A_{0\tau'}(0, 0, +) = \sqrt{2} A_{0\tau}(++-) \\ &= -\sqrt{\frac{2}{3}} A_{02'}(0\,0\,0) = \frac{b}{3\sqrt{2} f_\pi}\left(\frac{G_p}{G_s}\right) \end{aligned} \tag{167}$$

$$\begin{aligned} a_2(+-0) &= a_{\tau'}(0, 0, +) = -2a_\tau(++-) \\ &= a = -\frac{3m_\pi^2}{m_K^2 - m_\pi^2} = -0{\cdot}26 \end{aligned} \tag{168}$$

Except the last equalities in the above equations, which are the specific consequences of techniques of algebra of currents, the commutation relation (129b) and the Hamiltonian (128), the rest are the predictions of the $\Delta I = \frac{1}{2}$. The predictions of the $\Delta I = \frac{1}{2}$ rule are in good agreement with experiment[74]. To test the predictions of the algebra of currents, we note:

$$|\mathscr{E}| = a(K_1^0 \to \pi^+\pi^-),$$

the decay amplitude of $K_1^0 \to \pi^+\pi^-$. Experimentally[35]

$$\frac{1}{3\sqrt{2} f_\pi} |b| = (0{\cdot}81 \pm 0{\cdot}01) \times 10^{-6} \tag{169}$$

$$|A_{02}(+-0)| = |A_0| = (0{\cdot}89 \pm 0{\cdot}03) \times 10^{-6} \tag{170}$$

and

$$a_2(+-0) = -0{\cdot}24 \pm 0{\cdot}02 \tag{171}$$

Thus the prediction (165) for slope parameter is in very good agreement with experiment independent of the value of $|G_p/G_s|$. In view of (170) and (171) the prediction (165a) for the rates is also in agreement with experiment for $|G_p/G_s| \simeq 1$.

The predictions (165a) with $|G_p| \simeq |G_s|$ as well as (165b) are the same as one would also obtain[35,36] from the current-current form of $H_W{}^{NL}$ with octet dominance, except that in the model considered here $a(K_{10} \to \pi^+\pi^-)$ is not zero in the exact SU(3) limit as is the case for the current-current form of $H_W{}^{NL}$.

Appendix

We list here some of the many cross-section relations which have been obtained from the quark model. Because of the non-availability of the experimental data these relations have not been tested as yet. However many of these experiments are feasible and will be performed in the near future.

From the additivity alone the following cross-section relations follow immediately:

$$\sigma_t(\Lambda p) = \sigma_t(pn) + \sigma_t(K^- p) - \sigma_t(\pi^- p) \tag{1}$$

$$2\sigma_t(\Sigma^+ p) + \sigma_t(\Sigma^- p) = 2\sigma(pn) + \sigma(pp) + 3\sigma_t(K^- p) - 3\sigma_t(\pi^- p) \tag{2}$$

If we now assume the SU(2) symmetry, then from (1) and (2) we have

$$\sigma_t(\Lambda p) - \sigma_t(pp) = \sigma_t(K^- n) - \sigma_t(\pi^+ p) \tag{3}$$

$$\sigma_t(\Sigma^+ p) = \sigma_t(pp) + \sigma_t(K^- p) - \sigma_t(\pi^- p) \tag{4}$$

$$\sigma_t(\Sigma^- p) = 2\sigma(pn) - \sigma(pp) + \sigma(K^- p) - \sigma_t(\pi^- p) \tag{5}$$

From the SU(3) symmetry the following total cross-sections can be obtained:

$$\sigma_t(\Lambda p) = 2\sigma_t(pn) - \sigma_t(pp) \tag{6}$$

$$\sigma_t(\Sigma^+ p) = \sigma_t(pn) \tag{7}$$

$$\sigma_t(\Sigma^- p) = \sigma_t(\Xi^0 p) = 3\sigma(pn) - 2\sigma(pp) \tag{8}$$

For the production processes we mention the following relations (SU(3) is assumed):

$$\bar{\sigma}(K^+ n \to K^0 p) = 4\bar{\sigma}(K^- p \to \pi^0 \Lambda) - 2\bar{\sigma}(K^- p \to \pi^- \Sigma^+) \tag{9}$$

$$\bar{\sigma}(K^- p \to \pi^- Y_1{}^{*+}) = \tfrac{1}{3}\bar{\sigma}(K^+ p \to K^0 N^{*++}) \tag{10}$$

$$\bar{\sigma}(K^- p \to \pi^0 \Lambda) = \tfrac{3}{4}[\bar{\sigma}(K^- p \to \pi^- \Sigma^+) + \bar{\sigma}(K^- p \to \pi^- Y_1{}^{*+})] \tag{11}$$

For details the reader is referred to the original references[15,43,45].

References

1. Sakata, S., *Prog. Theor. Phys.* (*Kyoto*), **16,** 686 (1956).
2. Gell-Mann, M., and Ne'eman, Y., *The Eight-Fold Way*, Benjamin, New York, 1964.
3. Other fundamental particle models that do not require fractionally charged particles have been suggested. See for examples, Gell-Mann, M., *Phys. Lett.*, **8,** 214 (1964); Gursey, F., Lee, T. D., and Neunberg, M., *Phys. Rev.*, **135B,** 467 (1964); Lee, T. D., *Nuovo Cimento*, **35,** 933 (1965).
4. Gell-Mann, M., *Phys. Lett.*, **8,** 214 (1964).
5. Zweig, G., *CERN Report TH*-401 *and TH*-412 (1964), unpublished; *Symmetries in Elementary Particle Physics* (Ed. A. Zichichi), Academic Press, 1965, p. 192.
6. For details, Dalitz, R. H., 'Symmetries and the strong interactions', *Berkeley Conf. Report*, 1966, may be consulted.
7. Gell-Mann, M., 'Opening remarks', *XIII Intern. Conf. High Energy Phys., Berkeley*, 1966.
8. Beg, M. A. B., Lee, B. W., and Pais, A., *Phys. Rev. Letters*, **13,** 514 (1964).
9. Thirring, W. E., *Phys. Lett.*, **16,** 335 (1965).
10. Becchi, C., and Morpurgo, G., *Phys. Rev.*, **140B,** 687 (1965) **149,** 1284 (1966).
11. Anisovitch, Y., Anselm, A., Azimov, Y., Damlov, G., and Daytlor, I., *Phys. Lett.*, **16,** 194 (1965).
12. Soloviev, L., *Phys. Lett.*, **16,** 345 (1965).
13. Becchi, C., and Marpurgo, G., *Phys. Lett.*, **17,** 352 (1965).
14. Levin, E. M., and Frankfurt, L. L., *JETP Letters*, **2,** 65 (1965).
15. Lipkin, H. J., and Scheck, F., *Phys. Rev. Letters*, **16,** 7 (1966).
16. Barger, V., and Rubin, M. H., *Phys. Rev.*, **140B,** 1365 (1965).
17. Johnson, K., and Treiman, S. B., *Phys. Rev. Letters*, **14,** 189 (1965).
18. Sarker, A. Q., *Phys. Rev.* **161,** 1435 (1967).
19. Ruegg, H., and Volkov, D. V., *Nuovo Cimento*, **43,** 84 (1966).
20. The literature on non-chiral SU(3) X SU(3) is extensive. We refer the reader to following review articles or lecture notes where references to original literature can also be found: (i) Mathur, V. S., and Pandit, L. K., *Current Algebra: Application to Weak Interactions*, Rochester Preprint; (ii) Cabbibo, N., 'Weak interactions', *XIII Intern. Conf. High Energy Phys.*, Berkeley, 1966; (iii) Bell, J. S., 'Weak interactions and current algebra', *Lecture Notes, CERN Easter School, Noordemick*, June 1966; (iv) Amati, D., 'Current algebra', *CERN Report TH*-732 (December 1966); (v) Pietschman, H., *Selected Topics in Current Algebra*; (vi) Renner, B., 'Lectures on current algebra', *Rutherford Lab. Report* (1966); (vii) Papastamatiou, N. J., 'Some applications of current algebra', *Rutherford Lab. Report*, **130** (1966).
21. Gell-Mann, M., *Physics*, **1,** 63 (1964).
22. Kikkawa, K., *Progr. Theor. Phys.* (*Kyoto*), **35,** No. 2 (1966); Arafune, J., Imasaki, Y., Kikkawa, K., Matsuda, S., and Nakamura, K., *Phys. Rev.*, **143,** 1220 (1960).
23. Arnowitt, *Nuovo Cimento*, **40,** 985 (1965).
24. Riazuddin, and Mahanthappa, K. T., *Phys. Rev.*, **147,** 972 (1966).
25. Schwinger, J., *Phys. Rev.*, **135B,** 816 (1964).
26. Riazuddin and Fayyazuddin, *Phys. Rev.* **158,** 1447 (1967).

27. Bose, S. K., and Hara, Y., *Phys. Rev. Letters,* **17,** 409 (1966); see also Lai, C. S., *Phys. Rev.*, **155,** 1562 (1967); Rockmore, R., *Phys. Rev.*, **153,** 1490 (1967); Biswas, S. N., and Patil, S. H., *Phys. Rev.*, **155,** 1599 (1967).
28. Murashkin, M., and Glashow, S. L., *Phys. Rev.*, **132,** 482 (1963); Gupta, V., and Singh, V., *Phys. Rev.*, **135B,** 1442 (1964); Dullemond, C., MacFarlane, A. J., and Sudarshan, E. C. G., *Phys. Rev. Letters,* **10,** 423 (1963); Becchi, C., Ebrle, E., and Morpurgo, G., *Phys. Rev.*, **136B,** 808 (1964); Konuma, M., and Tomozawa, K., *Phys. Lett.*, **10,** 347 (1964).
29. Callan, C. G., and Treiman, S. B., *Phys. Rev. Letters,* **16,** 153 (1965).
30. Gillard, M. K., *Phys. Lett.*, **20,** 533 (1966); see also Gatts, R., Maiani, L., and Preparata, G., *Nuovo Cimento,* **41,** 622 (1966).
31. Coleman, S., and Glashow, S. L., *Phys. Rev.*, **134B,** 671 (1964).
32. Lee, B. W., *Phys. Rev.*, **140B,** 152 (1965).
33. Hara, Y., Nambu, Y., and Schechter, J., *Phys. Rev. Letters,* **16,** 380 (1960); Brown, L. S., and Sommafeld, C. M., *Phys. Rev. Letters,* **16,** 751 (1966); Badier, S., and Bouchiat, C., *Phys. Lett.*, **20,** 529 (1966); see also: Fayyazuddin and Riazuddin, *Nuovo Cimento,* **47,** 222 (1967); Itzykson, C., and Jacob, M., *Nuovo Cimento,* **48,** 655 (1967); Kumar, A., and Pati, J. C., *Phys. Rev. Letters,* **18,** 1230 (1967).
34. Suzuki, M., *Phys. Rev. Letters,* **16,** 212 (1966); Bose, S. K., and Biswas, S. M., *Phys. Rev. Letters,* **16,** 340 (1966); Nefkens, B. M. K., *Phys. Lett.*, **22,** 14 (1966); Weyers, J., Foldy, L. L., and Speiser, D. R., *Phys. Rev. Letters,* **17,** 1062 (1966). In these references current-current form of non-leptonic Hamiltonian $H_W{}^{NL}$ is used.
35. Hara, Y., and Nambu, Y., *Phys. Rev. Letters,* **16,** 875 (1966). In this reference both current-current form and the form (2) of the text is considered for $H_W{}^{NL}$.
36. Elias, D. K., and Taylor, J. C., *Nuovo Cimento,* **44,** 518 (1966) and **48,** 814 (1967). In this reference only current-current form of $H_W{}^{NL}$ is considered.
37. Abarbanel, H. D. I., *Phys. Rev.*, **153,** 1547 (1967).
38. Rosenfield, A. H., and others, *Rev. Mod. Phys.*, January (1967).
39. Gursey, F., Pais, A., and Radicati, L. A., *Phys. Rev. Letters,* **13,** 299 (1964).
40. Becchi, C., and Morpurgo, G., *Phys. Rev.*, **149,** 149 (1966).
41. Elitzur, M., Rubinstein, H. R., Stern, H., and Lipkin, H. J., *Phys. Rev. Letters,* **17,** 420 (1966).
42. Kokkedee, J. J. J., and Hove, L. Van, *Nuovo Cimento,* **42,** 711 (1966).
43. Chan, C. H., *Phys. Rev.*, **152,** 1244 (1966).
44. Galbraith, W., and others, *Phys. Rev.*, **138B,** 913 (1965); *BNL* 9410 preprint, (1965).
44a. Sawyer, R. F., *Phys. Rev. Letters,* **14,** 471 (1965).
45. Lipkin, H. J., Scheck, F., and Stern, H., *Phys. Rev.*, **152,** 1375 (1966).
46. Musgrave, B., and others, *Nuovo Cimento,* **35,** 735 (1965); Baltay, C., and others, *Phys. Rev.*, **140B,** 1027 (1965).
47. Miller, D. H., and others, *Phys. Rev.*, **140B,** 360 (1965); Wangler, T. P., and others, *Phys. Rev.*, **137B,** 114 (1965).
48. Smith, G. A., and others, *Phys. Rev.*, **123,** 2160 (1961); Goldhaver, G., and others, *Phys. Lett.*, **6,** 62 (1963).
49. Ferro-Luzzi, M., and others, *Nuovo Cimento,* **36,** 1101 (1965); **39,** 417 (1965).
50. Alexander, G., and others, *Phys. Rev. Letters,* **17,** 412 (1966).

51. Trilling, G. H., and others, *Phys. Lett.*, **19,** 427 (1965); *Proc. Sienna Conf., Bologna*, 1963, p. 75.
52. Kraemev, R., and others, *Phys. Rev.*, **36B,** 496 (1964).
53. Miller, D. H., and others, *Phys. Rev.*, **140B,** 360 (1965).
54. London, G., and others, *Phys. Rev.*, **143,** 1034 (1966).
55. Davis, R., and others, *Report at the Conf. High Energy Phys., Berkeley*, 1966.
56. Dalitz, R. H., 'Symmetries and the strong interactions', *Berkeley Conf. Report*, 1966.
57. Rubinstein, H., and Stern, H., *Phys. Lett.*, **21,** 447 (1966).
58. Harte, J., and others, *CERN preprints, No. TH.*701 *and TH.*697 (1966).
59. Baltay, C., and others, *Phys. Rev.*, **145,** 1103 (1966).
60. Bockmann, K., and others, *Nuovo Cimento*, **42A,** 954 (1966); Febel, T., and others, *Nuovo Cimento*, **38,** 12 (1965).
61. Swart, J. de, *Rev. Mod. Phys.*, **35,** 916 (1963).
62. Lee, B. W., *Phys. Rev. Letters*, **14,** 676 (1965); Lectures given at Brandeis Summer Institute of Physics, 1965.
63. Glashow, S. L., and Socolow, R. H., *Phys. Rev. Letters*, **15,** 329 (1965); see also Marshak, R. E., Okubo, S., and Wojtaszek, J. H., *Phys. Rev. Letters*, **15,** 463 (1965).
64. Yang, C. N., and Feldman, D., *Phys. Rev.*, **79,** 972 (1950).
65. Riazuddin, *Phys. Rev.*, **119,** 1184 (1959); see also Barger, V., and Kazes, E., *Nuovo Cimento*, **28,** 394 (1963).
66. Cini, M., Ferrari, E., and Galts, R., *Phys. Rev. Letters*, **2,** 7 (1954); Cottingham, W. N., *Ann. Phys. (N.Y.)*, **25,** 424 (1963).
67. Bose, S. K., and Marshak, R. E., *Nuovo Cimento*, **25,** 529 (1962); Goleman, S., and Schnitzer, H. J., *Phys. Rev.*, **136B,** 223 (1964); Socolow, R. H., *Phys. Rev.*, **137B,** 1221 (1965); Wajtaszek, J. H., Marshak, R. E., and Riazuddin, *Phys. Rev.*, **136B,** 1053 (1966).
68. Bjorken, J. D., *Phys. Rev.*, **148,** 1467 (1966); Harari, H., *Phys. Rev. Letters*, **17,** 1303 (1966); Okubo, S., *Phys. Rev. Letters*, **18,** 256 (1967).
69. Nieh, H. T., *Phys. Rev.*, **146,** 1020 (1966).
70. Sugawara, H., *Progr. Theor. Phys. (Kyoto)*, **31,** 213 (1964); Lee, B. W., *Phys. Rev. Letters*, **12,** 83 (1964).
71. Sugawara, H., *Phys. Rev. Letters*, **15,** 870, 997 (1965); Suzuki, M., *Phys. Rev. Letters*, **15,** 980 (1965).
72. Samios, N. P., *Proc. Intern. Conf. Weak Interactions*, October 1965, *Argonne Report* 7130.
73. Weinberg, S., *Phys. Rev. Letters*, **4,** 87, 585 (1960).
74. Sawyer, R. F., and Wali, K. C., *Nuovo Cimento*, **17,** 938 (1960).
75. Trilling, G. H., *Proc. Intern. Conf. Weak Interactions*, October 1965, *Argonne Report* 7130.

On the Bosons of Zero Baryon Number as Bound States of Quark–Antiquark Pairs

J. L. URETSKY

1 Introduction

The idea that a triplet of particles called 'quarks' might be the 'building blocks' of the hadrons[1] was proposed after the successes of SU(3) and larger symmetries[2] left unanswered the question, 'Why SU(3)?' From one viewpoint the question is a dull one, and the quark hypothesis gives rise, at best, to a simple computational scheme from which symmetries may be abstracted[3]. Such a viewpoint is not completely satisfactory, however, in that the symmetries that seem to operate within different classes of physical phenomena are different. Furthermore, the symmetries are broken in a very marked manner, and it is not always possible to describe the breaking in a simple way.

The quark model, on the other hand, seems to provide a common linkage among the various classes of phenomena that are unrelated by symmetry schemes in any straightforward manner. Furthermore, within the framework of the quark model it is possible to introduce symmetry breaking at a very primitive level, since once the quark hypothesis is adopted it is no longer necessary to assume any symmetry other than SU(2). In fact, one might take the rather extreme viewpoint that the successes of the SU(3) and higher symmetries only serve to teach us that the underlying dynamics of hadrons reflects the quark-triplet structure of the strong interaction universe.

This is not to say that one is required, at the present time, to postulate that the fundamental triplet must exist (in the usual sense) and that free quarks can be produced in a sufficiently expensive accelerator. Rather, the statement that is made is one concerning the structure of hadron *dynamics*. In particular, it is conceivable that this particular structure is only found in the dynamics of the hadrons of relatively low mass. From this viewpoint, wherein the apparent symmetry is a consequence of dynamics, the situation is reminiscent of the one that obtains in the theory of nuclear structure. There, it will be recalled, one sees in various regions of the periodic table the 'symmetries', respectively, of $\mathbf{L}\cdot\mathbf{S}$ coupling, $\mathbf{j}\cdot\mathbf{j}$ coupling, and the collective model, although no single one of these symmetries is characteristic of all nuclear physics.

The aim of the present work is to discuss a very simple independent quark model[4] of the hadrons without the *a priori* imposition of any symmetry other than isospin. The purpose is to investigate the ranges of phenomena to which this simple picture applies. Only the bosons of zero baryon number will be considered; it will be supposed that these have wave-functions that are primarily made up of quark-antiquark ($q\bar{q}$) pairs. It is clear at the outset, then, that an SU(3)-like structure will emerge as a trivial consequence of the fact that multiplets are constructed from a pair of quark triplets. Nevertheless, the SU(3) symmetry can be rather badly broken and not necessarily in a simple (e.g. octet dominant) manner. Realizing this possibility, it becomes of interest to reexamine some accepted dogma, namely, mass relations, in a way that will emphasize the abandonment of the assumption of an underlying symmetry. This is done in Section 2 where a model is developed for the purpose of extrapolating the boson mass spectrum.

In discussing the mass relations within the boson multiplets, the obvious picture to adopt is that of a quark and an antiquark moving non-relativistically (that is, with low velocity) in a very deep potential well[5]. If this picture is to be useful, then it is necessary that the parameters of the potential be determinable from a knowledge of the masses of a few low-lying states. Once the parameters are determined, it should be possible to predict the positions of new multiplets corresponding to the excitation of higher states in the potential. At some stage of excitation the simple picture should become inadequate either because the assumption of non-relativistic velocities or that of nearly pure $q\bar{q}$ composition is no longer valid. When the structure becomes sufficiently complicated then the usefulness of the quark picture will be lost.

One hopes, of course, that the usefulness of the model extends beyond the prediction of mass spectra. In particular, one wants to be able to understand strong decays as resulting from the transitions of bound quarks with the resulting emission of 'elementary' quanta*. If such a

* This view of the strong decays was also adopted by Becchi and Morpurgo and by Van Royen and Weisskopf[4].

description can be made to work, then our experience with nuclear physics tells us that the details of the bound state wave-functions may become important. In particular, the decay matrix elements may be sensitive to small admixtures of additional quark-antiquark pairs in the wave-function even when mass levels are satisfactorily obtained from the assumption of pure $q\bar{q}$ configurations. The strong decays are treated in Section 3. I shall not discuss weak and electromagnetic decays in this work*. However, in Section 4 I do discuss the mass differences of the charged and neutral π and ρ mesons.

The general philosophy, then, will be to build a model that resembles the nuclear shell model to the maximum possible extent. The 'forces' will be determined from phenomenology rather than symmetry principles, as will be the couplings to the external fields. There will be no attempt to justify the model on the basis of 'fundamental' principles in the present discussion. The reader is thereby free to characterize any successes of the model as purely accidental.

2 Mass Formulae

2.1 *The* 1*S-State*

Consider a $q\bar{q}$ pair bound in a potential of average strength U_0. If the quark velocities in the low-lying states are to be non-relativistic and the quark masses very large, then the potential is presumably 'flat-bottomed' in the manner of a square-well or harmonic oscillator. For reasonable and deep potentials of this sort the level structure is roughly independent of the shape, but it can always be adjusted by the introduction of L^2 dependent terms (where $\mathbf{L}$ is the orbital angular momentum operator). There is, therefore, little loss in generality in supposing that the potential is a square well†.

As is usual, we assume that the quark triplet consists of a nearly degenerate isotopic doublet, called (n_0, p_0) and an isotopic singlet of strangeness minus one called λ_0. The mass of the doublet will be denoted by m, that of the singlet by $m + \delta m$. Further, it will be supposed that the mass operator is very nearly diagonal in the number of strange quarks. In the approximation that the mass operator *is* diagonal we may then make the following ansatz for the spin and isotopic spin dependence of the masses of the s-wave bound state, namely,

$$M = M_0 + (a - \sigma_1 \cdot \sigma_2)(b + cY^2 - \tau_1 \cdot \tau_2)V_0 + n_\lambda \delta m \qquad (1)$$

* See instead Van Royen and Weisskopf[4], and references therein.

† Except that a harmonic oscillator gives (mass)^2 relationships in a natural way. It also gives quite different decay form-factors.

TABLE 1

MESONS, November 1966

	Symbol(J^P)	$I^G(J^P)C_n$ (underline = estab.)	Mass M (MeV)	Width Γ (MeV)	$M^2 \pm \Gamma M$[a] $(GeV)^2$	Partial Decay Modes: Mode	Fraction (%)	Q (MeV)	p or p_{max}[b] (MeV/c)	Nonets CP=−1 (0^-)	CP=+1 (0^+)	CP=+1 (1^-)	CP=+1 (1^+)	CP=+1 (2^+)
η(549)	$\eta(0^-)$	$0^+(0^-)+$	548.6 ±0.4	<0.01	0.301 <.000005	all neutral	73	} See Table S		η				
						$\pi^+\pi^-\pi^0+\pi^+\pi^-\gamma$	27							
σ, ε §														
ω(783)	$\phi'(1^-)$	$0^-(1^-)-$	783.4 ±0.7 S=1.8*	11.9 ±1.5	0.614 ±.009	$\pi^+\pi^-\pi^0$	≈ 90	369	328			ω		
						$\pi^+\pi^-$	seen (c)	504	366					
						$\pi^0\gamma$	9.7±0.8	648	380					
						η + neutral	< 1.5	234	199					
						$\pi^+\pi^-\gamma$	< 5	504	366					
						e^+e^-	0.012±.003	782	392					
						$\mu^+\mu^-$	< 0.10	572	377					
η'(958) or X^0	$\eta'(0^-)$	$0^+(0^-)+$	958.3 ±0.8	<4	0.918 <.004	ηππ	75 ± 3 S=1.8*	131	232	η'				
						$\pi^+\pi^-\gamma$(incl. $\rho^0\gamma$)	25 ± 3	679	458					
						for upper limits see footnote (f)								
H§														
φ(1019)	$\phi(1^-)$	$0^-(1^-)-$	1018.6 ±0.5	4.0 ±1.0 S=1.2*	1.039 ±.004	K^+K^-	48 ± 3	31	125			φ		
						K_LK_S	40 ± 3	23	107					
						$\pi^+\pi^-\pi^0$ (incl. ρπ)	12 ± 4	604	461					
						for upper limits see footnote (g)								
η_V(1050) → K_SK_S	$\eta(0^+)$	$0^+(0^+)+$	1050	50	1.10 ±.05	ππ	< 70	780	507		η_V			
Some data still favor large scattering length						$K\bar{K}$	> 30	54	167					
f(1250)	$\eta'(2^+)$	$0^+(2^+)+$	1254 ±12	117 ±15	1.57 ±.15	ππ	large	975	611					f
						$2\pi^+2\pi^-$	< 4	696	547					
						$K\bar{K}$	2.3±0.6	258	381					
D(1285)	η'(A)	$0^+(1^+)+$	1285 ±4	32 ±8	1.65 ±.04	$K\bar{K}\pi$ (mainly $\pi_V(1003)\pi$) only mode seen		154	304				D	
						$K^*\bar{K}+\bar{K}^*K$		-100						
						ππρ	not seen	256	356					
E(1420)	η (A)	$0^+(0^-)+$	1424 ±7	76 ±9	2.03 ±.11	$K^*\bar{K}+\bar{K}^*K$	50 ±10	38	157	?			E?	
						$\pi_V(1003)\pi$	50 ±10	284	338					
						ππρ	not seen	395	462					
K_sK_s§ ρρ f'(1500)	$\eta(2^+)$	$0^+(2^+)+$	1514 ±16	86 ±23	2.29 ±.13	ππ	< 14	1235	744					f'
						$K\bar{K}$	> 60	518	570					
						$K^*\bar{K}+\bar{K}^*K$	< 40	128	294					
						ηη	not seen	417	522					
$\pi^\pm$(140) π^0(135)	$\pi(0^-)$	$1^-(0^-)+$	139.58 134.98		0.019 0.018	See Table S				π				
$\rho^\pm$(760)	$\rho(1^-)$	$1^+(1^-)-$	778 (h)	160 (h)	0.605 ±.124	ππ	≈100	480	353			ρ		
						$\pi^\pm\pi^+\pi^-\pi^0$	< 0.2	206	243					
						$\pi^+\pi^-\pi^+\pi^-$	< 0.6	199	238					
						$\pi^0\gamma$	< 0.4	619	367					
ρ^0(760)			770 (h)	140 (h)	0.593 ±.108	$\eta\pi^\pm$	< 0.8	71	135					
						e^+e^-	.0065 +.011 −.005	759	380					
						$\mu^+\mu^-$	.0033 +.0016 −.0007	549	365					
δ(965)	?( )	1( )	963.1 ±4.2	<5	0.927 <.005	$\delta^\pm$ → 1 charged+neutral(s)	≈ 60							
						$\delta^\pm$ → ≥3 charged+neutral(s)	≈ 40							
π_V(1003)	$\pi(0^+)$	$1^-(0^+)+$	1003	70 ±15	1.006 ±.057	$K^\pm K^0$	large	11	75		π_V			
→ $K\bar{K}$						ηπ	see note in data listings	315	333					
may also be interpreted as due to large scattering length														
A1(1080)	$\pi(1^+)$	$1^-(1^+)+$	1079	130	1.16	ρπ	≈100	181	245					

(g) Empirical limits on fractions for other decay modes of φ(1019): $\pi^+\pi^- < 20\%$, $\eta\gamma < 8\%$, η + neutrals < 13%, $\pi^+\pi^-\gamma < 4\%$, $e^+e^- < 0.2\%$, $\mu^+\mu^- < 0.5\%$, $\omega\gamma < 5\%$, $\rho\gamma < 2\%$.

(h) m_ρ, Γ_ρ from p-wave fit to compiled spectrum of 2-4 GeV/c $\pi^+p \to \Delta^{++}\rho^0$, $|t| < 10\ m_\pi^2$, and comparison of $\rho^\pm - \rho^0$ in similar reactions. Results depend on background and t-cut, hence real errors unknown, but larger than statistical errors ~ 5 MeV. (See Matts Roos, Phys. Letters, to be publ.) Note contrast between Roos' p-wave fit vs. weighted average of published results (see listings). m(±) 778 vs. 757; m(0) 770 vs. 760, Γ(±) 160 vs. 132, Γ(0) 140 vs. 116. This demonstrates present uncertainty.

(i) Error on m_ρ taken to be 10 MeV.

						φπ	< 1.5	66	137
A2(1300)	π(2⁺)	$1^-(2^+)+$ (d)	1306 ±8 S=2.6*	81 ±8 S=1.4*	1.70 ±.11	ρπ $K\bar{K}$ ηπ η′π $\pi^+\pi^-\pi^0$(excl. ρπ)	93 ±3 3.8±1.3 2.9±2.4 S=1.5* < 1.5 < 17	408 314 618 208 892	417 425 527 276 616
π(1640) → 3π	π (A) ?	≥ 1^-(A)+ ?	1640 ±20	100 ±20	2.69 ±.16	3π [ρπ [fπ $K\bar{K}$	appears dominant < 40] ?] < 40	1235 746 251 644	792 636 319 652
ρ(1650) g→2π	ρ (V)	1^+(V) - S=1.4* →	1637 ±23	150 ±50	2.68 ±.24	2π 4π ρππ	observed probably observed	1358 1079 599	807 758 605
$R_1R_2R_3$ bumps suggest more structure in this peak									
$R_1R_2R_3$§									
S(1930) X^-	?(V) ? (e)	≥ 1^+( ) ?	1929 ±14	≤35	3.72 ≤.07	1 charged 3 charged >3 charged } +neutral(s)	6(+15/-6) 92(+ 8/-20) 2(+13/-2)		
T(2200) X^-	?( )	≥ 1	2195 ±15	≤13	4.82 ≤.03	1 charged 3 charged >3 charged } +neutral(s)	4(+11/-4) 94(+ 6/-19) 2(+13/-2)		
U(2380) X^-	?( )	≥ 1	2382 ±24	≤30	5.67 ≤.07	1 charged 3 charged >3 charged } +neutral(s)	30 ±10 45 ±15 25 ±10		
K^+(494) K^0(498)	K (0⁻)	1/2(0⁻)	493.78 497.7		0.244 0.248	See Table S			
K^*(890)	K (1⁻)	1/2(1⁻)	892.4 ±0.8	49.8 ±1.7 S=1.1*	0.796 ±.044	Kπ Kππ	≈100 < 0.2	259 119	288 216
κ(725)§ Kv(1080)§ K_c(1215)§ K_A(1320)	$m_0 - m_\pm$ = 3.5±1.8 S=1.2*								
	K (A) ?	1/2 (A) ?	1320 ±10	80 ± 20	1.742 ±.106	$K^*\pi$ Kρ } overlap Kω Kπ Kη	large probably seen, < 10 < 30 < 10	288 63 39 687 278	338 198 155 558 405
m and Γ values taken from Shen+. Appreciable discrepancies with other experiments. See note in data listings.									
K_V(1420)	K (2⁺)	1/2 (2⁺)	1411 ±5 S=1.8*	92 ±7 S=1.2*	1.991 ±.130	Kπ $K^*\pi$ Kρ Kω Kη	52 ± 5 36 ± 6 9 ± 5 1.0±1.7 2.1±3.0 } S=2.2*!	778 379 158 134 368	610 407 319 293 475
K_A(1800)	K (A)	1/2 (A)	1789 ±10	80 ±20	3.20 ±.14	Kπ $K^*\pi$ K_V(1420)π Kρ Remaining Kππ Kω	< 10 35 ± 12 8 ± 5 7 ± 5 40 ± 15 10 ± 3	1156 762 243 532 1021 508	819 664 315 630 801 616
$K^*_{3/2}$(1175)§ $K^*_?$(1270)§									

Multiplet labels in the right-hand columns: A_2; K; K^* (tentative); K_A (tentative); K_V.

§ The following bumps, excluded above, are listed among the data cards: σ (410), ε (700), H (975), K_SK_S (1440) and ρρ (1410), R_1, R_2, R_3(≈1700), κ (725) K_V(1080), K_c(1215), $K^*_{3/2}$(1175), $K^*_?$(1270).

* Quoted error includes scale factor $S = \sqrt{\chi^2/(N-1)}$. See footnote to Table S.

Footnotes continued in right margin.

	K	K^*	K_A	K_V
$m_8 = \sqrt{K^2 + \frac{K^2 - \pi^2}{3}}$ =	566.8 ±0.2	928.4 (i)±3.0	1391. ±13.	1444.3 ±6.9
$\sin^2\theta = \frac{\eta - m_8}{\eta - \eta'}$ =	0.033 ±0.001	0.414 (i)±0.013	0.25 ±0.10	0.29 ±0.06
θ =	10.4°	40.1°	29.7°	32.4°

(a) ΓM is the half-width of the resonance when plotted against M^2.
(b) For decay modes into ≥ 3 particles p_{max} is the maximum momentum that any of the particles in the final st[ate] can have. The momenta have been calculated using the averaged central mass values, without taking into a[ccount] the widths of the resonances.
(c) Reported values range between 1% and 10%, and depend on assumptions on ρ-ω interference.
(d) If A2 → both ρπ and $K\bar{K}$, then $J^P = 2^+$.
(e) S is ρ(V) if identified with $\pi^+\pi^-$ bump at 1910 MeV. See note on mesons.
(f) Empirical limits on fractions for other decay modes of η′(958): $\pi^+\pi^- < 7\%$, $3\pi < 7\%$, $4\pi < 1\%$, $6\pi < 1\%$, $\pi^+\pi^- e^+e^- < 0.6\%$, $\pi^0 e^+e^- < 1.3\%$, $\eta e^+e^- < 1.1\%$, $\pi^0\rho^0 < 4\%$, $\pi^0\omega < 8\%$.

[a] 'Permission of Professor Rosenfeld. Revisions of this table are available from the University of California, Lawrence Radiation Laboratory and *CERN*.' *Rev. Mod. Physics*, **39**, No. 1 (1967)—The Meson Table.

Here M_0 is twice the quark mass minus the average well-depth plus the kinetic energy of the $q\bar{q}$ pair. Also, n_λ is the number of strange quarks making up the boson, $\boldsymbol{\sigma}_i$ is the spin operator of the ith quark, $\boldsymbol{\tau}_i$ is the corresponding iso-spin operator, and Y is the total hypercharge of the boson.

There are eight bosons that are usually described as s-wave $q\bar{q}$ pairs (π, η, K, X^0, ρ, ω, K^*, ϕ) and six constants to be determined in (1) (from this point on the reader is advised to refer frequently to Table 1). It follows that two masses may be predicted from the present model. The input data is given in Table 2 together with the values of the six constants.

TABLE 2

Input data and values of constants in equation (1)

$^1S_0$		$^3S_1$
$\pi = 138$ (MeV)		$\rho = 775$ (MeV)
$K = 495$		$K^* = 892$
$\frac{1}{2}(\eta + X^0) = 754$		$\frac{1}{2}(\omega + \phi) = 901$
$a = 1{\cdot}142$	$b = -2{\cdot}25$	$c = 0{\cdot}224$
$V_0 = -49{\cdot}0$ MeV	$m = 109$ MeV	$M_0 = 798$ MeV

TABLE 3

Predictions from equation (1)

Particle	Predicted Mass	Experimental Mass	Quark Composition
η	559	549	$(\lambda_0\bar{\lambda}_0)$
X^0	950	958	$(N_0\bar{N}_0)_{I=0}$
ω	803	780	$(N_0\bar{N}_0)_{I=0}$
ϕ	1000	1019	$(\lambda_0\bar{\lambda}_0)$

We may now calculate the masses of the two $(\lambda_0\bar{\lambda}_0)$ mesons and the two $(N_0\bar{N}_0)$ iso-singlets. For the $(\lambda_0\bar{\lambda}_0)$ case we obtain 558 MeV for the spin singlet and 997 MeV for the spin triplet, respectively. Thus, the model forces us to identify the η- and ϕ-mesons as being made almost exclusively from $\lambda_0\bar{\lambda}_0$ quarks. For the ϕ this is essentially in accord with the SU(6) prediction, but for the η the result is contrary to the usual assumption. Nevertheless, I am unaware of any contradictory experimental evidence. For one consequence see the implication for $A_2 \rightarrow \eta\pi$ decay in Section 3.

The results so far are summarized in Table 3.

This is not yet the entire story, for we may expect that it is too much of an idealization to expect exact conservation of the number of strange quarks. In terms of a specific model, for example, we might suppose that there is a contribution from a contact interaction that permits the transition (see figure 1):

$$\lambda_0 + \bar{\lambda}_0 \leftrightarrow \bar{N}_0 + N_0$$

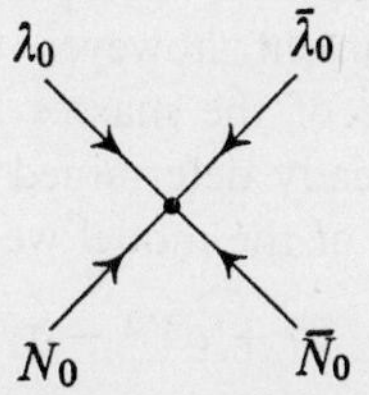

FIGURE 1 A model interaction for causing $\omega - \phi$ and $\eta - X^0$ 'mixing'

Then in addition to the diagonal mass operator of (1) there will also be a matrix mass operator that will mix each pair of iso-singlet states (η, X^0) and (ω, ϕ). Clearly, the additional mixing can only increase the iso-singlet mass differences of Table 2 so that, in particular, exact agreement with experiment can be obtained for the two mass differences. The success of the model is contained in the fact that quite good agreement with the experimental iso-singlet masses is obtained in the approximation where the number of λ quarks is conserved, and the discrepancy is in the direction that allows additional mixing.

In order to put numbers to the foregoing discussion we may add to the mass operator of (1) an off-diagonal term of the form

$$v_{\mathrm{i}} = \begin{pmatrix} 0 & v_{\mathrm{i}} \\ v_{\mathrm{i}} & 0 \end{pmatrix}$$

where the index 'i' denotes the spin state. It turns out, interestingly enough, that *both* the experimental singlet and triplet mass splittings are closely reproduced by the choice

$$v_{\mathrm{t}} \approx v_{\mathrm{s}} \approx 65 \text{ MeV}$$

The corresponding mixing angles (note that these are *not* the SU(3) mixing angles) turn out to be about 8° and 17°, respectively, for the spin singlet and spin triplet. The corresponding $SU(3)$ mixing angles* are 27° and 17°, if the η and ϕ are chosen to be the octet members in the limit of zero mixing.

2.2 *Higher states*

(i) *The* 2^+ *nonet*. We turn next to the excited states of the $q\bar{q}$ system. This portion of the discussion becomes rather speculative because additional experimental input is required and the experimental situation regarding higher boson resonances is still very murky[13].

There is reasonable confidence in the identification of a $J^P = 2^+$ nonet, and it may be supposed that it represents the excitation of the $q\bar{q}$ system to the 1P-state of the well so that the spectroscopic assignment is $^3P_2$. Independently of this assignment, however, we would expect the isospin and strangeness dependence of the masses to be given by (1) with the parameters b, c and δm already determined from the s-wave multiplets. That is to say, as a first test of the model we write

$$M(2^+) = M_0(2^+) - (b + cY^2 - \boldsymbol{\tau}_1 \cdot \boldsymbol{\tau}_2)V(2^+) + n_\lambda \delta m \tag{2}$$

with $M_0(2^+)$ and $V(2^+)$ to be determined. From (2) follow the two mass relations:

$$K^* = \tfrac{1}{4}[3A_2 + f] - \frac{c}{4}(A_2 - f) + \delta m = 1400 \pm 7 \text{ MeV} \tag{3a}$$

$$f' = \tfrac{1}{4}[3A_2 + f] + 2\delta m = 1511 \pm 7 \text{ MeV} \tag{3b}$$

in excellent agreement with the experimental K^* and f' masses which are, respectively, 1411 ± 5 MeV and 1514 ± 16 MeV. The f' was selected to be the $(\lambda_0\bar{\lambda}_0)$ member because its dominant decay is into $K\bar{K}$ (see decays). It is interesting to observe, from (3), that the mass differences in the 2^+ nonet are almost entirely due to the quark mass difference, δm. Thus, the successful feature of the model, so far, is that δm takes the same value for the 2^+ mesons as for the '36' s-wave mesons.

If we now take seriously the conjecture that the 2^+ nonet corresponds to a $^3P_2$ excitation, then we are invited to seek out the other three P-state nonets, namely, the $^3P_1(1^+)$, $^1P_1(1^+)$, and $^3P_0(0^+)$. It is at this point that experimental difficulties arise, since there are no 0^+ or 1^+ mesons that have been identified with really high confidence. There are, however, some candidates for these J^P assignments, so that one can attempt some

* Actually, the SU(3) mixing angles are not unambiguously defined by the $(\lambda_0\bar{\lambda}_0) \leftrightarrow (N_0\bar{N}_0)_{I=0}$ mixing angle. The ambiguity is in the direction of the displacement from the 'ideal' SU(3) mixing angle of about 34° which corresponds to pure $(\lambda_0\lambda_0)$ and $(n_0\bar{n}_0)$ states.

quantitative speculations by trying to fit likely candidates into the model.

Before doing this, it is necessary to generalize the mass formula, equation (1), to take into account forces that would not be contributing when the $q\bar{q}$ system is in an S-state. To this end one may generalize the definitions of M_0 and V_0 as follows:

$$M_0 \rightarrow M_0 + \Delta(n, L) + (\alpha_1 L^2 + 2\alpha_2 \mathbf{S} \cdot \mathbf{L} + \alpha_3 S_{12})V_0 \equiv M_J{}^{SL} \quad \text{(4a)}$$

$$V_0 \rightarrow V_0[1 + \beta_1 L^2 + 2\beta_2 \mathbf{S} \cdot \mathbf{L} + \beta_3 S_{12}] = V_J{}^{SL} \quad \text{(4b)}$$

Here **S** and **L** are, respectively, the total spin and orbital angular momentum operators and S_{12} is the tensor force operator. The matrix elements of S_{12} are (to fix the normalization)

$$\langle {}^3L_J | S_{12} | {}^3L_J \rangle = 1, \qquad J = L \quad \text{(5a)}$$

$$= -\frac{L+1}{2L-1}, \qquad J = L - 1 \quad \text{(5b)}$$

$$= -\frac{L}{2L+3}, \qquad J = L + 1 \quad \text{(5c)}$$

and

$$\langle {}^3L_J | S_{12} | {}^3L'_J \rangle = 3\left\{\frac{[J(J+1)]^{\frac{1}{2}}}{2J+1}\right\}, \; |L' - L| = 2, J = L \pm 1 \quad \text{(5d)}$$

Finally, $\Delta(n, L)$ denotes the difference in kinetic energy required for excitation from the $q\bar{q}$ state to the nL state. We shall suppose, provisionally, that the low-lying (at least) nL states are sufficiently separated that the mixing of states by the tensor force may be ignored. Of course, once the ordering and spacing of the states has been determined this assumption may be checked and corrected for.

(ii) *The Remaining* $1P$ *Nonets.* The next step is to determine the values of the parameters that have been introduced in (4). In order to do this it is necessary to identify a few members of the other $1P$ nonets. In accord with a rather long standing tradition, the two 1^+ isotriplets (see Table 1) are chosen to be the $A_1(^3P_1)$ and the $B(^1P_1)$. The spectroscopic assignments are determined by the known G-parity of the resonances. Neither the spin nor the intrinsic parity of either resonance has been measured, and in fact there is still some doubt as to their existence[7]. It is known[8], however, that decay angular distributions of the A_1 'enhancement' are consistent with the sequence $J^P = 1^+, 2^-, 3^+, \ldots$ At the moment A_1 and B are the only likely candidates available.

For the $J^P = 0^+$ iso-triplet there seem to be three available candidates (see Table 1 again). The δ-meson at 960 MeV is the lowest lying candidate. The so-called π_v enhancement in the $K\bar{K}$ system at 1003 is likely to be 0^+ on the basis of angular momentum barrier arguments. Finally, Montanet[7] has suggested identifying a recently observed 1250 MeV $K\bar{K}$ resonance ($I = 1$) as the needed object. In what follows I shall try two different choices: The δ-meson and $K\bar{K}$ (1250). Note that if the δ-meson is the correct choice then the π_v enhancement may be understood as resulting from the bound state just below the $K\bar{K}$ threshold.

We need one more resonance in order to fix all of the parameters in (4). Again the choice is somewhat arbitrary, namely, the 1285 MeV D_0 meson which is taken to be the iso-singlet ($N_0\bar{N}_0$) companion to the A_1. Further discussion of this choice will be found in the section on the decay widths (Section 3.5). The resulting values of the parameters are listed in Table 4 and the predictions for the masses in Table 5 where the input

TABLE 4

$\Delta(1P) + 2\alpha_1 = 394(426) \pm 10$ MeV	$1 + 2\beta_1 = 0 \pm 0{\cdot}015$
$\alpha_2 V_0 = 51{\cdot}5(27{\cdot}5) \pm 10$ MeV	$\beta_2 V_t = 1{\cdot}7 \pm 1{\cdot}7$ MeV
$\alpha_3 V_0 = 155(75) \pm 35$ MeV	$\beta_3 V_t = -48 \pm 5$ MeV

TABLE 5

$2^+(^3P_2)$	$1^+(^3P_1)$	$1^+(^1P_1)$	$0^+(^3P_0)$
$A_2 = \underline{1308}$	$A_1 = \underline{1079}$	$B = \underline{1208}$	$\delta = \underline{963(1270)}$
$f_0 = \underline{1254}$	$D_0 = \underline{1282}$	$X_0' = 1190$	$\omega'' = 612(916)\surd$
$K_v = 1401\surd$	$K^{*\prime} = 1250\surd$	$K' = 1310\surd$	$K^{*\prime\prime} = 965(1270)$
$f_0' = 1511\surd$	$\phi' = 1350$	$\eta' = 1420\surd$	$\phi'' = 1095\surd(1400)\surd$

masses are underlined. The parameter values and 0^+ masses corresponding to the choice $K\bar{K}(1250)$ are listed in parentheses. Predicted mass values corresponding to those of known resonances[6] are followed by a check mark.

Comments

(*a*) $1^+(^3P_1)$. The $K^{*\prime}$ predicted at 1250 MeV can possibly be identified with the so-called K_c (1215) for which there has been some weak evidence[6]. The dominant decay modes should probably be $K\rho$ and $K^*(890)\pi$. The predicted $\phi'(1350)$ might be expected to be a narrow $K\bar{K}\pi$ resonance (see Section 3).

(*b*) $1^+(^1P_1)$. The X_0' predicted at 1190 MeV should decay chiefly into $\rho + \pi$. The K' is tentatively identified with the $K_A(1320)$ of Table 1 provided, of course, that that particle continues to have no $K\pi$ decay mode. It is tempting to identify the η' with the E_0 meson despite the fact that one experiment[9] seems to prefer 0^- rather than 1^+ for the J^P values*.

(*c*) 0^+ with $\delta(965)$. The ω'' at 612, which is a $\pi\pi I = 0$ state, is an amusing compromise between the conjectured[6] $\sigma(410)$ and $\varepsilon(720)$. If $K^{*''}$ does in fact exist, at about 950 MeV it should be seen interfering with the neutral $K^*(890)$, and so far the evidence on this point seems to be negative. The K_SK_S enhancement[6] at 1050 MeV, which I have identified with $\phi''(1095)$ may well result simply from a large scattering length.

(*d*) 0^+ with $K^{\pm}K_0(1270)$. The mass of 916 MeV for ω'' is consistent with the recent $\pi - \pi$ scattering analysis of Walker and others[10]. At present there seems to be no evidence for $K^{*'}(1270)$ which has $K\pi$ as a possible decay mode. I have tentatively identified the ϕ'' with a K_SK_S enhancement reported[6] at 1440 MeV.

At this point alternative (d) seems to be the better choice although it leaves the δ-meson (and to some extent the π_v) shrouded in mystery. In the discussion of strong decays we shall again look at both alternatives with the hope of finding more criteria for choice.

(iii) *Higher multiplets.* As may be seen from Table 4 it is necessary to identify some one member of a higher multiplet in order to determine the value of α_1 and $\Delta(nL)$ in equation (4). In order to do this I shall adopt the suggestion[11] that the degenerate Regge trajectory containing the ρ and A_2 mesons is a straight line on a graph of (mass)^2 vs. L. Then the so-called $\rho(1650)$ and $S(1930)$ mesons should have the J^{PG} quantum numbers 3^{-+} and 4^{+-}, respectively, and should correspond to the $^3D_3$ and $^3F_4$ excitations of the $q\bar{q}$ system. Using the mass of $\rho(1650)$ as input, then, gives the result that

$$\alpha_1 V_0 = -205 \text{ MeV} \tag{6a}$$

$$M_q R^2 = 1{\cdot}28[1{\cdot}23] \text{ MeV}^{-1} \tag{6b}$$

where the M_qR^2 values correspond to the choice of $\delta(965)[K\bar{K}(1250)]$ as the $^3P_0$ isotriplet. In (6b) the quantity M_q is the quark mass and R the

* If E_0 is indeed a 0^- particle then it is probably a member of the $2S$ nonet. The positions of the other members are immediately determined since the $2S$ and $1S$ mass splittings are identical. Thus, the $2S$ nonets would be 0^-: π – 1010, η – 1420, K – 1370, X^0 – 1830 and 1^-: ρ – 1650, ω – 1655, K^* – 1760, ϕ – 1890. These might be expected to have quite small decay rates into PP, PV, or VV because of the orthogonality of the $1S$ and $2S$ wave-functions.

radius of the (effectively) infinite square well that binds the low-lying states. The consistency of the model may now be checked by calculating the mass of $S(1930)$ with the given parameters. The result, about 1820 MeV, is probably not in disagreement with the experimental mass considering the uncertainties in the input data. It is possible, however, that the discrepancy indicates the need for additional powers of L^2 in the potential (that is to say that the square well was not a good choice). Because of the uncertainty it is probably unwise to extrapolate very far (if, indeed, at all!) with (4), so only the $1D$ and $1F$ states will be discussed here.

It is now just a matter of arithmetic to calculate the masses of all the $1D$ and $1F$ mesons, and the results are shown in Table 6. Only for the

TABLE 6

Corresponding Member of 0^- nonet	Singlet States			
	π	X^0	K	η
$J^P = 2^-(^1D_2)$	2720	1100	2320	2540 (MeV)
$J^P = 3^+(^1F_3)$	very high	850	very high	2540 (MeV)

Corresponding Member of 1^- nonet	Triplet States			
	ρ	ω	K^*	ϕ
$J^P = 3^-(^3D_3)$	1650 (input)	1520	1720	1840
$2^-(^3D_2)$	1315	1460	1470	1570
$1^-(^3D_1)$	1090(1360)	880(1170)	1130(1400)	1250(1530)
$4^+(^3F_4)$	1950	1715	1990	2110
$3^+(^3F_3)$	1590	1650	1650	1820
$2^+(^3F_2)$	1170(1540)	920(1290)	1190(1570)	1320(1690)

cases $J = L - 1$ does it make any difference which of the two sets of input data is used, and as before the set corresponding to $K\bar{K}(1270)$ as the 0^+ isotriplet is shown in parentheses. In addition, all the $1F$ energies have been increased 140 MeV over the values calculated in order for the 'ρ'

with $J^P = 4^+$ to have mass 1950 MeV. In a sense, then, this is also an input number.

Comments

(*a*) The singlet masses are dominated by the V_J^{SL} term which, in this model, grows like $L(L + 1)$ with a large coefficient. The results are so dependent upon the assumed L^2 dependence that they should probably be ignored, especially in view of the low mass, high spin X^0 predicted. If the $\pi(1640)$ of Table 1 turns out to have $J^P = 2^-$ then it may be identified with the $^1D_2$ state to provide additional input. At the moment, however, I am correlating $\pi(1640)$ with $\rho(1590)$, $J^P = 3^+$, in the table.

(*b*) The K^*'s 890, 1420 and 1800 of Table 1 fall on a straight line on an L vs. (mass)^2 plot just as do $\rho(760)$, A_2, $\rho(1650)$, S, T, U. This suggests that the K^* predicted at 1720 correlates with $K^*(1800)$, $J^P = 3^-$ and that the next K^* on the (degenerate) trajectory lies near 2 GeV. This prediction need have little to do with the validity of the quark model.

(*c*) The $J = L - 1$ predictions seem to be uncomfortably low unless the numbers in parentheses, based upon $K\bar{K}(1250)$ as input, are used.

(*d*) There are some interesting degeneracies predicted. For example, the model gives a second f^0 meson ($^3F_2$) degenerate with the usual one which was assumed to be $^3P_2$. The splitting from the tensor interaction is readily calculated and turns out to be small (about 20 MeV) compared with the quoted width. Depending upon the unperturbed mass difference of the two states, the presence of a doublet might be difficult to ascertain experimentally. Also predicted is a $1^- K\bar{K}$ state, $\phi(1530)$, lying very near the f' and $1^- K^*$ near the $2^+ K^*(1405)$. It should be recalled that these predictions are sensitively dependent upon the correct identification of one member of the $^3P_0$ nonet.

3 The Strong Decays

This section is devoted to the calculation of meson decays in a manner that is analogous to the calculation of electromagnetic transition rates in nuclear physics. Since the end result of a (strong) mesonic decay is mesons it is necessary that they play a dual role. From one viewpoint they are bound states of quark-antiquark systems as has already been discussed. On the other hand at least some of the mesons are 'elementary quanta' that are emitted as a quark makes a transition in its potential well. Accordingly, the elementary interaction for a free nucleon type of quark to emit a pion of type i will be taken to be

$$(2\pi)^{-3/2}(2\omega)^{-1/2} g\bar{u}(p')\gamma_5 \gamma \cdot k \tau^i u(p)\, e^{ik\cdot x} \qquad (7)$$

where $\bar{u}$, u are free quark Dirac spinors, k_μ is the meson 4-momentum, τ^i is the isospin operator and ω is the energy of the emitted pion. An axial vector interaction has been chosen (in preference to pseudoscalar) largely because of its current popularity.

For bound quarks it is hopefully permissible to neglect terms in expression (7) that depend upon the velocity of the quarks within the well. Expression (7) then reduces to just

$$(2\pi)^{-3/2}(2\omega)^{-1/2} g \boldsymbol{\sigma} \cdot \mathbf{k} \tau^i \exp(\mathrm{i}\, k \cdot x) \tag{8}$$

Defining a dimensionless coupling constant by

$$g = G/M_\rho$$

where M_ρ is the mass of the ρ-meson, the decay matrix element for the process

$$a \to b + \pi^i$$

is

$$M(a \to b + \pi^i) = (2\pi)^{-3/2}(2\omega)^{-1/2}(G/M_\rho) \int \mathrm{d}^3 r \Psi_b^* [\boldsymbol{\sigma}_1 \cdot \mathbf{k} \tau_1^i \, \mathrm{e}^{\mathrm{i}\mathbf{k}\cdot\mathbf{r}} - \\ - \boldsymbol{\sigma}_2 \cdot \mathbf{k} \tau_2^i \, \mathrm{e}^{-\mathrm{i}\mathbf{k}\cdot\mathbf{r}}] \Psi_a \tag{9}$$

where Ψ_a, Ψ_b are the wave-functions of the initial and final $q\bar{q}$ bound states. In keeping with previous comments the radial wave-functions are taken to be those appropriate to an infinite square well. The subscripts '1' and '2' refer, respectively, to the quark and antiquark, and the relative (minus) sign guarantees conservation of G-parity in the decay process.

In order to describe K-meson emission the isospin operators must be augmented by four new operators denoted by Λ_+, Λ_-, Λ_0, $\overline{\Lambda}_0$. These are defined sufficiently for our purposes by the relations*

$$\Lambda_+ p = -\lambda \qquad \Lambda_0 n = -\lambda \qquad \overline{\Lambda}_0 \lambda = -n \qquad \Lambda_- \lambda = p \tag{10a}$$

$$\Lambda_- \bar{p} = +\bar{\lambda} \qquad \overline{\Lambda}_0 \bar{n} = -\bar{\lambda} \qquad \Lambda_+ \bar{\lambda} = \bar{p} \qquad \overline{\Lambda}_0 \bar{\lambda} = \bar{n} \tag{10b}$$

$$\Lambda_+ n = \Lambda_0 p = \text{etc.} = 0 \tag{10c}$$

where etc. stands for all the combinations not written down. Clearly Λ_-, for example, is the operator that describes K^- emission from a quark and may be written as $\frac{1}{2}(\lambda_4 + \mathrm{i}\, \lambda_5)$ in the language of Gell-Mann. The meson wave-functions, expressed in terms of their quark constituents, are recorded in Table 7 in order to avoid any confusion regarding phase conventions.

* The Λ's are obviously SU(3) shift operators.

TABLE 7

Meson wave-functions

$\pi^+ = -p\bar{n}$	$K^+ = p\bar{\lambda}$	$\bar{K}^0 = \bar{\lambda}n$
$\pi^0 = \frac{1}{\sqrt{2}}[p\bar{p} - n\bar{n}]$	$K^0 = n\bar{\lambda}$	$K^- = -\lambda\bar{p}$
$\pi^- = n\bar{p}$	$X^0 = +\frac{1}{\sqrt{2}}(p\bar{p} + n\bar{n})$	$\eta_0 = +\lambda\bar{\lambda}$

With these conventions it is convenient to define the operator for K^+ emission from a quark to be

$$-gx\Lambda_+\boldsymbol{\sigma}\cdot k_K$$

with g the same as in (8) and x a positive constant to be determined.

Having made this preparation let us turn to the calculation of the 2-body and quasi-2-body decays.

3.1 *Decay of the ρ meson and K**(890)*

For the decay of $\rho^+ \to \pi^+\pi^0$ suppose first that the π^0 is the emitted quantum so that the 'nuclear' transition is $\rho^+ \to \pi^+$. The matrix element is

$$M(\rho^+ \to \pi^+ + \pi^0) = g(2\omega_0)^{-1/2}(2\pi)^{-3/2}\int d^3r j_0(kr/2)P_0(\hat{k}\cdot\hat{r})\Psi_f{}^+ \times \times [\boldsymbol{\sigma}_1\tau_1{}^0 - \boldsymbol{\sigma}_2\tau_2{}^0]\cdot\mathbf{k}_0\Psi_i \quad (11)$$

where the subscripts on ω_0 and $\mathbf{k}_0$ denote the charge of the emitted pion. It is supposed now that the radius of the binding potential is small enough so that it is a reasonable approximation to replace $j_0(kr/2)$ by unity. (This will be confirmed by the calculation of f_0 decay.) Since the ρ and π are both described by the same spatial wave-functions the integral gives unity, leaving (it is assumed throughout that the decaying particle is at rest):

$$\begin{aligned} M(\rho^+ \to \pi^+ + \pi^0) &= \frac{g}{(2\pi)^{3/2}}\frac{k_0}{\sqrt{(2\omega_0)}}\langle\pi^+|\sigma_{12}\tau_{10} - \sigma_{22}\tau_{20}|\rho^+\rangle \\ &= \frac{g}{(2\pi)^{3/2}}\frac{k_0}{\sqrt{(2\omega_0)}}\langle p\bar{n}|\tau_{10} + \tau_{20}|p\bar{n}\rangle \\ &= \frac{2g}{(2\pi)^{3/2}}\frac{k_0}{\sqrt{(2\omega_0)}} \end{aligned} \quad (12)$$

where the z-axis is defined along the direction of $\mathbf{k}_0$. In the same way

$$M(\rho^+ \to \pi^0 + \pi^+) = -\frac{2g}{(2\pi)^{3/2}}\frac{k_+}{\sqrt{(2\omega_+)}} \quad (13)$$

I now adopt the *convention** that the total decay matrix element is the average of the difference (since the pions are in an anti-symmetrical spatial state) of (12) and (13). As long as the discussion is confined to boson decays the averaging procedure (as opposed to taking the unaveraged difference of the two matrix elements) simply scales all the coupling constants in a consistent manner. However, it appears from other work[4] that it is necessary to average in order to describe ρ decay and $N^*(1385)$ decay with the same coupling of the pion to the quarks. An intuitive justification follows from the observation that in a field theoretical formalism the ρ decay matrix element may be written as proportional to

$$M \sim \tfrac{1}{2}[\langle \pi^+ | j_0(0) | \rho^+ \rangle - \langle \pi^0 | j_+(0) | \rho^+ \rangle] \tag{14}$$

where the j's are the pion currents.

The decay width is given by

$$\begin{aligned}\Gamma(\rho \rightarrow \pi\pi) &= (\tfrac{1}{3}) 4\pi^2 k\omega |M|^2 \\ &= \left(\frac{4}{3}\right)\left(\frac{G^2}{4\pi}\right)\left(\frac{k}{M_\rho}\right)^2 k = 96{\cdot}5 \left(\frac{G^2}{4\pi}\right) \text{MeV}\end{aligned} \tag{15}$$

remembering the definition of G. Using the current value of 160 MeV (Table 1) for the width, one finds that

$$\frac{G^2}{4\pi} = 1{\cdot}65 \tag{16}$$

This is the prototype calculation for all of the subsequent ones, hence the very detailed discussion that accompanied it. The calculation of $K^* \rightarrow K\pi$ is quite similar except that a new coupling constant, gx, is used to describe K-emission and the phase space is proportional to $2k\omega_\pi\omega_K/M_{K^*}$ instead of the $\frac{1}{2}k\omega$ in (15). The result is

$$\Gamma(K^* \rightarrow K\pi) = 2\left(\frac{G^2}{4\pi}\right)\left(\frac{\omega_K}{M_{K^*}}\right)\left(\frac{k}{M_\rho}\right)^2 k \times \tfrac{1}{4}\left[1 + \frac{x}{\sqrt{2}}\left(\frac{\omega_\pi}{\omega_x}\right)^{\frac{1}{2}}\right]^2 \tag{17}$$

Setting this expression equal to the value of 50 MeV (Table 1) and solving for x one finds

$$x = 1{\cdot}04 \tag{18}$$

The result that the π and K coupling constants are nearly equal was, of course, to be expected from the success of SU(3) symmetry in relating these two decays.

* In agreement with that used by Van Royen and Weisskopf[4]. Note, however, that of the strong decays they treat only $\rho \rightarrow 2\pi$.

3.2 *ϕ decay and the $\omega - \phi$ mixing angle*

In view of the discussion on mass spectra the ϕ wave-function must be written

$$\phi = (\lambda\bar{\lambda}) \cos \theta + \frac{1}{\sqrt{2}} (p\bar{p} + n\bar{n}) \sin \theta \equiv \phi_0 \cos \theta + \omega_0 \sin \theta \quad (19)$$

In the discussion of mass formulae θ was found to be about 17. The partial width for $\phi \rightarrow K\bar{K}$ is given by

$$\Gamma(\phi \rightarrow K\bar{K}) = \frac{2}{3}\left(\frac{G^2}{4\pi}\right) x^2 \cos^2\theta \left(\frac{k}{M_\rho}\right)^2 k \left[1 - \frac{\tan \theta}{\sqrt{2}}\right]^2$$
$$\approx \begin{pmatrix} 2{\cdot}0 \\ 5{\cdot}3 \end{pmatrix} \text{MeV} \quad (20)$$

where the two values correspond to a $\begin{pmatrix}\text{positive}\\ \text{negative}\end{pmatrix}$ value of θ. The current Rosenfeld value is $3{\cdot}5 \pm 0{\cdot}9$ MeV which is not in gross disagreement. However, the $\phi \rightarrow \rho\pi$ rate provides a much more sensitive test of the mixing angle, and this may be calculated when the coupling of the ρ to the quarks is determined. I shall return to this point and show that θ is probably much smaller than the value determined from the mass formula.

3.3 *Decays in the 2^+ nonet*

The matrix element for emission of a pion accompanied by a transition to an S-wave $q\bar{q}$ state from a state with quantum numbers $SLJM$ is

$$M = \frac{8}{(2\pi)^{3/2}} \frac{k}{\sqrt{(2\omega)}} \mathrm{i}^l(2l + 1)\langle S', m_{S'}; L = 0| P_l \left[\sigma_{1z}\tau_1{}^i - \right.$$
$$\left. -(-1)^l \sigma_{2z}\tau_2{}^i \right] |SL; JM\rangle R_{fi} \quad (21)$$

where the radial matrix element is

$$R_{fi} = \left(\frac{2}{\pi^2}\right) |j_{l-1}(\alpha_l\pi)|^{-1} \int_0^\pi x^2 \mathrm{d}\, x j_0(x) j_l(\alpha_l x) j_l \left[\left(\frac{kR}{2\pi}\right) x\right] \quad (22)$$

and $\alpha_l\pi$ is the appropriate zero of $j_l(z)$. In the case where $L = 1$

$$R_{fi}(kR) \equiv R_{1P;1S}(kR) \rightarrow {\cdot}088(kR) \quad (23)$$

in the limit of small kR. It is important to note that the 'form-factor' is a 'universal' function that depends only upon the initial and final *orbital* states. As a particular consequence, it is to be expected that if the well-radius R is small enough then all of the p-wave $q\bar{q}$ bosons (0^+, 1^+, 2^+) will have decay rates proportional to k^5. This means that, in terms of the relative angular momenta of the emitted mesons, the s-wave and d-wave amplitudes show the same momentum dependence. It is interesting that Capps[12] finds a rather similar behaviour resulting from an $SU(6)_W$ 'bootstrap' model.

For the convenience of the reader the formulae needed to evaluate the expression in brackets in (21) are tabulated in the Appendix. The result for the case of f_0 decay into two pions is

$$\Gamma(f_0 \to \pi\pi) = \left(\frac{12}{5}\right)\left(\frac{G^2}{4\pi}\right)\left(\frac{k}{M_\rho}\right)^2 k|R_{1P;2S}(kR)|^2 \tag{24}$$

Setting this equal to the current experimental value of 120 MeV (Table 1) and making use of a numerical evaluation of the radial matrix element determines the value of R. The result is* (the prediction of the preceding section that f_0 is a doublet is ignored here):

$$R \approx 0{\cdot}82f \tag{25}$$

This value is small enough so that the small kR approximation, (23), may be used for all of the other decays of the $1Pq\bar{q}$ states with an error of less than about 5%. It is also easy to go back and verify that the small kR approximation was justified in the discussion of the decays of the $1S$ states.

The other rates for a 2^+ meson to decay into two pseudoscalars may be calculated readily with the information now at hand. When there is a vector meson in the final state, however, it is necessary to include the possibility that the vector meson was emitted by the $q\bar{q}$ system making a transition to the state describing the other meson in the decay process. In order to do this one must introduce the non-relativistic coupling of a vector meson to the $q\bar{q}$ system. This is taken to be (in the case of the ρ meson)

$$(2\pi)^{-3/2}(2\omega)^{-1/2}\tau^i[g_E\mathbf{k}\cdot\boldsymbol{\epsilon} - \mathrm{i}\, g_M\boldsymbol{\sigma}\cdot(\mathbf{k}\times\boldsymbol{\epsilon})]e^{i\mathbf{k}\cdot\mathbf{x}} \tag{26}$$

where $\boldsymbol{\epsilon}$ is the polarization of the emitted vector particle. The overall sign is the same for both quark and antiquark emission.

* This was obtained by a numerical evaluation of the matrix element. The radius would be ~10% smaller if the approximation $R_{1P;2S}(kR) \sim kR$ had been used. I am indebted to R. Arnold for assistance with the numerical evaluation.

For the process $A_2 \to \rho\pi$, then, the rate is given by

$$\Gamma(A_2 \to \rho\pi) = \left(\frac{48}{5}\right)\left(\frac{G^2}{4\pi}\right)\left(\frac{k}{M_\rho}\right)^2\left(\frac{\omega_\rho}{MA_2}\right) k \cdot \frac{1}{4} \left[1 + \left(\frac{\omega_\pi}{\omega_\sigma}\right)^{\frac{1}{2}}\left(\frac{g_M}{g}\right)\right]^2 |R_{1P;1S}|^2 \quad (27)$$

Equating this expression to the experimental value (Table 1) of about 75 MeV gives the result that

$$g_M \approx 2g \quad (28)$$

We may now calculate all of the 2^+ decay rates with the results shown in Table 8. In doing this calculation it is supposed that $K^*(890)$ and ρ are symmetrically coupled to the quarks in the same sense that K and π were found to be. On the other hand, since the observed ω and η rates are very small, experiment does not provide a useful guide to these couplings. Consequently, these rates are estimated by considering pion emission only

TABLE 8
2^+ decay rates

Decay		Theoretical Width (MeV)	Measured Width (MeV)
$A_2 \to$	$\rho\pi$	Input (determines g_M)	75
	$K\bar{K}$	3·7	3·1 ± 1·0
	$\eta\pi$	0·9(0 = 8°) 2·4(0 = 13°)[a]	2·4 ± 1·9
	$X_0\pi$	2·3[b]	<1·2
$f_0 \to$	$\pi\pi$	Input (determines R)	120
	$K\bar{K}$	2·0	2·8 ± 0·7
$K^*(1430) \to$	$K\pi$	45	48 ± 10
	$K^*\pi$	17[a]	33 ± 7
	$K\rho$	5·9	8 ± 4
	$K\omega$	0·7[a]	1·0 ± 1·7
	$K\eta$	6·4[a]	2·1 ± 3·0
$f_0' \to$	$\pi\pi$	0	<11
	$K\bar{K}$	31	>52 ± 13
	$KK^*(890) + K^*(890)\bar{K}$	3·5[b]	<35

a π or K 'emission' only
b Assumes 'universal' 1^- coupling

without averaging amplitudes. It is interesting that most of the experimental rates are reasonably well approximated by supposing that only pseudoscalar emission takes place. The $A_2 \rightarrow \rho\pi$ rate seems to come out a bit small, but there are indications that the presently quoted 'experimental' width may be too large[13].

In general the results of the calculation are very close to those of Glashow and Socolow[14] who, however, required three pieces of input data rather than the two needed here. Probably the most important difference is the small branching ratio (~10%) predicted in Table 8 for $f_0' \rightarrow K^*\bar{K} + \bar{K}^*K$. It is heartening to me that present (and preliminary) indications[15] are that the branching ratio is much smaller than the 40–60% predicted by other authors[14,16]. On the other hand the prediction of a very small total width for the f_0' seems to provide a most glaring discrepancy with *observation* (but not with other calculations!) unless an alternative (and dominant) decay mode can be invented.

It should be remarked, to conclude this section, that the prediction of zero rate for $f_0' \rightarrow \pi\pi$ follows from the assumption that there is no $(f_0 - f_0')$ mixing. The assumption is, of course, a gratuitous one and a small amount of mixing could easily be tolerated without otherwise affecting the model. It is also interesting that the small rate for $A_2 \rightarrow \eta\pi$ can only be obtained if the η is nearly pure $(\lambda_0\bar{\lambda}_0)$ in its quark composition, thereby verifying the prediction of the mass formula.

3.4 *Decay of the* 1^+ *and* 0^+ *nonets*

Since the membership in the 0^+ and the two 1^+ nonets is highly speculative I consider here a few of the possibilities.

(i) *The* $A_1(1080)$. The decay rate $A_1 \rightarrow \rho\pi$ is now completely determined and predicted to be about 8 MeV. The quoted experimental number is 130 ± 40 MeV although there are experimental difficulties in resolving the A_1 peak[7,17] and it may be much narrower. Assuming that A_1 exists and has a width somewhat closer to 100 MeV, one can look for other decay modes. The obvious candidate could be a $J^P = 0^+ I^G = 0^+$ meson if the mass is small enough since the decay rate would go as k^3 rather than k^5. In fact, a quick calculation gives the result that the partial width for $A_1 \rightarrow \varepsilon_0(740) + \pi$, if $\varepsilon_0(740)$ is a $1P$ $q\bar{q}$ state with the requisite quantum numbers, would be about 50 MeV. If $\varepsilon(740)$ does not exist then the present model argues against[18] the existence of a broad A_1.

(ii) $D_0(1285)$. The D_0 has been seen[6] as a $K\bar{K}\pi$ 'enhancement' with the $K\bar{K}$ invariant mass showing a peak near threshold (1003 MeV). As a first guess one might suppose that the process is $D_0 \to \delta + \pi$ where δ is the 0^+ isotriplet. However, the width for this process (experimentally ~30 MeV) comes out to be about 450 MeV! As a consequence, the model suggests that either D_0 is misidentified or the 0^+ isotriplet lies well above 1 GeV. If the alternate identification (see the discussion of mass spectra) of the 0^+ member is made (1270 MeV) then quasi 2-body decays seem to be forbidden. In the absence of a calculation this would account for the narrow width.

(iii) $K^{*\prime}(1250)$. It will be recalled that the existence of such an object was predicted by the mass formulae. The main 2-body decay mode would be either $K^*(890) + \pi$ or $K^{*\prime\prime}(0^+) + \pi$ if $K^{*\prime\prime}(0^+)$ is as light as 965 MeV (Table 3). The two partial widths turn out to be, respectively, 4 MeV and 50 MeV. Again, with the preferred (higher) masses for the 0^+ nonet, a very narrow K^* seems to be indicated. The results are qualitatively similar if the $K^{*\prime}(1^+)$ mass is increased to 1320 MeV where there is a known 1^+K^* resonance (Table 1).

(iv) $\phi'(1350)$. The $\phi'(1350)$ is the $\lambda_0\bar{\lambda}_0$ member of the A_1 nonet. This is predicted to be very narrow ($\lesssim$1 MeV) since there are no (quasi) 2-body decays available unless $K^{*\prime\prime}(0^+)$ has a mass less than 850 MeV.

(v) *The B multiplet* ($^1P_1$). In the decays of this multiplet into 1^- and 0^- a new coupling constant makes its appearance for the first time, the g_E in expression (26). In particular the $\omega\pi$ decay of the B-meson depends upon the g_E that describes the coupling of the ω-meson to a quark. It will be recalled that there is no reason, in the present model, why this must be the same as the ρ-meson coupling. Nevertheless, in view of the paucity of experimental information it is convenient to assume equality of all the vector meson coupling constants in order to arrive at quantitative estimates of the widths.

The partial rate for B decay into $\omega + \pi$, then, may be written

$$\Gamma(B \to \omega\pi) = 8\left(\frac{G^2}{4\pi}\right)\left(\frac{k}{M_\rho}\right)^2\left(\frac{\omega_\omega}{M_B}\right)k \cdot \frac{1}{4} \left[1 + \left(\frac{\omega_\pi}{\omega_\omega}\right)^{\frac{1}{2}}\frac{g_E}{g}\right]^2 |R_{1P;1S}|^2 \approx 120 \text{ MeV} \quad (29)$$

from whence it turns out that

$$g_E/g \approx 5{\cdot}1 \quad (30)$$

Assuming that $K^*(1310)$ *belongs to the B multiplet* one obtains

$$\Gamma[K'(1310) \rightarrow K^*(890) + \pi] \approx 55 \text{ MeV} \tag{31a}$$

$$\Gamma[K'(1310) \rightarrow K\rho] \approx 6{\cdot}5 \text{ MeV} \tag{31b}$$

$$\Gamma[K'(1310) \rightarrow K\omega] \approx 2{\cdot}1 \text{ MeV} \tag{31c}$$

which does not seem to be in severe contradiction with presently available (and poorly determined) experimental numbers given in Table 1.

The reason that I emphasized the assumption that $K'(1310)$ belongs to a particular multiplet is because one must consider possible mixing of the strange members of the two $J^P = 1^+$ nonets. This is an interesting point that will now be discussed.

Consider a decay $A \rightarrow V + P$ where A, V, and P are, respectively, axial vector, vector, and pseudoscalar mesons. From the viewpoint where the pseudoscalar is the emitted quantum (with momentum $\mathbf{k}$) the transition matrix element is either $\langle {}^3S_1|\boldsymbol{\sigma}\cdot\mathbf{k}|{}^3P_1\rangle$ or $\langle {}^3S_1|\boldsymbol{\sigma}\cdot\mathbf{k}|{}^1P_1\rangle$ depending upon which nonet is considered. It is easy to verify that the z-component of the total angular momentum can only be ± 1 in the first case and zero in the second, where the z-axis was chosen in the direction $\mathbf{k}$. Two conclusions follow immediately:

(*a*) The $V + P$ intermediate state cannot mix the strange members of the two nonets[19] (the non-strange members cannot mix because they have opposite g-parity).

(*b*) If $V \rightarrow P + P$, then the decay angular distribution *of the V* with respect to the direction $\mathbf{k}$ is $\sin^2\theta$ or $\cos^2\theta$, respectively, for the ${}^3P_1$ and ${}^1P_1$ axial vector mesons. *This appears to be the only experimental distinction between the strange members of the two nonets.*

The $V + P$ decays, however, do not tell the whole story. Consider now $A \rightarrow S + P$ where S is a ${}^3P_0$ scalar meson. In this case both of the axial vector mesons must decay from states with zero component of angular momentum along the $\mathbf{k}$-axis so the $S + P$ intermediate states will cause mixing. In particular, if the two axial vector K^* mesons are nearly degenerate in the absence of coupling through $S + P$ decay, then the physical states will contain approximately equal admixtures from the two nonets. In the limit of exact degeneracy, then, the decay angular distribution referred to in (*b*) will be flat. There are preliminary indications[20] that this is in fact the situation in the decay sequence

$$\begin{array}{l} K^*(1320) \rightarrow K^*(890) + \pi \\ \qquad\qquad\quad \hookrightarrow K + \pi \end{array} \tag{32}$$

Turning next to the $(\lambda_0\bar{\lambda}_0)$ member of the B nonet (predicted mass of 1420 MeV, see Table 5) we encounter a now familiar situation. The partial width for $V + P$ decay $(\bar{K}K^* + \bar{K}^*K)$ is only about 3 MeV. If we identify the predicted meson with $E_0(1420)$ then the observed width for this decay is about 35 MeV. The strong suppression of the $V + P$ decay rate is brought about by the k^5 phase space. The situation is not improved by taking seriously the suggestion[9] that E_0 has quantum numbers $J^P = 0^-$ rather than 1^+. In this case one would assign the E_0 to the nonet corresponding to the $2S$ excitation of the $q\bar{q}$ pair. Since the initial and final radial states would then be orthogonal the decay matrix element would be proportional to k^7, thereby providing even more suppression. For the time being, then, E_0 decay will be left as a mystery not unravelled by the model under discussion.

The remaining member of this nonet is $X_0'(1190)$ which can decay into $\rho + \pi$. Since the kinematics are nearly identical to that for $B \rightarrow \omega + \pi$ the ratio of the two rates is just given by isospin factors. The prediction, therefore, is

$$\Gamma(X_0' \rightarrow \rho\pi)/\Gamma(B \rightarrow \omega\pi) \approx 3$$

(vi) *The* $^3P_0$ *Mesons*. The partial widths for 0^+ decay into two pseudoscalars are summarized in Table 9 for the two alternative mass spectra of Table 5.

TABLE 9

Rates for $0^+ \rightarrow P + P$

Decay		Predicted Width (MeV)
$\rho''(965)$	$\rightarrow K\bar{K}$	0
$\rho''(1270)$	$\rightarrow K\bar{K}$	7
$\omega''(612)$	$\rightarrow \pi\pi$	5
$\omega''(916)$	$\rightarrow \pi\pi$	60
$K^{*''}(965)$	$\rightarrow K\pi$	3
$K^{*''}(1270)$	$\rightarrow K\pi$	30
$\phi''(1095)$	$\rightarrow K\bar{K}$	~0
$\phi''(1400)$	$\rightarrow K\bar{K}$	60

The present experimental situation is such that one can say very little concerning these predictions. It has already been suggested that the low-mass alternative would result in undesirably large widths for the 1^+ mesons, so we may focus attention upon the higher mass choices. Here the most uncomfortable prediction is that of a relatively narrow width for the $I = 0$, $J^P = 0^+$ π–π resonance. This may be compared with the phase shift analysis of Walker and others[10], who would seem to prefer a width of the order of 500 MeV! If this 'experimental' result is correct then major modifications of the present model are indicated. Also, I am unable to find indications of any $K\pi$ resonances between $K^*(890)$ and $K^*(1420)$.

As for the other predictions concerning the higher mass alternative, they are perhaps not too unreasonable. The predicted small width of $\rho''(1270)$ would explain why the question of its existence[7] is still a matter for debate. There is some evidence[6] for a $K_S K_S$ enhancement near 1400 MeV, although the concurrent presence of $\rho\rho$ is an embarrassment if one insists that ϕ'' is pure $(\lambda_0\bar{\lambda}_0)$. Nevertheless, the fact that the quoted total width was 90 MeV leaves room for hope that this is the $\phi''(1400)$ of the table.

3.5 *The decay* $\phi \rightarrow \rho + \pi$

With the ρ meson coupling to the quarks (g_M) now known from A_2 decay it is a simple matter to calculate the rate for $\phi(1020) \rightarrow \rho + \pi$. The result is

$$\Gamma(\phi \rightarrow \rho\pi) = 175 \sin^2\theta_v \text{ MeV} \tag{33}$$

Since the quoted experimental value is $0{\cdot}48 \pm 0{\cdot}13$ MeV it is necessary to take $\theta_v \approx 3°$, a considerably smaller value than the 17° suggested by the mass formula. Using the smaller value, however, one finds

$$\Gamma(\phi \rightarrow K\bar{K}) = 3{\cdot}96 \text{ or } 3{\cdot}43 \tag{34}$$

depending upon the sign of θ_v. Either value is in good agreement with the value quoted in Table 1.

3.6 $\omega \rightarrow \rho + \pi$ *decay*

Although the calculation of ω decay will not be described here in detail the result is well known in the model where $\omega \rightarrow \rho \rightarrow \pi$ with pion emission accompanying each arrow. The predicted width comes out to be about 10 MeV compared with the current experimental value of $10{\cdot}8 \pm 1{\cdot}3$ MeV.

3.7 *Decay of* $\rho(1650)$

Assuming $\rho(1650)$ is the $J^P = 3^-$, $I^G = 1^+$ meson corresponding to the $^3D_3$ excitation of the $q\bar{q}$ system, the rate for $\rho(1650) \rightarrow \pi\pi$ is readily calculated. The result is $\Gamma[\rho(1650) \rightarrow \pi\pi] \approx 10$ MeV. The actual value is not yet known. This result is included in order to exhibit the predictive power of the model.

4 Mass Differences of $\pi^\pm - \pi^0$, $\rho^\pm\rho^0$

The electromagnetic mass difference of the charged and neutral pions is[21]

$$\Delta M_\pi = \frac{1}{2}\left\langle \frac{e^2}{r} \right\rangle - \frac{1}{2}(\mu_p - \mu_n)^2 \cdot \frac{16\pi}{3}\,|\psi(0)|^2 \langle \boldsymbol{\sigma}_1 \cdot \boldsymbol{\sigma}_2 \rangle_s \tag{34}$$

where $\psi(0)$ is the value of the $1S$ wave-functions at zero separation, and $\langle \boldsymbol{\sigma}_1 \cdot \boldsymbol{\sigma}_2 \rangle_s$ denotes the spin expectation value (equal to -3 for pions). In accordance with deductions made from the baryon magnetic moments I shall take μ_p, μ_n, the moments of the p_0, n_0 quarks, equal to $+\frac{2}{3}(2{\cdot}79)$ and $-\frac{1}{3}(2{\cdot}79)$ nuclear magnetons. The expectation value of r^{-1} for ground state square-well wave-functions is readily found to be

$$\langle 1/r \rangle = 1{\cdot}23 \left(\frac{2}{\pi R}\right) \tag{35}$$

so that

$$m(\pi^\pm) - m(\pi^0) = \frac{1}{\pi}\frac{e^2}{R}\left\{1{\cdot}23 + \left[\frac{2\pi}{eR}(\mu_p - \mu_n)\right]^2\right\} = 3{\cdot}6 \text{ MeV} \tag{36}$$

if $R = 0{\cdot}82f$ as deduced from f_0 decay. The experimental value of 4·6 MeV would have been attained for a radius 10% smaller, and this is within the error of the determination of R. Now, substituting $+1$ instead of -3 for $\langle \boldsymbol{\sigma}_1 \cdot \boldsymbol{\sigma}_2 \rangle_s$ and using the smaller value of R, we find the electromagnetic mass splitting for the ρ mesons. The result is

$$m(\rho^\pm) - m(\rho^0) = -0{\cdot}55 \text{ MeV} \tag{37}$$

It seems unlikely that this prediction will be confronted by experiment in the near future.

The interested reader is encouraged to extend this calculation to the K and K^* multiplets. He should remember, however, that in the present

21

'non-symmetry' model the λ quark and neutron quark are not required to have the same magnetic moment.

5 Conclusions

I have presented in this paper a model of the bosons in which they are viewed as bound states of a quark and an antiquark moving in a very deep potential. Some degree of symmetry has been built into the model in two different ways. First, invariance under the isotopic spin transformations has been assumed to hold for two members of the quark triplet. Second, the binding potential has been assumed to be 'mostly' independent of the spin or isotopic spin state of the bound pair, provided that the binding energy is comparable to the rest mass of the constituent particles. Beyond these two assumptions there was left room for an arbitrary degree of difference between the interactions and properties of the quark singlet and the quark doublet. In this fashion it became possible for an apparent symmetry such as U(3) or U(6) to manifest itself in connection with the low-lying states but to fail to provide a 'global' description of boson systematics.

To the extent that such behaviour is exhibited, the model becomes of interest in its own right independently of its success in describing the real world. Consider, for example, the Gell-Mann–Okubo formula for the pseudoscalar *octet*. It reads

$$M_\pi = [4m_K{}^2 - 3m_\eta{}^2]^{1/2} = 164 \text{ MeV} \tag{38}$$

which is close enough to the measured value of 139 MeV so that one is willing to consider that SU(3) symmetry must be important to boson dynamics. On the other hand, consider the same formula applied to the corresponding members of the 3^- nonet listed in Table 6. In this case the result would be

$$m_\rho = 2215 \text{ MeV}$$

instead of the 'experimental' value of 1650 MeV, an error of about 35 % in mass (or ~80 % in mass²). Yet both nonets are contained within the model that has been constructed here. It should be clear from this example that one should use caution in trying to identify higher multiplets by fitting them to the Gell-Mann–Okubo mass formula.

It is clear, then, that a model of a badly broken 'symmetry' has been constructed with quarks as the building blocks. In comparing the model with real life one finds some comforting successes. The mass difference between the λ quark and nucleon quarks that describes the pseudo-scalar and vector mass splittings also works for the tensor (2^+) nonet. Most of the known strong decay rates are reproduced with relatively few parameters as input. And it is perhaps non-trivial that a very reasonable value for the $\pi^\pm - \pi^0$ mass difference is obtained using only the nucleon magnetic moments, the ratio of the f_0 and ρ meson decay rates, and the quark model as inputs.

Yet it should also be apparent that these successes do not, in themselves, provide a justification for the model, for almost all of them may be achieved by other means. The major importance of the non-relativistic quark approach lies in its potential for extrapolation of mass spectra and decay rates. Only with an underlying dynamics at hand could one make the mass predictions that appear in Table 6 and, having made them, go on to estimate the corresponding widths as was done, by way of example, in Section 2.7 for $\rho(1650)$. For this reason it seems of major importance to determine the extent to which the simple-minded viewpoint advocated here does provide such dynamics. The question of incorporating the model into the framework of present day theoretical understanding, uncomfortable as that question is to most of us, is to my mind secondary.

It is a bit disappointing, therefore, that no clear-cut test of the model seems to be available at the present time. To be sure, some difficulties have already presented themselves. For example, the predicted widths of the low-lying members of the $1P$ multiplets (such as the A_1) seem to be much too narrow because of the fifth power dependence upon the decay momenta. Further, although both the square-well parameter MR^2 and the square-well radius R were determined in the text I have been too embarrassed to quote the resultant mass M of the supposedly very massive quarks. Suffice it to say that the mass parameters and the decay parameters do not seem to be consistent each with the other and also with the non-relativistic quark picture. We have already seen indications of this in the determinations of the $\eta - X_0$ and $\phi - \omega$ mixing angles.

At the moment, it is hard to take these difficulties seriously, remembering the uncertain state of our experimental information. In addition there is the possibility, indeed the likelihood, that the quark game was played badly in some of its finer details. For example, if I had used an oscillator potential instead of a square-well the decay matrix elements would have acquired a form-factor with a Gaussian dependence upon the decay momentum, as assumed by Mitra and Srivastava[16]. This would have done much to increase the width of A_1 relative to, say, the A_2. In addition, of course, an oscillator potential would have given mass relationships in terms of $(\text{mass})^2$. Whether this is desirable remains to be seen.

This, then, is the model. As implemented here it is probably wrong in many of its details and conceptually it is ugly and uncomfortable to live with. Yet some of its successes are astonishing and it seems inescapable that the essential features of its mathematical structure must survive. It seems appropriate, then, to ignore aesthetic considerations for the time being while bearing in mind the advice of the contemporary philosopher[22] who said:

> 'We must undoubtedly criticize wrong ideas of every description. It certainly would not be right to refrain from criticism, look on while wrong ideas spread unchecked and allow them to monopolize the field. Mistakes must be fought and wrong ideas checked whenever they crop up. However, such criticism should not be dogmatic, and the metaphysical method should not be used, but efforts should be made to apply the dialectical method. What is needed is scientific analysis and convincing argument.'

Acknowledgments

I am indebted to David Horn for a series of remarks that stimulated my interest in the quark model, to Harry Lipkin for some illuminating comments, and to Richard Arnold and Stanley Fenster for their interest, encouragement and incisive remarks. I am also deeply grateful to Prof. S. Tzitzeica for the hospitality of the Romanian Institute of Atomic Physics where the writing of this work was begun, and to Prof. Hadi Aly of the American University, Beirut, for the hospitality of his institution and for encouraging me to record the calculations described herein.

Appendix

Although the formulae recorded here are well known to every nuclear physicist, they may prove useful to an occasional reader.

What is needed for the calculation of most of the decay amplitudes is the matrix element either of $P_l(\hat{k}\cdot\hat{r})\boldsymbol{\sigma}\cdot\mathbf{k}$ or of $P_l(\hat{k}\cdot\hat{r})\boldsymbol{\sigma}\cdot\mathbf{k}\times\boldsymbol{\epsilon}$. Consider, therefore, an arbitrary vector $\mathbf{A}$ whose *spherical* components (A_+, A_-, A_0) are given by

$$A_\pm = \mp(A_x \pm \mathrm{i}\,A_y);\; A_0 = A_z$$

Then the matrix elements of $\boldsymbol{\sigma}\cdot\mathbf{A}P_L(\hat{k}\cdot\hat{r})$ from an arbitrary initial state of a quark-antiquark coupled to spin S, orbital angular momentum L, and total angular momentum J with z-component M to a state with $L' = 0$, described by quantum numbers $S' = J'$, M' are tabulated below.

1. $S = 1$, $S' = 1$

(*a*) $J = L+1$: $\dfrac{1}{2L+1}\left\{(-1)^{M-M'}M\sqrt{\left(\dfrac{L+2}{2}\right)} - M'\sqrt{(L+1)}\delta_{M0}\right\} a_{M-M'}$

(*b*) $J = L$: $\dfrac{|M|}{2\sqrt{(2L+1)}}(-1)^{M-M'}a_{M'-M}$

(*c*) $J = L-1$: $\dfrac{1}{2L+1}\left\{(-1)^{M-M'}M\sqrt{\left(\dfrac{L-1}{2}\right)} - M'\sqrt{L}\delta_{M0}\right\} a_{M-M'}$

2. $S = 1$, $S' = 0$

(*a*) $J = L+1 \pm \dfrac{1}{2L+1}\left\{\sqrt{\left(\dfrac{L+2}{2}\right)}\,\delta_{|M|,2} + 2\sqrt{(L+1)}\delta_{M,0}\right\} a_M$

(*b*) $J = L \pm \dfrac{M}{[2(2L+1)]^{\frac{1}{2}}} a_M$

(*c*) $J = L-1 \mp (-1)^M \dfrac{1}{2L+1}\left\{\sqrt{\left(\dfrac{L-1}{2}\right)}\,\delta_{|M|,2} + \sqrt{L}\delta_{M0}\right\} a_M$

3. $S = 0$, $S' = 1$ $\qquad \pm (-1)^{M'}\dfrac{1}{\sqrt{(2L+1)}} a_{M'}$

4. $S = 0$, $S' = 0$ $\qquad$ Forbidden

When a $\pm$ or $\mp$ sign is indicated the upper sign is to be used if $\boldsymbol{\sigma}$ operates on a quark, the lower sign if it operates on an antiquark.

References

1. Gell-Mann, M., *Phys. Lett.*, **8,** 214–15 (1964).
2. A useful reference is Harari, H., in *Lectures in Theoretical Physics*, Vol. VIIIb, University of Colorado Press, Boulder, 1966, p. 345.
3. See, for example, Okubo, S., and Marshak, R. E., in *Symmetry Principles at High Energy II* (report of second Coral Gables Conference), W. H. Freeman and Co., San Francisco and London, 1965, p. 128.
4. There is an extensive literature on quark models. An early advocate was Dalitz, R. H. See the *Proc. Oxford. Intern. Conf. on Elementary Particles*, the Rutherford High Energy Laboratory, 1966, p. 157. A few of the other references known to me are: Lipkin, H. J., *Proc. 1967 Rehoboth Conf. High Energy Phys. and Nucl. Struct.* (to be published); Morpurgo, G., *Physics*, **2,** 2 (1965); Becchi, C., and Morpurgo, G., *Phys. Rev.*, **140B,** 687 (1965); Badier, S., Orsay preprint (Jan. 1967); Various combinations of

Rubenstein, H. R., Talmi, I., and collaborators in *Phys. Lett.*, **22,** 208, 210 (1966), *Phys. Lett.*, **23,** 693 (1966), *Phys. Rev. Letters*, **17,** 41, 420 (1966) and Weizmann Institute preprints; Oktay Sinanoglu, *Phys. Rev.*, **145,** 1205 (1966); Feld, B., MIT preprint (March, 1967); Royen, R. Van, and Weisskopf, V. F., to be published; Iizuka, Jugaro, *Prog. Theor. Phys. Suppl.*, Nos. 37–8, 21 (1966); Minamikawa, T., Miura, K., and Miyamoto, Y., *Prog. Theor. Phys.*, Nos. 37–8, 56 (1966); Ishida, S., and Roman, P., to be published.

5. Such a 'realistic' model has also been treated by Sinanoglu, *Phys. Rev.*, **145,** 1205 (1966). After completion of the present work Dr. B. Maglič showed me a recent (February 1967) University of Kansas preprint by Kwak, N., and Wong, K. W., who investigated a harmonic oscillator model for the $I = 1$ boson spectrum.
6. Rosenfeld, A. H., and others, *Rev. Mod. Phys.*, **39,** 1 (1967).
7. Montanet, L., 'Les resonances', *CERN preprint CERN/TC/Physics 66–23*, 26 August 1966.
8. Allard, J. F., and others, *Phys. Lett.*, **19,** 431 (1965).
9. Baillon, P., and others, *CERN preprint CERN/TC/Physics, 66–24 of 10/10/66* (to be published in *Nuovo Cimento*).
10. Walker, W. D., Carroll, J., Garfinkel, A., and Oh, B. Y., *Phys. Rev. Letters*, **18,** 630 (1967).
11. Cline, D., *Nuovo Cimento*, **XLVA,** 750 (1966).
12. Capps, Richard H., preprint. I am grateful to Dr. Capps for pointing out to me that the momentum dependence of the rate does *not* specify uniquely the relative orbital angular momentum of the final meson pair.
13. While this manuscript was being completed Dr. Maglič kindly informed me that the A_2 resonance now seems to be split into two peaks, each with a width of about 20 MeV (Chikovani, G., and others, to be published). One might wonder if this is a $^3P_2$, $^3F_2$ doublet similar to the one predicted for the f_0. A width of 20 MeV, by the way, corresponds to $g_M \approx 0$.
14. Glashow, Sheldon L., and Socolow, Robert H., *Phys. Rev. Letters*, **15,** 329 (1967). The predictions of this paper were based on SU(3).
15. Private communication from Dagan, S., and Hwang, C.
16. In particular Elitzur, M., Rubinstein, H. R., Stein, H., and Lipkin, H. J., reference 4; Rubinstein, H. R., and Talmi, I., reference 4; and Mitra, A. N., and Srivastava, P. P., Trieste preprint IC/67/31. I am indebted to the latter two authors especially for an advance copy of their paper.
17. Allard, J. F., and others, *Nuovo Cimento*, **46A,** 737 (1966).
18. But apparently not that of Mitra and Srivastava who assume a Gaussian form-factor (as would be obtained from a harmonic oscillator potential).
19. See also Lipkin, H. J., 'Intrinsic orbital angular momentum and *W* spin', (to be published in *Phys. Rev.*) for a discussion of this selection rule from the *W* spin viewpoint.
20. I am indebted to Wangler, Dr. T. P., for this information.
21. See also Minamikawa and others, *Prog. Theor. Phys.*, Nos. 37–8, 56 (1966).
22. Mao Tse-Tung, *Quotations from Chairman Mao Tse-Tung*, Bantam Books, Inc., New York, 1967, p. 29.

Inelastic N/D Equations and Regge Theory

R. L. WARNOCK

1 Introduction

In the literature there are two kinds of applications of the N/D method. First, as in the original work of Chew and Mandelstam[1], the method is used in the construction of dynamical models[2]. Second, there are applications of a more theoretical type. For instance the N/D integral equation provides a method of analytic continuation of partial wave amplitudes $A(l, s)$ from high to low values of Re l. It also provides a means of extending Levinson's theorem outside the domain of potential scattering[4]. Applications of the N/D method in the second category may prove to be of some lasting value, while it is not clear at present whether the N/D dynamical models are capable of giving more than a vague sketch of the physical world. The defects of the models are due to our ignorance about the physical forces to be put into the equations, and not to mathematical defects of the N/D approach. From a mathematical viewpoint the method is very good. It is related to techniques in the theory of singular integral equations developed since Riemann and Hilbert, and still actively studied by mathematicians in Russia and elsewhere. Much of this mathematical theory fits in nicely with scattering problems. For instance a solution of the 'Hilbert problem in several unknown functions' published by Plemelj in 1908 is easily adapted to give an existence proof for the many-channel N/D representation[5].

In the following I am concerned with both the practical and the theoretical aspects of N/D equations. The discussion will be for physical values of the angular momentum l in the direct channel, but with account taken of Regge asymptotic behaviour in the limit of infinite direct channel energy.

In Section 2 the inelastic N/D equation is derived, and a generalized form of Levinson's theorem is proved. The equation takes as given data the term in the partial wave amplitude arising from left-hand singularities and also the elasticity function η. The discussion is somewhat more general than those of earlier publications[4], in that requirements on the behaviour at infinity of partial wave amplitudes are relaxed, and zeros of the elasticity function at physical energies are allowed. A new and possibly advantageous method of handling the Castillejo–Dalitz–Dyson (CDD) poles is described.

Section 3 is concerned with the behaviour of partial waves at infinity in Regge theory. To deduce this behaviour some assumptions are necessary. I hope that the assumptions are reasonable, but I wish to emphasize that Regge theory without such assumptions does not make a statement about partial waves at infinity. In Section 3 a Regge behaviour modified by logarithmic factors is also considered. With that hypothesis a very elementary derivation of a Froissart-type bound is given.

In Section 4 the Fredholm character of the N/D equations in Regge theory is demonstrated. Here there is no discussion of the construction of a Regge theory from first principles. Rather, the properties of the N/D kernel are deduced after assuming that the partial wave amplitude has the high-energy behaviour determined in Section 3. This is possible because the kernel is expressed directly in terms of the partial wave amplitudes themselves. The Fredholm property of the equation holds even in one delicate case of a CDD pole at infinity.

In Regge theory the behaviour of phase shifts is such that whenever CDD poles are required, one of them may be placed at infinity in such a way that only one free parameter enters in place of the usual two. The preceding statements hold if the elasticity η is free of zeros at physical energies. In the accidental case of a zero of η the equation is not of Fredholm type, but is effectively equivalent to one that is. The latter equation may have practical value as an approximation to the case in which η almost vanishes. With regard to regularity the N/D equation with η describing the inelasticity has an advantage over the Chew–Mandelstam equation in which the inelasticity parameter is the ratio R of total to elastic partial wave cross-sections. The R equation is marginally singular in Regge theory[7].

Section 5 describes a tentative proposal for dynamical calculations with Regge trajectory exchange. The term trajectory exchange refers to a sum of Regge pole terms from which the left-hand singularities and the elasticity (at high energies only) are computed. In this model the N/D equation has the same regularity properties as in a correct Regge theory. Some attention is given to the matter of direct channel resonances, which can be introduced in our scheme through CDD poles when the resonances are not dynamically generated by the trajectory exchange mechanism. These poles give an economical description of important effects of the channels not treated explicitly in the single-channel integral equation. Another important point has to do with dips in the η function which occur at resonances. There is an interesting requirement of consistency between the width of the dip and width of the associated resonance, the latter width coming out of the solution of the integral equation. The model is largely phenomenological; it will involve several parameters that must be fitted to experiments. On the other hand, it is almost free of 'mystery parameters' like the cut-offs required in non-Regge bootstrap models with higher spins. The model is closer to nuclear reaction theory than to the ideal 'Reggeized bootstrap', since at this stage the ambition of making it symmetric under crossing is given up.

The discussion throughout is for the case of zero spin, equal mass scattering. I will not claim that everything goes through in the same way for higher spins and unequal masses, but the techniques for working out the general cases are available. In general one can use helicity amplitudes free of kinematic singularities[8], and the matrix N/D method with arbitrary inelasticity[9]. Many-channel analogs of certain items (for instance the special CDD pole at infinity in Section 2) have not been discovered up to now.

2 Inelastic N/D Equations and Levinson Relation for Real Part of the Phase Shift

The partial wave amplitude is denoted $A_l(z) = A_l^*(z^*)$. Henceforth the subscript l is dropped and $A_\pm(\nu)$ denotes the limit of A as z approaches the physical cut from above (below). The limit from above satisfies the unitarity relation

$$2\mathrm{i}\,\rho(\nu)A_+(\nu) = \eta(\nu)\exp\,[2\mathrm{i}\,\delta(\nu)] - 1,\ \nu \geqslant 0$$
$$\nu = q^2/m^2,\quad \rho(\nu) = [\nu/(\nu+1)]^{1/2},\quad \eta^2(\nu) \leqslant 1 \tag{1}$$

Here q is the momentum in the centre of mass frame and m is the particle mass. The factor η will be called the elasticity, and δ, the real part of the phase, will usually be called simply the phase shift. The inelastic threshold $\nu = \nu_1$ is assumed to lie higher than the elastic threshold. In the elastic region $0 \leqslant \nu \leqslant \nu_1$ we have $\eta \equiv 1$.

The amplitude A is to fulfill the following conditions:

(i) $A(z)$ is analytic except for the stable particle poles in the z plane with cuts $(-\infty, -1]$, $[0, \infty)$;

(ii) $A(z) = 0(|z|^n)$ at infinity, for some integer n;

(iii) $\eta(\nu)$ and $\delta(\nu)$ are Hölder-continuous in any finite interval; i.e., $|\eta(\nu) - \eta(\mu)| \leqslant c|\nu - \mu|^{\alpha}$, $\alpha > 0$, and similarly for δ;

(iv) $\delta(\nu)$ is bounded at $\nu = \infty$. (2)

In support of these assumptions it may be mentioned first that (2i) is supposed to hold in every order of perturbation theory[10]. If (2i) does not hold in general it may very well be possible to adapt the methods to more complicated situations (for example to a situation where complex singularities are present). Since $A_+(\nu) = 0(1)$ by unitarity, (2ii) is an assumption of fairly uniform behaviour at infinity which may seem plausible enough but is nevertheless unproved on a basis of physical principles.

Kinoshita[11] has been able to find conditions under which A obeys a dispersion relation with one subtraction. He assumes a weak bound on the amplitude at infinity (which in itself would not rule out infinitely many subtractions) and also a limitation on the number of zeros of the left cut discontinuity. For N/D equations one can get along with less uniformity in behaviour at infinity than Kinoshita obtains, but it does not seem easy to weaken his conditions so as to obtain just what one needs and not more[12].* Assumption (2iii) holds if our usual ideas of analyticity and threshold behaviour are correct since then $\eta(\nu)$ and $\delta(\nu)$ are values of functions $\eta(z)$ and $\delta(z)$ which are analytic in a neighbourhood of the physical cut (the neighbourhood is partly on the physical sheet and partly on the unphysical sheet reached by continuation through the cut). There are branch points of η and δ at particle thresholds, but at worst such branch points are of the type $(z - \bar{\nu})^{1/2}$ which means that they do not prevent Hölder-continuity. An exception to this claim of continuity occurs if $\eta(\nu)$ has a zero and is defined so that $\eta(\nu) \geqslant 0$. Then $\delta(\nu)$ jumps by $\pi/2$ at the zero. One avoids this difficulty by defining η so that it changes sign at a point where it has a zero. Then if $A_+(\nu)$ has a continuous derivative at the zero (which will be assumed), δ is Hölder-continuous. The boundedness of the phase required in (2iv) holds in Regge theory under the assumptions of Section 3, and also in a theory with Regge asymptotic behaviour modified by logarithmic factors (Section 3). Also, Eden[13] gives some general assumptions under which the phase is bounded. In both Section 3 and in Eden's examples the phase tends to an integer multiple of π. Hence in the following particular attention is given to the example in

* Note added in proof: It is possible to avoid assumption (2ii) entirely, and to weaken greatly assumption (2i). Instead of (2i) it is sufficient to have analyticity only in a neighborhood of the physical cut. In other words, the N/D equation (16) holds with only the analyticity properties that have been proved from axiomatic field theory, provided that the phase shift is bounded. The clue to this generalization, which will be published elsewhere, is simply to *define* $B(Z)$ as the difference between the amplitude $A(Z)$ itself and a right-cut integral (cf. (15)).

which the phase approaches a constant at infinity, although for *N/D* equations it is sufficient to have a merely bounded phase which could oscillate indefinitely.

From (2i, 2ii) it follows that $A(z)$ obeys a dispersion relation with $n+1$ subtractions. In the right cut integral partial fraction identities may be used to remove all but one subtraction because of the unitarity bound $\text{Im}\, A_+(\nu) = 0(1)$. If the subtraction point ν_0 is placed in the gap, $-1 < \nu_0 < 0$, and the constant $a = A(\nu_0)$ is defined, the dispersion relation has the form

$$A(z) = a + (z - \nu_0)\left[A^{\text{L}}(z) + \sum_{i=1}^{n_b} \frac{\gamma_i}{\nu_i - z} + \frac{1}{\pi}\int_0^\infty \frac{\text{Im}\, A_+(\nu)\, \text{d}\nu}{(\nu - \nu_0)(\nu - z)}\right] \tag{3}$$

where A^{L} is a polynomial plus an integral over the left cut with $n+1$ subtractions. The poles represent stable particles. If the angular momentum is 1 or greater, one may take $\nu_0 = 0$ and thereby have $a = 0$. It is useful to split off a part of the right cut integral which depends only on the elasticity; namely the part coming from the term $(1 - \eta)/2\rho$ in the following

$$\text{Im}\, A_+ \equiv \frac{\eta \sin^2 \delta}{\rho} + \frac{1 - \eta}{2\rho} \tag{4}$$

By adding that part to A^{L} a convenient function B is defined:

$$B(z) = (z - \nu_0)\left[A^{\text{L}}(z) + \frac{1}{\pi}\int_0^\infty \frac{1 - \eta(\nu)}{2\rho(\nu)(\nu - \nu_0)(\nu - z)}\text{d}\nu\right] \tag{5}$$

The method proceeds from the factorization $A(z) = N(z)/D(z)$, where D meets the following requirements:

(i) $D(z)$ is meromorphic in the plane with cut $(0, \infty)$, and $D(z) = D^*(z^*)$;
(ii) $D(z) = 0(|z|^m)$ at infinity for some integer m;
(iii) $D_-(\nu) = \exp\,[2\text{i}\, \delta(\nu)]D_+(\nu)$, $\nu \geqslant 0$. (6)

One example of such a D function is well known:

$$\mathscr{D}(z) = \exp\left[-\frac{z - \nu_0}{\pi}\int_0^\infty \frac{\delta(\nu)\text{d}\nu}{(\nu - \nu_0)(\nu - z)}\right] \tag{7}$$

The property (6i) of this function is obvious, and the Hölder-continuity of δ allows one to verify the Hilbert barrier condition (6iii); [it also allows application of the rule $1/(x \pm \text{i}\, 0) = P(1/x) \mp \text{i}\, \pi\delta(x)$].

The bound (6ii) is proved in appendix A of reference 6. If D is any function satisfying the three conditions (6), then $D/\mathscr{D}$ is a rational function. That is true because the only singularities of $D/\mathscr{D}$ in the finite plane are poles, and $1/\mathscr{D}$ is bounded by a power at infinity[6]. Hence the general D has the form

$$D(z) = \mathscr{D}(z)R(z) \tag{8}$$

where $R(z) = R^*(z^*)$ is rational. If the phase shift tends to a constant $\delta(\infty)$, then the function $\mathscr{D}$ behaves as $z^{\delta(\infty)/\pi}$ at infinity apart from a possible logarithmic factor. The precise statement is[6]

$$\mathscr{D}(z) = z^{\delta(\infty)/\pi}\, e^{\lambda(z)} \tag{9}$$

where $|\lambda(z)| < \varepsilon \ln |z|$ for any $\varepsilon > 0$ and all z for which $|z| > r(\varepsilon)$.

Formulae for the discontinuities of N and D on the right cut are obtained by putting $\exp(2i\,\delta) = D_+^*/D_+$ in (1):

$$2\rho \operatorname{Re} N_+ = -(1+\eta)\operatorname{Im} D_+, \quad 2\rho \operatorname{Im} N_+ = (1-\eta) \operatorname{Re} D_+ \tag{10}$$

According to (10), the right-hand cut of N begins at the inelastic threshold. At this point the usual procedure is to write dispersion relations for N and D, and substitute the relation for D into that for N. All unknown functions may then be eliminated save the one function $\operatorname{Im} D_+$ which obeys a linear integral equation, the kernel and right side being functions of $\operatorname{Re} B_+(\nu)$. When there is inelasticity this method is awkward, and it requires an unnecessarily strong bound on $A(z)$ at infinity in complex directions. Also, it is not applicable in a new treatment of CDD poles given below.

The case where the phase shift tends to a constant will be treated first, and two classes of amplitudes will be distinguished according to phase shift behaviour at infinity:

$$\text{class A: } [\delta(\infty)/\pi] \leqslant -n_b; \quad \text{class B: } [\delta(\infty)/\pi] > -n_b \tag{11}$$

The symbol $[x]$ denotes the largest integer contained in x. We first derive the integral equation for class A. It is an equation for the imaginary part of a particular D function chosen as follows:

$$D(z) = c \prod_{i=1}^{n_b} (1 - z/\nu_i)\mathscr{D}(z), \quad D(\nu_0) = 1, \quad c = \text{constant} \tag{12}$$

According to (7) and (11) we have $D(z) = 0(|z|^{1-\varepsilon})$ for some $\varepsilon > 0$. Therefore D has the Cauchy representation

$$D(z) = 1 + \frac{z - \nu_0}{\pi} \int_0^\infty \frac{\operatorname{Im} D_+(\nu)\, d\nu}{(\nu - \nu_0)(\nu - z)} \tag{13}$$

From (5) and (13) we form an auxiliary function:

$$\Lambda(z) = \frac{1}{z - \nu_0}[N(z) - B(z)D(z)] + \frac{1}{\pi}\int_0^\infty \frac{\text{Im } D_+(\nu) \text{ Re } B_+(\nu)\, d\nu}{(\nu - \nu_0)(\nu - z)} \tag{14}$$

The desired integral equation is obtained by taking the real part of $\Lambda(\nu + i\,0)$, after showing that $\Lambda(z) = a/(z - \nu_0)$. To demonstrate the latter we first note that the integral of (14) converges because $\text{Im } D_+(\nu) = 0(\nu^{1-\varepsilon})$ and $\text{Re } B_+(\nu) = 0(\ln \nu)$. The bound on Re B_+ comes from the dispersion relation (3), which yields

$$\text{Re } B_+(\nu) = \text{Re } A_+(\nu) - a - (\nu - \nu_0)\left[\sum_{i=1}^{nb} \frac{\gamma_i}{\nu_i - \nu} + \frac{P}{\pi}\int_0^\infty \frac{\eta(\mu)\sin^2\delta(\mu)\, d\mu}{(\mu - \nu_0)(\mu - \nu)}\right] \tag{15}$$

The first three terms of (15) are bounded, while the last term is $0(\ln \nu)$ according to a result of reference 6, appendix A. By using the relations (10) it is easy to check that Λ has zero discontinuity over either cut. Furthermore Λ is bounded by a power at infinity and therefore is a rational function. It has a single pole with residue a at $z = \nu_0$, and $\lim \Lambda_+(\nu) = 0$, $\nu \to \infty$. Hence $\Lambda(z) = a/(z - \nu_0)$ as claimed. Evaluation of Re $\Lambda_+(\nu)$ from (14) and (13) yields the N/D integral equation:

$$\eta(\nu)\phi(\nu) = \frac{a + C(\nu)}{\nu - \nu_0} + \frac{1}{\pi}\int_0^\infty \frac{C(\nu) - C(\mu)}{\nu - \mu}\rho(\mu)\phi(\mu)\, d\mu$$

$$\phi(\nu) = \frac{-\text{Im } D_+(\nu)}{(\nu - \nu_0)\rho(\nu)}, \quad C(\nu) = \text{Re } B_+(\nu) \tag{16}$$

We wish to emphasize that (16) is a *necessary* condition on the particular D function (12) of a class A amplitude. Thus, if a class A solution of the dispersion relation exists, (16) must have a solution whether the equation is of Fredholm type, is singular, or whatever. The question of finding solutions of the dispersion relation by solving equations such as (16) will be discussed later. For the moment we are concerned only with finding necessary conditions on *particular* D functions. It goes without saying that different choices of the rational function R in (8) will lead to different integral equations.

Next we take the case of class B amplitudes and choose a D function as follows:

$$D(z) = \prod_{i=1}^{nb}(1 - z/\nu_i)\mathscr{D}(z)/P(z), \quad D(\nu_0) = 1 \tag{17}$$

The polynomial $P(z)$ is to have the minimum degree allowing a dispersion relation with one subtraction. Define the integer n_c by the equation

$$[\delta(\infty)/\pi] = -(n_b - n_c) \tag{18}$$

In class B one has $n_c > 0$. The polynomial P may always be chosen so that D has the representation[14]

$$D(z) = 1 + (z - \nu_0)\left[A + \sum_i \frac{c_i}{\mu_i - z} + \frac{1}{\pi}\int_0^\infty \frac{\text{Im } D_+(\nu)\, d\nu}{(\nu - \nu_0)(\nu - z)}\right] \tag{19}$$

where the number of poles (counting the possible pole at infinity from the term Az) is equal to n_c. The real constant A is non-zero if and only if δ/π tends to an integer from *below*; i.e., $\delta(\nu) - \pi n \leqslant 0$ for all ν greater than some $\bar{\nu}$. Each μ_i is either positive and such that $\sin \delta(\mu_i) = 0$ or else negative and such that $A(\mu_i) = 0$. As a consequence Im D_+ is continuous at a point $\mu_i > 0$; the poles occur only in Re D_+. If A is non-zero, $D(z)$ is asymptotic to Az in directions not parallel to the real axis. If $A \neq 0$ the convergence of the integral in (19) is a very delicate matter. The convergence, which is established by a theorem of Herglotz[4], may happen only by virtue of weakly vanishing logarithmic factors. Hence in applying the Λ method with the D function (19) we cannot be sure of convergence of the integral since Re $B_+(\nu)$ might increase at infinity [we know only the upper bound Re $B_+(\nu) = 0(\ln \nu)$]. The method still goes through, however, if the integrand in (14) is multiplied by an extra factor $(z - \nu_0)/(\nu - \nu_0)$ to ensure convergence.

Thus we adopt the definition

$$\Lambda(z) = \frac{1}{z - \nu_0}[N(z) - B(z)D(z)] + \frac{1}{\pi}\int_0^\infty \frac{\alpha(z, \nu)\text{Im } D_+(\nu)\text{Re } B_+(\nu)}{(\nu - \nu_0)(\nu - z)}\, d\nu \tag{20}$$

where $\alpha = 1$ if the integral converges with $\alpha = 1$, and otherwise $\alpha(z, \nu) = (z - \nu_0)/(\nu - \nu_0)$. Taking account of the poles of Λ we find that the following integral equation replaces (16):

$$\eta(\nu)\phi(\nu) = \frac{a + C(\nu)}{\nu - \nu_0} + AC(\nu) + \Lambda(\infty) - \sum_i c_i \frac{C(\mu_i) - C(\nu)}{\mu_i - \nu} + \\ + \frac{1}{\pi}\int_0^\infty \frac{C(\nu) - \alpha(\nu, \mu)C(\mu)}{\nu - \mu}\rho(\mu)\phi(\mu)\, d\mu \tag{21}$$

One has $\alpha = 1$ and $\Lambda(\infty) = 0$ if $A = 0$, since in that case there is the bound $D(z) = 0(|z|^{1-\epsilon})$. In Section 4 we shall find that in Regge theory $A \neq 0$, $\alpha = 1$, and $\Lambda(\infty) = -AC(\infty)$. Thus in Regge theory there are

$2n_c - 1$ real parameters associated with the poles of D: $2(n_c - 1)$ of the parameters c_i, μ_i plus the constant A.

The poles of the D function (19) are called Castillejo–Dalitz–Dyson (CDD) poles, although it must be realized that not all of these poles necessarily have the same interpretation as the CDD poles of a purely elastic problem (for instance a problem like that originally discussed by Castillejo, Dalitz and Dyson[15]).

In an inelastic problem there is the new feature that CDD poles are associated with zeros of the S-matrix element $S(z) = 1 + 2i\,\rho(z)A(z)$. There will be a number of 'inelastic CDD poles' equal to the net number of zeros of S that enter the upper half of the physical sheet through the inelastic cut as the interaction is increased in strength from zero to its final value. (The net number is the number that enters minus the number that leaves.) This matter is discussed by various authors with various viewpoints[16] and is taken up again from a viewpoint that I prefer in a forthcoming publication.

It is clear that the choice of location of the CDD poles is arbitrary[17]. We could have taken them all at real points below threshold, at complex points, or even at points μ_i above threshold for which $\sin \delta(\mu_i) \neq 0$. Then one would have integral equations (different from but analogous to equation 21) as necessary conditions on the corresponding D functions. We shall see that it is not necessary to consider all such equations. We find particular equations obeyed by certain D functions. We have an equation for each possible type of phase shift behaviour, and these équations will be sufficient for anything one wishes to do with the N/D method.

The poles in (19) were located so as to minimize the number of parameters occurring in the integral equation. The same choice makes the kernel regular at points $\nu = \mu_i > 0$, as a result of Im D_+ being continuous at such points. Minimization of the number of parameters is clearly important when the N/D equation is being used to construct solutions of the dispersion relation. Then the CDD parameters are part of the data put into the equations, and there is an advantage in making the set of parameters as small and as non-redundant as possible. Had poles been placed at points $\nu = \mu_i < 0$ such that $A(\mu_i) \neq 0$, there would have been a greater number of parameters. Poles at such points occur in N as well as D, and therefore residues of the poles in N appear as additional parameters. If poles had been placed at points $\nu = \mu_i > 0$ such that $\sin \delta(\mu_i) \neq 0$, a principal value definition of the integral in (19) at $\nu = \mu_i$ would have been necessary. There would be a consequent complication in the integral equation. If the phase tends to an integral multiple of π from below one does not necessarily have a number n_c of points where either $A(\mu_i) = 0$ or

$\sin\delta(\mu_i) = 0$; the number may be only $n_c - 1$. In that case one applies a theorem of Herglotz to show that the integral representation (19) still converges with one subtraction if the last CDD pole is put at infinity[4].

Before turning to the case in which the phase shift may oscillate at infinity, it is useful to give a new method of treating class B. The method is to put all of the poles of D at infinity, and use a Λ function with an appropriate number of subtractions. This trick has advantages:

(i) it is simpler, since it does not require the argument of reference 4 to establish existence of the points μ_i;

(ii) it is applicable to the case of oscillating phase and to the many-channel problem. In these latter problems the existence of points analogous to the μ_i is not assured, so if the poles are taken at finite points one encounters difficulties such as unnecessary parameters occurring as residues of N function poles.

The D function will be chosen just as it was in class A (cf. equation 12). Suppose first that δ/π does not tend to an integer from below. Since $D(z) = 0(|z|^{n_c+1-\varepsilon})$ for some $\varepsilon > 0$, we may write a Cauchy representation of $D(z)(z-\nu_0)^{-n_c-1}$. The latter gives

$$D(z) = 1 + \sum_{n=1}^{n_c} d_n(z-\nu_0)^n + \\ + \frac{(z-\nu_0)^{n_c+1}}{\pi}\int_0^\infty \frac{\operatorname{Im} D_+(\nu)\,d\nu}{(\nu-\nu_0)^{n_c+1}(\nu-z)} \tag{22}$$

where the real constants d_n are the coefficients of the Taylor development about $z = \nu_0$: $d_n = D^{(n)}(\nu_0)/n!$. If δ/π does tend to an integer from below, $D(z)$ has the following representation which is proved by using the factorization of D as a Herglotz function times a rational function[4]:

$$D(z) = 1 + \sum_{n=1}^{n_c-1} d_n(z-\nu_0)^n + \\ + (z-\nu_0)^{n_c}\left[A + \frac{1}{\pi}\int_0^\infty \frac{\operatorname{Im} D_+(\nu)\,d\nu}{(\nu-\nu_0)^{n_c}(\nu-z)}\right] \tag{23}$$

When δ/π does not go to an integer from below, the Λ function is chosen as follows:

$$\Lambda(z) = \frac{1}{(z-\nu_0)^{n_c+1}}[N(z) - B(z)D(z)] + \frac{1}{\pi}\int_0^\infty \frac{\operatorname{Im} D_+(\nu)\operatorname{Re} B_+(\nu)}{(\nu-\nu_0)^{n_c+1}(\nu-z)}\,d\nu \tag{24}$$

Again Λ is a rational function. It vanishes at infinity and has a pole of order $n_c + 1$ with principal residue $a = A(\nu_0)$ at $z = \nu_0$. Therefore we write Λ as

$$\Lambda(z) = \frac{a}{(z - \nu_0)^{n_c+1}} + \sum_{n=1}^{n_c} \frac{a_n}{(z - \nu_0)^n} \tag{25}$$

The residues a_n are real constants because all the functions appearing in Λ are real at $z = \nu_0$. From Re $\Lambda_+(\nu)$ we get the integral equation

$$\eta(\nu)\phi(\nu) = \frac{a + C(\nu)}{(\nu - \nu_0)^{n_c+1}} + \sum_{n=1}^{n_c} \left[\frac{a_n}{(\nu - \nu_0)^n} + C(\nu)\frac{d_n}{(\nu - \nu_0)^{n_c+1-n}}\right] + $$
$$+ \frac{1}{\pi}\int_0^\infty \frac{C(\nu) - C(\mu)}{\nu - \mu}\rho(\mu)\phi(\mu)\,\mathrm{d}\mu \tag{26}$$
$$\phi(\nu) = \frac{-\mathrm{Im}\ D_+(\nu)}{\rho(\nu)(\nu - \nu_0)^{n_c+1}}$$

There are $2n_c$ constants a_n, d_n just as there are $2n_c$ constants c_i, μ_i in the corresponding case with (21). When δ/π tends to an integer from below, the Λ function is

$$\Lambda(z) = \frac{1}{(z - \nu_0)^{n_c}}[N(z) - B(z)D(z)] + \frac{1}{\pi}\int_0^\infty \frac{\alpha(z, \nu)\mathrm{Im}\ D_+(\nu)\mathrm{Re}\ B_+(\nu)}{(\nu - \nu_0)^{n_c}(\nu - z)}\,\mathrm{d}\nu \tag{27}$$

where again $\alpha = 1$ if the integral converges with $\alpha = 1$, and $\alpha(z, \nu) = (z - \nu_0)/(\nu - \nu_0)$ otherwise. The corresponding integral equation is

$$\eta(\nu)\phi(\nu) = \frac{a + C(\nu)}{(\nu - \nu_0)^{n_c}} + \sum_{n=1}^{n_c-1}\left[\frac{a_n}{(\nu - \nu_0)^n} + C(\nu)\frac{d_n}{(\nu - \nu_0)^{n_c+1-n}}\right] +$$
$$+ AC(\nu) + \Lambda(\infty) + \frac{1}{\pi}\int_0^\infty \frac{C(\nu) - \alpha(\nu, \mu)C(\mu)}{\nu - \mu}\rho(\mu)\phi(\mu)\,\mathrm{d}\mu \tag{28}$$
$$\phi(\nu) = \frac{-\mathrm{Im}\ D_+(\nu)}{\rho(\nu)(\nu - \nu_0)^{n_c}}$$

If $\Lambda(\infty) = -AC(\infty)$ as in Regge theory, the number of constants in (28) is $2n_c - 1$ as in the corresponding case with (21).

It is now easy to treat the situation where the phase does not approach a limit. Since we assume that the phase is bounded, it follows from appendix A of reference 6 that $\mathscr{D}(z)$ is bounded by a power. Define D as in (12). If $D(z)(z - \nu_0)^{-1}$ has a convergent Cauchy representation, then the situation is just as in class A, and Im D_+ satisfies (16), in which no CDD parameters enter. In the contrary event, let n_c be the smallest positive integer such that $D(z)(z - \nu_0)^{-n_c-1}$ has a convergent Cauchy representation. Then Im D_+ satisfies (26) in which there are $2n_c$ parameters.

To prove the Levinson relation we have only to add the following hypotheses to the conditions already laid down in (2):

(i) The phase shift tends to a limit $\delta(\infty)$;
(ii) The homogeneous N/D equation has no non-trivial solution. (29)

The homogeneous N/D equation is defined in the obvious way; i.e. it is the following equation:

$$\eta(\nu)\phi(\nu) = \frac{1}{\pi}\int_0^\infty \frac{C(\nu) - \alpha(\nu, \mu)C(\mu)}{\nu - \mu}\,\rho(\mu)\phi(\mu)\,\mathrm{d}\mu \tag{30}$$

Under conditions (2) and (29), Levinson's relation holds in the form

$$[\delta(\infty)/\pi] = -(n_b - n_c) \tag{31}$$

where n_c, the number of CDD poles, is defined in the way already described. In fact, (31) follows from the definition of the CDD poles if the amplitude is of class B. It only remains to show that in class A one has $[\delta(\infty)/\pi] = -n_b$; i.e. the case $[\delta(\infty)/\pi] < -n_b$ is impossible. If the latter inequality holds, then there are two D functions which both have the representation (13) because of the asymptotic behaviour (9):

$$D_1(z) = c\prod_{i=1}^{n_b}(1 - z/\nu_i)\mathscr{D}(z) = 0(|z|^{-\varepsilon})$$
$$D_2(z) = (z/\nu_0)D_1(z) = 0(|z|^{1-\varepsilon}) \tag{32}$$

The corresponding functions ϕ_1, ϕ_2 (where $\phi_i = -\text{Im}\, D_{+i}/(\nu - \nu_0)\rho$) both satisfy the inhomogeneous N/D equation (16). Hence the difference $\phi_1 - \phi_2$ satisfies the homogeneous equation, contrary to hypothesis.

The physical meaning of hypothesis (29ii) is not as clear as it should be, but one can at least note that (29ii) holds in all the usual models. If the N/D equation is of Fredholm type as it is in Regge theory under the assumptions of the following section, then we know that when (29ii) fails there are two possibilities: either (*a*) the inhomogeneous N/D equation has no solution, or (*b*) it has infinitely many solutions. Either possibility seems unsatisfactory; (*a*) means that there are no solutions of the partial-wave dispersion relations of the sort we want (bounded phase shift, etc.), while (*b*) gives an infinite family of solutions parametrized by constants which are unrelated to the CDD constants such as μ_i, c_i, A. Situation (*b*) would mean an ambiguity in the solution of partial wave dispersion relations of a new sort lacking the physical interpretation that may be given the CDD ambiguity. Also, situation (*b*) is quite exceptional in that it requires orthogonality between the inhomogeneous term and every solution of the adjoint homogeneous equation.

The N/D integral equations were derived as necessary conditions on particular D functions. The questions of existence and explicit construction of amplitudes having these D functions were put aside. We now turn to the problem of constructing amplitudes by means of the N/D equations. First we note that a solution of the dispersion relation that meets conditions (2) may be represented in terms of an appropriate Λ function; for instance from (20) we have

$$A(z) = \frac{N(z)}{D(z)} = B(z) + \frac{(z - \nu_0)}{D(z)}\left[\Lambda(z) - \frac{1}{\pi}\int_0^\infty \frac{\alpha(z, \nu)\,\mathrm{Im}\,D_+(\nu)\,\mathrm{Re}\,B_+(\nu)\,\mathrm{d}\nu}{(\nu - \nu_0)(\nu - z)}\right] \tag{33}$$

(This expression is appropriate if our first method of treating CDD poles is used. With the second method, $A(z)$ is constructed from (24) or (27).) In (33), Λ has the value

$$\Lambda(z) = -\sum_i c_i \frac{C(\mu_i)}{\mu_i - z} + \frac{a}{z - \nu_0} + \Lambda(\infty) \tag{34}$$

where $c_i = \Lambda(\infty) = 0$ if the amplitude is of class A. Every solution of the dispersion relation (meeting conditions 2) may be represented as in (33) with $D(z)$ constructed from a solution of the appropriate integral equation (one uses (13) and (16) for class A amplitudes, and (19) and (21) for class B). On the other hand a particular solution of the integral equation, when substituted in (33), does not necessarily yield a solution of the dispersion relation. The function fails to be a solution if $D(z)$ has ghost zeros; i.e. zeros producing poles of A which cannot be interpreted as stable particle poles. It is *only* in the presence of ghosts that the function constructed from (33) fails to be a solution. This is seen by noting that apart from possible ghosts the function has the correct singularities. Its left singularities are explicitly correct, arising as they do from the input term $B(z)$. On the right-hand cut the unitarity condition is satisfied. A simple calculation from (33) and the integral equation yields immediately the formula

$$A_+ = -\eta\,\frac{\mathrm{Re}\,D_+\,\mathrm{Im}\,D_+}{\rho|D_+|^2} + \mathrm{i}\left[\frac{1 - \eta}{2} + \eta\,\frac{(\mathrm{Im}\,D_+)^2}{\rho|D_+|^2}\right] \tag{35}$$

This agrees with (1) if δ is defined by $D_+ = |D_+|\exp(-\mathrm{i}\,\delta)$. The amplitude (33) is unitary only if Im D_+ satisfies the N/D integral equation.

In deciding whether a zero of D corresponds to a genuine stable particle we have only two general criteria. First, the zero must lie at a real energy below the elastic threshold, and second, the residue of the corresponding pole in A must have the appropriate sign. According to potential theory and field theory the pole should have the form

$$\frac{R}{\nu_i - \nu}, \quad R(-1)^l > 0 \tag{36}$$

It is not clear that these two criteria are always sufficient to distinguish ghosts from particles. There might be unusually diabolical ghosts which would look like particles with respect to these criteria. Within any particular model it should be possible to tell which are the real particles by physical arguments.

The factorization (8) gives one an effective method for calculating the number of zeros in a function D constructed from a solution of the integral equation. Such a D has this factorization provided only that its phase is bounded, which we assume for the moment. The poles of D are part of the input to the integral equation, so we know the number of poles of R. Thus by finding the asymptotic behaviour of $D/\mathscr{D}$ as ν goes to infinity we can discover the number of zeros of R which is also the number of zeros of D. In a practical calculation one could first look on the real axis for stable particle zeros of D and then check for the presence of additional zeros of D by the method described. This seems to be a practical procedure for testing for ghosts. The method has been mentioned before in the literature[18], but it has not been applied as far as I know in numerical calculations. Many authors who carry out calculations either neglect to check for ghosts or else do so in a relatively difficult and uninformative way.

To finish these remarks it is worth pointing out that if (16) has a unique L^2 solution, and if a class A amplitude having $\delta(\infty) = -\pi n_b$ exists*, then there is only one such amplitude and its stable particle parameters γ_i, ν_i are completely determined by the input $a + C(\nu)$. To prove this we note that the function $\phi = -\text{Im } D_+/(\nu - \nu_0)\rho$ constructed from the class A phase shift by means of (12) certainly satisfies the integral equation (16). Furthermore $\phi(\nu) = 0(\nu^{-1+\varepsilon})$ for any $\varepsilon > 0$ according to (9); hence $\phi \varepsilon L^2$. Any class A amplitude with $\delta(\infty) = -\pi n_b$ gives an L^2 solution of (16), and since by hypothesis there is only one of the latter, there is at most one class A amplitude. This is the situation expected in a Regge theory under the assumptions of Section 3; there one has $\delta(\infty) = -\pi n_b$

* Notice the absence of the integral part symbol around δ. Here we are concerned with the typical case in which phases tend to integral multiples of π.

for class A amplitudes and the integral equation has a unique L^2 solution (barring the accidental case in which η has a zero, or that in which the homogeneous equation has a non-zero solution). If a ghost appears in the amplitude (33) when D is constructed from (13) and (16), then there is no class A solution of the dispersion relation.

3 Partial Waves at Infinity in Regge Theory

We begin by evaluating the contribution of the Pomeranchuk trajectory to the partial wave amplitude at infinity. Later we argue that the contribution of secondary trajectories and of Mandelstam branch points vanishes more rapidly than that of the Pomeranchuk trajectory. The latter is written

$$B(t)\,\frac{1+\mathrm{e}^{-\mathrm{i}\pi\alpha(t)}}{\sin\pi\alpha(t)}\,s^{\alpha(t)}+\bar{B}(u)\,\frac{1+\mathrm{e}^{-\mathrm{i}\pi\alpha(u)}}{\sin\pi\alpha(u)}\,s^{\alpha(u)} \tag{37}$$

The partial wave amplitude is

$$A_l(s)=\frac{1}{s-s_0}\int_{-(s-s_0)}^{0}A(s,t)P_l\left(1+\frac{2t}{s-s_0}\right)\mathrm{d}t \tag{38}$$

Since the region of integration goes to infinity as s increases, the contribution of a Regge pole term cannot be evaluated without some sort of assumption on the behaviour of trajectories and residue functions at large negative t. If we make such an assumption there is still the question of uniformity with respect to t of the approximation of the amplitude by Regge poles at a given s. We deal with these questions by simply assuming that the leading contribution to the partial wave amplitude is given by integration over fixed but arbitrary regions of the 2-momentum transfers: $-T\leqslant t\leqslant 0$, $-U\leqslant u\leqslant 0$. This restriction to forward and backward peaks seems plausible, since it is suggested both by experiment and by our intuitive ideas of peripheralism. Thus we have to find the large s behaviour of the expression

$$\frac{1}{s-s_0}\int_{-T}^{0}B(t)\,\frac{1+\mathrm{e}^{-\mathrm{i}\pi\alpha(t)}}{\sin\pi\alpha(t)}\,s^{\alpha(t)}P_l\left(1+\frac{2t}{s-s_0}\right)\mathrm{d}t$$
$$+\frac{(-1)^l}{s-s_0}\int_{-U}^{0}\bar{B}(u)\,\frac{1+\mathrm{e}^{-\mathrm{i}\pi\alpha(u)}}{\sin\pi\alpha(u)}\,s^{\alpha(u)}P_l\left(1+\frac{2u}{s-s_0}\right)\mathrm{d}u \tag{39}$$

We assume as usual that

$$\alpha'(t)\geqslant 0,\quad t\leqslant 0,\quad \alpha(0)\leqslant 1 \tag{40}$$

We shall go beyond (40) in assuming that the trajectory slope is non-zero at $t = 0$:

$$\alpha'(0) > 0 \tag{41}$$

Experimental evidence at present is consistent with a more or less flat Pomeranchuk trajectory, and one theoretical viewpoint[19] is that it is perfectly flat; i.e. the Pomeranchuk singularity is a fixed pole in the angular momentum plane, $\alpha'(t) \equiv 0$. Although this view takes us back to a traditional diffraction picture, it is difficult to reconcile in a natural way with the usual ideas of S-matrix theory[20]. Therefore we prefer to suppose that there is some slope, however small, and that there is nothing special about the point $t = 0$ that would make $\alpha'(0) = 0$. We shall see that assumptions (40) and (41) imply forward and backward peaks which shrink within the intervals $[-T, 0]$, $[-U, 0]$ as s increases. It should not be thought that this shrinkage justifies our restriction of the integration to the fixed intervals. Such a justification requires a stronger statement about both α and B, $\bar{B}$ at large momentum transfers.

The functions $B(t)$ and $\alpha(t)$ are believed to be real-analytic for $\mathrm{Re}\, t < 4m^2$, but for the following we require only that they be real and have continuous second derivatives for $-T \leqslant t \leqslant 0$. If $\alpha(\tau) = 0, -2, -4, \ldots, -T < \tau < 0$ we naturally demand that $B(\tau) = 0$ since there can be no singularity of the scattering amplitude in the physical region.

First we analyse the integral

$$I_n(s) = \int_{-T}^{0} t^n s^{\alpha(t)}\, \mathrm{d}t \tag{42}$$

where n is a non-negative integer. Since $\alpha'(0) > 0$ and $\alpha'(t) \geqslant 0$, it is more or less obvious that the leading behaviour of I_n at large s is determined by the integration over an infinitesimal region near $t = 0$. Therefore, the reader not interested in the following tedium can obtain the same results that we obtain in general by substituting a linear function for $\alpha(t)$ in (42), and performing the integration explicitly. Writing (42) as

$$I_n = s^{\alpha(0)} \int_{-T}^{0} t^n \exp\left\{t \ln s \left[\frac{\alpha(t) - \alpha(0)}{t}\right]\right\} \mathrm{d}t \tag{43}$$

we notice that the quantity in brackets ([ ]) has a positive lower bound because of (40) and (41):

$$\frac{\alpha(0) - \alpha(t)}{-t} > \varepsilon > 0 \tag{44}$$

That is to say there is some line $\alpha(0) + \varepsilon t$ below which $\alpha(t)$ must remain. Hence

$$
\begin{aligned}
|I_n| &\leqslant s^{\alpha(0)} \int_{-T}^{0} (-t)^n \, \mathrm{e}^{\varepsilon t \ln s} \, \mathrm{d}t \\
&= \frac{s^{\alpha(0)}}{[\varepsilon \ln s]^{n+1}} \int_0^{\varepsilon T \ln s} x^n \, \mathrm{e}^{-x} \, \mathrm{d}x \\
&= 0 \left[\frac{s^{\alpha(0)}}{(\ln s)^{n+1}} \right]
\end{aligned}
\tag{45}
$$

This upper bound on I_n may now be used to show that I_n is actually asymptotic to a constant times $s^{\alpha(0)}(\ln s)^{-n-1}$. In (43) make the substitution $x = -t \ln s$:

$$
\begin{aligned}
I_n &= \frac{-s^{\alpha(0)}}{(-\ln s)^{n+1}} \int_0^{T \ln s} x^n \exp\left\{ \ln s \left[\alpha\left(\frac{-x}{\ln s}\right) - \alpha(0) \right] \right\} \mathrm{d}x \\
&= \frac{(-1)^n s^{\alpha(0)}}{(\ln s)^{n+1}} J(s)
\end{aligned}
\tag{46}
$$

Since by (45), $J(s)$ is a bounded function, we can show that it approaches a finite limit if we can show that its derivative tends to zero; i.e. that it does not oscillate indefinitely. Computation of the derivative gives

$$
\begin{aligned}
\frac{\mathrm{d}J}{\mathrm{d}s} = \frac{1}{s} \Bigg\{ & T^{n+1} \ln^n s \, s^{\alpha(-T)-\alpha(0)} + \\
& + \int_0^{T \ln s} x^n \exp\left\{ \ln s \left[\alpha\left(\frac{-x}{\ln s}\right) - \alpha(0) \right] \right\} \cdot \left[\left[\alpha\left(\frac{-x}{\ln s}\right) \right. \right. \\
& \left. \left. - \alpha(0) + \frac{x}{\ln s} \alpha'\left(\frac{-x}{\ln s}\right) \right] \right] \mathrm{d}x \Bigg\}
\end{aligned}
\tag{47}
$$

The integral in (47) converges exponentially as s goes to infinity, since $\ln s[\alpha(-x/\ln s) - \alpha(0)] \leqslant -\varepsilon x$ by (44). Also $\alpha(-T) - \alpha(0) < 0$, so $|\mathrm{d}J/\mathrm{d}s| = 0(s^{-1})$. Hence we have the asymptotic statement

$$
I_n(s) \sim (-1)^n c_n \frac{s^{\alpha(0)}}{(\ln s)^{n+1}}, \quad c_n > 0 \tag{48}
$$

The constants c_n are clearly independent of T for $T > 0$. Since

$$
c_n(T) = \lim_{s \to \infty} \int_0^{T \ln s} x^n \exp\left\{ \ln s[\alpha(-x/\ln s) - \alpha(0)] \right\} \mathrm{d}x \tag{49}
$$

we have

$$c_n(T_1) - c_n(T_2) = \lim_{s\to\infty} \int_{T_2 \ln s}^{T_1 \ln s} x^n \exp\left\{\ln s[\alpha(-x/\ln s) - \alpha(0)]\right\} dx$$
$$= 0 \tag{50}$$

We may evaluate c_n by taking T so small that the linear approximation for α is valid. Thus

$$c_n = \int_0^\infty x^n \, e^{-\alpha'(0)x} \, dx = [\alpha'(0)]^{-n-1} \tag{51}$$

If $\phi(t)$ is any piecewise continuous function we clearly have also

$$\left| \int_{-T}^{0} \phi(t) t^n s^{\alpha(t)} \, dt \right| \leqslant M \int_{-T}^{0} (-t)^n s^{\alpha(t)} \, dt = 0 \left[\frac{s^{\alpha(0)}}{(\ln s)^{n+1}} \right] \tag{52}$$

A similar argument applied to the integral

$$K(s) = \int_{-T}^{-T_1} \phi(t) t^n s^{\alpha(t)} \, dt, \quad T > T_1 > 0 \tag{53}$$

yields

$$|K| \leqslant \frac{M s^{\alpha(0)}}{[\varepsilon \ln s]^{n+1}} \int_{\varepsilon T_1 \ln s}^{\varepsilon T \ln s} x^n \, e^{-x} \, dx \tag{54}$$

The integral in (54), call it $L(s)$, clearly vanishes as s goes to infinity. One may obtain its behaviour by l'Hospital's theorem:

$$\lim_{s\to\infty} \frac{L(s)}{(\ln s)^n s^{-\varepsilon T_1}} = \lim_{s\to\infty} \frac{dL/ds}{-\varepsilon T_1 (\ln s)^n s^{-1-\varepsilon T_1}} = (\varepsilon T_1)^n \tag{55}$$

Thus

$$|K(s)| = 0 \left[\frac{s^{\alpha(0) - \varepsilon T_1}}{\ln s} \right] \tag{56}$$

The partial integral K vanishes faster by at least a power than the total integral I_n. As was claimed above, the asymptotically dominant part of I_n comes from an arbitrarily small range of integration below $t = 0$. This is the shrinkage of the forward peak.

Returning to the Regge pole projection (39), we examine first the part of the integral over a range $[-T_1, 0]$ where T_1 is chosen so that

$\alpha(t)$ has no zero for $-T_1 \leqslant t < 0$. By Taylor's theorem with remainder we may write

$$\frac{1 + e^{-i\pi\alpha(t)}}{\sin \pi\alpha(t)} = r(t) - i$$

$$r(t) = r^*(t) = r(0) + r'(0)t + r''(\theta t)t^2/2, \quad 0 < \theta(t) < 1$$

$$B(t) = B^*(t) = B(0) + B'(\phi t)t, \quad 0 < \phi(t) < 1 \tag{57}$$

where $r''(\theta t)$ and $B'(\phi t)$ are bounded for $-T_1 \leqslant t \leqslant 0$. If we have a conventional Pomeranchuk trajectory with $\alpha(0) = 1$, then

$$r(0) = 0, \quad r'(0) = -\frac{\pi}{2}\,\alpha'(0) \tag{58}$$

With $\alpha(0) = 1$ we use (57), (58), (48) and (52) to obtain the result

$$\frac{1}{s - s_0}\int_{-T_1}^{0} B(t)\,\frac{1 + e^{-i\pi\alpha(t)}}{\sin \pi\alpha(t)}\,s^{\alpha(t)}P_l\left(1 + \frac{2t}{s - s_0}\right)dt \sim$$

$$\sim \frac{B(0)}{\alpha'(0)}\left(\frac{\pi}{2\ln^2 s} - \frac{i}{\ln s}\right) + C(s)$$

$$\operatorname{Re} C(s) = 0(\ln^{-3} s), \quad \operatorname{Im} C(s) = 0(\ln^{-2} s) \tag{59}$$

If $\alpha(\tau) = 0, -2, -4, \ldots$ for $-T \leqslant \tau < -T_1$ we may apply Taylor's theorem about the point $t = \tau$ to the ratio $B(t)/\sin \pi\alpha(t)$ and use (56) to show that the integral over $[-T, -T_1]$ vanishes faster by a power than the leading term (59).

A secondary Regge trajectory $\alpha_i(t)$ of either signature and with $\alpha_i(0) < 1$ will give instead of (59) the asymptotic behaviour

$$-\frac{s^{\alpha_i(0)-1}B_i(0)}{\alpha_i'(0)\ln s}\left[-r_i(0) + i\right] + C(s), \quad |C(s)| = 0[s^{\alpha_i(0)-1}\ln^{-2} s] \tag{60}$$

All of these arguments apply as well to the u channel Regge poles. Presently we shall argue that the Mandelstam branch points contribute less by at least one power of $1/\ln s$ than the leading terms of (59). We also make the plausible assumption that the partial wave projection of the background integral in the Watson–Sommerfeld representation contributes less than (59). Thus we have the result

$$A_l(s) \sim \frac{1}{\alpha'(0)}\,[B(0) + (-1)^l \bar{B}(0)]\left(\frac{\pi}{2\ln^2 s} - \frac{i}{\ln s}\right) \tag{61}$$

if $\alpha(0) = 1$. If $\alpha(0) < 1$, as has sometimes been suggested, then we have $A_l(s) = 0(s^{-\varepsilon})$ for some $\varepsilon > 0$. We shall not be much concerned with that possibility, so from here on $\alpha(0) = 1$ is understood. The behaviour of N/D equations is, of course, closer to being singular if $\alpha(0) = 1$.

The behaviours of the phase shift δ and the elasticity factor η are calculated easily from (61)*:

$$\begin{aligned}
\delta_l(\nu) &\sim \pi n_l - (\pi/2)\gamma_l \ln^{-2}\nu \\
\eta_l(\nu) &\sim 1 - 2\gamma_l \ln^{-1}\nu \\
\eta_l &= \text{integer} \\
\gamma_l &= -[1/\alpha'(0)][B(0) + (-1)^l \bar{B}(0)] \geqslant 0
\end{aligned} \tag{62}$$

The non-negative property of γ_l is obtained from unitarity. For analysis of N/D equations we shall also need bounds on the first two derivatives of δ_l and η_l. Assuming that the leading terms in the derivatives of A_l are obtained from the derivatives of (39) one finds that one can differentiate the right and left sides of (62) and still have valid asymptotic relations. Thus

$$\begin{aligned}
\delta_l'(\nu) &= 0(\nu^{-1} \ln^{-3}\nu), \quad \delta_l''(\nu) = 0(\nu^{-2} \ln^{-3}\nu) \\
\eta_l'(\nu) &= 0(\nu^{-1} \ln^{-2}\nu), \quad \eta_l''(\nu) = 0(\nu^{-2} \ln^{-2}\nu)
\end{aligned} \tag{63}$$

Of course, the derivatives of δ_l and η_l are continuous according to our assumptions.

How will branch points in the crossed channel angular momentum affect our results? The question is indefinite, since the theory of the branch points is in an incomplete state. Mandelstam[21] has argued that a certain set of non-planar graphs involving multiple exchange of Regge trajectories, roughly speaking, would produce branch points at positions related to the positions of the exchanged trajectories. Gribov, Pomeranchuk and Ter-Martirosyan[22] and Polkinghorne[23] agree with Mandelstam's conclusions and in addition determine the type of the branch points.

An n-fold exchange of the trajectory $\alpha(t)$ is supposed to provide a branch point $\alpha_n(t)$ as follows:

$$\alpha_n(t) = n\alpha(t/n^2) - n + 1 \tag{64}$$

* In (62) it is assumed that $\eta = 1$ at $\nu = \infty$ rather than $\eta = -1$. According to the definition of η in Section 2, it is possible to have $\eta = -1$ if η has a zero at some physical energy. If there is one such zero, then $\delta_l \sim (n_l + 1/2)\pi - (\pi/2)\gamma_l \ln^{-2}\nu$, $\eta_l \sim -1 + 2\gamma_l \ln^{-1}\nu$.

If $\bar{A}(\lambda, t)$ is the t-channel partial wave amplitude, then the branch point α_n is of the logarithmic type

$$\bar{A}(\lambda, t) = \alpha(\lambda, t)[\alpha_n(t) - \lambda]^{n-2} \ln [\alpha_n(t) - \lambda] + \beta(\lambda, t) \tag{65}$$

where α and β are regular in λ near $\lambda = \alpha_n$. By the Watson–Sommerfeld transformation it is immediately possible to verify that the asymptotic contribution of these branch points at large s is

$$\sum_{n=2}^{\infty} b_n(t) \frac{1 + e^{-i\pi\alpha_n(t)}}{\sin \pi\alpha_n(t)} \frac{s^{\alpha_n(t)}}{(\ln s)^{n-1}} \tag{66}$$

for the case in which the branch points occur in the positive signature amplitude $\bar{A}^{(+)}(\lambda, t)$. As a consequence of the regularity of $\alpha(\lambda, t)$, $b_n(t)$ should be a real function for negative t. The convergence properties of the series (66) are apparently unknown, but it would seem that the Regge theory would be useless for determining the large s behaviour unless there is convergence for some appreciable region of negative t. We assume that to get the asymptotic contribution to partial waves it is correct to integrate the series term by term over the intervals $[-T, 0]$, $[-U, 0]$. The exchange of two or more Pomeranchuk trajectories will give the right-most branch points, so we evaluate the partial waves of (66) with $\alpha(t)$ identified as the Pomeranchukon. Then the analysis above shows that each term of (66) gives a term like (61), but modified with an additional factor $(\ln s)^{n-1}$ in the denominator. We conclude tentatively that the contribution of branch points is down by a factor $1/\ln s$ in comparison with that of the Pomeranchuk pole.

Instead of the standard Regge model, we may try a similar but more general asymptotic behaviour as considered by van Hove[24] and others; viz.,

$$A(s, t) \sim C(t) s^{\alpha(t)} (\ln s)^{\beta(t)} (\ln \ln s)^{\gamma(t)} \ldots \tag{67}$$

where there may be any finite number of factors with compounded logarithms, and where $\alpha, \beta, \gamma, \ldots$ etc., are real functions of t for $-T \leqslant t \leqslant 0$. The exponent $\alpha(t)$ is to satisfy $\alpha'(t) \geqslant 0$, $\alpha'(0) > 0$.

We proceed as we did before to project partial waves, and suppose that the ratio of real to imaginary part of the partial wave tends to zero. The latter condition means that $C(0) = -i|C(0)|$ so we write

$$C(t) = t\mathscr{A}(t) - i\,\mathscr{B}(t) \tag{68}$$

where $\mathscr{A}$ and $\mathscr{B}$ are real, $\mathscr{A}(0)$ is finite, and $\mathscr{B}(0) \neq 0$. With assumptions about differentiability of $C(t)$ like those made previously, we have instead of (61) the result

$$A_l(s) \sim \frac{1}{\alpha'(0)} [s^{\alpha(0)-1}(\ln s)^{\beta(0)}(\ln \ln s)^{\gamma(0)} \ldots] \times$$
$$\times \left\{ -\frac{1}{\alpha'(0) \ln^2 s} [\mathscr{A}(0) + (-1)^l \bar{\mathscr{A}}(0)] - \frac{i}{\ln s} [\mathscr{B}(0) + (-1)^l \bar{\mathscr{B}}(0)] \right\} \tag{69}$$

To satisfy unitarity Im A_l must be bounded. By choosing l so that $\mathscr{B}(0) + (-1)^l \bar{\mathscr{B}}(0) \neq 0$, we see that $\alpha(0) \leqslant 1$. If $\alpha(0) = 1$ then $\beta(0) \leqslant 1$. If $\alpha(0) = \beta(0) = 1$, then $\gamma(0) \leqslant 0$ and so on if there are higher compounded logarithms. Thus we have the bound

$$|A(s, 0)| = 0(s \ln s) \tag{70}$$

which is closer by one power of ln s than that of Froissart[25]. Needless to say the assumptions we have made in deriving (70) are much less satisfactory than those of Froissart and Martin[25], but the derivation does have the good feature of showing the role of unitarity in a very clear way.

From (69) and unitarity it follows that sin δ tends to zero. In fact

$$\delta_l(\nu) = \pi n_l + 0(\ln^{-1}\nu) \tag{71}$$

The asymptotic value of η_l depends on the parameters of (69). If Im A_l tends to zero, then η_l tends to 1; more generally, η_l goes to a constant less than one. For other models in which δ_l tends to an integral multiple of π the reader is referred to the book by R. J. Eden[13].

4 Regularity of *N/D* Equations

For the regularity question the first step is to find the asymptotic behaviour of the quantities $\mathscr{D}_+(\nu)$ and $C(\nu) = \text{Re}\, B_+(\nu)$ as defined in (7) and (5). For this we use the following theorem[26]: if $\phi(x) = 0(x^{-1} \ln^{\alpha} x)$, $\phi'(x) = 0(x^{-2} \ln^{\alpha} x)$, and $\phi'(x)$ is continuous, then

$$tP \int_{x_0}^{\infty} \frac{\phi(x)\,dx}{t - x} = \int_{x_0}^{t} \phi(x)\,dx + 0(\ln^{\alpha} t), \quad t \to \infty \tag{72}$$

Equation (72) may be applied to the exponent of $\mathscr{D}_+(\nu)$ with $\phi(\nu) = \delta(\nu)/(\nu - \nu_0)$. Here and in the rest of this section we assume the Regge behaviour (62, 63) for δ and η. According to (72) and (62), (63), the real part of the exponent of $\mathscr{D}_+(\nu)$ has the form

$$n\int^{\nu}\frac{\mathrm{d}x}{x}+0(1)\sim n\ln\nu \tag{73}$$

Hence

$$\mathscr{D}_+(\nu)\sim c\nu^n,\quad \delta(\infty)=n\pi \tag{74}$$

where c is a non-zero constant.

The approach of δ to its limit is sufficiently rapid (as $\ln^{-2}\nu$) to give a precise power behaviour of $\mathscr{D}$. Had the approach been as $\ln^{-1}\nu$, there would have been a power of $\ln\nu$ as a factor on the right-hand side of (74).

By using (4) and (5) in the partial wave dispersion relation (3), we obtain an expression for the force function $C(\nu)$ in terms of physical quantities:

$$C(\nu)=\frac{\eta(\nu)\sin 2\delta(\nu)}{2\rho(\nu)}-a-(\nu-\nu_0)\left[\sum_{i=1}^{n_b}\frac{c_i}{\nu_{bi}-\nu}+\frac{P}{\pi}\int_0^{\infty}\frac{\eta(\mu)\sin^2\delta(\mu)\,\mathrm{d}\mu}{\rho(\mu)(\mu-\nu_0)(\mu-\nu)}\right] \tag{75}$$

The theorem (72) shows that the last term of the expression has the behaviour $b+0(\ln^{-3}\nu)$ where b is a constant. Hence

$$C(\nu)=C(\infty)-(\pi/2)\gamma\ln^{-2}\nu+0(\ln^{-3}\nu)$$

$$C(\infty)=-a+\sum_{i=1}^{n_b}c_i-\lim_{\nu\to\infty}\frac{\nu}{\pi}P\int_0^{\infty}\frac{\eta(\mu)\sin^2\delta(\mu)\,\mathrm{d}\mu}{\rho(\mu)(\mu-\nu_0)(\mu-\nu)} \tag{76}$$

We shall also need the behaviour of $C'(\nu)$ which is obtained from a theorem resembling (72): if $\psi(x)=0(\ln^{\alpha}x)$, $\psi'(x)=0(x^{-1}\ln^{\alpha-1}x)$, $\psi''(x)=0(x^{-2}\ln^{\alpha-1}x)$, and $\psi''(x)$ is continuous, then

$$\frac{\mathrm{d}}{\mathrm{d}t}\left[t\,P\int_{x_0}^{\infty}\frac{\psi(x)\,\mathrm{d}x}{x(t-x)}\right]=0\left(\frac{\ln^{\alpha}t}{t}\right),\quad t\to\infty \tag{77}$$

To prove (77) we note that the left side is equal to

$$\frac{\mathrm{d}}{\mathrm{d}t}P\int_{x_0/t}^{\infty}\frac{\psi(tu)\,\mathrm{d}u}{u(1-u)}=$$

$$=\frac{\mathrm{d}}{\mathrm{d}t}\left[\int_{x_0/t}^{\infty}\frac{\psi(tu)-\psi(t)}{u(1-u)}\mathrm{d}u+\psi(t)P\int_{x_0/t}^{\infty}\frac{\mathrm{d}u}{u(1-u)}\right] \tag{78}$$

By direct calculation one finds that the second term of (78) is $0(t^{-1}\ln^{\alpha} t)$. The derivative may be taken under the integral sign of the first term, since the differentiated integrand is continuous in both t and u. Thus the first term of (78) is

$$\int_{x_0/t}^{\infty} \frac{\mathrm{d}}{\mathrm{d}t}\left[\frac{\psi(tu)-\psi(t)}{u(1-u)}\right] \mathrm{d}u + \frac{1}{t}\left[\frac{\psi(x_0)-\psi(t)}{1-x_0/t}\right] \tag{79}$$

In (79) the part of the integral for small u may be treated as follows, with $\varepsilon < 1$:

$$\begin{aligned}\int_{x_0/t}^{\varepsilon} \frac{u\psi'(tu)-\psi'(t)}{u(1-u)}\,\mathrm{d}u &= \frac{1}{t}\int_{x_0}^{\varepsilon t}\frac{\psi'(x)\,\mathrm{d}x}{1-x/t} - \psi'(t)\int_{x_0/t}^{\varepsilon}\frac{\mathrm{d}u}{u(1-u)}\\ &= \frac{1}{t}\int_{x_0}^{\varepsilon t}\psi'(x)\left(1+\frac{x}{t-x}\right) + 0(t^{-1}\ln^{\alpha} t)\\ &= \frac{-\psi(x_0)}{t} + \frac{1}{t^2}\int_{x_0}^{\varepsilon t}\frac{x\psi'(x)\,\mathrm{d}x}{1-x/t} + 0(t^{-1}\ln^{\alpha} t)\end{aligned} \tag{80}$$

The integral in the last line of (80) is less in magnitude than

$$\frac{M}{t^2(1-\varepsilon)}\int_{x_0}^{\varepsilon t} \ln^{\alpha-1}x\,\mathrm{d}x = 0(t^{-1}\ln^{\alpha} t) \tag{81}$$

To find the behaviour of the integral in (81) we computed the limit of its ratio with $t\ln^{\alpha-1}t$ by l'Hospital's rule. To handle the integral $\int_{\varepsilon}^{\infty}$ in (79) the following identity is useful:

$$\frac{\mathrm{d}}{\mathrm{d}t}\left[\frac{\psi(tu)-\psi(t)}{u-1}\right] = \frac{1}{t(u-1)}\int_{t}^{tu}[\psi'(v)+v\psi''(v)]\,\mathrm{d}v \tag{82}$$

By using the assumed bounds on ψ' and ψ'' in the last line of (82), one finds a convenient bound on the derivative:

$$\left|\frac{\mathrm{d}}{\mathrm{d}t}\left(\frac{\psi(tu)-\psi(t)}{u-1}\right)\right| \leqslant \frac{M}{t}\left|\frac{\ln^{\alpha}tu - \ln^{\alpha}t}{u-1}\right| \tag{83}$$

Hence

$$\begin{aligned}\left|\int_{\varepsilon}^{\infty}\frac{\mathrm{d}u}{u}\frac{\mathrm{d}}{\mathrm{d}t}\left[\frac{\psi(tu)-\psi(t)}{u-1}\right]\right| &\leqslant M\frac{\ln^{\alpha}t}{t}\int_{\varepsilon}^{\infty}\left|\frac{(1+\ln u/\ln t)^{\alpha}-1}{u-1}\right|\frac{\mathrm{d}u}{u}\\ &= 0(t^{-1}\ln^{\alpha} t)\end{aligned} \tag{84}$$

We now have all terms bounded by $t^{-1}\ln^{\alpha} t$ except for the terms $\pm\psi(x_0)/t$ in (79) and (80), which cancel. The theorem (77) follows.

To apply the result (77) to the integral appearing in the expression (75) for $C(\nu)$, we take $\psi(\nu) = \eta \sin^2 \delta = 0(\ln^{-4}\nu)$. Then according to (63) $\psi'(\nu) = 0(\nu^{-1}\ln^{-5}\nu)$, $\psi''(\nu) = 0(\nu^{-2}\ln^{-5}\nu)$, and ψ'' is continuous. It follows that

$$C'(\nu) = 0(\nu^{-1}\ln^{-3}\nu) \tag{85}$$

since the integral and the two other terms in C' separately are $0(\nu^{-1}\ln^{-3}\nu)$.

We now turn to the question of regularity (i.e. Fredholm character) of the N/D equation (16). First assume that the elasticity η has no zero. We regard $\phi(\nu) = -[\eta(\nu)/\rho(\nu)]^{1/2} \operatorname{Im} D_+(\nu)/(\nu - \nu_0)$ as the unknown function, so that the kernel is symmetrical:

$$\phi(\nu) = \left[\frac{\rho(\nu)}{\eta(\nu)}\right]^{1/2} \frac{a + C(\nu)}{\nu - \nu_0} + \frac{1}{\pi}\int_0^\infty \left[\frac{\rho(\nu)\rho(\mu)}{\eta(\nu)\eta(\mu)}\right]^{1/2} \frac{C(\nu) - C(\mu)}{\nu - \mu}\phi(\mu)\,\mathrm{d}\mu \tag{86}$$

We have assumed that η and δ satisfy the Hölder continuity condition on every finite interval, from which it follows that $C(\nu)$ is also Hölder-continuous (reference 6, appendix B). Then the kernel has at worst a weak-singularity[27] at $\nu = \mu$, so that as far as the point $\nu = \mu$ is concerned, the Fredholm theory applies to (86). The only other questionable points are at infinity. Since ρ and η both tend to 1 at infinity, the Fredholm theory applies if the following integrability conditions are met:

$$\int_{\bar{\nu}}^\infty \mathrm{d}\nu \left[\frac{a + C(\nu)}{\nu - \nu_0}\right]^2 < \infty \tag{87a}$$

$$\int_{\bar{\nu}}^\infty \int_{\bar{\nu}}^\infty \mathrm{d}\nu\,\mathrm{d}\mu \left[\frac{C(\nu) - C(\mu)}{\nu - \mu}\right]^2 < \infty \tag{87b}$$

The lower limit $\bar{\nu}$ is arbitrary, but it is chosen greater than 1 for convenience. Since C is bounded, (87a) clearly holds. To handle the double integral the following bound is effective:

$$\begin{aligned}\left|\frac{C(\nu) - C(\mu)}{\nu - \mu}\right| &= \left|\int_0^1 C'(\mu + u(\nu - \mu))\,\mathrm{d}u\right| \\ &\leqslant M\left|\int_0^1 \frac{\ln^{-3}[\mu + u(\nu - \mu)]}{\mu + u(\nu - \mu)}\,\mathrm{d}u\right| \\ &= M\left|\frac{\ln^{-2}\nu - \ln^{-2}\mu}{\nu - \mu}\right|\end{aligned} \tag{88}$$

In the μ integral one may put $\mu = \nu u$ in order to see that our double integral is bounded by

$$\int_{\bar{\nu}}^{\infty} \frac{d\nu}{\nu} \int_{\bar{\nu}/\nu}^{\infty} du \left[\frac{\ln^{-2}\nu - \ln^{-2}(\nu u)}{1-u}\right]^2 \tag{89}$$

The u integral of (89) is now to be broken into two parts, which may be bounded at large ν as follows:

$$\int_{\varepsilon}^{\infty} = \ln^{-4}\nu \int_{\varepsilon}^{\infty} du \left[\frac{1-(1+\ln u/\ln \nu)^{-2}}{1-u}\right]^2 = 0(\ln^{-4}\nu) \tag{90}$$

$$\int_{\bar{\nu}/\nu}^{\varepsilon} \leqslant M \int_{\bar{\nu}/\nu}^{\varepsilon} du\, [\ln^{-2}\nu - \ln^{-2}(\nu u)]^2 \tag{91}$$

The division point ε is less than 1, and ν is taken so large that $1 + \ln u/\ln \nu$ in (90) cannot vanish in the region where $\ln u$ is negative. To evaluate the right side of (91) consider

$$\int_{a/\nu}^{\varepsilon} \ln^{-n}(\nu u)\, du = \nu^{-1} \int_{a}^{\nu\varepsilon} \ln^{-n} x\, dx, \quad n = 2, 4 \tag{92}$$

The behaviour of the x integral is ascertained through l'Hospital's rule by computing the limit of its ratio with $\nu \ln^{-n}\nu$. In fact that limit is a non-zero constant, from which it follows that (91) is $0(\ln^{-4}\nu)$. Thus for the ν integral we have convergence at least as

$$\int^{\infty} \frac{d\nu}{\nu \ln^4 \nu} < \infty \tag{93}$$

Since the bounds we have used are all rather close, there is not much leeway. The L^2 property of the kernel is a fairly delicate matter, depending as it does on logarithms for convergence.

In the case of class B amplitudes, where the D function requires CDD poles, the integral equation is (21). In Regge theory the α factor is 1, because convergence of the integral in (20) with $\alpha = 1$ follows from the known convergence of the integral in (19). This is true because $\text{Re}\, B_+(\nu) = C(\nu)$ tends to a constant $C(\infty)$. The constant $\Lambda(\infty)$ that appears in the integral equation is easily calculated as $\lim \Lambda_+(\nu)$ from the definition (20) if one applies (61), (76) and (19). The result is

$$\Lambda(\infty) = -AC(\infty) \tag{94}$$

Hence if $A \neq 0$ the inhomogeneous term of the integral equation contains $A[C(\nu) - C(\infty)]$ which is not square-integrable according to (76). One can redefine the unknown function, however, so that the inhomogeneous term becomes L^2 and the kernel remains L^2. After multiplication by $\nu^{-1/2}\ln(\nu-\nu_0)$ the equation takes the form

$$\phi(\nu)=\frac{1}{\eta(\nu)}\frac{\ln(\nu-\nu_0)}{\nu^{1/2}}\left[\frac{a+C(\nu)}{\nu-\nu_0}+A[C(\nu)-C(\infty)]+\right.$$

$$\left.+\sum_i c_i\frac{C(\mu_i)-C(\nu)}{\mu_i-\nu}\right]+\frac{1}{\pi}\int_0^\infty \mathrm{d}\mu\,\frac{\rho(\mu)}{\eta(\nu)}\left(\frac{\mu}{\nu}\right)^{1/2}\frac{\ln(\nu-\nu_0)}{\ln(\mu-\nu_0)}\left[\frac{C(\nu)-C(\mu)}{\nu-\mu}\right]\phi(\mu) \tag{95}$$

The L^2 property of the inhomogeneous term is now obvious, but we must check up on the kernel using the method of equations (88) ff. Instead of (89) we have

$$\int_{\bar{\nu}}^\infty \mathrm{d}\nu\,\frac{\ln^2(\nu-\nu_0)}{\nu}\int_{\bar{\nu}/\nu}^\infty\frac{u\,\mathrm{d}u}{\ln^2(\nu u-\nu_0)}\left[\frac{\ln^{-2}\nu-\ln^{-2}(\nu u)}{1-u}\right]^2 \tag{96}$$

In the region $\varepsilon\leqslant u<\infty$ we can employ the bound $\ln^{-2}(\nu u-\nu_0)\leqslant\ln^{-2}(\bar{\nu}u-\nu_0)$ if we take $\bar{\nu}$ so that $\bar{\nu}\varepsilon>\nu_0$. Thus in place of (90) one has

$$\int_\varepsilon^\infty\leqslant\ln^{-4}\nu\int_\varepsilon^\infty\frac{u\,\mathrm{d}u}{\ln^2(\bar{\nu}u-\nu_0)}\left[\frac{1-(1+\ln u/\ln\nu)^{-2}}{1-u}\right]^2=0(\ln^{-4}\nu) \tag{97}$$

Since $u\ln^{-2}(\nu u-\nu_0)$ is bounded, (91) still holds. Hence the u integral is $0(\ln^{-4}\nu)$ as before, and the ν integral converges at least as

$$\int^\infty \mathrm{d}\nu\,\nu^{-1}\ln^{-2}\nu<\infty$$

We conclude that if δ and η have the behaviour of (62) and (63), and η has no zero, then the N/D integral equation is of Fredholm type in the space L^2. This statement is clearly true for (26) and (28), as well as for (16) and (21) discussed above.

Now suppose that the elasticity vanishes at one finite point $\bar{\nu}$: $\eta(\bar{\nu})=0$. By defining η so that it changes sign as ν passes through $\bar{\nu}$, we have made sure that the N/D integral equation has the same algebraic form as in the case of non-vanishing η. Since η occurs in the denominator, the kernel is not L^2 at $\nu=\bar{\nu}$. We can find a related equation which does have an L^2 kernel and inhomogeneous term. Suppose that the amplitude is of class A; then (16) implies

$$0=\frac{a+C(\bar{\nu})}{\bar{\nu}-\nu_0}+\frac{1}{\pi}\int_0^\infty\frac{C(\bar{\nu})-C(\mu)}{\bar{\nu}-\mu}\rho(\mu)\phi(\mu)\,\mathrm{d}\mu \tag{98}$$

because $\phi(\mu)$ is bounded at $\mu = \bar{\nu}$. If (98) is multiplied by $(\bar{\nu} - \nu_0)/(\nu - \nu_0)$ and subtracted from (16), we have an integral equation with better behaviour at $\nu = \bar{\nu}$:

$$\phi(\nu) = \frac{1}{\eta(\nu)} \frac{C(\nu) - C(\bar{\nu})}{\nu - \nu_0}$$
$$+ \frac{1}{\pi} \int_0^\infty \frac{1}{\eta(\nu)} \left[\frac{C(\nu) - C(\mu)}{\nu - \mu} - \frac{\bar{\nu} - \nu_0}{\nu - \nu_0} \frac{C(\bar{\nu}) - C(\mu)}{\bar{\nu} - \mu} \right] \rho(\mu)\phi(\mu)\, \mathrm{d}\mu \quad (99)$$

The purpose of the factor $(\bar{\nu} - \nu_0)/(\nu - \nu_0)$ is to provide good behaviour of the modified kernel and inhomogeneous term as ν tends to infinity. In (99) the kernel and inhomogeneous term are bounded at $\nu = \bar{\nu}$, provided C has a continuous first derivative at that point. The latter condition is met if η has a continuous second derivative near $\nu = \bar{\nu}$, which we assume. Then the question of Fredholm character of (99) is essentially the same as that for the original equation. The squares of the two terms in the kernel are clearly L^2, and the cross term is seen to be L^2 by applying Schwarz's inequality.

Any solution of the original equation is also a solution of the new equation (99) if $\eta(\bar{\nu}) = 0$. But when can one be sure that any solution of (99) is also a solution of (16)? The question is easily answered if (16) has an L^2 solution, and (99) a unique L^2 solution. In that case, the solution of (99) is also a unique L^2 solution of (16). If an amplitude with $\delta(\infty) = -\pi n_b$ exists, then the corresponding D function provides an L^2 solution of (16) since $\phi(\nu) = 0(\nu^{-1+\varepsilon})$, all $\varepsilon > 0$. Hence we can find the amplitude by solving (99), provided only that the homogeneous form of the latter has no non-trivial solution. Similar statements apply to class B amplitudes.

A zero of η right on the physical cut is to be considered a case of 'measure zero'—purely an accident. The S-matrix element $S = 1 + 2i\,\rho A$ might have a zero close to the physical cut, however, in which case η might become quite small at a physical energy. In that case (99) might be a useful idealization of the actual situation. Equation (16) is probably difficult to solve numerically when η nearly vanishes. In approximating (16) by (99) one would have an easier numerical problem.

5 Regge Trajectory Exchange and Properties of the Elasticity Function

Probably the main interest in N/D equations has to do with their role in dynamical models. Does the discussion of the preceding sections have any value toward the construction of such models, or does it apply only to a hypothetical complete Regge theory which we do not know how to construct? In this Section, I would like to suggest a form of the Regge trajectory exchange model in which the discussion of Section 4 is directly relevant.

It is conceivable that a good approximation to the left-hand singularities of $A(z)$ is given by the left-hand singularities of a small number of Regge pole terms. Such an approximation, which can take several forms, is what we shall mean by the 'Regge trajectory exchange model'. For about five years this idea has been pursued in one form or another by several physicists. Chew and others, in particular, have devoted a good deal of work to the subject[28]. Recently Bali, Chew and Chu[29] have proposed that a Mandelstam iteration with a certain kind of cut-off might prove to be a better basis for a Reggeized bootstrap theory. These latter authors have advanced some arguments against the trajectory exchange model. Indeed the latter was not very successful in a quantitative way in the particular formulations studied. I think, however, that the results will depend strongly on the detailed formulation of the dynamical equations. The model outlined here differs substantially from those studied earlier, and there will no doubt be corresponding differences in its predictions. It is more phenomenological, more flexible and less ambitious than the models of Chew and others[28].

In our proposal the main part of the left-hand singularity function $A^L(z)$ is given by an approximate partial wave projection of leading Regge pole terms associated with t- and u-channel trajectories. The projection is approximate in that the integration over $\cos\theta$ is restricted to fixed regions of t and u, as in Section 3: $-T \leqslant t \leqslant 0$, $-U \leqslant u \leqslant 0$. The projection is from the full Regge pole term, not from its asymptotic form. For instance, we take the term arising from the left singularities of the following function:

$$\frac{1}{s-s_0}\int_{-T}^{0} \mathrm{d}t\; \theta(t+s-s_0) P_l\left(1+\frac{2t}{s-s_0}\right)\frac{\beta(t)}{\sin \pi\alpha(t)} \times \\ \times \left[P_{\alpha(t)}\left(-1-\frac{2s}{t-t_0}\right) \pm P_{\alpha(t)}\left(1+\frac{2s}{t-t_0}\right)\right] \qquad (100)$$

The sum of such terms from the leading t- and u-trajectories will be denoted by $A_R{}^L(z)$. We hope that T may be chosen small enough so that the trajectory and reduced residue are well approximated by rather simple functions that can be partially inferred from experiment. At the same time we would like T to be large enough so that the restriction of the full range of integration $-(s-s_0) \leqslant t \leqslant 0$ to the fixed interval $-T \leqslant t \leqslant 0$

is not a serious error. The discontinuity over the left cut obtained in this way may not agree in detail with that of the untruncated projection of the Regge term, but that need not be a difficulty if the truncated and untruncated left singularity terms $A_R{}^L$ agree closely when they are evaluated in the physical region $s \geqslant s_0$. The N/D equation has the nice property of referring only to the physical region, so it is necessary to approximate $A^L(z)$ only in that region.

The elasticity function η will also be computed from our Regge trajectory exchange, but only at sufficiently high energies. According to (61) the sum of the Regge terms can be regarded as a unitary scattering amplitude at high energies. From this unitary amplitude one can read off a value of η for input to the N/D equation. What to use for η at lower energies is a problem that will require some study. One simple model that seems reasonable is that of 'resonance saturation' of the integral in (5); i.e. we would take $\eta = 1$ in the low energy region except near resonances where $1 - \eta$ is expected to have a sharp peak. The peak is seen from the unitarity formula

$$\eta^2 = 1 - 4\rho_1 \sum_{i \neq 1} |T_{+1i}|^2 \rho_i \tag{101}$$

which expresses the elasticity of channel 1 in terms of a sum over transitions to and from all other open channels. All of the T-matrix elements T_{+1i} possess a common resonance pole, so we get a dip in η^2 at the physical resonance energy. For a narrow resonance $\eta^2 - 1$ should have a Breit–Wigner form describable by two parameters:

$$1 - \eta^2 = \frac{\gamma}{(s - s_*)^2 + \Gamma^2/4}, \quad 0 < \gamma \leqslant \Gamma^2/4 \tag{102}$$

There is a potentially useful consistency requirement between the input width Γ and the width of the resonance computed from the solution of the N/D equation. By demanding equality of these widths, it may be possible to achieve a more accurate treatment of the resonance narrowing due to inelastic processes. Up to now, resonance narrowing has been studied in two ways: either in coupled channel models with a small number of channels (two or three) computed explicitly[30], or else by means of our equation (16) but with a function lacking the resonance dip[31]. In the former method it is likely that the number of channels is insufficient to give an accurate picture of the narrowing, especially for a high-energy resonance like the ρ in π-π scattering. The second method is probably unsatisfactory if the constant γ in (102) is appreciable. Since η occurs in the denominator of the kernel one can expect a strong effect on the solution of the integral equation if there is a pronounced dip in η.

Specification of η does not amount in general to a complete description of the effect of inelastic channels. The inelastic channels can give rise to CDD poles, as was noted following (21), and the associated CDD parameters must be specified.

If the residue of a CDD pole is sufficiently small there will be a resonance or stable particle close to the energy at which the pole is located. Here I refer to the poles at finite energies μ_i in the first formulation (19) of CDD poles. By means of these poles account is taken of any resonances not generated by the Regge trajectory exchange in the N/D equation. This picture reminds one of the rather successful model of Barger and Cline[32] and others in which the Regge pole terms of the invariant amplitudes are simply added to Breit–Wigner amplitudes for direct channel resonances. The sum is fitted directly to experimental data. The fits are often satisfactory down to surprisingly low energies. The Barger–Cline model has the defect of involving Breit–Wigner amplitudes evaluated at energies far away from their resonance maxima. Our model will predict a certain non-Breit–Wigner behaviour away from the resonance positions. The success of the Barger–Cline model, and the prospects of improving it by ensuring unitarity, correct analyticity, and a better treatment of resonances, may be construed as encouragement for undertaking the numerical calculations required in our proposal.

The quantitative mathematical interplay of the CDD poles and the dips in the η function is a subject worth studying in itself apart from the particular model of Regge trajectory exchange.

To check on the matter of Fredholm character of the N/D integral equation, we note first that our approximate partial wave projection $A_R(z)$ of the leading Regge terms satisfies a once-subtracted dispersion relation:

$$A_R(z) = A_R(\nu_0) + \frac{z - \nu_0}{\pi}\left[\int_{-\infty}^{-1} + \int_0^{\infty}\right] \frac{\operatorname{Im} A_{R+}(\nu)\, \mathrm{d}\nu}{(\nu - \nu_0)(\nu - z)} \tag{103}$$

The input to the N/D equation in the trajectory exchange model is

$$C(\nu) = \operatorname{Re} B_{R+}(\nu) = \frac{\nu - \nu_0}{\pi}\left[\int_{-\infty}^{-1} \frac{\operatorname{Im} A_{R+}(\mu)\, \mathrm{d}\mu}{(\mu - \nu_0)(\mu - \nu)} + \right.$$

$$\left. + P\int_0^{\infty} \frac{[1 - \eta(\mu)]\, \mathrm{d}\mu}{2\rho(\mu)(\mu - \nu_0)(\mu - \nu)}\right] \tag{104}$$

At sufficiently large positive ν it is possible to write Im A_{R+} in the form

$$\operatorname{Im} A_{R+} = \frac{\eta \sin^2 \delta_R}{\rho} + \frac{1-\eta}{\rho}, \quad \eta^2 \leqslant 1 \tag{105}$$

where δ_R is a real function. The η functions of (104) and (105) coincide for large ν, say for all ν greater than some $\bar{\nu}$. The values of η for $\nu < \bar{\nu}$ in (104) are immaterial for the question of Fredholm character of the integral equation provided we require that η is Hölder-continuous on any finite interval. According to (62) the first and second terms of (105) are $0(\ln^{-4}\nu)$ and $0(\ln^{-1}\nu)$, respectively, so the integrands of the right cut integrals in (103) and (104) are asymptotically equal. Their difference vanishes as $\ln^{-4}\nu$. From this we can now show that $C(\nu)$ and Re $A_{R+}(\nu)$ have the same asymptotic behaviour up to an additive constant; i.e.,

$$C(\nu) = C(\infty) - (\pi/2)\gamma \ln^{-2}\nu + 0(\ln^{-3}\nu) \tag{106}$$

$$\operatorname{Re} A_{R+}(\nu) = -(\pi/2)\gamma \ln^{-2}\nu + 0(\ln^{-3}\nu), \quad \nu \to \infty \tag{107}$$

This follows from the asymptotic theorem stated in (72) as is seen by rewriting (103) and (104) as follows:

$$\operatorname{Re} A_{R+}(\nu) = \frac{\nu-\nu_0}{\pi}\left\{\int_{-\infty}^{-1} \frac{\operatorname{Im} A_{R+}(\mu)\,\mathrm{d}\mu}{(\mu-\nu_0)(\mu-\nu)} + P\int_{\bar{\nu}}^{\infty} \frac{[1-\eta(\mu)]\,\mathrm{d}\mu}{2\rho(\mu)(\mu-\nu_0)(\mu-\nu)}\right\} + \left[A_R(\nu_0) + \frac{\nu-\nu_0}{\pi}\int_0^{\bar{\nu}} \frac{\operatorname{Im} A_{R+}(\mu)\,\mathrm{d}\mu}{(\mu-\nu_0)(\mu-\nu)} + \frac{\nu-\nu_0}{\pi} P\int_{\bar{\nu}}^{\infty} \frac{\eta(\mu)\sin^2\delta_R(\mu)\,\mathrm{d}\mu}{(\mu-\nu_0)(\mu-\nu)}\right] \tag{108}$$

$$C(\nu) = \frac{\nu-\nu_0}{\pi}\left\{\int_{-\infty}^{-1} \frac{\operatorname{Im} A_{R+}(\mu)\,\mathrm{d}\mu}{(\mu-\nu_0)(\mu-\nu)} + P\int_{\bar{\nu}}^{\infty} \frac{[1-\eta(\mu)]\,\mathrm{d}\mu}{2\rho(\mu)(\mu-\nu_0)(\mu-\nu)}\right\} + \frac{\nu-\nu_0}{\pi}\int_0^{\bar{\nu}} \frac{[1-\eta(\mu)]\,\mathrm{d}\mu}{2\rho(\mu)(\mu-\nu_0)(\mu-\nu)} \tag{109}$$

According to (72) the second square bracket in (108) has the form $c + 0(\ln^{-3}\nu)$, $c =$ constant, at large ν. On the other hand we know from (61) that Re $A_{R+}(\nu)$ has the form (107), so the first square bracket of (108) has the form

$$-c - (\pi/2)\gamma \ln^{-2}\nu + 0(\ln^{-3}\nu)$$

By comparison with (109) the desired result (106) follows. The first derivatives of C and Re A_{R+} may be compared in a similar way by means of the methods of Section 4. As in Section 4 we find $C'(\nu) = 0(\nu^{-1}\ln^{-3}\nu)$

which together with (106) implies the Fredholm property of the integral equation.

An advantage of the scheme just described over the usual non-Regge bootstrap models is that it avoids the mysterious and probably meaningless cut-off procedures. It shares the defect of those models of not furnishing the correct threshold behaviour of partial waves without an *ad hoc* modification. A modification is easily made by applying the N/D procedure to the function

$$G(z) = \frac{P(z)}{z^l} A(z) = \frac{N(z)}{D(z)} \tag{110}$$

where $P(z)$ is a polynomial of order l with real coefficients. Since $G(0)$ computed as $N(0)/D(0)$ will normally be finite and non-zero, we achieve the correct threshold behaviour $A(\nu) \sim c\nu^l$, $\nu \to 0$. The only modifications in the N/D equations are expressed by the replacements

$$\begin{aligned} &\rho(\nu) \to \theta(\nu)\rho(\nu), \quad \theta(\nu) = \nu^l/P(\nu) \\ &\text{Im } A_+(\nu)|_L \to \theta^{-1}(\nu)\text{Im } A_+(\nu)|_L \end{aligned} \tag{111}$$

where L denotes the left cut, $\nu \leqslant -1$. Since $\theta(\nu)$ tends to a constant at large ν, the discussion of regularity of the integral equation is the same as before. The drawback of this procedure is that in general it introduces poles in $A(z)$ at the zeros of $P(z)$. If the input to the N/D equations were really correct, and correct threshold behaviour thereby guaranteed, these poles would actually be absent. Thus some kind of measure of the adequacy of the input is given by the magnitudes of the residues of these poles. One could attempt to make the poles an approximate representation of some known left-hand singularities (other than those of the trajectory exchange terms); for instance the left singularities associated with s-channel Regge trajectories[28]. In any case, one can expect a dependence of the results on the choice of the polynomial P, and that is an ambiguity which is more or less serious depending on the accuracy of the input data.

In the work of Chew, Jones, Teplitz and Collins, part of the programme was to compute the output trajectories (that is, to determine s-channel trajectories $\alpha(s)$ from the equation $D[\alpha(s), s] = 0$) and to achieve consistency between input and output. We do not advocate such a bold approach in connection with the scheme just outlined, since it would appear that our knowledge of the elasticity function η is too limited to make possible a meaningful continuation in l. Therefore our scheme is not a complete Reggeized bootstrap, but only an attempt to find an approximate description of the left-hand singularities, the elasticity function, and the direct channel resonances. Regge trajectories and residues are merely taken as input. Since we favour flexibility in the choice of η, and inclusion of CDD poles, any conclusions that were made about the validity of the trajectory exchange model in the Chew–Jones framework must now be reexamined. We think that this is worth doing before returning to the Mandelstam iteration, which has difficulties of its own. The type of iteration with cut-off proposed in reference 29 has the property of completely damping out the elastic term in the unitarity condition at energies above the s-strip boundary. This seems contrary to a proper Regge theory, in which the elastic term vanishes faster than the inelastic term by only one power of $\ln s$.

Acknowledgements

I am very grateful to Professors H. H. Aly and A. B. Zahlan for the opportunity to visit the American University of Beirut, Lebanon. I also wish to thank the International Atomic Energy Agency and Professors A. Salam and P. Budini for their hospitality and support of my work at the International Centre for Theoretical Physics, Trieste. A correction to the manuscript of this paper was kindly provided by Dr. M. Stihi, Bucharest.

References

1. Chew, G. F., and Mandelstam, S., *Phys. Rev.*, **119**, 467 (1960).
2. There are, of course, innumerable papers in which the N/D method is applied to the bootstrap model. For a review see Zachariasen, F., in *Strong Interactions and High Energy Physics* (Ed. R. G. Moorhouse), Plenum Press, New York, 1964. For more recent work see articles in *Phys. Rev.* and *Nuovo Cimento* by the following authors, among others: Balázs, Ball, Capps, Carruthers, Chan Hong-mo, Cutkosky, Dashen, Diu, Frautschi, Hara, Hwa, Martin, Patil, Rajasekaran, Rubenstein, Sawyer, Singh, Suzuki, Udgaonkar, Wali, Warnock, Wong. For N/D methods incorporating Regge theory, see reference 30.
3. Mandelstam, S., *Ann. Phys.*, **21**, 302 (1963).
4. Warnock, R. L., *Phys. Rev.*, **131**, 1320 (1963).
5. Warnock, R. L., *Nuovo Cimento*, **50**, 894 (1967), and **52A**, 637 (1967). For mathematical theory related to the N/D method, see Muskhelishvili, N. I., *Singular Integral Equations*, P. Noordhoff Ltd., Groningen, 1953; Plemelj, J., *Problems in the Sense of Riemann and Klein*, Interscience, New York, 1964; Vekua, N. P., *Systems of Singular Integral Equations*, P. Noordhoff Ltd., Groningen, 1967; Pogorzelski, W., *Integral Equations and their Applications*, Vol. I, Pergamon Press, New York, 1966.

6. Frye, G., and Warnock, R. L., *Phys. Rev.*, **130,** 478 (1963).
7. Atkinson, D., and Contogouris, A. P., *Lawrence Radiation Laboratory Report UCRL*-17364 to be published in *J. Math. Phys.*
8. Ling-lie Chan Wang, *Phys. Rev.*, **142,** 1187 (1966).
9. Warnock, R. L., *Phys. Rev.*, **146,** 1109 (1966).
10. Taylor, J. G., *Nuovo Cimento*, **22,** 92 (1961); Nakanishi, N., *Phys. Rev.*, **126,** 1225 (1962).
11. Kinoshita, T., *Phys. Rev.*, **154,** 1438 (1967).
12. I wish to thank Professor Kinoshita for correspondence about this point.
13. Eden, R. J., *High Energy Collisions of Elementary Particles*, Cambridge University Press, London, 1967.
14. A proof that the polynomial P may be chosen so as to yield the representation (19) is given in Section 3 of reference 4. In the proof there is a misstatement about the signs of residues of particle poles following equation (3.19). Instead of $R_i > 0$ one should have $(-1)^l R_i > 0$. If there is a stable particle (bound state) then the phase is negative just above threshold, and consequently $\varepsilon = 1$ in the notation of reference 4. Thus we obtain the required number $n_b - \varepsilon$ of zeros of A using only the inequality $(-1)^l R_i > 0$. Also in reference 3 the statement about square-integrability of $C(\nu)$ just before equation (3.18) is false but irrelevant in view of the present paper. In equation (3.18) there is a misprint: $N\text{-}BD$ must be divided by $z - \nu_0$. In equation (3.15), Re $B(z)$ should read Re $B(\nu)$.
15. Castillejo, L., Dalitz, R. H., and Dyson, F. J., *Phys. Rev.*, **101,** 453 (1956).
16. A list of references is to be found in footnote 14 of reference 9. See also Atkinson, D., Dietz, K., and Morgan, D., *Ann. Phys.*, **37,** 77 (1966).
17. In a particular field-theoretical model there may be natural locations of the CDD poles corresponding to poles of the determinant $\det[(E - H)/(E - H_0)]$ where E is the energy and H and H_0 are the perturbed and unperturbed Hamiltonians. See Vaughn, M. T., Aaron, R., and Amado, R. D., *Phys. Rev.*, **124,** 1258 (1961). This determinant can differ, however, from our D function by a multiplicative rational function at least.
18. Sugawara, M., and Kanazawa, A., *Phys. Rev.*, **126,** 2251 (1962).
19. Present data are consistent with a vanishing Pomeranchuk slope. Van Hove, L., *Proc. XIII Intern. Conf. High Energy Phys., Berkeley*, 1966, University of California Press, Berkeley, 1967.
20. Oehme, R., in *Strong Interactions and High Energy Physics* (Ed. R. G. Moorhouse), Plenum Press, New York, 1964. By postulating a particular kind of moving branch point, it is possible to reinstate a fixed pole with $J = 1$: Oehme, R., *Phys. Rev. Letters*, **18,** 1222 (1967). At present this postulate seems artificial.
21. Mandelstam, S., *Nuovo Cimento*, **30,** 1127, 1148 (1963).
22. Gribov, V. N., Pomeranchuk, I. Ya., and Ter-Martirosyan, K. A., *Phys. Rev.*, **139B,** 184 (1965).
23. Polkinghorne, J. C., *J. Math. Phys.*, **6,** 1960 (1965); Osborne, P., and Polkinghorne, J. C., *Nuovo Cimento*, **47,** 526 (1967).
24. Van Hove, L., in *High Energy Physics and Elementary Particles*, International Atomic Energy Agency, Vienna, 1965.
25. Froissart, M., *Phys. Rev.*, **123,** 1053 (1961); Martin, A., *Phys. Rev.*, **129,** 1432 (1963).

26. This theorem is proved in the appendix of reference 4. The hypotheses of the theorem were mis-stated in reference 4 in that the condition $\phi'(x) = 0(x^{-2} \ln^{\alpha} x)$ was omitted. An attempt was made to deduce this condition from $\phi(x) = 0(x^{-1} \ln^{\alpha} x)$ which is only possible if certain oscillations at infinity are forbidden. The error occurs in the following places in reference 4: appendix, equation (2.17), before equation (2.19), footnote 18. In reference 6 the error shows up after equation (III.13), in footnote 26, and in appendix D. The result of appendix D is applied following equation (A8). These papers are corrected by adding the assumption $\phi'/\phi = 0(x^{-1})$ where necessary, and that causes very little change in the conclusions.
27. Mikhlin, S., *Integral Equations*, Pergamon Press, Oxford, 1964.
28. Chew, G. F., *Phys. Rev.*, **129,** 2363 (1963), **130,** 1264 (1963), **140B,** 1427 (1965); Chew, G. F., and Jones, C. E., *Phys. Rev.*, **135B,** 208 (1964); Jones, C. E., *Phys. Rev.*, **135B,** 214 (1964); Teplitz, D. C., and Teplitz, V. L., *Phys. Rev.*, **137B,** 142 (1965); Collins, P. D. B., and Teplitz, V. L., *Phys. Rev.*, **140B,** 663 (1965); Collins, P. D. B., *Phys. Rev.*, **142,** 1163 (1966), **157,** 1432 (1967); Chew, G. F., and Teplitz, V. L., *Phys. Rev.*, **136B,** 1154 (1964).
29. Bali, N. F., Chew, G. F., and Chu, S.-Y., *Phys. Rev.*, **150,** 1352 (1966); Bali, N. F., *Phys. Rev.*, **150,** 1358 (1966).
30. Zachariasen, F., and Zemach, C., *Phys. Rev.*, **128,** 849 (1962); Fulco, J. R., Shaw, J. L., and Wong, D. J., *Phys. Rev.*, **137B,** 1242 (1965).
31. Coulter, P. W., and Shaw, G. L., *Phys. Rev.*, **138B,** 1273 (1965), **141,** 1419 (1966).
32. Barger, V., and Cline, D., *Phys. Rev. Letters*, **16,** 913 (1966); Barger, V., *Proc. Symp. Regge Poles*, Argonne National Laboratory, December 1966, unpublished.

Some Aspects of High Energy Potential Scattering

H. H. ALY

1 Introduction

In 1958 Mandelstam put forward a very useful conjuncture which assumed that the scattering amplitude for the pion-nucleon system is essentially an analytic function in both energy and momentum transfer variables, with singularities located on the real axis of these two variables. This conjuncture led to a simple integral representation of the scattering amplitude.

Since then, a vast number of papers appeared along these lines in the hope of finding a satisfactory theoretical framework for the strongly interacting particles.

It must, however, be emphasized that the Mandelstam integral representation can only be given for the first few orders of perturbation theory. No general proof has been given in the case of covariant 2-particle scattering.

Since there are various difficulties in investigating the 2-particle relativistic scattering processes, the more successful techniques of potential theory were looked on as a testing ground in the hope of gaining an insight into some features of relativistic scattering. Both regular (e.g. Yukawa type) and singular [$V(r) \sim gr^{-n}$; $n > 3$] potential models were considered.

In these notes we shall discuss various relevant problems arising in the relativistic scattering of hadrons in the context of non-relativistic scattering by regular and singular potentials.

The interest in these potentials derives from various indications that the interaction responsible for high energy elementary particles with each other may be more singular in nature than the Yukawa type potentials. In quick succession, the renormalizable $\lambda\phi^4$-field theory (in the ladder approximation) seems to be reduced to a non-relativistic analog of r^{-2} potential[1], while a more singular potential like r^{-4} seems to correspond to the four bosons interaction

$$(g_1\bar{\psi}(\gamma_5, 1)\psi\phi + g_2\bar{\psi}(\gamma_5, 1)\psi\phi^2)$$

As a further example we also notice (in the ladder approximation) that the four fermion interactions correspond to the short range potential r^{-6}.

It has, however, been recently shown that a better analogy between quantum field theory (Bethe–Salpeter equation) and potential scattering may in fact be discussed in terms of the Lippman–Schwinger equation (L-S equation) rather than the Schrödinger equation[2]. For completeness we shall first consider very briefly what has been learnt so far in terms of the Schrödinger equation[3], and then proceed to compare the picture obtained from the L-S equation.

In what follows, we shall give some general criteria for writing down dispersion relations for the 2-particle scattering amplitude due to singular potentials. This will be preceded by giving the bounds on the scattering amplitudes as discussed by Martin. Finally we shall briefly discuss some analyticity aspects of the residue function of the Regge poles, both in the case of regular and singular potentials. Regge trajectories for the potential r^{-4} are also briefly discussed.

2 Upper Bounds on Partial Wave Amplitudes, *f(s)*

We shall follow Martin[4] in calculating the bounds imposed on the partial wave scattering amplitudes by considering first the case of purely repulsive potential $V(V)$, with a finite range R, given as

$$V_e = \mathrm{e}^{-\varepsilon/r}V(r) \tag{1}$$

then the radial Schrödinger equation is written as

$$\phi''(r) + \left[k^2 - \frac{l(l+1)}{r^2} - V_e(r)\right]\phi(r) = 0 \tag{2}$$

where the normalized ϕ in the physical region is

$$\phi = \phi_{01} + \mathrm{tag}\ \delta_l\eta_{0l}, \quad \text{for } r \geqslant R \tag{3}$$

For the case of $l > kR$, both ϕ_{0l} and η_{0l} are positive Bessel functions whenever R is to the left of the turning point of the Bessel equation. Martin shows that

$$0 < \phi(R) < \phi_{0l}(R) \tag{4}$$

Due to the fact that the effective potential, i.e. $V_{\text{eff}}(e, r)$, dominates the Hamiltonian, it follows that

$$0 \leqslant \phi(r) < \phi_{0l}(R)$$

and

$$\text{tag}\, \delta_{\text{Born}}(l, e) < \text{tag}\, \delta(l, e) < 0 = -\frac{1}{k}\int_0^R [\phi_{0l}(r)]^2 V_e \, \mathrm{d}r \tag{5}$$

and we obtain the inequality

$$-\frac{1}{k}\int_0^R [\phi_{0l}(r)]^2 V(r)\, \mathrm{d}r < \text{tag}\, \delta(l, e) < 0 \tag{6}$$

As a final remark we can easily show that the $\lim\limits_{\text{as } e \to 0} \delta(l, e)$ exists, leading to the conclusion that in the case of singular potential scattering all partial wave amplitudes are well defined and bounded.

Similar arguments are repeated for the case of $k^2 < 0$, and the following bounds are obtained by Martin:

$$\left.\begin{aligned} |f(l, e)| &< \frac{1}{|k|}\int_0^R |\phi_{0l}(r)|^2 V_e(r)\, \mathrm{d}r \\ |f(l, e)| &< \frac{1}{|k|}\int_0^R |\phi_{0l}(r)|^2 V(r)\, \mathrm{d}r \end{aligned}\right\} \tag{7}$$

Having established these bounds on the partial wave amplitudes, we may obtain in a similar fashion, with the help of (6) and (7) in the physical region $k^2 > 0$, an upper bound on the forward scattering amplitude $f_e(k^2, 0)$,

$$|f_e(k^2, \cos\theta = 1)| < \frac{1}{k}\sum_0^{l=kR} (2l+1) + \frac{1}{k^2}\sum_{kR}^{\infty}\int (2l+1)|\phi_{0l}(r)|^2 V(r)\, \mathrm{d}r \tag{8}$$

We change the sum and the integral signs; we write the Born approximation for the total scattering amplitude:

$$(A_{\text{B}})_{\text{tot}} \simeq \frac{1}{k^2}\int V(r)\left[\sum_0^{\infty}(2l+1)|\phi_{0l}(r)|^2\right] \mathrm{d}r \tag{9}$$

where

$$\sum_0^{\infty}(2l+1)|\phi_{0l}(r)|^2 = \tfrac{1}{2}\int_{-1}^{+1} k^2 r^2\, \mathrm{e}^{ikr\cos\theta}\, \mathrm{e}^{-ikr\cos\theta}\, \mathrm{d}(\cos\theta) = k^2 r^2 \tag{10}$$

It is now evident that

$$|f_e(k^2, \cos\theta = 1)| < \frac{(kR + 1)^2}{k} + \int_0^R r^2 V(r)\, \mathrm{d}r \tag{11}$$

Equation (11) shows that a bound independent of e and growing like k as $k \to \infty$, can be imposed on the forward scattering amplitude. This bound was obtained with the help of unitary condition

$$|\mathrm{e}^{\mathrm{i}\delta_l(k)} \sin \delta_l(k)| < 1$$

In a similar manner a bound independent of e and with a finite value for a finite negative k^2 can be obtained for the partial wave amplitudes for $k^2 < 0$. Here, however, we must point out that although the S-matrix is not unitary the inequality given by (7) still holds, so we write:

$$f_e(k^2 < 0, \cos\theta = 1) < \frac{1}{|k^2|} \int V(r) \sum_0^\infty (2l + 1)|\phi_{0l}(r)|^2\, \mathrm{d}r \tag{12}$$

Again we may use the expansion of the plane wave in partial wave terms as given by (10), provided we replace k^2 by $|k^2|$. Then we get in a similar manner

$$|f_e(k^2 < 0, \cos\theta = 1)| < \exp(2|k|R) \int_0^R V(r) r^2\, \mathrm{d}r \tag{13}$$

3 Dispersion Relations for Singular Potential

As was pointed out before, the Mandelstam representation was proved for the 2-particle covariant Bethe–Salpeter equation for ladder graphs. Otherwise it has only been proved up to the fourth order.

The analytic structure of the scattering amplitude can be studied explicitly in some exactly solvable models in potential scattering. The properties of the scattering amplitude can then be used to study a wide class of both regular and singular potentials. In what follows we shall show for what class of short range singular potentials one can write down a dispersion relation for the partial wave scattering amplitude with a finite number of subtractions. First, we shall show this with the help of the Phragmén–Lindelöf theorem.

Let us then first recall that the partial wave expansion of the total scattering amplitude is

$$f(k^2, \cos\theta) = \frac{1}{2\mathrm{i}\,k} \sum_0^\infty (2l + 1) f_l P_l(\cos\theta) \tag{14}$$

In what follows we shall consider only the *s*-wave partial amplitude $f(k^2)$ which satisfies the inequality

$$|f(k^2)| \leqslant \text{const.}, \quad k^2 \geqslant 0 \tag{15}$$

This inequality is due to the unitarity condition, and the function $f(k^2)$ is taken to be analytic in the upper (Im $k^2 > 0$) and the lower (Im $k^2 < 0$) half planes with a cut running from 0 to $+\infty$ and from $-\infty$ to $-\lambda_0 < 0$ along the real axis. Furthermore we shall neglect the possible finite number of poles.

We now look at the analytic structure of $f(k^2)$ at large negative k^2 and as $|k^2| \to \infty$ in any complex direction. Indeed as a consequence of the partial wave projection we expect that $f(k^2)$ has a left-hand cut along the negative real axis. This behaviour is evident both in the case of generalized Yukawa type potentials and also for more singular type potentials[5], for example $e^{-\mu r}/r^2$.

With the help of the Phragmén–Lindelöf theorem and some other related theorems[6], we notice that the asymptotic behaviour of $f(k^2)$ for $k^2 \to \pm\infty$ and $|k^2| \to \infty$ along any given ray in the complex k^2-plane are strongly related to each other. Consequently some sufficient conditions for the validity of the Cauchy formula for $f(k^2)$ with a finite number of subtractions can indeed be given.

As the singularity of the potential in configuration space becomes strong (r^{-n}, $n > 3$), it has been proved that the scattering amplitude has an essential singularity[7] as $k^2 \to \infty$.

(1) We shall first consider the class of generalized Yukawa potential, i.e.

$$V(r) = \int_m^\infty \sigma(\mu)\, e^{-\mu r}/r \, d\mu \tag{16}$$

where $\sigma(\mu)$ is a suitable weight distribution. Yukawa type potentials, such as the ones given by (16), can be continued for complex values of the coordinate variable r in the half plane Re(r) > 0. This fact follows from the properties of Laplace transforms which are analytic in the half plane of convergence.

It is clear that

$$\lim_{k^2 \to +\infty} f(k^2) = 0 \quad \text{since } \delta(\infty) = 0 \tag{17}$$

This holds for all complex k^2, as $|k^2| \to \infty$, provided that $f(k^2)$ does not increase faster than some exponential function on a sequence of semi-circles of radius $|k^2| = R_n$ of which $R_n \to \infty$ as $n \to \infty$ in the upper half plane, which is a consequence of the Phragmén–Lindelöf theorem.

If, in addition, $f(k^2)$ has a definite limit for $k^2 \to -\infty$, this limit must equal zero and we can write down a dispersion relation for $f(k^2)$ with at most one subtraction. This last statement is also true under the less restrictive assumption, i.e.

$$\lim_{k^2 \to -\infty} |f(k^2)| < \text{const.} \tag{18}$$

when the existence of the $\lim f(k^2)$ is not claimed. Indeed, in this case the inequality (15) is true on the whole of the real axis, and due to the Phragmén–Lindelöf theorem it is true for all complex values of $s = k^2$. We can then unite

$$\frac{T(s) - T(s_0)}{s - s_0} = \frac{1}{\pi} \int_0^\infty \frac{ds' \operatorname{Im} T(s')}{(s' - s_0)(s' - s)} - \frac{1}{\pi} \int_\beta^\infty \frac{ds' \operatorname{Im} T(-s')}{(s' - s_0)(s' - s)} \tag{19}$$

4 Total Scattering Amplitude

The potential $V(v) = gr^{-4}$ is the only known potential more singular than the centrifugal term r^{-2} for which the radial Schrödinger equation can be exactly solved for all values of k^2 and all partial waves. Recently this potential had received considerable attention[8]. Without reproducing the results already published, the wave-functions are obtained in terms of the standard Mathieu functions and their parameters. The work of various authors has recently been concerned with the general analytic properties of the partial and total scattering amplitude for this and various other singular potentials.

We shall write the standard partial wave expansion and its contour integral in the complex λ-plane:

$$\begin{aligned} f(k^2, \cos\theta) &= \frac{1}{2\mathrm{i}\,k} \sum_{l=0}^{\infty} (2l+1) P_l(\cos\theta)[S_l(k) - 1] \\ &= -\frac{1}{2k} \int_c \frac{\lambda\, d\lambda}{\cos \pi\lambda} P_{\lambda-\frac{1}{2}}(-\cos\theta)[S(\lambda, k) - 1] \end{aligned} \tag{20}$$

where $\lambda = l + \frac{1}{2}$.

The contour C must contain the positive real axis. To study the analytic properties of $f(k^2, \cos\theta)$, we must open out the contour C. The case of the singular potentials is quite different from that for the regular ones, since in the latter case one can make the contour run along the imaginary axis.

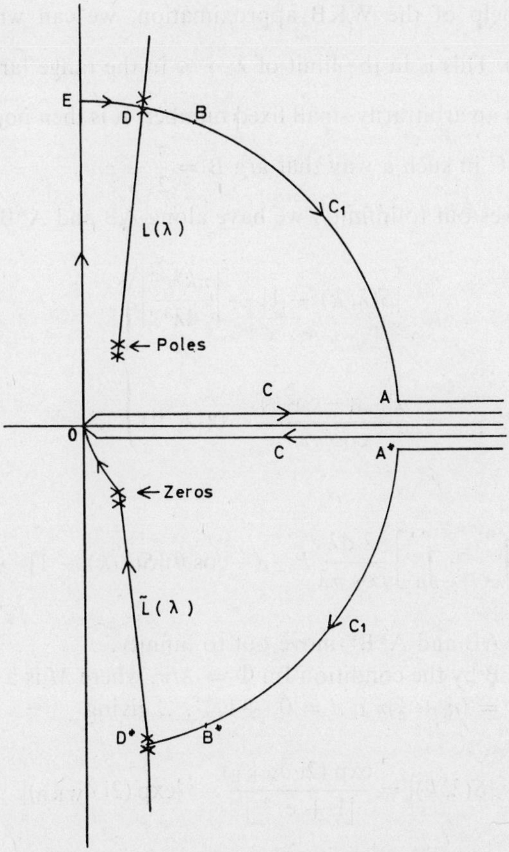

FIGURE 1 The contours C and C_1 in the λ-plane[9]

We shall then carefully consider the contour given in figure 1 and restrict ourselves to values of $0 < \theta < 2\pi$. We write:

$$f(k^2, \cos\theta) = -\frac{1}{2k}\int_{C_1} \frac{\lambda\,d\lambda}{\cos\pi\lambda} P_{\lambda-\frac{1}{2}}(-\cos\theta)[S(\lambda, k) - 1]$$

$$= -\frac{\pi i}{k}\sum_m \frac{\lambda m}{\cos\pi\lambda_m} P_{\lambda_m-\frac{1}{2}}(-\cos\theta)\gamma_m \quad (21)$$

With the help of the contour given in figure 1, it was proved by Domby and Jones[9] that as AE and A*D* are moved out to infinity the parts of the integral in (21) taken along these arcs tend to zero. Their argument may be summarized below.

With the help of the WKB approximation, we can write $S(\lambda, k) \to \exp(2i\,\delta_{WKB})$. This is in the limit of $\lambda \to \infty$ in the range $|\arg \lambda| \leqslant \frac{\pi}{2} - \varepsilon$; $\varepsilon > 0$ which is an arbitrarily small fixed number. It is then possible to select a point B on C in such a way that $\arg B = \frac{\pi}{2} - \varepsilon$.

As AB moves out to infinity, we have along AB and A*B*

$$\left.\begin{aligned} &|S(\lambda, k) - 1| \to \left|\frac{\pi k^2 g^2}{4\lambda^3}\right| \\ \text{and} \\ &\frac{P_{\lambda-\frac{1}{2}}(-\cos\theta)}{\cos\pi\lambda} \leqslant 0(|\lambda^{-\frac{1}{2}}|) \end{aligned}\right\} \tag{22}$$

resulting in

$$\left[\int_{A^*}^{B^*} + \int_{B}^{A}\right] \frac{\lambda\, d\lambda}{\cos\pi\lambda} P_{\lambda-\frac{1}{2}}(-\cos\theta)[S(\lambda_1 k) - 1] \to 0 \tag{23}$$

provided that AB and A*B* move out to infinity.

We define EB by the condition $\operatorname{Im}\Phi = M\pi$, where M is a large positive integer and $\Phi = (n + \frac{1}{2})\pi\, i$, $n = 0, \pm 1, \ldots$, giving

$$|S(\lambda, k)| = \frac{\exp(2i\,\delta_{WKB})}{|1 + e^{-2\phi}|} \leqslant |\exp(2i\,\delta_{WKB})| \tag{24}$$

Since the δ_{WKB} is small for large λ, the S-matrix is bounded along EB. Also for the real value of Φ which is greater than zero, we notice that

$$\tfrac{1}{2} \leqslant \left|\frac{S(\lambda, k)}{\exp(2i\,\delta_{WKB})}\right| \leqslant 1 \tag{25}$$

Finally by unitarity, $S(\lambda, k)$ is also bounded along B*D*. We then conclude that $P_{\lambda-\frac{1}{2}}(-\cos\theta)|\cos\pi\lambda$ is exponentially small when $|\operatorname{Im}\lambda|$ is large and

$$\left[\int_{B^*}^{D^*} + \int_{E}^{B}\right] \frac{\lambda\, d\lambda}{\cos\pi\lambda} P_{\lambda-\frac{1}{2}}(-\cos\theta)[S(\lambda, k) - 1] \to 0 \tag{26}$$

as EB and B*D* move out to infinity.

Combining (21), (22) and (26), we have

$$f(k^2, \cos\theta) = -\frac{1}{2k}\int_{C_2}\frac{\lambda\,d\lambda}{\cos\pi\lambda}P_{\lambda-\frac{1}{2}}(-\cos\theta)[S(\lambda, k)-1]$$

$$= -\frac{\pi\,i}{k}\sum_{m=0}^{\infty}\frac{\lambda_m}{\cos\pi\lambda_m}P_{\lambda_m-\frac{1}{2}}(-\cos\theta)\gamma_m \qquad (27)$$

where C_2 lies along $L(\lambda)$ in the fourth quadrant and along the positive imaginary axis. The range of θ is $0 < \theta < 2\pi$. In the limit of large λ we write

$$\frac{P_{\lambda-\frac{1}{2}}(-\cos\theta)}{\cos\pi\lambda} = 0\left\{\frac{\lambda^{-\frac{1}{2}}\exp[\pm i\,\lambda(\pi-\theta)]}{\exp[\pm i\,\pi\lambda]}\right\} \qquad (28)$$

Where $\pm$ are chosen to give the larger terms in both the numerator and the denominator. We notice that when one sums over the poles, $\lambda_m = \lambda_{m_1} + i\,\lambda_{m_2}$ where $\frac{\lambda_{m_2}}{\lambda_{m_1}} \to \infty$, as $m \to \infty$. We can then write:

$$\frac{P_{\lambda_m-\frac{1}{2}}(-\cos\theta)}{\cos\pi\lambda_m} = 0\left\{\frac{\lambda_m^{-\frac{1}{2}}\exp[\pm(\lambda_{m_1}\theta_2-\lambda_{m_2})(\pi-\theta_1)]}{\exp(\pi\lambda_{m_2})}\right\} \qquad (29)$$

A few remarks can be added here: (i) the above expression decreases exponentially at infinity, since $\lambda_{m_2}/\lambda_{m_1} \to \infty$ for any value of θ_2 if $0 < \theta_1 < 2\pi$; (ii) the infinite sum over Regge poles converges if $0 < \text{Re}\,\theta < 2\pi$ and $0 \leqslant \text{Im}\,\theta < \infty$.

It is not difficult to see therefore that (27) is an analytic representation of the scattering amplitude $f(k^2, \cos\theta)$ in the cut plane.

5 Behaviour of Wave Equating at the Origin

Before we compare the wave-functions for two particles interacting relativistically (i.e. the solution near the origin of the Bethe–Salpeter equation, and the Lippman–Schwinger equation), we shall first briefly summarize what has been so far learnt from the comparison of the B-S equation and the non-relativistic Schrödinger equation.

The covariant description of the 2-body problem in the B-S equation is the fact that the studies of the two scalar bosons interactions, either via a vector particle or through the exchange of two scalar particles, lead to non-Fredholm type integral equations. This implies that all the Fredholm traces diverge at large momenta. This difficulty can, however, be removed by giving a proper treatment of the wave-functions at small distances. This problem is extensively treated by Bastai and others[3].

5.1 *The non-relativistic Schrödinger equation*

We write the radial Schrödinger equation as

$$\left[\frac{d^2}{dr^2} + k^2 - \frac{l(l+1)}{r^2} - V(r)\right]\phi(r) = 0$$

In this equation we shall consider the potential $V(r)$ to be pure power of the form:

$$V(r) \sim gr^{-n} \tag{30}$$

We shall use the power n to classify the behaviour of the wave-function $\phi(r)$ at small distances as follows:

(1) $n < 2$ (this is commonly referred to as regular potential) and $\phi(r) \sim r^{l+1}$.

In this case the $\phi(r)$ is dominated by the kinetic energy part of the Hamiltonian.

(2) $n = 2$ separates the regular from singular potentials and we notice that $\phi(r)$ still has a modified power behaviour or

$$\phi(r) \sim r^{\bar{l}+1}; \quad \bar{l} = +\tfrac{1}{2} + [(l + \tfrac{1}{2})^2 - g]^{\frac{1}{2}} \tag{31}$$

where g is the strength of the potential.

(3) $n > 3$, we notice that the potential term dominates the centrifugal term and one can no longer establish a power behaviour for $\phi(r)$ as $r \to 0$.

It is interesting to note that a somewhat similar classification can also be given in the case of the covariant 2-particle B-S equation.

5.2 *The relativistic* B-S *equation*

The differential form for the B-S equation is

$$(\Box_1 - \mu^2)(\Box_2 - \mu^2)\Phi(\bar{X}_1\bar{X}_2) = V[(\bar{X}_1 - \bar{X}_2)]\Phi(\bar{X}_1\bar{X}_2) \tag{32}$$

In (32), $V(\bar{X}_1 - \bar{X}_2)$ is the 'relativistic potential', responsible for the exchange of one or more particles.

In order to compare the 'wave-functions' $\Phi(\bar{X}_1\bar{X}_2)$ as the 4-dimension distance

$$R^2 = (\bar{X}_1 - \bar{X}_2)^2 \to 0$$

the comparison must take into account the contribution of both the potential and the kinetic terms which contain 4-order derivatives in R. This fact leads one to modify any power behaviour of the Φ at small values of R and test the form $\Phi \sim R^{n-4}$. We now look more carefully at the behaviour of Φ at small values of R.

(1) $n < 4$. We notice here that the kinetic term dominates the Hamiltonian at small distances implying that a plane wave representation gives the correct behaviour on the light cone and the problem is similar to potential scattering in obtaining eigenvalues. The scattering process can be represented by a Fredholm type integral equation.

(2) $n = 4$. Here we have an equal contribution from the kinetic and the potential terms and one can still write a power behaviour of the solution as $\Phi \sim R^n$ except that n becomes a function of the potential strength.

Here we no longer have a Fredholm type integral equation for the 2-particle scattering.

(3) $n > 4$. Here the singular potential dominates and no power behaviour is to be found for Φ in terms of R.

These classifications are well known and serve to a better understanding of the relativistic B-S equation. We shall now go over to compare the Lippman–Schwinger equation with the B-S equation[3]. We look at the problem more closely in terms of three classes of field theoretic interactions where (I) only finite quantities appear, (II) renormalizable interactions and (III) non-renormalizable interactions.

We write the B-S equation in the form

$$M(p, p'; q) = V(p, p'; q) + \int \frac{d^4k \; V(p, k, q)M(k, p', q)}{[(k+q)^2 - m^2][(k-q)^2 - m^2]} \tag{33}$$

In (33), $V(p, p'; q)$ is the relativistic potential while $q \pm p, q \pm p'$ are the initial and final momenta of the particles.

We notice that for the super-renormalizable, i.e. class (I) interaction, the potential tends to zero for large momentum transfer and external masses. While class (II) (for spinless particles exchange) the potential tends to a constant. Finally, in class (III), the potential may increase either (a) as a power or (b) at least exponentially in momentum transfer of external masses.

The relativistic B-S equation has a well-defined integral equation under iteration only for the super-normalizable case, i.e. class (I) potentials.

J. G. Taylor[10] has carefully discussed the use of the symmetric differential operator

$$\mathscr{D} = \sum_{i,\mu=1}^{4} p_{i_\mu} \partial / \partial p_{i_\mu} \tag{34}$$

Here p_{i_μ} are the components of the external momenta $P_1, P_2, P_3, P_4 = q \pm p, q \pm p'$, connection with the potential of class (II) arising in the B-S equation from a renormalizable field theory. When this differential operator $\mathscr{D}^{-1}\mathscr{D}$ is applied on (33) we obtain a differentio-integral equation containing the value of the scattering amplitude at zero momenta for all particles. We notice that the value of the zero momenta scattering amplitude cannot be determined from the differentio-integral equation, but plays the role of the renormalized charge.

In the previous discussion, we note that the comparison between the relativistic B-S equation and the non-relativistic Schrödinger equation has led to a lot of speculation with regard to the non-renormalizable field theories. We would like to ask, then, what information one would obtain from the comparison of the B-S equation with the Lippman–Schwinger equation? Here again we shall classify the L-S equation as follows:

Class (I): regular potentials behaving like r^{-n} ($n < 2$ when $r \to 0$),
Class (II): centrifugal type potential or $V(r) = r^{-n}$ ($n = 2$ as $r \to 0$), and
Class (III): (a) $V(r) = r^{-n}$ ($n > 2$ as $r \to 0$),
(b) $V(r) = A(r)\, r^{-n}$ where $A(r)$ can have for example an essential singularity $e^{a/r}$ or a branch point $\ln r$ as $r \to 0$.

We follow closely some previous work which was done to compare the B-S and L-S equations[2].

With the help of the operator $\mathscr{D}^{-1}\mathscr{D}$, we can extend the B-S equation of class (I) to class (II). It is then conjectured that we can also, with the help of $\mathscr{D}^{-n}\mathscr{D}^{n}$ for $n > 1$, accomplish step 2 (i.e. from class II to class III for the B-S equation).

With the help of $\mathscr{D}^{-n}\mathscr{D}^{+n}$, it is possible to accomplish steps (3) and (4), i.e. to go from class (I) to class (II) and from class (II) to class (III) in the Lippman–Schwinger equation. The Lippman–Schwinger equation is

$$f(\mathbf{p}, \mathbf{p}'; \mathbf{k}^2) = \tilde{V}(|\mathbf{p} - \mathbf{p}'|) + \int \frac{\tilde{V}(|p - q|) f(q, p', k^2)}{(k^2 - q^2)} \, d^3q \qquad (35)$$

where

$$\tilde{V}(|\mathbf{p}|) = \int e^{i\mathbf{p}\cdot\mathbf{r}} V(|\mathbf{r}|) \, d^3\mathbf{V}$$

We then look at the asymptotic behaviour of the potential $\tilde{V}(|\mathbf{p}|)$ in the limit of $|\mathbf{p}| \to \infty$ or in configurational space as $r \to 0$. This behaviour is

$$\tilde{V}(|\mathbf{p}|) \sim |\mathbf{p}|^{n-3} \qquad (36)$$

When this behaviour is inserted in L-S equation (35) we get for $n < 2$, a well-behaved equation on iteration, while for $n = 2$, we have a logarithmic divergence on iteration.

This logarithmic divergence may be removed in a manner similar to that of the B-S equation (see for example reference 10).

We replace the operator $\mathscr{D}$ by the operator **d** which is defined as acting on any function of the three sector **p** by

$$(\mathbf{d_p}f)(\mathbf{p}) = \left[\frac{\mathrm{d}}{\mathrm{d}\lambda} f(\lambda\hat{p})\right]_{\lambda=|\mathbf{p}|} \tag{37}$$

where $\hat{p} = \mathbf{p}/|\mathbf{p}|$ is the unit vector along **p**. Similarly

$$(\mathbf{d_p}^n f)(\mathbf{p}) = \left[\frac{\mathrm{d}^n}{\mathrm{d}\lambda^n} f(\lambda\,\hat{p})\right]_{\lambda=|\mathbf{p}|}$$

We note that (35) becomes

$$f(\mathbf{p}, \mathbf{p}'; \mathbf{k}^2) = \tilde{V}(|\mathbf{p} - \mathbf{p}'|) - \tilde{V}(|\mathbf{p}'|) + f(0, \mathbf{p}'; \mathbf{k}^2)$$
$$+ \int_0^{|\mathbf{p}|} \mathrm{d}\lambda \left\{\frac{\mathrm{d}_{\mathbf{p}''} V(|\mathbf{p}'' - \mathbf{q}|) f(\mathbf{q}, \mathbf{p}'; \mathbf{k}^2)}{(\mathbf{k}^2 - \mathbf{q}^2)}\right\}_{\mathbf{p}''=\lambda\mathbf{p}} \mathrm{d}^3 q \tag{38}$$

We define

$$U(\mathbf{p}, \mathbf{p}') = \tilde{V}(|\mathbf{p} - \mathbf{p}'|) - \tilde{V}(|\mathbf{p}'|) + f(0, \mathbf{p}') \tag{39}$$

Since $\mathrm{d}_{\mathbf{p}} U(\mathbf{p}, \mathbf{p}') = \mathrm{d}_{\mathbf{p}} \tilde{V}(|\mathbf{p} - \mathbf{q}|)$, then we may rewrite (38) as

$$f(\mathbf{p}, \mathbf{p}'; k^2) = U(\mathbf{p}, \mathbf{p}') + \int^{|\mathbf{p}|} \mathrm{d}\lambda \, \frac{\mathrm{d}_{\mathbf{p}''} U(\mathbf{p}'', \mathbf{q}) f(\mathbf{q}, \mathbf{p}', \mathbf{k}^2)}{(\mathbf{k}^2 - \mathbf{q}^2)} \mathrm{d}^3\mathbf{q} \tag{40}$$

We note that iteration of (38) for potential of class (II) is convergent at every step. This is due to the differentiation acting on $\tilde{V}(|\mathbf{p}'' - \mathbf{q}|)$ under the integral sign of (40), together with integration over finite range of the λ-variable.

In conclusion we may regard (40) as the correct equation extending the L-S equation to the r^{-2}-type potential in as similar manner as possible to the renormalized B-S equation for $\lambda\phi^4$-theory.

6 Residue Functions of Regge Poles

To illustrate the concept of maximal analyticity, Chew[11] considered, as a prototype model, scattering by potentials of the Yukawa type with a range m^{-1} and strength g. These two parameters can be matched to the simple pole in the scattering amplitude $A(s, t)$ at the invariant momentum $t = m^2$, and to its constant residue β respectively, by writing

$$\beta = \lim_{t \to m^2} (m^2 - t) A(s, t) \tag{41}$$

Where $t = -2q_s^2(1 - \cos\theta_s)$; q_s and θ_s being respectively the C-M system momentum and scattering angle in the s-reaction.

This interpretation can be further understood to imply the existence of a single boson elementary particle of mass m communicating with the channels of t-reaction according to the coupling constant.

Until now there have been only very few models in potential scattering where one can explicitly construct the Regge poles trajectories and calculate their residue functions. In this note we consider in some detail the analytic properties of the residue function $\beta(k)$ for a wide class of both regular (Coulomb and Yukawa type potentials) and singular (gr^{-n}, $n > 3$) potentials where the explicit form of $\beta(k)$ can be calculated.

6.1 *The Coulomb potential* gr^{-1}

The exact location of the Regge poles are well known for the Coulomb potential and the residue functions are given by[12]

$$\beta_n(k) = \frac{(-1)^n}{n!\Gamma(-n + g\,i/\sqrt{|k^2|})}, \quad \text{where } n = 0, 1, 2, \ldots \tag{42}$$

Since the Γ-function in (42) is a non-vanishing mesomorphic in the entire z-plane with simple poles at $-n + g\,i\sqrt{|k^2|} = 0, -1, \ldots, \beta_n(k)$ will have only zeros at the location of the poles. $\beta_n(k)$ will also vanish as $k \to \pm\infty$. It also vanishes at a number of points $l_{n,m} = \frac{1}{2}(-m - n - 2)$ on the negative real axis in the complex l-plane.

In the neighbourhood of small values of k^2, $\beta_n(k)$ behaves as

$$\beta(k) \sim \frac{(-1)^n}{n!} e^{-|z(k)|^2 \ln |z(k)|}$$

$$\text{where } z(k) = -n + \frac{gi}{\sqrt{|k^2|}} \tag{43}$$

We also notice that for $k^2 \to \infty$ the Γ-function has simple poles for negative integral value of n, hence $\beta_n(k)$ vanishes.

The signs of the residue alternate for different values of n. These features are also common to the residue function of the Yukawa potential.

6.2 *The Yukawa type potential*

For this class of potential the residue function $\beta(k)$ at a pole $\lambda_0(k)$ is given by[13]

$$\beta(k) = \frac{e^{i\pi\lambda_0}k/\lambda_0}{\displaystyle\int_0^\infty dr\ r^{-2}f_-^2} \tag{44}$$

where the Jost solution f_- is defined by

$$\bar{f}(\lambda, k; r) \equiv k^{\lambda-\frac{1}{2}}\,e^{i\pi/2}[\lambda - \tfrac{1}{2}]f(\lambda, k; r) \tag{45}$$

Since $\bar{f}$ is finite generally non-zero as $k \to 0$, the threshold behaviour is

$$\beta(k) \simeq 0(k^{2l_0+1}) \tag{46}$$

For $\lambda_0 > -\frac{1}{2}$, $\beta(k)$ is an analytic function in k except at points where the poles are not simple. Again $\beta(k)$ is analytic for $k < 0$ in the region where λ_0 is physical and it does not vanish. But for a non-physical value of λ, β may not be analytic as the poles move to infinity.

We may conclude by saying that when the poles of the S-matrix move through a negative integral value of λ, $\beta(k) \to 0$.

We notice that the threshold behaviour given in (46) can be compared to that of the Coulomb potential.

In a recent work Müller and Schilcher[14] calculated among various other things the residue functions of the Regge poles for Yukawa potential and found that

$$\beta_n(k) = \frac{(-1)^n}{n![\Gamma-n + 2i\operatorname{Im}\alpha_n(k)]} \tag{47}$$

where $\operatorname{Im}\alpha_n(k) = l$ and $l + n + 1 = -\dfrac{\Delta(k)}{2k}$, and $\Delta(k)$ is an expansion in descending powers of k, $n = 0, 1, 2, \ldots$.

The function $\Delta(k)$ is obtained from the secular equation of the eigenvalue problem and is known up to and including the term k^{-7} (see Müller reference 14).

We notice that (47) is independent of any high energy approximation and thus is valid also for other approximations of $\alpha_n(k)$.

For $\operatorname{Im}\alpha_n(k)$ to be very small, the case of resonances which occur just above threshold, we may expand the Γ-function and write

$$\beta_n(k) \simeq -\frac{2i}{(n!)^2}\operatorname{Im}\alpha_n(k) \tag{48}$$

In the high energy region,

$$\beta_n(k) \simeq -\frac{\mathrm{i}}{(n!)^2}\frac{M_0}{k}; \; k^2 \to \infty \text{ and } M_0 \text{ is a constant} \tag{49}$$

We notice that the residue function is zero for $k^2 < 0$, where $\alpha_n(k)$ is real. Above threshold it increases from zero but with increasing energy it decreases and becomes zero again for $k^2 \to \infty$.

The behaviour of $\beta(k)$ is similar to that in the case of the Coulomb potential.

6.3 *Singular power potentials gr^{-n}*

Here we give the *S*-matrix and the Regge poles for an exactly solvable potential model gr^{-4} which has been discussed by various authors[9]:

$$S(\lambda, k) = \frac{e^{i\pi\lambda}}{\cos \pi\beta + \mathrm{i}\,(e^{-2\Phi} + \sin^2 \pi\beta)^{\frac{1}{2}}} \tag{50}$$

where $\cos \pi\beta = 1 - 2\Delta(0) \sin^2 \pi\lambda/2$ and $\Delta(0)$ is some determinant, function of λ and k. While Φ is

$$\Phi = \Phi_0 + 0[(\lambda^2 + 2\mathrm{i}\, kg)^{-\frac{1}{4}}] \tag{51}$$

We notice that the *S*-matrix has a pole if

$$e^{-2\Phi} = -1 \tag{52}$$

i.e.

$$\Phi = (m + \tfrac{1}{2})\pi\, \mathrm{i}, \quad m = 0, \pm 1, \pm 2, \ldots$$

The Regge poles thus lie along a line $L(\lambda)$ defined by

$$\operatorname{Re} \Phi(\lambda. k) = 0 \tag{53}$$

It can be readily seen that (50) gives the residue function at each pole as

$$\beta_m(k) = \lim_{\lambda \to \lambda_m} (\lambda - \lambda_m) S(\lambda_1 k) \tag{54}$$

and for large m,

$$\beta_m(k) = \frac{1}{4 \ln \left[\dfrac{2\lambda_m}{\sqrt{(\mathrm{i}\, gk)}}\right]} \tag{55}$$

For a general power gr^{-n} potential the residue functions read

$$\beta_m(k) = \frac{n-2}{2n \ln \left[\dfrac{2\lambda_m}{(\mathrm{i}\, g)^{\frac{1}{2}} k^{1-2/n}}\right]} \tag{56}$$

We shall restrict our discussion here to the exactly solvable model gr^{-4} since the mth *Regge trajectory* was discussed in detail[8].

$\beta_m(k)$ in (55) has a simple pole for $2mk < 0$ at $k = -\mathrm{i}\left(\frac{2\lambda_m}{g^{\frac{1}{2}}}\right)^2$.

We notice here that $\beta_m(k) \to 0$ as $k \to 0$ and ∞.

6.4 *Regge trajectories*

The poles of the *S-Matrix* give the trajectories in the plane of complex angular momentum. There trajectories can also be traced with the help of the secular equation of the eigenvalue problem.

Since the potential model gr^{-4} can be transformed into a Mathieu equation, where the eigenvalues are known and can be calculated by the condition that the solutions be periodic functions of periods π, 2π, we shall use this fact. We must remember however that only under this periodicity condition can one relate these eigenvalues to those of the Schrödinger equation.

The *modified Mathieu equation* is

$$\phi''(Z) - [\lambda - 2h^2 \cosh 2Z]\phi(Z) = 0 \tag{57}$$

The periodicity conditions here imply that $\phi(Z) \to 0$ at the points where $\phi(\mathrm{i}\, Z)$ is zero, i.e. $\mathrm{i}\, Z = p(h, q)$, or where

$$r = r_a\, \mathrm{e}^{-\mathrm{i}p(h,q)},\ r_a = (g^{\frac{1}{2}}/\mu^2 k)^{\frac{1}{2}}$$

p being a real function of h for real h.

The threshold behaviour of the trajectories may easily be inferred from the *Mathieu equation*, which for $h^2 \to 0$, i.e. energy $k^2 \to 0$, we write,

$$\phi'' + \lambda\phi \sim 0 \tag{58}$$

so that for $h^2 \to 0$,

$$\lambda \sim n^2, \quad n = 0, \pm 1, \pm 2, \ldots \tag{59}$$

or

$$\lambda \sim n^2 + 0(h^2)$$

The eigenvalues corresponding to the expression of the S-matrix may be given by

$$\lambda_{2n} \equiv a_{2n}, \quad \lambda_{2n+1} \equiv D_{2n+1}$$
$$\lambda(-2n-1) \equiv b_{2n+1}, \quad \lambda(-2n-2) \equiv b_{2n+2} \tag{60}$$

For large value of m and small values of $|h^2|$ the eigenvalues are given by

$$a_m(h^2) = m^2 + \frac{1}{2(m^2-1)}h^4 + \frac{(5m^2+7)}{3^2(m^2-1)^3(m^2-4)}h^8 + \frac{9m^4+58m^2+29}{64(m^2-1)^5(m^2-4)(m^2-9)}h^{12} + \cdots \quad (61)$$

For more detail see reference (8).

We also give the high energy expansion for a_m as

$$a_m(h^2) = -2h^2 + 2hq - \frac{1}{2^3}(q^2+1) - \frac{1}{2^7h}q(q^2+3) - \frac{1}{2^{12}h^2}(5q^4+34q^2+9) - \frac{1}{2^{17}h^3}q(33q^4+410q^2+405) + \cdots \quad (62)$$

We show in the schematic trajectory below (figure 2) the mth Regge

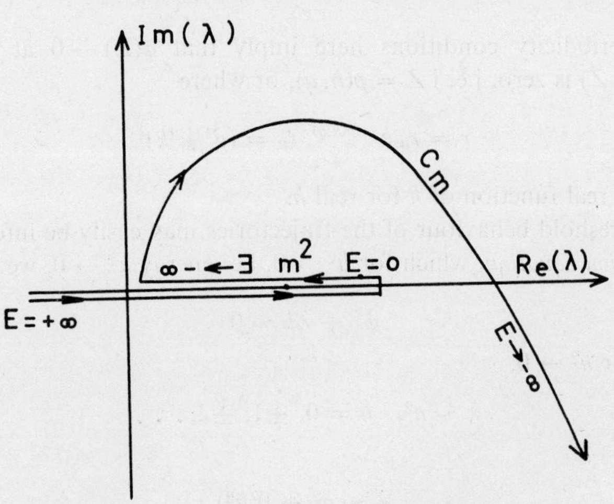

FIGURE 2 Schematic trajectory of the mth Regge pole in the complex λ-plane for the attractive potential gr^{-4} (reference 14)

pole trajectories. The low energy expansion (61) and the high energy expansion (62) are used. But we notice that only large values of m seem to account for the ground state for this potential which occurs at $E = -\infty$.

The trajectory is given in the complex λ-plane and in order to stay in the physical sheet we consider only the case of $\mathrm{i}\,k > 0$. This means that the physical sheet in the E-plane is mapped into the upper half of the k-plane, and this again is mapped into the first quadrant of the $\sqrt{k}$-plane and so on.

References

1. Contogouris, A. P., *Nuovo Cimento*, **37,** 1712 (1965); Furlan, G., and Mahoux, G., *Nuovo Cimento*, **36,** 215 (1965); Aly, H. H., and Müller, H. J. W., *J. Math. Phys.*, **8,** 367 (1967).
2. Aly, H. H., and Taylor, J. G., *A New Field Theory Approach to Singular Potential*, American Univ. of Beirut preprint (1967), to appear in *J. Math. Phys.*
3. Bastai, A., Bertochi, L., Furlan, G., Fubini, S., and Tonin, M., *Nuovo Cimento*, **30,** 1512 (1963).
4. Martin, A., *Preludes in Theoretical Physics*, North-Holland, Amsterdam, 1966.
5. Aly, H. H., and Wit, R., *Z. Physik*, **201,** 129 (1967).
6. Meiman, N. N., *Soviet Phys. JEPT* (*English Transl.*), **19,** 706 (1964).
7. Jabbur, R. J., *Phys. Rev.*, **138B,** 1527 (1965); Aly, H. H., and Müller, H. J. W., *J. Math. Phys.*, **8,** 367 (1967).
8. Spector, R. M., *J. Math. Phys.*, **5,** 1185 (1964); Aly, H. H., and Müller, H. J. W., *J. Math. Phys.*, **7,** 1 (1966).
9. Dombey, N., and Jones, R. H., *Analyticity Properties of the Scattering Amplitudes for Singular Potentials*, Univ. of Kent at Canterbury preprint (1967).
10. Taylor, J. G., *Nuovo Cimento Suppl.*, 1, 2, 3, 4 and 5 (1963).
11. Chew, G. F., *Strong Interaction Physics*, W. A. Benjamin, Inc., 1964.
12. Singh, V., *Phys. Rev.*, **127,** 632 (1962).
13. Newton, R. G., *The Complex j-Plane*, W. A. Benjamin, Inc., 1964, Chap. 10.
14. Müller, H. J. W., and Schilcher, K., *High-energy Scattering for Yukawa Potential*, Amer. Univ. of Beirut preprint (1967).

Perturbation Approach for Regular Interactions

H. J. W. MÜLLER

1 Introduction

Recently various difficulties in quantum field theory have led to a considerable revival of interest in the analytic properties of scattering amplitudes in the frame of quantum mechanics and potential theory. This interest is largely due to the fact that in potential theory many problems are solvable either exactly or else approximately by well-known methods, whereas fully relativistic theories often lead to insurmountable mathematical difficulties. Moreover, potential theory is flexible enough to serve as an invaluable simple prototype of almost any kind of model theory.

For a long time it was customary, however, to argue that even in potential theory only very few highly idealized examples were solvable exactly and that others had to be treated by numerical methods. Perturbation theory, although used for deriving first or second order approximations, was not generally considered as giving true series solutions, from which even analytic properties could be inferred.

In various investigations the author has shown recently that this disregard—if not contempt—of perturbation theory is unjustified. The present article is a first attempt to present a unified picture of a perturbative treatment of almost any type of non-singular interaction—unified, because we attack all these problems with one and the same perturbation approach. It is for this reason, moreover, that we devote the introductory section to the underlying mathematical method. Once this is mastered and understood, it can be applied to almost any type of regular interaction. This is shown in the following sections, where we apply our method to perturbed Coulomb interactions and to perturbed harmonic oscillators.

2 Perturbation Formalism

2.1 *An introductory example: the Mathieu equation*

We consider first as an introductory example of our perturbation formalism the simple case of *S*-wave scattering by a periodic potential. The radial Schrödinger equation may then be written in the form

$$y'' + (\lambda - 2h^2 \cos 2x)y = 0 \tag{1}$$

where y is the radial wave function, λ the energy k^2 in units $\hbar = c = 1$ (taking the reduced mass $m = 1/2$), and h^2 is a coupling parameter, which is positive for a repulsive and negative for an attractive interaction. Equation (1) is well known in the mathematics of periodic differential equations[1,2], where it is known as the Mathieu equation. We consider separately the two cases of large and small values of $|h^2|$. Moreover, since in ordinary scattering theory the potential cos $2x$ (as also its modified form cosh $2x$) is rather unrealistic, our main emphasis in this section will be on a simple formulation of the mathematical eigenvalue problem defined by the differential equation (1) together with suitable boundary conditions.

We consider first the simple case[3] of small $|h^2|$, although the perturbation method was originally[4] first applied to the case of large $|h^2|$.

Now, for an equation of the form

$$y'' + f(x)y = 0 \tag{2}$$

the characteristic numbers or eigenvalues are largely determined by the behaviour of $f(x)$ (x real) in the range for which this function is positive and the solutions therefore oscillatory. Approximations to the eigenvalues can therefore be found by solving the simpler equation obtained by replacing $f(x)$ by an approximation valid over this range. For the Mathieu equation ($-\pi \leqslant x \leqslant \pi$)

$$y'' + f[x)y = y'' + [\lambda + 2h^2 \cos 2(x \pm \tfrac{1}{2}\pi)]y = 0 \tag{3}$$

we see that near $h^2 = 0$, $x \simeq \mp \dfrac{\pi}{2}$

$$y'' + (\lambda + 2h^2)y \simeq 0 \tag{4}$$

and hence

$$\lambda = \nu^2 + 0(h^2) \quad \text{and} \quad y = \frac{\cos}{\sin} \nu x + 0(h^2) \tag{5}$$

where ν is an integer ($n = 0, 1, 2, \ldots$) or approximately an integer depending on the boundary conditions imposed on y. We shall not be concerned here with the particular boundary conditions imposed on the periodic solutions of the Mathieu equation. However, supposing a boundary condition B imposed on y yields

$$B(y) = 0 \tag{6}$$

one can use this relation to obtain—by Taylor expansion around $\nu = n$, for instance—an expansion of the form

$$\nu^2 = n^2 + 0(h^2) \tag{7}$$

The eigenvalue problem will thereby be solved uniquely. If, in particular, solution of (6) for ν yields $\nu = n$, then ν itself is already an integer, and (7) reduces to a triviality. In potential problems, as we shall show later, this is often the case if the solution of the radial Schrödinger equation is required to vanish at infinity.

We now rewrite the first of the conditions (5) in the form

$$\lambda + 2h^2 = \nu^2 + \Delta(h^2) \cdot h^2 \tag{8}$$

Choosing as a first approximation to y the expression

$$y^{(1)} = y_\nu = \cos \nu x \tag{9}$$

we see that y_ν leaves uncompensated terms in the Mathieu equation amounting to

$$R^{(1)} \equiv h^2(2 + 2 \cos 2x - \Delta) \cos \nu x \tag{10}$$

since the equation may be written as

$$y'' + \nu^2 y = h^2(2 + 2 \cos 2x - \Delta)y \tag{11}$$

It is convenient in our formalism to set

$$R^{(1)} = h^2[(\nu, \nu + 2)y_{\nu+2} + (\nu, \nu)y_\nu + (\nu, \nu - 2)y_{\nu-2}] \tag{12}$$

where the parenthetical expressions denote coefficients defined by

$$\left.\begin{aligned}(\nu, \nu + 2) = 1 = (\nu, \nu - 2)\\ (\nu, \nu) = 2 - \Delta\qquad\end{aligned}\right\} \tag{13}$$

It is now clear that a term $\mu y_{\nu+2t}$ on the right-hand side of (11) may be removed by adding a further contribution

$$-[\mu/4t(\nu + t)]\, y_{\nu+2t} \tag{14}$$

to the previous approximation except, of course, when $t = 0$. This follows from the fact that $y_{\nu+2t}$ is a solution of the equation

$$y''_{\nu+2t} + \nu^2 y_{\nu+2t} = -4t(\nu + t)\, y_{\nu+2t} \tag{15}$$

Thus the terms $y_{\nu\pm 2}$ in (12) may be removed by adding to $y^{(1)}$ the contribution

$$y^{(2)} = -h^2\left[\frac{(\nu, \nu+2)}{4\cdot 1\cdot(\nu+1)}\,y_{\nu+2} + \frac{(\nu, \nu-2)}{4(-1)(\nu-1)}\,y_{\nu-2}\right]$$
$$= \frac{h^2}{4}\left[\frac{\cos(\nu-2)x}{(\nu-1)} - \frac{\cos(\nu+2)x}{(\nu+1)}\right] \tag{16}$$

In its turn $y^{(2)}$ leaves uncompensated terms on the right-hand side of (11) amounting to

$$R^{(2)} = -h^4\left\{\frac{(\nu, \nu+2)}{4\cdot 1\cdot(\nu+1)}\,[(\nu+2, \nu+4)\,y_{\nu+4} + (\nu+2, \nu+2)\,y_{\nu+2} + (\nu+2, \nu)\,y_\nu]\right.$$
$$\left. + \frac{(\nu, \nu-2)}{4(-1)(\nu-1)}\,[(\nu-2, \nu-4)\,y_{\nu-4} + (\nu-2, \nu-2)\,y_{\nu-2} + (\nu-2, \nu)y_\nu]\right\} \tag{17}$$

so the next contribution $y^{(3)}$ becomes

$$y^{(3)} = (-h^2)^2\left[\frac{(\nu, \nu+2)(\nu+2, \nu+4)}{4\cdot 1\cdot(\nu+1)\cdot 4\cdot 2\cdot(\nu+2)}\,y_{\nu+4}\right.$$
$$+ \frac{(\nu, \nu+2)(\nu+2, \nu+2)}{4\cdot 1\cdot(\nu+1)\cdot 4\cdot 1\cdot(\nu+1)}\,y_{\nu+2} + \frac{(\nu, \nu-2)(\nu-2, \nu-2)}{4(-1)(\nu-1)\cdot 4(-1)(\nu-1)}\,y_{\nu-2}$$
$$\left. + \frac{(\nu, \nu-2)(\nu-2, \nu-4)}{4(-1)(\nu-1)\cdot 4(-2)(\nu-2)}\,y_{\nu-4}\right] \tag{18}$$

Similarly higher contributions may readily be obtained. Then, for

$$y = y^{(1)} + y^{(2)} + y^{(3)} + \cdots \tag{19}$$

to be a solution of the Mathieu equation, the sum of the coefficients of the terms in y_ν (e.g. in (12) and (17))—left uncompensated so far—must vanish. Hence

$$0 = (\nu, \nu) - h^2\left[\frac{(\nu, \nu+2)}{4\cdot 1\cdot(\nu+1)}\,(\nu+2, \nu) + \frac{(\nu, \nu-2)}{4(-1)(\nu-1)}\,(\nu-2, \nu)\right] + \cdots \tag{20}$$

This equation determines the unknown function Δ of (8) and hence λ. One finds:

$$\lambda = \nu^2 + \frac{h^4}{2(\nu^2-1)} + \frac{(5\nu^2+7)h^8}{32(\nu^2-1)^3(\nu^2-4)} + \cdots \tag{21}$$

Our simple and straightforward formalism is thus seen to yield easily any number of terms of the expansions of the solutions and eigenvalues. Moreover, the coefficients of these expansions follow a definite pattern—which we shall examine in more detail later on—so that 'complete' series for the solutions and eigenvalues are obtained.

We next consider the case of large values of $|h^2|$ and show that the same formalism leads to simple asymptotic expansions in h for both the solutions and eigenvalues.

Again we consider the Mathieu equation in the form (1) or (2). But now we observe that the real-valued function $f(x) = \lambda - 2h^2 \cos 2x$ attains a maximum positive value at $x = \mp \pi/2$, so that it is convenient to rewrite (1) as

$$y'' + [\lambda + 2h^2 \cos 2\,(x \pm \tfrac{1}{2}\pi)]\,y = 0 \tag{22}$$

We confine ourselves for the moment to that region of x, for which $|2(x \pm \frac{1}{2}\pi)| \ll 1$. Then $\cos 2\,(x \pm \frac{1}{2}\pi) \simeq 1 - 2(x \pm \frac{1}{2}\pi)^2$. Changing the variable in (22) to $z = 2h^{1/2}(x \pm \frac{1}{2}\pi)$, the equation reduces to

$$\mathrm{d}^2y/\mathrm{d}z^2 + \left(\frac{\lambda + 2h^2}{4h} - \frac{z^2}{4}\right) y = 0 \tag{23}$$

The solutions of this equation are normalizable parabolic cylinder functions, provided

$$\frac{\lambda + 2h^2}{2h} = q \tag{24}$$

where q is either exactly or approximately an odd integer depending on the boundary conditions imposed on y (in fact, it is exactly an odd integer if y is required to vanish for a large, though not infinite value of $|z|$). The rough asymptotic behaviour of λ for large $|h^2|$ is thus given by

$$\lambda = -2h^2 + 2hq + \frac{\Delta}{8} \tag{25}$$

where Δ is an (as yet) undetermined asymptotic expansion in h.

Substituting (25) into (1), we have the equation

$$y'' + 2h\left(q + \frac{\Delta}{16h} - 2h \cos^2 x\right) y = 0 \tag{26}$$

It is convenient here to separate off the behaviour of y for $h \to \infty$, i.e. a factor $\exp(\pm 2h \sin x)$. Then, writing

$$y = A(x) \cdot \mathrm{e}^{\pm 2h \sin x} \tag{27}$$

the function A is seen to satisfy the differential equation

$$A'' \pm 4h \cos x \cdot A' + 2h \left(q \mp \sin x + \frac{\Delta}{16h} \right) A = 0 \tag{28}$$

We now choose

$$y = A(x) \cdot e^{2h \sin x} \tag{29}$$

The equation for A may then be written in the form

$$\cos x \cdot A' + \tfrac{1}{2}(q - \sin x)A = -\frac{1}{2^6 h}(16A'' + 2\Delta A) \tag{30}$$

The solution of the first-order equation

$$\cos x \cdot A_q' + \tfrac{1}{2}(q - \sin x)A_q = 0 \tag{31}$$

is

$$A_q(x) = \frac{\cos^{\frac{1}{2}(q-1)}(\frac{1}{4}\pi + \frac{1}{2}x)}{\cos^{\frac{1}{2}(q+1)}(\frac{1}{4}\pi - \frac{1}{2}x)} \tag{32}$$

so that the first approximation to A is

$$A^{(1)} = A_q \tag{33}$$

This first approximation leaves uncompensated on the right-hand side of (30) the terms

$$\begin{aligned} R^{(1)} &= -\frac{1}{2^6 h}(16A_q'' + 2\Delta A_q) \\ &= -\frac{1}{2^6 h}[(q, q+4)A_{q+4} + (q, q)A_q + (q, q-4)A_{q-4}] \end{aligned} \tag{34}$$

where

$$\left.\begin{aligned} (q, q+4) &= (q+1)(q+3) \\ (q, q) &= 2(q^2 + 1 + \Delta) \\ (q, q-4) &= (q-1)(q-3) \end{aligned}\right\} \tag{35}$$

Since, however,

$$\cos x \cdot A_{q+4t}' + \tfrac{1}{2}(q - \sin x)A_{q+4t} = -2tA_{q+4t} \tag{36}$$

a term μA_{q+4t} on the right-hand side of (30) can be cancelled out by adding a new contribution $(-\mu/2t)A_{q+4t}$ to A, except, of course, when $t = 0$. Thus the terms in (34) involving $A_{q\pm4}$ can be removed by adding $A^{(2)}$ to $A^{(1)}$, where

$$A^{(2)} = \frac{1}{2^7 h}\left[\frac{(q, q+4)}{1} A_{q+4} + \frac{(q, q-4)}{-1} A_{q-4}\right] \tag{37}$$

In its turn $A^{(2)}$ leaves uncompensated

$$R^{(2)} = -\frac{1}{2^{13}h^2}\left\{\frac{(q, q+4)}{1}(q+4, q+8)A_{q+8} + \frac{(q, q+4)}{1}(q+4, q+4)A_{q+4} + \left[\frac{(q, q+4)}{1}(q+4, q) + \frac{(q, q-4)}{-1}(q-4, q)\right]A_q + \frac{(q, q-4)}{-1}(q-4, q-4)A_{q-4} + \frac{(q, q-4)}{-1}(q-4, q-8)A_{q-8}\right\} \quad (38)$$

The next contribution $A^{(3)}$ therefore becomes

$$A^{(3)} = \frac{1}{2^{14}h^2}\left[\frac{(q, q+4)}{1}\frac{(q+4, q+8)}{2}A_{q+8} + \frac{(q, q+4)}{1}\frac{(q+4, q+4)}{1}A_{q+4} + \frac{(q, q-4)}{-1}\frac{(q-4, q-4)}{-1}A_{q-4} + \frac{(q, q-4)}{-1}\frac{(q-4, q-8)}{-2}A_{q-8}\right] \quad (39)$$

For

$$A = A^{(1)} + A^{(2)} + A^{(3)} + \cdots \quad (40)$$

to satisfy (30), the sum of the terms in A_q (e.g. in equations (34) and (38))—left uncompensated so far—must vanish. Thus

$$0 = -\frac{(q, q)}{2^6 h} - \frac{1}{2^{13}h^2}\left[\frac{(q, q+4)}{1}(q+4, q) + \frac{(q, q-4)}{-1}(q-4, q)\right] + \cdots \quad (41)$$

an equation determining Δ and hence the eigenvalue λ. The region of validity of the series (40) is given by

$$|\cos(\tfrac{1}{4}\pi \pm \tfrac{1}{2}x)| \gg h^{-1/2} \quad (42)$$

for sufficiently large values of h.

Comparing now the eigensolutions of the Mathieu equation for the two cases of small and large $|h^2|$, we see that in each case the solution is obtained as a linear combination of (suitably defined) unperturbed basic functions, which may be considered as the base vectors of an infinite dimensional Hilbert space (properties such as square-integrability can be defined, also for Mathieu functions, but we shall not be concerned with these here). We may write therefore (apart from normalizing constants)

$$y = y_\nu + \sum_{i=1}^{\infty} h^{2i} \sum_{\substack{j=-i \\ j\neq 0}}^{i} p_i(\nu, j) y_{\nu+2j} \tag{43}$$

and

$$y = e^{2h \sin x} \left[A_q + \sum_{i=1}^{\infty} \frac{1}{(2^7 h)^i} \sum_{\substack{j=-i \\ j\neq 0}}^{i} P_i(q, j) A_{q+4j} \right] \tag{44}$$

where p, P denote the respective coefficients.

We shall not pursue the case of the Mathieu equation any further in the present context. Solutions of the type (44) over the entire range of the independent variable, as well as their analytic continuation, normalization and eigenvalues have been discussed in detail by Dingle and Müller[4]. Alternative forms of these solutions important for a generalization of the perturbation approach have also been discussed[5]. Finally, we remark that the solutions and eigenvalues of the Mathieu equation as given above are by no means devoid of physical meaning. The Schrödinger equation for the strongly singular potential $1/r^4$, for instance, may be rewritten in the form of a Mathieu equation[3,6,7,8]. Also, although $V(r) = \cosh r$ does not represent a realistic potential, the Mathieu equation suggests the method of solution of S-wave Schrödinger equations for potentials expandable in powers of $\cosh r$ or related functions. The Mathieu equation therefore belongs to the class of Schrödinger equations which contain a potential expandable in even powers of r. In fact, we shall find in a later section that many of the general properties of the scattering and eigensolutions of the Schrödinger equation for realistic potentials—such as the anharmonic oscillator[9] and Gauss potentials[40]—are also characteristic of Mathieu functions, so that even from this angle the Mathieu equation deserves particular attention.

2.2 *The potential* $V(r) = -g(1 - \alpha \cosh r)^{-1}$

It is well known that the Schrödinger equation can be solved exactly and in closed form only for a very small number of simple potentials. Perturbation theory was only rarely used to extend the number of analytically solvable radial Schrödinger equations. However, the author has shown recently that almost any type of non-singular potential permits a perturbation-theoretical solution of the radial Schrödinger equation by the above method. Nevertheless, several potentials such as the Gauss potential, for which not even approximate analytic solutions were known before, lead (at least in first investigations) to series solutions which are far less satisfactory than those for the Yukawa potential, for instance. It is thus highly desirable to invent and analyse other potentials with an analogous physical behaviour for which the Schrödinger equation can be solved much more easily. A potential of this type, worthy of separate detailed investigation[10] is

$$V(r) = -g[1 - \alpha \cosh r]^{-1} \tag{45}$$

where $\alpha = \cos \gamma < 1$. In the vicinity of the origin this potential behaves as a constant (for $\alpha \neq 1$) together with a small harmonic disturbance; at infinity it dies off to zero, while at some point in between it changes from an attractive to a repulsive interaction.

We consider the case of S-wave scattering and confine ourselves for the moment to that region of r, for which the potential is attractive, i.e. $\cosh r < 1/\alpha$. The radial Schrödinger equation may then be written

$$\frac{d^2\psi}{dr^2} + (k^2 + g)\psi = -g \sum_{j=1}^{\infty} (\alpha \cosh r)^j \psi \tag{46}$$

We now assume that both α and g are sufficiently small so that the following series solutions all converge. By (5) we see that

$$k^2 + g = \nu^2 + 0(g) \tag{47}$$

or, if we require ψ to vanish at infinity for $k^2 < 0$, we may set

$$k^2 + g = -n^2 + g\Delta \tag{48}$$

where Δ is an (as yet) undetermined expansion in powers of g, and $n = 1, 2, 3, \ldots$. Substituting (48) into (46), we have

$$\psi'' - n^2\psi = -g(\Delta + \alpha \cosh r)\psi - g \sum_{j=2}^{\infty} (\alpha \cosh r)^j \psi \tag{49}$$

As a regular solution of the equation

$$\psi_n'' - n^2\psi_n = 0 \tag{50}$$

we choose

$$\psi_n = \sinh nr \tag{51}$$

so that to 0(o) in g the solution of (49) is

$$\psi^{(0)} = \psi_n \tag{52}$$

Clearly this first approximation leaves uncompensated on the right-hand side of (49) terms amounting to

$$A^{(0)} = -g\Delta\psi_n - g\sum_{j=1}^{\infty}(\alpha\cosh r)^j\psi_n \tag{53}$$

By repeated application of the relation

$$\cosh r\cdot\sinh nr = \tfrac{1}{2}[\sinh(n+1)r + \sinh(n-1)r] \tag{54}$$

one finds

$$(\cosh r)^j\cdot\sinh nr = \frac{1}{2^j}\sum_{k=0}^{j}\binom{j}{k}\sinh(n+j-2k)r \tag{55}$$

The expression (53) may therefore be written

$$A^{(0)} = -g\Delta\psi_n - g\sum_{j=1}^{\infty}(\alpha/2)^j\sum_{l=0}^{j}\binom{j}{l}\psi_{n+j-2l} \tag{56}$$

or

$$A^{(0)} = -g\Delta\psi_n - g\sum_{j=-\infty}^{\infty}\beta_j\psi_{n+j} \tag{57}$$

where (for $\alpha = \cos\gamma$)

$$\left.\begin{aligned} \beta_0 &= \frac{1-\sin\gamma}{\sin\gamma} \\ \text{and for } j\neq 0,\quad \beta_j = \beta_{-j} &= \frac{(1-\sin\gamma)^j}{\sin\gamma(\cos\gamma)^j} \end{aligned}\right\} \tag{58}$$

For $\cos\gamma < 1$, as we assumed,

$$\left|\frac{1-\sin\gamma}{\cos\gamma}\right| < 1 \tag{59}$$

so that for given $\cos\gamma < 1$

$$|\beta_j| > |\beta_{j+1}| \tag{60}$$

It is therefore plausible and legitimate to regroup a sum of terms containing factors β_j in the order of increasing j and to group product terms such as $\beta_j\beta_{j'}$ with $\beta_{j+j'}$ (since $\beta_{j+j'} = \sin\gamma\cdot\beta_j\beta_{j'}$).

We now rewrite (57) as

$$A^{(0)} = \sum_{j=-\infty}^{\infty} (n, n+j)\psi_{n+j} \tag{61}$$

where for $j = 0$:

$$(n, n) = -g(\Delta + \beta_0)$$

and for $j \neq 0$:

$$(n, n+j) = -g\beta_j \tag{62}$$

Setting

$$D_n \equiv \frac{\mathrm{d}^2}{\mathrm{d}r^2} - n^2 \tag{63}$$

we see that

$$D_n\psi_{n+j} = j(2n+j)\psi_{n+j} \tag{64}$$

so that a term $\mu\psi_{n+j}$ on the right-hand side of (49) can be cancelled out by adding to the previous approximation the term

$$\frac{\mu}{j(2n+j)}\,\psi_{n+j}$$

except, of course, when $j = 0$. The case $0 = 2n + j$ has a different significance and limits effectively the range of applicability of the ensuing expansions, as we shall indicate in a later section—so in our perturbation formalism this case does not play the same role as $j = 0$. The next contribution $\psi^{(1)}$ to $\psi^{(0)}$ is therefore given as

$$\psi^{(1)} = \sum_{\substack{j=-\infty,\\ j\neq 0}}^{\infty} \frac{(n, n+j)}{j(2n+j)}\,\psi_{n+j} \tag{65}$$

In its turn this contribution leaves uncompensated

$$A^{(1)} = \sum_{\substack{j=-\infty\\ j\neq 0}}^{\infty} \frac{(n, n+j)}{j(2n+j)} \sum_{j'=-\infty}^{\infty} (n+j, n+j+j')\psi_{n+j+j'} \tag{66}$$

and yields the next contribution

$$\psi^{(2)} = \sum_{\substack{j=-\infty\\ j\neq 0}}^{\infty} \frac{(n, n+j)}{j(2n+j)} \sum_{\substack{j'=-\infty\\ j'\neq -j}}^{\infty} \frac{(n+j, n+j+j')}{(j+j')(2n+j+j')}\,\psi_{n+j+j'} \tag{67}$$

Higher contributions clearly follow along the same line. The sum $\psi = \psi^{(0)} + \psi^{(1)} + \psi^{(2)} + \cdots$, however, will be a solution of (49) only if the sum of the coefficients of ψ_n (e.g. in $A^{(0)}$, $A^{(1)}$)—left uncompensated so far—is set equal to zero; i.e.

$$0 = (n, n) + \sum_{\substack{j=-\infty \\ j\neq 0}}^{\infty} \frac{(n, n+j)}{j(2n+j)}(n+j, n)$$
$$+ \sum_{\substack{j=-\infty \\ j\neq 0}}^{\infty} \frac{(n, n+j)}{j(2n+j)} \sum_{\substack{j'=-\infty \\ '\neq -j}}^{\infty} \frac{(n+j, n+j+j')}{(j+j')(2n+j+j')}(n+j+j', n) \tag{68}$$
$$+ \cdots$$

This is the equation determining Δ and hence the eigenenergy k^2.

It is now convenient to regroup successive terms in ψ and (68) according to the approximations suggested by the inequality (60). The solution ψ then becomes

$$\psi = \psi_n + \left[\frac{(n, n+1)}{1\cdot(2n+1)}\psi_{n+1} + \frac{(n, n-1)}{(-1)\cdot(2n-1)}\psi_{n-1}\right]$$
$$+ \left[\frac{(n, n+2)}{2(2n+2)}\psi_{n+2} + \frac{(n, n-2)}{(-2)(2n-2)}\psi_{n-2}\right.$$
$$+ \frac{(n, n+1)(n+1, n+1)}{1\cdot(2n+1)\cdot 1\cdot(2n+1)}\psi_{n+1}$$
$$\left.+ \frac{(n, n-1)(n-1, n-1)}{(-1)(2n-1)(-1)(2n-1)}\psi_{n-1}\right] + \cdots \tag{69}$$

or, for corresponding coefficients $P_i(n, j)$:

$$\psi = \psi_n + \sum_{i=1}^{\infty} \sum_{\substack{j=-i \\ j\neq 0}}^{i} P_i(n, j)\psi_{n+j} \tag{70}$$

Solving (68) for Δ, we obtain

$$\Delta = -\beta_0 - \frac{2g\beta_1^2}{(2n-1)(2n+1)} + 0(\beta_2^2) \tag{71}$$

or, by (48),

$$k^2 + g = -n^2 - g\beta_0 - \frac{2g^2\beta_1^2}{(2n-1)(2n+1)} + \cdots \tag{72}$$

We observe that the denominators of successive terms of both the solution (69) and the eigenvalue (72) contain factors which vanish for half-integral and integral values of n. At first sight one might think, therefore, that these expansions are completely useless. However, a closer look shows that this is by no means the case. A similar phenomenon occurs in the solutions and eigenvalues of the Mathieu equation, as can be seen by comparing (69) and (72) with (19) and (21). Without going into extensive details, we simply remark here that the first three terms of (69) constitute a valid approximation for $n > 1$, and in general, if m is the largest value of n for which a certain number of terms for an expansion of this type becomes infinite, this number of terms is a valid approximation for values of n larger than m. For a detailed discussion of this problem and in particular of the convergence properties of these expansions in the case of the Mathieu equation, we refer to Meixner and Schäfke[2] (Section 2.25). We shall return to this point again at a later stage in connection with anharmonic oscillator potentials[10].

So far the analysis in this section proceeded along the same lines as in Section 2.1. We now extend the eigenvalue problem to a scattering problem. This implies passing from negative to positive values of k^2 by analytic continuation. A unique analytic continuation is defined by the eigenvalue expansions (72). All we have to do is to reverse this series to calculate n^2 and to replace n^2 by ν^2, where ν is a non-integral parameter of the theory (a close look at equation (72) shows that all the coefficients are functions of n^2 only but not of n). Then

$$\begin{aligned}\nu^2 &= -k^2 - g - g\beta_0 + 2g^2\beta_1{}^2/(4k^2 + 1) + \cdots \\ \nu &= +ik + \cdots\end{aligned} \tag{73}$$

and the corresponding scattering solution is

$$\psi_R(k, r) = N\left[\psi_\nu(k, r) + \sum_{i=1}^{\infty} \sum_{\substack{j=-i \\ j\neq 0}}^{i} P_i(\nu, j)\psi_{\nu+j}(k, r)\right] \tag{74}$$

where N is a constant factor. We now observe that ψ_R is a regular solution satisfying the boundary condition

$$\psi_R(k, 0) = 0 \tag{75}$$

In order to obtain the scattering matrix and hence the phase-shift, it is necessary to consider solutions of the radial Schrödinger equation for $r \to \infty$. A glance at equations (50) to (55) shows that if $\psi_n = \sinh nr$ is replaced by

$$\psi_n{}' = e^{nr} \tag{76}$$

the form of the solution (69) remains unchanged, and also the same expansion for the eigenvalues is obtained as before. This new solution (called the Jost solution)

$$\psi^{(+)}(k, r) = \psi_\nu{}'(k, r) + \sum_{i=1}^{\infty} \sum_{\substack{j=-i \\ \neq 0}}^{i} P_i(\nu, j)\psi_{\nu+j}{}'(k, r) \tag{77}$$

however, satisfies a different boundary condition:

$$\psi^{(+)}(k, r) \simeq e^{ikr} \quad \text{for } r \to \infty \tag{78}$$

The S-wave Jost function $f(-k)$ is therefore given by

$$f(-k) = \psi^{(+)}(k, 0) \tag{79}$$

and the scattering matrix S by

$$S = e^{2i\delta_0} = f(k)/f(-k) \tag{80}$$

where for k real

$$f(k) = f^*(-k) \tag{81}$$

(the asterisk * implying complex conjugation).

In order to obtain the proper S-matrix it is convenient to replace ψ_ν' of (77) by the expression

$$e^{\nu r} = (2\cosh r)^\nu F\left(\frac{1-\nu}{2}, -\frac{\nu}{2}, -\nu+1; \frac{1}{\cosh^2 r}\right) \tag{82}$$

By (79) it then follows that

$$f(-k) = F\left(\frac{1-\nu}{2}, -\frac{\nu}{2}, -\nu+1; 1\right)\cdot 2^\nu$$

$$+\sum_{i=1}^{\infty}\sum_{\substack{j=-i\\ j\neq 0}}^{i} 2^{\nu+j}P_i(\nu, j)F\left(\frac{1-\nu-j}{2}, -\frac{\nu+j}{2}, -\nu-j+1; 1\right)$$

$$= 2^\nu\cdot\frac{\Gamma(1-\nu)\Gamma(\frac{1}{2})}{\Gamma\left(\frac{1-\nu}{2}\right)\Gamma\left(1-\frac{\nu}{2}\right)}\left[1+\sum_{i=1}^{\infty}\sum_{\substack{j=-i\\ j\neq 0}}^{i} P_i(\nu, j)\right] \tag{83}$$

The S-matrix may therefore be written

$$S = 2^{\nu^*-\nu}\cdot\frac{\Gamma(1-\nu^*)\Gamma\left(\frac{1-\nu}{2}\right)\Gamma\left(1-\frac{\nu}{2}\right)}{\Gamma(1-\nu)\Gamma\left(\frac{1-\nu^*}{2}\right)\Gamma\left(1-\frac{\nu^*}{2}\right)}$$

$$\cdot\left[\frac{1+\sum_{i=0}^{\infty}\sum_{\substack{j=-i\\ j\neq 0}}^{i} P_i(\nu^*, j)}{1+\sum_{i=0}^{\infty}\sum_{\substack{j=-i\\ \neq 0}}^{i} P_i(\nu, j)}\right] \tag{84}$$

The quotient of sums in this expression can be shown to cancel out under fairly general conditions. The remaining factors show that S does in fact possess poles at the points $\nu, \nu^* = 1, 2, 3, \ldots$—in complete agreement with our earlier eigenvalue condition (48).

We leave the discussion of this potential here; a more detailed discussion will be given elsewhere. This simple example, however, should suffice to indicate the procedure we shall use in later sections for much more realistic interactions.

2.3 *Formal development of the perturbation method*

For a deeper understanding of our formalism it is best to apply it next to a general differential equation of the second order[11]. We consider therefore an equation of the form

$$D\phi + (E - V)\phi = 0 \tag{85}$$

where D is a differential operator and E a parameter. It is immaterial here whether D does or does not contain a first order derivative—in fact, with certain restrictions on the function V, the analysis below would be valid even if D contained only the derivative of the first order but not of the second.

We also assume that V can be reexpressed as the sum of two terms

$$V = V_0 + v \tag{86}$$

such that if correspondingly

$$E = E_0 + \epsilon \tag{87}$$

(85) may be written

$$D\phi + (E_0 - V_0)\phi = (v - \epsilon)\phi \tag{88}$$

The subdivision (86) is thereby chosen such that the equation

$$D\phi + (E_0 - V_0)\phi = 0 \tag{89}$$

is exactly solvable and has the eigenvalue $E_0 = E_m$ and eigensolution $\phi = \phi_m$, where m is an additional integral or near-integral parameter. Then, if

$$|\epsilon| \ll |E| \quad \text{and} \quad |v - \epsilon| \ll |V_0 - E_0|$$

(the latter in a limited region of the independent variable),

$$\phi^{(1)} = \phi_m \tag{90}$$

represents a first approximation to the solution ϕ of (85).

The next step is to use recursion formulae to reexpress $(v - \epsilon)\phi_m$ as a linear combination of ϕ_{m+s}:

$$(v - \epsilon)\phi_m = \sum_s (m, m + s)\phi_{m+s} \tag{91}$$

where it is advantageous to consider the coefficient of ϕ_{m+s} as a 'step' from m to $m + s$. Since, however,

$$\begin{aligned} D\phi_{m+s} + (E_m - V_0)\phi_{m+s} \\ &= D\phi_{m+s} + (E_{m+s} - V_0)\phi_{m+s} + (E_m - E_{m+s})\phi_{m+s} \\ &= (E_m - E_{m+s})\phi_{m+s} \end{aligned} \tag{92}$$

the second approximation $\phi^{(2)}$, for which we may set

$$\phi^{(2)} = \sum_s c_s \phi_{m+s}$$

is

$$\phi^{(2)} = \sum_{s\neq 0} \frac{(m, m + s)}{(E_m - E_{m+s})} \phi_{m+s} \tag{93}$$

This may be seen as follows:

$$\begin{aligned} (D + E_m - V_0)(\phi^{(1)} + \phi^{(2)}) \\ &= (D + E_m - V_0)\phi^{(2)} = \sum_s c_s(E_m - E_{m+s})\phi_{m+s} \\ &= (v - \epsilon)\phi^{(1)} = \sum_s (m, m + s)\phi_{m+s} \end{aligned} \tag{94}$$

and hence

$$c_{s\neq 0} = (m, m + s)/(E_m - E_{m+s}) \tag{95}$$

To ensure that $\phi = \phi^{(1)} + \phi^{(2)}$ is actually a solution of (85) to a second approximation, the coefficient of the term $(m, m)\ \phi_m$ in (94) must vanish; i.e.

$$(m, m) = 0 \text{ to a first approximation} \tag{96}$$

Repeating this procedure with $\phi^{(2)}$ instead of $\phi^{(1)}$, we obtain the third contribution

$$\phi^{(3)} = \sum_{s\neq 0} \frac{(m, m + s)}{(E_m - E_{m+s})} \sum_{r\neq 0} \frac{(m + s, m + r)}{(E_m - E_{m+r})} \phi_{m+r} \tag{97}$$

together with

$$(m, m) + \sum_{s\neq 0} \frac{(m, m + s)(m + s, m)}{(E_m - E_{m+s})} = 0 \tag{98}$$

This procedure may be repeated as often as desired. We find

$$\begin{aligned}\phi &= \phi^{(1)} + \phi^{(2)} + \phi^{(3)} + \cdots \\ &= \phi_m + \sum_{s\neq 0} \frac{(m, m+s)}{(E_m - E_{m+s})} \phi_{m+s} \\ &\quad + \sum_{s\neq 0} \sum_{r\neq 0} \frac{(m, m+s)(m+s, m+r)}{(E_m - E_{m+s})(E_m - E_{m+r})} \phi_{m+r} + \cdots \end{aligned} \tag{99}$$

and

$$\begin{aligned} 0 &= (m, m) + \sum_{s\neq 0} \frac{(m, m+s)(m+s, m)}{(E_m - E_{m+s})} \\ &\quad + \sum_{s\neq 0} \sum_{r\neq 0} \frac{(m, m+s)(m+s, m+r)(m+r, m)}{(E_m - E_{m+s})(E_m - E_{m+r})} \\ &\quad + \cdots \end{aligned} \tag{100}$$

Equation (99) is the desired eigensolution and (100) the secular equation from which ϵ and hence E may be determined.

3 Perturbed Coulomb Interactions

3.1 *The Yukawa potential*

In the theory of the scattering of particles the most important potential is the Yukawa potential. The predominance of this potential is on the one hand due to the fact that it is the only known potential which can be directly associated with the exchange of a particle by the scattering particles, while on the other hand it appears to play a singular role in mathematical analysis, where the Mandelstam representation of the scattering amplitude, for instance, has so far been shown to exist only for Yukawa-like interactions. Here we shall use extensively the fact that at high energies the Yukawa potential represents a mildly perturbed Coulomb interaction, the latter being a well-known potential for which the eigenvalues and eigensolutions are simple and known in closed form.

The eigenvalues and eigensolutions for the Yukawa potential were first discussed by Lovelace and Masson[12], Mandelstam[13] and Bethe and Kinoshita[14]. The present author[15,16] then used the perturbation formalism of the preceding chapter to derive high-energy asymptotic expansions for the eigenvalues and eigensolutions of the Schrödinger equation, as well as corresponding expressions for the S-matrix and its residues at the Regge poles. Explicit evaluation of several expansion coefficients later revealed (cf. Müller and Schilcher[17]) useful simplifications. We review this work here and then discuss several consequences and applications.

3.2 *Asymptotic behaviour of the complex angular momentum at high energies*

We start with the radial Schrödinger equation in the form

$$\psi''(r) + \left[k^2 - \frac{l(l+1)}{r^2} - V(r)\right]\psi(r) = 0 \tag{101}$$

where $E = k^2$ and $\hbar = c = 1 = 2m$, m being the reduced mass of the 2-particle system. We now set

$$\psi(r) = \mathrm{e}^{\mathrm{i}kr} r^{l+1} \chi(r) \tag{102}$$

and

$$\kappa = \mathrm{i}k, \quad z = -2\mathrm{i}kr, \quad \dot{\chi} = \mathrm{d}\chi/\mathrm{d}r \tag{103}$$

Then

$$z\ddot{\chi} + (2l + 2 - z)\dot{\chi} - \left[l + 1 + \frac{z}{(2\kappa)^2} V\left(-\frac{z}{2\kappa}\right)\right]\chi = 0 \tag{104}$$

Next we assume that the potential V represents a superposition of Yukawa potentials

$$V(r) = \int_{m_0>0}^{\infty} \mathrm{d}\mu\, \sigma(\mu)\, \mathrm{e}^{-\mu r}/r \tag{105}$$

which may be expanded in ascending powers of r; i.e. we have

$$V(r) = \sum_{i=-1}^{\infty} M_{i+1}(-r)^i = \sum_{i=-1}^{\infty} M_{i+1}\left(\frac{z}{2\kappa}\right)^i \tag{106}$$

so that

$$\int_{m_0>0}^{\infty} \sigma(\mu)\mu^n\, \mathrm{d}\mu < C < \infty \text{ for all } n \tag{107}$$

For the real, energy-independent potentials we discuss first, all M_i are real and independent of k.

Following Regge we now consider l, the quantum number of angular momentum, to be a complex parameter. What then is the behaviour of this parameter at high energies, if ψ is required to satisfy the boundary condition of a bound-state wave-function? This question has been discussed by many authors[12,13,14]. A rough approximation to the asymptotic behaviour of l may be obtained as follows. Setting

$$\chi(z) = \sum_{i=0}^{\infty} C_i z^i$$

one finds a recursion formula for the coefficients C_i:

$$(i + 1)(2l + 2 + i)C_{i+1} = (i + l + 1)C_i + \sum_{j=0}^{i} \frac{C_{i-j}M_j}{(2\kappa)^{j+1}} \tag{108}$$

If l is now chosen such that for $k^2 < 0$

$$i + l + 1 + \frac{M_0}{2\kappa} = 0\left(\frac{1}{\kappa^2}\right) \tag{109}$$

for $i = 0, 1, 2, \ldots$, we see that

$$C_{i+1}/C_i = 0(1/\kappa^2)$$

The relationship

$$C_{i+n}/C_i = 0(1/\kappa^2)$$

is readily seen to hold for $n = 1, \ldots, i$. Thereafter (due to the factor $2l + 2 + i$ on the left-hand side of equation 108) it is of order $1/\kappa$. Thus in the limit $|\kappa| \to \infty$, χ represents effectively a polynomial, so that in this limit (and its neighbourhood) a bound state or Regge pole can be defined. For $k^2 < 0$ we may therefore write

$$l + n + 1 = -\frac{\Delta(\kappa)}{2\kappa} \tag{110}$$

where Δ is as yet an undetermined expansion in descending powers of κ, its first approximation being $\Delta = +M_0$.

3.3 *The regular solution and the asymptotic expansion for the angular momentum*

We now use the formalism of Section 2 to solve (104). Substituting (110) into (104), we have the fundamental equation

$$D_n\chi = \frac{1}{2\kappa}(M_0 - \Delta(\kappa))\chi + \frac{1}{2\kappa}\sum_{i=1}^{\infty}\left(\frac{z}{2\kappa}\right)^i M_i\chi \tag{111}$$

where

$$D_n = z\mathrm{d}^2/\mathrm{d}z^2 + (b - z)\,\mathrm{d}/\mathrm{d}z - a \tag{112}$$

and

$$\left.\begin{aligned} a &= l + 1 + \frac{\Delta(\kappa)}{2\kappa} = -n \\ b &= 2l + 2 = -2n - \frac{\Delta(\kappa)}{\kappa} \end{aligned}\right\} \tag{113}$$

Since, however, the right-hand side of (111) is of order $1/\kappa$, we have to a first approximation $\chi = \chi^{(1)}$ and

$$D_n\chi^{(1)} = 0 \tag{114}$$

Equation (114) is seen to be a confluent hypergeometric equation. A particular solution of this equation (which leads to the regular solution ψ of the radial Schrödinger equation) is the function

$$\Phi(a, b; z) = 1 + \frac{a}{b}\frac{z}{1!} + \frac{a(a+1)}{b(b+1)}\frac{z^2}{2!} + \cdots \tag{115}$$

Here again we observe that if we require ψ to be a normalizable bound-state wave-function, the Kummer series (115) must be broken off after a finite number of terms—and this means setting $a = -n$, which is precisely our earlier approximation (110).

In the next step of our perturbation formalism we require the recurrence relation[18] of the confluent hypergeometric functions (115). We write this relation in the following form:

$$z\Phi(a) = (a, a+1)\Phi(a+1) + (a, a)\Phi(a) + (a, a-1)\Phi(a-1) \tag{116}$$

where $\Phi(a) \equiv \Phi(a, b; z)$ and

$$\left.\begin{aligned}(a, a+1) &= a\\ (a, a) &= b - 2a\\ (a, a-1) &= a - b\end{aligned}\right\} \tag{117}$$

By repeated application of (116) we obtain

$$z^m\Phi(a) = \sum_{j=m}^{-m} S_m(a, j)\Phi(a+j) \tag{118}$$

with the coefficients S_m satisfying the following recurrence relation

$$\begin{aligned}S_m(a, r) = {} & S_{m-1}(a, r-1)\cdot(a+r-1, a+r)\\ & + S_{m-1}(a, r)\cdot(a+r, a+r)\\ & + S_{m-1}(a, r+1)\cdot(a+r+1, a+r)\end{aligned} \tag{119}$$

The boundary conditions for this difference equation are

1. $S_0(a, 0) = 1$ and all $S_0(a, i \neq 0) = 0$;
2. $S_m(a, r) = 0$ for $|r| > m$.

The first approximation $\chi^{(1)} = \Phi(a)$ therefore leaves uncompensated terms on the right-hand side of (111) amounting to

$$\frac{1}{2\kappa}[a, a]_1\Phi(a) + \sum_{i=1}^{\infty}\frac{1}{(2\kappa)^{i+1}}\sum_{j=-i}^{i}[a, a+j]_{i+1}\Phi(a+j) \tag{120}$$

where

$$\left.\begin{aligned}[a, a]_1 &= M_0 - \Delta(\kappa)\\ [a, a+j]_{i+1} &= M_i S_i(a, j), \quad 0 \leqslant |j| \leqslant i\end{aligned}\right\} \tag{121}$$

We now observe that

$$D_n\Phi(a+j) = j\Phi(a+j) \tag{122}$$

Thus a term $\dfrac{1}{(2\kappa)^{i+1}}[a, a+j]_{i+1}\Phi(a+j)$ in (120) can be cancelled out by adding to the previous approximation the contribution

$$1/(2\kappa)^{i+1}\cdot[a, a+j]_{i+1}/j\cdot\Phi(a+j)$$

except, of course, when $j = 0$. The next contribution to $\chi^{(1)}$ is therefore

$$\chi^{(2)} = \sum_{i=1}^{\infty}\frac{1}{(2\kappa)^{i+1}}\sum_{\substack{j=-i\\ j\neq 0}}^{i}\frac{[a, a+j]_{i+1}}{j}\Phi(a+j) \tag{123}$$

However $\chi = \chi^{(1)} + \chi^{(2)}$ represents an approximation to χ only if simultaneously the sum of the coefficients of terms in $\Phi(a)$ is set equal to zero; i.e.

$$[a, a]_1 + \sum_{i=1}^{\infty}\frac{[a, a]_{i+1}}{(2\kappa)^{i+1}} = 0$$

Proceeding in this manner and reordering terms according to powers in $1/\kappa$, we find that the solution χ may be written

$$\chi(a, b; z) = \Phi(a) + \sum_{i=2}^{\infty}\frac{1}{(2\kappa)^i}\sum_{\substack{j=-i+1\\ j\neq 0}}^{i-1}P_i(a, j)\Phi(a+j) \tag{124}$$

where

$$\begin{aligned}P_2(a, 1) &= \frac{[a, a+1]_2}{1}, \quad P_2(a, -1) = \frac{[a, a-1]_2}{-1}\\ P_3(a, 2) &= \frac{[a, a+2]_3}{2}\\ P_3(a, 1) &= \frac{[a, a+1]_3}{1} + \frac{[a, a+1]_2}{1}\frac{[a+1, a+1]_1}{1}\end{aligned} \tag{125}$$

etc. These coefficients again follow from a recurrence relation:

$$tP_r(a, t) = \sum_{i=1}^{r}\sum_{j=-i+1}^{i-1}[a+t-j, a+t]_i\cdot P_{r-i}(a, t-j) \tag{126}$$

subject to the boundary conditions:

1. $P_0(a, 0) = 1$ and $P_0(a, t \neq 0) = 0$;
2. $P_r(a, t) = 0$ for $|t| \geqslant r$;
3. $P_r(a, 0) = 0$ for $r > 1$.

The subsidiary condition determining Δ and hence l may also be expressed in terms of these coefficients (cf. Müller and Schilcher[17]) by a formalism previously developed for Mathieu[5] and other functions[19,20]. Here we restrict ourselves to the expansion

$$0 = \frac{1}{2\kappa}[a, a]_1 + \frac{1}{(2\kappa)^2}[a, a]_2 + \frac{1}{(2\kappa)^3}[a, a]_3 + \frac{1}{(2\kappa)^4}\left\{[a, a]_4 + \frac{[a, a+1]_2}{1}[a+1, a]_2 + \frac{[a, a-1]_2}{-1}[a-1, a]_2\right\} + \cdots \tag{127}$$

The solution $\psi(r) = e^{ikr}r^{l+1}\chi(r)$ is now seen to be the required bound-state eigensolution which is regular at the origin $r = 0$.

3.4 *The complex angular momentum*

Solving (127) for $\Delta \equiv \Delta_n$ we obtain[15]

$$\Delta_n(\kappa) = M_0 - \frac{1}{2\kappa^2}[n(n+1)M_2 + M_0M_1] - \frac{(2n+1)}{4\kappa^3}M_0M_2 + \frac{1}{8\kappa^4}[3M_4(n-1)n(n+1)(n+2) + 2M_3M_0(3n^2+3n-1) + 6M_2M_1n(n+1) + 2M_2M_0^2 + 3M_1^2M_0] + \frac{(2n+1)}{8\kappa^5}[3M_4M_0(n^2+n-1) + 3M_3M_0^2 + M_2^2n(n+1) + 4M_2M_1M_0] + 0(1/\kappa^7) \tag{128}$$

Grouping the term M_1 (of the potential) together with k^2, we obtain an expansion for l with a larger region of validity (cf. Müller[16]):

$$l = -n - 1 + \frac{M_0}{2h} - \frac{n(n+1)M_2}{4h^3} + \frac{(2n+1)M_0M_2}{8h^4}$$

$$+ \frac{1}{16h^5}[3M_4(n-1)n(n+1)(n+2) + 2M_3M_0(3n^2+3n-1)$$
$$+ 2M_2M_0^2]$$
$$- \frac{1}{16h^6}[3M_4M_0(2n^3+3n^2-n-1) + 3M_3M_0^2(2n+1)$$
$$+ M_2^2n(n+1)(2n+1)]$$
$$- \frac{1}{16h^7}[10M_6(n-2)(n-1)n(n+1)(n+2)(n+3)$$
$$+ 2M_5M_0\langle 5n(n+1)(3n^2+3n-10)+12\rangle$$
$$+ 2M_4M_0^2(6n^2+6n-11) + 4M_3M_0^2$$
$$+ 10M_3M_2n(n+1)(3n^2+3n-4)$$
$$+ M_2^2M_0(18n^2+18n-7)] + \cdots \tag{129}$$

where $h = \sqrt{(\kappa^2 + M_1)}$. For $|M_1/\kappa^2| < 1$ and $\sqrt{\kappa^2} = -\kappa$ the denominators may be expanded in descending powers of κ and again yield the expansion (128) for Δ. Expansion (129) is valid for $k^2 < 0$ or $k^2 > M_1 > 0$.

If we reverse (129) and calculate n in terms of l, we find that Δ may also be expressed as

$$\Delta_l(\kappa) = M_0 - \frac{1}{2\kappa^2}[l(l+1)M_2 + M_0M_1]$$

$$+ \frac{1}{8\kappa^4}[3(l-1)l(l+1)(l+2)M_4 + 2M_3M_0(3l^2+3l-1)$$

$$+ 6l(l+1)M_2M_1 + 3M_2M_0^2 + 3M_1^2M_0] + 0(1/\kappa^6) \tag{130}$$

Thus: for $k^2 < 0$ the expansions (128), (129) determine the eigenvalues counted in terms of the discrete quantum number n. The same relations define a unique analytic continuation into the region $k^2 > 0$ if n is considered to be a function of l and k. The corresponding expansion for n substituted into the regular eigensolution obtained above defines uniquely the regular scattering solution for $k^2 > 0$.

3.5 *The generalized Balmer formula*

The expansion (129) may also be reversed to yield the discrete energy eigenvalues. One finds[21]

$$\kappa^2 + M_1 = \frac{M_0^2}{4(l+n+1)^2}\left\{1 - 4n(n+1)\frac{M_2}{M_0}(l+n+1)^2 \right.$$
$$+ 4(2n+1)\frac{M_2}{M_0^3}(l+n+1)^3$$
$$+ \frac{4}{M_0^6}[3M_4M_0(n-1)n(n+1)(n+2) + 2M_3M_0^2(3n^2+3n-1)$$
$$- 3M_2^2n^2(n+1)^2 + 2M_2M_0^3]\cdot(l+n+1)^4$$
$$- \frac{24(2n+1)}{M_0^6}[M_4M_0(n^2+n-1) + M_3M_0^2 - M_2^2n(n+1)]$$
$$\left.\cdot(l+n+1)^5 + \cdots\right\} \qquad (131)$$

(one more term is given in the literature[21]). This formula is readily seen to be a generalization of the Balmer formula for the discrete energy-levels of the hydrogen atom. In fact, in atomic physics, where the Yukawa potential is known as the Debye–Hückel potential, the additional terms represent the screening effect of the electron-cloud surrounding the atom. For a more detailed discussion and numerical examples demonstrating the usefulness of (131) see Müller[21].

3.6 *The Jost solutions*

We have seen before that to obtain the S-matrix, we need the Jost solutions which behave like $\exp(\pm ikr)$ for $r \to \infty$. So we define a solution $\psi^{(+)}$ by the condition

$$\psi^{(+)}(l, k; r) \simeq e^{ikr} \quad \text{for} \quad |r| \to \infty \qquad (132)$$

As before we set

$$\psi^{(+)}(l, k; z) = c \cdot e^{-z/2} z^{l+1} \chi^{(+)}(l, k; z) \qquad (133)$$

where c is an appropriately chosen constant:

$$c = 1/\Gamma(a) \qquad (134)$$

The fact that $a = -n$ by (113) need not bother us here, since n is now—in the case of positive energies—considered to be a non-integral parameter. $\chi^{(+)}$, of course, is again a solution of equation (111). In view of the boundary condition (132), however, we now choose for the first approximation to $\chi^{(+)}$ a different solution of the confluent hypergeometric equation. The particular choice we make is[15]

$$\chi^{(+)(1)} = \Gamma(a)\Psi(a, b; z) \equiv \tilde{\Psi}(a, b; z) \tag{135}$$

where

$$\Psi(a, b; z) = \frac{\Gamma(1-b)}{\Gamma(a-b+1)}\,\Phi(a, b; z) + \Gamma(b-1)/\Gamma(a)\cdot z^{1-b}\Phi(a-b+1, 2-b; z) \tag{136}$$

The solution (135) was chosen so that it again obeys a simple recurrence relation similar to (116). One finds:

$$z\tilde{\Psi}(a) = (a, a+1)^*\tilde{\Psi}(a+1) + (a, a)^*\tilde{\Psi}(a) + (a, a-1)^*\tilde{\Psi}(a-1) \tag{137}$$

where

$$\left.\begin{aligned} (a, a+1)^* &= a - b + 1 \equiv \alpha \\ (a, a)^* &= b - 2a = \beta - 2\alpha, \qquad \beta = 2 - b \\ (a, a-1)^* &= a - 1 = \alpha - \beta \end{aligned}\right\} \tag{138}$$

In a manner completely analogous to our previous procedure we may now define coefficients

$$S_m^*(a, j), \quad [a, a+j]_i^*, \quad P_i^*(a, j)$$

and thus obtain a solution

$$\chi^{(+)}(a, b; z) = \tilde{\Psi}(a, b; z) + \sum_{i=2}^{\infty} \frac{1}{(2\kappa)^i} \sum_{\substack{j=-i+1 \\ j\neq 0}}^{i-1} P_i^*(a, j)\tilde{\Psi}(a+j, b; z) \tag{139}$$

The subsidiary condition

$$0 = \frac{1}{2\kappa}\,[a, a]_1^* + \frac{1}{(2\kappa)^2}\,[a, a]_2^* + \cdots \tag{140}$$

(cf. equation 127) leads again to the same eigenvalue expansion as before.

Next, substituting (139) for $\chi^{(+)}$ into (133) and taking $V = 0$, we obtain

$$\psi^{(+)}(l, k; r) = \mathrm{e}^{\mathrm{i}\frac{\pi}{2}(l+1)}\left(\frac{\pi kr}{2}\right)^{\frac{1}{2}} H^{(1)}_{l+\frac{1}{2}}(kr) \tag{141}$$

where

$$\left(\frac{\pi kr}{2}\right)^{\frac{1}{2}} H_n^{(1)}(kr) \simeq \mathrm{e}^{\mathrm{i}(kr - \frac{1}{2}n\pi - \frac{1}{4}\pi)} \tag{142}$$

for $r \to \infty$. $\psi^{(+)}(l, k; r)$ is therefore the Jost solution with the required asymptotic behaviour (132).

3.7 *The S-matrix*

The S-matrix for all l and k is defined as

$$S(l, k) = \exp[2i\delta(l, k)] = f(l, k)/f(l, -k) \tag{143}$$

where the Jost function $f(l, -k)$ follows from the solution $\psi^{(+)}$ by the relation[22]

$$f(l, -k) = \lim_{r\to 0} \frac{\Gamma(l+1)}{\Gamma(2l+1)}(-2ikr)^l \psi^{(+)}(l, k; r) \tag{144}$$

Substituting in (144) for $\psi^{(+)}$, we find

$$f(l, -k) = \frac{\Gamma(l+1)}{\Gamma\left[l+1+\dfrac{\Delta(\kappa)}{2\kappa}\right]}\left[1 + \sum_{i=2}^{\infty} \frac{1}{(2\kappa)^i} \sum_{\substack{j=-i+1 \\ j\neq 0}}^{i-1} P_i^*(a, j)\right] \tag{145}$$

for $\operatorname{Re} l > -1/2$ (and for $\operatorname{Re} l < -1/2$ by analytic continuation). Evaluating the sum on the right-hand side (cf. Müller and Schilcher[17]), the Jost function becomes

$$f(l, -k) = \frac{\Gamma(l+1)}{\Gamma\left[l+1+\dfrac{\Delta(\kappa)}{2\kappa}\right]}\left[1 + \frac{(n+1)M_1}{2\kappa^2} + \frac{M_0 M_3}{4\kappa^3} + \cdots\right] \tag{146}$$

Here, of course, n is a function of l and k as pointed out before. Substituting for n the corresponding expansion in descending powers of κ (obtained by reversing equation 110), the Jost function is found to be given by[17]

$$\begin{aligned} f(l, -k) = \frac{\Gamma(l+1)}{\Gamma\left[l+1+\dfrac{\Delta_l(\kappa)}{2\kappa}\right]} \Bigg\{ 1 &- \frac{M_1 l}{2\kappa^2} + \frac{1}{3(2\kappa)^4} \\ &\times [9(2l-1)M_0M_2 + 3l(l+3)M_1^2 + 2l(l-1)(8l+11)M_3] \\ &+ \frac{1}{60(2\kappa)^6}[60M_0^2M_1^2 - 1200(l-1)M_0^2M_3 \\ &- 120(3l^2+17l-10)M_0M_1M_2 - 100(22l^3-3l^2 \\ &- 55l+15)M_0M_4 + 40l(l+2)(l-5)M_1^3 \\ &- 40l(l-1)(7l^2+59l+70)M_1M_3 \\ &+ 5l(3l^3-226l^2-39l+310)M_2^2 \\ &- 8l(l-1)(l-2)(128l^2+479l+411)M_5] + 0(1/\kappa^8) \Bigg\} \end{aligned} \tag{147}$$

We observe that at least to $0(1/\kappa^6)$ the expansion on the right of (147) is an even function of κ. It is plausible to conjecture that the complete expansion is an even function of κ, although we have no general proof. Similarly we assume (cf. equation 130) that

$$\Delta_l(\kappa) = \Delta_l(-\kappa) \tag{148}$$

provided the potential is independent of k. By (143) the S-matrix is seen to be

$$S(l, k) = \frac{\Gamma\left[l + 1 + \dfrac{\Delta_l(\kappa)}{2\kappa}\right]}{\Gamma\left[l + 1 - \dfrac{\Delta_l(\kappa)}{2\kappa}\right]} \tag{149}$$

Taking the logarithm of this expression and expanding the gamma-functions around $l + 1$, one obtains the high-energy expansion for the phase-shift[17]

$$\delta(l, k) = -\frac{1}{2k} M_0\psi(l + 1) + \frac{1}{48k^3}\left\{M_0{}^3\psi^{(2)}(l + 1) - 12[l(l + 1)M_2 + M_0M_1]\psi(l + 1)\right\} + 0(1/k^5) \tag{150}$$

where

$$\psi(x) = \frac{\mathrm{d}}{\mathrm{d}x}\ln\Gamma(x) \quad \text{and} \quad \psi^{(2)} = \frac{\mathrm{d}^2\psi(x)}{\mathrm{d}x^2} \tag{151}$$

Thus at infinitely high energies the phase-shift decreases to zero as one expects for a proper regular potential.

3.8 *Regge poles, trajectories, residues*

Since $\Gamma(z)$ is a non-vanishing meromorphic function of z with simple poles at $z = -n$, $n = 0, 1, 2, \ldots$, we see that the only poles of the S-matrix (149) are given by

$$l + 1 + \frac{\Delta(\kappa)}{2\kappa} = -n$$

which is precisely our earlier eigenvalue condition (110). The contours of $l \equiv \alpha_n(\kappa)$ considered as complex functions of κ are called the Regge trajectories. It is hardly necessary to point out the obvious similarity between our results here and those for the Coulomb potential. The Regge trajectories for the latter are in fact the asymptotes of those for the Yukawa potential. For further details and discussions of Regge trajectories for the Coulomb potential we refer to the paper of Singh[23]; Yukawa trajectories have been plotted and analysed by several authors—for instance Lovelace and Masson[12] and Ahmadzadeh, Burke and Tate[24]. We recall, however, that one of the unphysical features of the Coulomb potential is the infinite number of bound states it produces for any given value of n. This number may even be calculated[21] from the condition $k^2 < 0$ in the case of Yukawa potentials, where it is finite; i.e. (cf. equation 131)

$$M_1 - \frac{M_0^2}{4(l+n+1)^2} < 0 \text{ (to first order)}$$

or

$$0 \leqslant l < \frac{1}{2}\sqrt{\left(\frac{M_0}{M_1}\right)} - n - 1 \tag{152}$$

If the right-hand side of this inequality lies between n_1 and $n_1 + 1$, where n_1 is an integer, approximately n_1 bound states may be expected (we are not considering exchange potentials).

The residues at the poles of the S-matrix in the plane of complex l can also be calculated easily. One readily finds[17]

$$\beta_n(\kappa) = \frac{(-1)^n}{n!\Gamma[-n + 2\mathrm{i}\,\mathrm{Im}\,\alpha_n(\kappa)]} \tag{153}$$

for the residue at the nth Regge pole. This expression is remarkable as it is completely independent of the deriving perturbation formalism. Moreover, it is valid also for other approximations of $\alpha_n(\kappa)$ apart from our asymptotic expansion (129). In particular, if $\mathrm{Im}\,\alpha_n(\kappa)$ is small—as for resonances which occur just above threshold, we may expand the gamma function in (153) and obtain

$$\beta_n(\kappa) \simeq -\frac{2\mathrm{i}}{(n!)^2}\,\mathrm{Im}\,\alpha_n(\kappa) \tag{154}$$

—an approximation which is also valid for $|\kappa| \to \infty$.

3.9 *Half-widths and phase-shifts*

For practical analysis it is useful to have an approximate relation between the half-width of a resonance and the phase-shift near the resonance energy. This expression may be obtained as follows.

Suppose $s = 4(k^2 + 1)$ is the square of the total energy in the s-channel of a scattering process for two scalar particles (e.g. π-mesons) both having the same mass $m = 1$. The half-width of a resonance on the Regge trajectory α_n is then

$$\Gamma/2 = (\text{Im}\ \alpha_n)_{E_R}/(\text{d Re}\ \alpha_n/\text{d}E)_{E_R}, \quad E = k^2 \tag{155}$$

where E_R is the resonating energy. This formula is well known from elementary Regge pole analysis. Also

$$\delta(l, k) = -\arg f(l, -k) = -\sum_{n=0}^{\infty} \arg\left[l - \alpha_n(k)\right] \tag{156}$$

where the second equality follows readily from the Weierstrass product theorem. Thus, expanding α_n in the neighbourhood a resonance, we have

$$\alpha_n(E) \simeq l + \left(\frac{\text{d Re}\ \alpha_n}{\text{d}E}\right)_{E=E_R} \cdot (E - E_R) + \text{i}\,(\text{Im}\ \alpha)_{E=E_R} \tag{157}$$

and so

$$\begin{aligned}\delta &\simeq -\tan^{-1}\left[\text{Im}\ \alpha_n(k)/(E - E_R)\left(\frac{\text{d Re}\ \alpha}{\text{d}E}\right)_{E=E_R}\right] \\ &= -\tan^{-1}\left(\frac{\Gamma/2}{E - E_R}\right) \simeq -\frac{\pi}{2} + \frac{E - E_R}{\Gamma/2}\end{aligned} \tag{158}$$

and

$$\frac{\Gamma}{2} \simeq \frac{1}{4}\left[\frac{\text{d}\delta(s)}{\text{d}s}\right]^{-1}_{E=E_R} \tag{159}$$

If, furthermore, the resonance occurs at $s = m^2$ or $k^2 = q_R^2$, the 'reduced width' Γ_{red} is defined by

$$\frac{2q_R^3\Gamma_{\text{red}}}{m} = \frac{\Gamma}{2} \tag{160}$$

Hence

$$\Gamma_{\text{red}} = \frac{m}{8q_R^3}\left[\frac{\text{d}\delta}{\text{d}s}\right]^{-1}_{s=m^2} \tag{161}$$

3.10 *Eigenvalues of the Lippmann–Schwinger kernel*[11]

Considerable insight into various aspects of potential theory is gained from a method developed by Weinberg[25]. Here we shall touch only one sideline which is of particular interest in connection with the previous results of this section.

Weinberg[25] starts from the Lippmann–Schwinger equation for the T-matrix element and then considers the eigenvalues of its kernel. Thus

$$T(E) = V + T(E)K \tag{162}$$

where

$$K = GV, \quad G = 1/(E - H_0) \tag{163}$$

is the kernel and H_0 the free Hamiltonian. Here we assume again that V represents a superposition of Yukawa potentials which can be expanded in ascending powers of r. The eigenvalues η_n and eigenvectors $|\psi_n\rangle$ of the kernel are then defined by

$$K(E)|\psi_n(E)\rangle = \eta_n(E)|\psi_n(E)\rangle \tag{164}$$

where $|\psi_n\rangle$ is assumed to be a Hilbert vector of finite norm.

For square-integrable kernels (e.g. if V is a Yukawa potential) Weinberg[25] has shown that the iterated Born expansion of the Lippmann–Schwinger equation, i.e.

$$T(E) = V + VGV + \cdots \tag{165}$$

converges at the energy $E \equiv k^2$ if

$$|\eta_n(E)| < 1 \text{ for all } n \tag{166}$$

This condition is both necessary and sufficient for short-range regular potentials. The Born series diverges at the energy E if for some n, $|\eta_n(E)| > 1$. Poles in the plane of complex E, i.e. bound states and resonances, lie on the unit circles where $|\eta_n| = 1$. The eigenvalues η_n plotted as functions of E therefore lead to trajectories (called Weinberg trajectories) which yield considerable insight into the analytic nature of the scattering amplitude.

We observe now that (164) may be rewritten as

$$\left[H_0 + \frac{1}{\eta_n(E)} V\right] |\psi_n(E)\rangle = E|\psi_n(E)\rangle \tag{167}$$

Thus the eigenvalue of the Lippmann–Schwinger kernel may be regarded as a number by which the interaction has to be divided in order to produce at the energy E a bound state (where η_n, of course, is real and equal to 1). The radial wave equation corresponding to (167) is

$$\left[\frac{\mathrm{d}^2}{\mathrm{d}r^2} - \frac{l(l+1)}{r^2} - \frac{V(r)}{\eta_n}\right] \psi_n(r, k^2) = -k^2\psi_n(r, k^2) \tag{168}$$

Since the eigenvalues of this equation are now known in the form of series expansions, a similar expansion is readily found for η_n. Replacing the M_i in (106) by

$$M_i = -2^{i+1}U_i \tag{169}$$

one finds[11]

$$\begin{aligned}\eta_n(\kappa = \mathrm{i}k) = {} & \frac{U_0}{(l+n+1)\kappa} + \frac{2U_1}{\kappa^2} \\ & + \frac{1}{\kappa^3}\left[2(l+n+1)\left(2U_2 + \frac{U_1{}^2}{U_0}\right) + 2(2n+1)U_2 - \frac{2n(n+1)U_2}{l+n+1}\right] \\ & + \frac{1}{\kappa^4}\left[8(l+n+1)^2\left(U_3 + \frac{2U_1U_2}{U_0}\right) - 4U_3(3n^2+3n-1)\right. \\ & + 4(l+n+1)(2n+1)\left(3U_3 + \frac{2U_1U_2}{U_0}\right) \\ & \left. - 8n(n+1)\frac{U_1U_2}{U_0}\right] + 0(1/\kappa^5)\end{aligned} \tag{170}$$

For $n = 0$, $l = 0$ these coefficients agree with expressions obtained by Warburton[26].

The eigenvalues (170) can be used for many purposes—e.g. for the calculation of scattering phase-shifts. It is easy to show that

$$f(l, -k) = \prod_n (1 - \eta_n) \tag{171}$$

so that (by equation 156)

$$\delta = -\sum_n \arg(1 - \eta_n) = +\sum_n \arg(1 - \eta_n^*) \tag{172}$$

Rewriting this expression as

$$\delta = \sum_n \tan^{-1}[\operatorname{Im}\eta_n/\operatorname{Re}(1 - \eta_n)] \tag{173}$$

and expanding for large energies, one obtains the approximation

$$\delta_l{}^{\mathrm{Born}} = \sum_n \operatorname{Im}\eta_n \tag{174}$$

Equation (172) may therefore be written

$$\delta(l, k) = \delta_l{}^{\mathrm{Born}} - \sum_n [\arg(1 - \eta_n) + \operatorname{Im}\eta_n] \tag{175}$$

—a formula which is particularly useful for computational work since δ_l^{Born} is also known in other forms—e.g. for $V = A\,\mathrm{e}^{-\mu r}/r$:

$$\delta_l^{\text{Born}} = \frac{A}{2k} Q_l\left(1 + \frac{\mu}{2k^2}\right) \tag{176}$$

where

$$\left.\begin{aligned} Q_0(x) &= \tfrac{1}{2}y \\ Q_1(x) &= \tfrac{1}{2}xy - 1, \text{etc.} \end{aligned}\right\} \quad \text{for} \quad y = \ln\left(1 + \frac{x}{1-x}\right) \tag{177}$$

Astonishingly accurate results can be obtained from the few terms given in (170) if this expansion is first converted into a Padé approximant. In view of the limited space available here, we do not discuss these questions in the present context. For details the reader is referred to the literature (Müller[11]).

3.11 *An example: π-π scattering*

As an example of a realistic Yukawa interaction we consider π-π scattering[27,28]. Here the equivalent potential may be calculated as suggested by Balázs[29]. The Schrödinger equation-potential is assumed to have the general form

$$V(r, k^2) = -\frac{1}{\pi}\int_{t_0}^{\infty} \mathrm{d}t' v(t', k^2)\,\frac{1}{r}\,\mathrm{e}^{-r\sqrt{t'}} \tag{178}$$

where v can be constructed from the given absorptive part A_t of the pion-pion amplitude in the crossed channel

$$v(t, k^2) = 2s^{-1/2}A_t(t, s) - \frac{1}{\pi}\int_0^{\infty} \mathrm{d}k'^2\,\frac{\alpha(k'^2, t)}{k'^2 - k^2} \tag{179}$$

Here k is the magnitude of the 3-momentum in the barycentric system of the s-channel, $s = 4(k^2 + 1)$ is the square of the total energy (taking $m_\pi = 1$) and α is the double-spectral function which is determined by the unitarity condition; $t = -2k^2(1 - \cos\theta_s)$ is the momentum transfer in the direct channel. To a first approximation the integral term in (179) may be omitted and A_t may be approximated by a small number (L) of partial waves in the t-channel. Writing the scattering amplitude for isospin I in the s-channel and isospin I' in the t-channel in the form

$$A^I(t, s) \simeq \sum_{I'=0}^{2} \beta_{II'} \sum_{l=0}^{L} (2l + 1)A_l^{I'}(t)P_l(\cos\theta_t) \tag{180}$$

we have

$$A_t^I(t, s) \simeq \sum_{I'=0}^{2} \beta_{II'} \sum_{l=0}^{L} (2l + 1)\,\mathrm{Im}\,A_l^{I'}(t)P_l(\cos\theta_t) \tag{181}$$

where

$$\cos\theta_t = 1 + \frac{2s}{t-4} \tag{182}$$

and

$$\beta = \begin{pmatrix} \frac{1}{3} & 1 & \frac{5}{3} \\ \frac{1}{3} & \frac{1}{2} & -\frac{5}{6} \\ \frac{1}{3} & -\frac{1}{2} & \frac{1}{6} \end{pmatrix} \tag{183}$$

is the well-known crossing matrix.

We now neglect everything except the P-wave, so that the input-potential is equivalent to a vector ρ-meson exchange in the crossed channel. Then, approximating this ρ-meson by a resonance of zero width, the potential (178) becomes

$$V(r, k^2) = -24\beta_{I1}\Gamma_{\text{in}}s^{-1/2}(s + 2q_{\text{R}}^2)\,\frac{1}{r}\,\mathrm{e}^{-mr} \tag{184}$$

where Γ_{in} is the reduced input width, $q_{\text{R}}^2 = \frac{1}{4}m^2 - 1$, and m is the mass of the ρ-meson in units $m_\pi = 1$.

Expanding (184) in rising powers of r and comparing with (106), we find

$$M_i = 6\beta_{I1}\Gamma_{\text{in}}\frac{8k^2 + m^2 + 4}{\sqrt{(k^2+1)}} \cdot \frac{m^i}{i!} \tag{185}$$

If m and Γ_{in} are known, the Regge trajectories may be calculated by (129). Here we take $m = 5{\cdot}3m_\pi$, the experimentally observed mass, and $\Gamma_{\text{in}} = 0{\cdot}5$, a bootstrap value[27], which is approximately three times the observed width. For $I = 1$ the threshold of only one trajectory lies in the right of the complex l-plane and there gives rise to the ρ-meson at $l = 1$. For $I = 0$ we see from (183) that the coupling is twice as strong as for $I = 1$; the Regge trajectories are therefore expected to move further into the right of the l-plane. The calculations show that this is the case: two trajectories now give rise to physical states at $l = 2$, and may be identified with the P and P' trajectories of even signature. For $I = 2$, β_{21} is negative and the potential repulsive. Thus no physical states are found. For numerical details see Müller[28].

3.12 *Energy-dependent potentials*

The above example has shown that even the first approximation of a realistic interaction is strongly energy-dependent, although in its r-dependence the potential is of the Yukawa type. Reexamining now our earlier work, we find that the S-matrix of a potential of the type (106), in which the M_i are functions of k, may be written

$$S = \frac{\Gamma\left[l + 1 + \dfrac{\Delta_l(\kappa)}{2\kappa}\right]}{\Gamma\left[l + 1 - \dfrac{\Delta_l(-\kappa)}{2\kappa}\right]} \cdot \frac{\Phi(-\kappa)}{\Phi(\kappa)} \tag{186}$$

where

$$\Phi(\kappa) = 1 - \frac{M_1(\kappa)l}{2\kappa^2} + \frac{1}{3(2\kappa)^4}\,[9(2l-1)M_0(\kappa)M_2(\kappa) + 3l(l+3)M_1{}^2(\kappa) + 2l(l-1)(8l+11)M_3(\kappa)] + \cdots \tag{187}$$

Thus even additional bound states and resonances may be expected from the condition $\Phi(\kappa) = 0$ in the low-energy region. For a potential such as (184) the expansion (187) will certainly converge asymptotically at least in a high-energy region since for $|k| \to \infty$ all $M_i \propto k$. In fact, a behaviour of this type, for which

$$V(r, k^2) \propto k \quad \text{for} \quad |k| \to \infty \tag{188}$$

is expected from many other arguments—see, for instance, Nambu and Sugawara[30]. We shall not enter into a detailed investigation of these points here and instead consider another similar example due to Serber[31] (for additional discussions see also Aly, Lurié and Rosendorff[32]).

Serber found empirically that the high-energy diffraction maximum in proton-proton scattering could be reproduced in the frame of an optical model, in which the scattering medium is assumed to be purely absorptive. More specifically Serber assumed in his model that the wave number k of the free incoming wave changes in the scattering region to k' with

$$k_1 \equiv k' - k = \mathrm{i}\eta\,\frac{\mathrm{e}^{-\Lambda r}}{r} \tag{189}$$

Thus, assuming the validity of the non-relativistic Schrödinger equation,

$$\left.\begin{aligned} &\psi'' + \mathbf{k}^2\psi = 0 \quad \text{where} \quad V = 0 \\ \text{and} \qquad &\\ &\psi'' + (\mathbf{k} + \mathbf{k}_1)^2\psi = 0 \quad \text{where} \quad V \neq 0 \end{aligned}\right\} \tag{190}$$

so that

$$(\mathbf{k} + \mathbf{k}_1)^2 = E - V, \quad E = k^2 \tag{191}$$

or

$$\mathbf{k}_1 = \mathbf{k}\left[\left(1 - \frac{V}{E}\right)^{\frac{1}{2}} - 1\right] \simeq \mathbf{k}(-V/2E) \quad \text{for} \quad |V|E| \ll 1 \tag{192}$$

If, moreover, absorption is assumed to take place in the scattering region, we see from

$$e^{ik'r} = e^{i(k+k_1)r}$$

that the modulus of this factor decreases provided k_1 possesses a positive and thus V a negative imaginary part for $k^2 > 0$. Equating (192) and (189) we see that

$$V = -2ik\eta\, e^{-\Lambda r}/r \tag{193}$$

is the appropriate non-relativistic Schrödinger potential (Im $V < 0$ is in fact the absorptivity condition). Accordingly we shall call

$$V_s(r) = 2\kappa V(r), \quad \kappa = ik \tag{194}$$

where

$$V(r) = \sum_{i=-1}^{\infty} M_{i+1}(-r)^i$$

a Serber potential. Calculating the Regge trajectories again as before, we obtain[33]

$$\begin{aligned}
\alpha_n(\kappa) = {} & -n - 1 - M_0 + \frac{M_0 M_1}{\kappa} - \frac{1}{2\kappa^2}[3M_0M_1^2 + 2M_2M_0^2 \\
& - (2n+1)M_2M_0 - n(n+1)M_2] \\
& - \frac{1}{2\kappa^3}\Big\{M_0M_3[-2M_0^2 + 3(2n+1)M_0 + (3n^2+3n-1)] \\
& + M_1M_2[-10M_0^2 + 4(2n+1)M_0 + 3n(n+1)] \\
& + M_1^3M_0[(2n+1)M_0 + (n^2+n-5)]\Big\} \\
& - \frac{1}{8\kappa^4}\Big\{M_4[8M_0^4 - 24(2n+1)M_0^3 - 2M_0^2(6n^2+6n-11) \\
& + 6M_0(2n^3+3n^2-n-1) + 3n(n^3+2n^2-n-2)] \\
& + 2M_3M_1M_0[28M_0^2 - 36(2n+1)M_0 - (30n^2+30n-9)] \\
& + M_1^4M_0[4M_0^2 - 22(2n+1)M_0 - (18n^2+18n-35)] \\
& + M_1^2M_2[-20(2n+1)M_0^3 - 6(2n^2+2n-25)M_0^2 \\
& + 4(2n+1)(n^2+n-12)M_0 + 2n(n+1)(n^2+n-15)] \\
& + M_2^2[32M_0^3 - 30(2n+1)M_0^2 - (18n^2+18n-7)M_0 \\
& + 2n(n+1)(2n+1)]\Big\} + 0(1/\kappa^5)
\end{aligned} \tag{195}$$

A noteworthy feature of this result is that the limit of α at infinite energy is no longer an integer, but strongly dependent on the coupling constant M_0.

From Regge pole analysis we know that the differential cross-section of the non-relativistic scattering amplitude for the s-channel ($s = k^2$) is given by

$$\frac{d\sigma}{d\Omega} = |F(t, s)|^2, \quad F(t, s) \simeq B(s) \cdot t^{\alpha_m(s)} \tag{196}$$

for large negative momentum transfers (i.e. $|t| \to \infty$). The gradient of the plot $\log d\sigma/d\Omega$ versus $\log |t|$ is therefore $g = 2\alpha_m(s)$. For moderately large values of s we may replace $\alpha_m(s)$ by $\alpha_m(\infty)$, so that

$$g \simeq -2 - 2M_0 \tag{197}$$

for the leading trajectory of the Serber potential (194). Substituting for M_0 the value calculated by Serber[31] in his optical model, i.e. $M_0 = 1$, we have $g = -4$, compared with $g = -5$, the approximate empirical gradient measured by Serber[31].

Any model of this type, of course, has numerous shortcomings. The use of the non-relativistic Schrödinger equation at high energies, for instance, is questionable. It is easy to see, however, that relativistic wave equations such as the Dirac and Klein–Gordon equations are effectively energy-dependent Schrödinger equations, if V is (for instance) a simple energy-independent Yukawa potential. We close this chapter therefore with a brief discussion of the Klein–Gordon equation[34]. For a similar discussion of the Dirac equation see Müller[35].

3.13 *The Klein–Gordon equation*

The equation for the radial part $\psi(r)/r$ of the wave-function for relativistic potential scattering is[34]

$$\psi''(r) + \left[(E - V)^2 - m^2 - \frac{l(l+1)}{r^2}\right] \psi(r) = 0 \tag{198}$$

We set

$$\left.\begin{aligned}
V &= \sum_{i=-1}^{\infty} M_{i+1}(-r)^i, \quad N_i = \sum_{j=0}^{i} M_{i-j}M_j \\
E^2 &= m^2 + k^2, \quad \kappa = ik, \quad z = -2ikr \\
\psi(r) &= e^{ikr} r^{l+1} z^s H(z) \\
s &= -(l + \tfrac{1}{2}) + \sqrt{[(l + \tfrac{1}{2})^2 - N_0]} \\
a &= -n = \sqrt{[(l + \tfrac{1}{2})^2 - N_0]} + \frac{1}{2} + \frac{E}{\kappa} M_0 - \frac{\Delta(\kappa)}{2\kappa} \\
b &= 2\sqrt{[(l + \tfrac{1}{2})^2 - N_0]} + 1 = -2n - \frac{2E}{\kappa} M_0 + \frac{\Delta(\kappa)}{\kappa}
\end{aligned}\right\} \tag{199}$$

The solutions H are again found to have the general form [writing $\Phi(a) \equiv \Phi(a, b; z)$]

$$H(z) = \Phi(a) + \sum_{i=1}^{\infty} \frac{1}{(2\kappa)^i} \sum_{\substack{j=i \\ j\neq 0}}^{-i} P_i(a, j)\Phi(a + j) \tag{200}$$

and the S-matrix becomes

$$S(l, \kappa) = \frac{\Gamma\left\{\sqrt{[(l + \frac{1}{2})^2 - M_0^2]} + \frac{1}{2} + \frac{E}{\kappa} M_0 - \frac{\Delta(+\kappa)}{2\kappa}\right\}}{\Gamma\left\{\sqrt{[(l + \frac{1}{2})^2 - M_0^2]} + \frac{1}{2} - \frac{E}{\kappa} M_0 + \frac{\Delta(-\kappa)}{2\kappa}\right\}} \tag{201}$$

—again a simple closed form (for all M_i real and independent of K), the function Δ being given by[34]

$$\begin{aligned}
\Delta(\kappa) = {} & 2M_0M_1\left[1 + \left(\frac{E}{\kappa}\right)^2\right] \\
& + \frac{1}{\kappa}\left(\frac{E}{\kappa}\right)\left\{n(n + 1)M_2 + (2n + 1)M_0M_2\left(\frac{E}{\kappa}\right)\right. \\
& \left. - (2M_0^2M_2 + 3M_0M_1^2)\left[1 + \left(\frac{E}{\kappa}\right)^2\right]\right\} + \cdots \\
= {} & 2M_0M_1\left[1 + \left(\frac{E}{\kappa}\right)^2\right] \\
& + \frac{1}{\kappa}\left(\frac{E}{\kappa}\right)\left\{M_2l(l + 1) - 3M_0M_1^2\left[1 + \left(\frac{E}{\kappa}\right)^2\right]\right. \\
& \left. - 3M_0^2M_2\left[1 + \left(\frac{E}{\kappa}\right)^2\right]\right\} \\
& - \frac{1}{\kappa^2}\left\{l(l + 1)(M_0M_3 + M_1M_2)\left[1 + 3\left(\frac{E}{\kappa}\right)^2\right]\right. \\
& - \left(\frac{E}{\kappa}\right)^2[6M_0^3M_3 + 9M_0^2M_1M_2 + M_0M_3 + 4M_0M_1^3] \\
& - \left(\frac{E}{\kappa}\right)^4[5M_0^3M_3 + M_0^2M_1M_2 + 4M_0M_1^3] \\
& - M_0^2(M_0M_3 + M_1M_2) \\
& \left. + 2M_0M_1M_2\left(\frac{E}{\kappa}\right)\left[1 + \left(\frac{E}{\kappa}\right)^2\right]\sqrt{[(l + \tfrac{1}{2})^2 - M_0^2]}\right\} + \cdots
\end{aligned} \tag{202}$$

Thus, even relativistic equations can be solved by our perturbation method. Regge poles, phase-shifts, residues again follow along the lines of our previous calculations.

4 Perturbed Harmonic Oscillators

4.1 *The anharmonic oscillator*

The simplest and best-known of all regular potentials for which the Schrödinger equation is exactly solvable in closed form are the Coulomb and harmonic oscillator potentials. In the preceding section we used the formalism of Section 2 to solve the Schrödinger equation for almost any type of potential which can remotely be considered a perturbed Coulomb interaction. In the present section we outline some methods—still based on our perturbation procedure—to deal with potentials which can be considered as perturbed harmonic oscillator interactions. Several authors[36,37] have pointed out recently that these two types of scattering problems are not entirely independent of each other—in fact, in the words of Fivel[36], one can be regarded as the dual of the other. This relationship between the two problems is based on a set of substitutions which transforms the radial Schrödinger equation for a Yukawa potential into another equation of the type of the radial Schrödinger equation for a 3-dimensional anharmonic oscillator. We do not discuss this duality here; for a detailed investigation the reader is referred to the paper of Aly, Müller and Schilcher[38]. We simply remark that this substitution procedure has definite limitations; it is not possible, for example, to use it in order to derive the *S*-matrix for a Gauss potential from that for the Yukawa potential.

In the present context we restrict ourselves to potentials which can be expanded in ascending even powers of r. Many well-known examples fall into this category, such as the Gauss potential and the potential of Section 2.2. It is obvious that if the anharmonic terms are only to perturb the harmonic oscillator without destroying its essential features, the energy can play only a secondary role as a perturbing parameter in our formalism; the proper perturbing parameter has to be roughly proportional to the coupling constant of the anharmonic terms.

The procedure we discuss in this section has several undesirable features, the chief difficulty being a clumsiness of expansion coefficients due to several essential series expansions. Nevertheless we believe it represents a first attempt to obtain analytic results for anharmonic interactions. Additional details are given elsewhere[9].

4.2 *The radial Schrödinger equation*

We now consider the radial wave equation in the following more general form:

$$\left[\frac{d^2}{dr^2} + k^2 - \frac{L(L+1)}{(\sinh r)^2} + \sum_{i=0}^{\infty} f_i r^{2i}\right] \psi_l(k, r) = 0 \tag{203}$$

If, in particular,

$$f_i = A_{2i} - N_{2i} \tag{204}$$

and

$$A_{2n} = L(L+1)C_{2n} \tag{205}$$

and

$$C_{2n} = \sum_{i=-1,1,3,\ldots}^{2n+1} \frac{4[2^i - 1][2^{2n-i} - 1](-1)^{n+1}}{(i+1)!(2n-i+1)!} B_{\frac{1}{2}(i+1)} B_{\frac{1}{2}(2n-i+1)} \tag{206}$$

the B's being Bernouilli's numbers, (203) becomes

$$\left[\frac{d^2}{dr^2} + k^2 - \frac{L(L+1)}{r^2} - V'(r)\right] \psi_l(k, r) = 0 \tag{207}$$

where

$$V'(r) = \sum_{i=0}^{\infty} N_{2i} r^{2i} \tag{208}$$

is an even-power potential. Here we have also used the expansion

$$\frac{1}{r^2} = \frac{1}{\sinh^2 r} - \sum_{n=0}^{\infty} C_{2n} r^{2n} \tag{209}$$

The additional substitutions

$$r^2 = \sum_{i=1}^{\infty} e_{i-1} \sinh^{2i} r \tag{210}$$

where $e_0 = 1$, $e_1 = -1/3$, $e_2 = 8/45$, $e_3 = -4/35$, $e_4 = 128/1575$, etc., and

$$\sum_{i=0}^{\infty} f_i r^{2i} = \sum_{i=0}^{\infty} G_i \sinh^{2i} r \tag{211}$$

where

$$G_0 = f_0, \quad G_1 = f_1, \quad G_2 = f_2 + f_1 e_1, \ldots \tag{212}$$

convert (203) into

$$\frac{d}{du}\left[(1-u^2)\frac{d\Psi}{du}\right]+\left[\Lambda-\frac{\mu^2}{1-u^2}-4h^2(1-u^2)\right]\Psi$$
$$=\sum_{i=2}^{\infty}G_i(u^2-1)^i\Psi \tag{213}$$

where we have set

$$u=\cosh r,\quad \psi=(\sinh r)^{1/2}\Psi \tag{214}$$

and

$$k^2+G_0=-\Lambda-\tfrac{1}{4},\quad G_1=-4h^2,\quad L=\mu-\tfrac{1}{2} \tag{215}$$

We now use (213) as the fundamental equation of our subsequent analysis.

We observe first that for $G_i(i \geqslant 2)=0$ the equation reduces to a spheroidal wave equation and for all $G_i=0$ to the equation for generalized spherical harmonics. It is useful therefore to set

$$\Lambda \equiv \nu(\nu+1) \tag{216}$$

In the limit of all $G_i \to 0$, we have

$$\Psi \to \mathscr{P}_\nu{}^\mu(u),\quad \mathscr{Q}_\nu{}^\mu(u) \tag{217}$$

the functions on the right being generalized spherical harmonics of the first and second kind respectively. As in the preceding chapters it is convenient to use this property of Ψ, in order to obtain first the approximate behaviour of Λ for $G_i \simeq 0$ under the normal bound-state boundary conditions. For this reason we now consider the case of all $G_i(i \geqslant 2)=0$ in more detail.

4.3 *The potential* $[\sinh r]^{-2}$

Setting all $G_i(i \geqslant 2)$ in (213) equal to zero we obtain an equation which is equivalent to an *S*-wave radial Schrödinger equation containing the potential $+L(L+1)[\sinh r]^{-2}$. This last-named potential has already been discussed in the literature[39]; however, we now present a different and perhaps more elegant derivation of the relevant *S*-matrix.

The radial wave-function is seen to be given as

$$\psi(r)=(\sinh r)^{1/2}$$
$$\times \text{ linear combinations of generalized spherical harmonics}$$

In particular we have the regular solution

$$\psi_{\mathrm{R}}(r) = (\sinh r)^{1/2}\mathscr{P}_{\nu}{}^{\mu}(\cosh r) \tag{218}$$

If

$$L = -p, p > 1 \quad \text{and} \quad \mu = -p + \tfrac{1}{2} < -\tfrac{1}{2} \tag{219}$$

this solution behaves as

$$\psi_{\mathrm{R}} \simeq r^{p}/[2^{p-\frac{1}{2}}\Gamma(p + \tfrac{1}{2})] \tag{220}$$

for $r \to 0$. Its analytic continuation for $r \to \infty$ is obtained from the formula

$$\mathscr{P}_{\nu}{}^{\mu}(u) = \frac{2^{\mu}}{\sqrt{\pi}}(u^2 - 1)^{\mu/2}\Bigg[\frac{1}{\mathscr{S}^{\nu+\mu+1}}\frac{\Gamma(-\nu - \frac{1}{2})}{\Gamma(-\nu - \mu)}\cdot {}_2F_1\left(\tfrac{1}{2} + \mu, \nu + \mu + 1; \nu + \frac{3}{2}; \frac{1}{\mathscr{S}^2}\right)$$
$$+ \frac{1}{\mathscr{S}^{\mu-\nu}}\frac{\Gamma(\nu + \frac{1}{2})}{\Gamma(\nu - \mu + \frac{1}{2})}\,{}_2F_1\left(\tfrac{1}{2} + \mu, \mu - \nu; \tfrac{1}{2} - \nu; \frac{1}{\mathscr{S}^2}\right)\Bigg] \tag{221}$$

where $\mathscr{S} \equiv u + \sqrt{(u^2 - 1)}$ provided $\mathrm{Re}\,(u) > 1$, $\mathscr{S}$ real and > 1 and $2\nu \neq \pm 1, \pm 3, \pm 5, \ldots$. Substituting (221) into (218) we obtain (choosing $\nu = \mathrm{i}k - \frac{1}{2}$ for $G_0 = 0$)

$$\psi(r, p) \simeq \frac{1}{2\mathrm{i}k}\left[\frac{\Gamma(1 + \mathrm{i}k)}{\Gamma(p + \mathrm{i}k)}\,\mathrm{e}^{\mathrm{i}kr} - \frac{\Gamma(1 - \mathrm{i}k)}{\Gamma(p - \mathrm{i}k)}\,\mathrm{e}^{-\mathrm{i}kr}\right] \tag{222}$$

in the limit $r \to \infty$. Hence the S-matrix is

$$S = \mathrm{e}^{2\mathrm{i}\delta_0(k)} = \frac{\Gamma(p - \mathrm{i}k)\Gamma(1 + \mathrm{i}k)}{\Gamma(p + \mathrm{i}k)\Gamma(1 - \mathrm{i}k)} \tag{223}$$

in agreement with the expression in the literature[39].

We note now that the poles of (223) are given by

$$p - \mathrm{i}k = -n, \quad 1 + \mathrm{i}k = -n' \tag{224}$$

where $n, n' = 0, 1, 2, \ldots$. Only the first of these conditions depends on the potential and is therefore physical. Hence the energy eigenvalues are given by

$$k^2 = -(p + n)^2 = -(\mu - \tfrac{1}{2} - n)^2 \tag{225}$$

4.4 *Solution of the original equation*

We now want to solve (213) by our perturbation procedure. For this reason we set

$$\Lambda - 4h^2 = \nu_0(\nu_0 + 1) - 4\Delta h^2 \tag{226}$$

where Δ is an (as yet) unknown function in terms of the G_i. ν_0 however, follows by (219) from (224) as

$$\nu_0 = \mu - 1 - n \quad \text{or} \quad -\mu + n \tag{227}$$

Thus for physical (integral) values of L, ν_0 is half-integral.

Substituting (226) into (213), we obtain

$$D_{\nu_0}\Psi = -4h^2(u^2 - \Delta)\Psi + \sum_{i=2}^{\infty} G_i(u^2 - 1)^i\Psi \tag{228}$$

where

$$D_{\nu_0} \equiv \frac{\mathrm{d}}{\mathrm{d}u}\left[(1 - u^2)\frac{\mathrm{d}}{\mathrm{d}u}\right] + \nu_0(\nu_0 + 1) - \frac{\mu^2}{1 - u^2} \tag{229}$$

Again we consider negative values of $L < -1$; results for positive L then follow by the substitutions

$$L \to -L - 1, \quad \nu_0 \to -\nu_0 - 1$$

A zeroth approximation to Ψ is therefore

$$\Psi^{(0)}(u) = \mathscr{P}_{\nu_0}{}^{\mu}(u) \tag{230}$$

and for negative L this solution is regular at $r = 0$.

Proceeding in the standard manner of our perturbation method, we next express

$$\mathscr{R}^{(1)} \equiv -4h^2(u^2 - \Delta)\mathscr{P}_{\nu_0}{}^{\mu}(u) + \sum_{i=2}^{\infty} G_i(u^2 - 1)^i\mathscr{P}_{\nu_0}{}^{\mu}(u) \tag{231}$$

as a linear combination of $\mathscr{P}^{\mu}_{\nu+2j}$. Straightforward manipulations yield

$$\mathscr{R}^{(1)} = \sum_{i=1}^{\infty}\sum_{j=-i}^{i} [\nu_0, \nu_0 + 2j]_{2i}\mathscr{P}^{\mu}_{\nu_0+2j}(u) \tag{232}$$

where for $|j| \leqslant i$, $i = 2, 3, \ldots$

$$[\nu_0, \nu_0 + 2j]_{2i} = G_i T_{2i}(\nu, 2j) \tag{233}$$

except for $i = 1$, $j = 0$, in which case

$$[\nu_0, \nu_0]_2 = G_1[S_2(\nu_0, 0) - \Delta] \tag{234}$$

Here the coefficients S_m are defined by

$$u^m\mathscr{P}_{\nu_0}{}^{\mu}(u) = \sum_{i=-m,\,-m+2,\ldots}^{\ldots,\,(m-2),\,m} S_m(\nu_0, i)\mathscr{P}^{\mu}_{\nu_0+i}(u) \tag{235}$$

those for $m = 1$,

$$S_1(\nu_0, 1) = (\nu_0 + 1 - \mu)/(2\nu_0 + 1) \tag{236}$$
$$S_1(\nu_0, -1) = (\nu_0 + \mu)/(2\nu_0 + 1)$$

being given by the recurrence relation for Legendre functions. The T's are more complicated:

$$T_{2i}(\nu_0, 2i) = \binom{i}{i}(-1)^{2i}S_{2i}(\nu_0, 2i)$$

$$T_{2i}(\nu_0, 2i-2) = \binom{i}{i}(-1)^{2i}S_{2i}(\nu_0, 2i-2) + \binom{i}{i-1}(-1)^{2i-1}S_{2i-2}(\nu_0, 2i-2)$$

. . . .

$$T_{2i}(\nu_0, -2i) = \binom{i}{i}(-1)^{2i}S_{2i}(\nu_0, -2i) \tag{237}$$

Again it is possible to derive recurrence relations, etc., for these coefficients; for further details the reader is referred to a separate publication[9].

At this stage we have to ask ourselves what we mean by a perturbation formalism in the present case. At first sight the expression (232) does not appear to contain a parameter ε, say, which is such that successive terms of the solution can be written in ascending powers of ε so that an expansion is obtained which is valid at least near $\varepsilon = 0$. The mathematics is not (unfortunately) as simple as that. However, it is both plausible and legitimate to assume that the coefficients G_i are such that successive terms yield decreasing contributions (in magnitude). We assume therefore that

$$|G_i| > |G_{i+1}| \tag{238}$$

and that

$$|G_{i+j}| \text{ is roughly of the same order as } |G_i| \cdot |G_j| \tag{239}$$

These conditions are satisfied, for instance, in the case of a Gauss potential for small angular momenta and coupling constants. They also define a definite ordering according to the magnitudes of contributions of successive terms of our expansions. It is legitimate, therefore, under the above restrictions to proceed in our standard manner and then to rearrange various contributions according to (238), (239). In this way we obtain[9]

$$\Psi(u) = \mathscr{P}_{\nu_0}{}^{\mu}(u) + \sum_{i=1}^{\infty} \sum_{\substack{j=-i \\ j\neq 0}}^{i} P_{2i}(\nu_0, 2j)\mathscr{P}^{\mu}_{\nu_0+2j}(u) \tag{240}$$

where

$$P_2(\nu_0, 2) = \frac{[\nu_0, \nu_0 + 2]_2}{-2(2\nu_0 + 3)}, \quad P_2(\nu_0, -2) = \frac{[\nu_0, \nu_0 - 2]_2}{2(2\nu_0 - 1)}$$

$$P_4(\nu_0, 4) = \frac{[\nu_0, \nu_0 + 2]_2}{-2(2\nu_0 + 3)} \cdot \frac{[\nu_0 + 2, \nu_0 + 4]_2}{-4(2\nu_0 + 5)}$$

$$P_4(\nu_0, 2) = \frac{[\nu_0, \nu_0 + 2]_4}{-2(2\nu_0 + 3)} + \frac{[\nu_0, \nu_0 + 2]_2}{-2(2\nu_0 + 3)} \cdot \frac{[\nu_0 + 2, \nu_0 + 2]_2}{-2(2\nu_0 + 3)} \tag{241}$$

etc. The denominators in these expressions again follow from the fact that

$$D_{\nu_0} \mathscr{P}^{\mu}_{\nu_0+i}(u) = -i(2\nu_0 + i + 1)\mathscr{P}^{\mu}_{\nu_0+i}(u) \tag{242}$$

The corresponding subsidiary condition from which Δ and hence the eigenvalues may be determined is

$$0 = [\nu_0, \nu_0]_2 + \left\{[\nu_0, \nu_0]_4 + \frac{[\nu_0, \nu_0 + 2]_2}{-2(2\nu_0 + 3)}[\nu_0 + 2, \nu_0]_2 \right. \\ \left. + \frac{[\nu_0, \nu_0 - 2]_2}{2(2\nu_0 - 1)}[\nu_0 - 2, \nu_0]_2\right\} + \cdots \tag{243}$$

We thus find

$$\begin{aligned} k^2 &= N_0 - N_2 - (\nu_0 + \tfrac{1}{2})^2 + \frac{2}{5} l(l + 1) \\ &- \frac{G_1}{(2\nu_0 - 1)(2\nu_0 + 3)}[(2\nu_0^2 + 2\nu_0 - 1) - 2\mu^2] \\ &+ \frac{G_1^2}{(2\nu_0 - 3)(2\nu_0 - 1)^3(2\nu_0 + 3)^3(2\nu_0 + 5)}[-2\mu^4(20\nu_0^2 + 20\nu_0 + 33) \\ &+ 4\mu^2(12\nu_0^4 + 24\nu_0^3 - \nu_0^2 - 13\nu_0 + 15) \\ &- 2(4\nu_0^6 + 12\nu_0^5 - 3\nu_0^4 - 26\nu_0^3 + 2\nu_0^2 + 17\nu_0 - 3)] \\ &- \frac{G_2}{(2\nu_0 - 3)(2\nu_0 - 1)(2\nu_0 + 3)(2\nu_0 + 5)}[6\mu^4 + 2\mu^2(2\nu_0^2 + 2\nu_0 - 15) \\ &+ 2(3\nu_0^4 + 6\nu_0^3 - 11\nu_0^2 - 14\nu_0 + 12)] \\ &+ 0(G_3, G_1 \cdot G_2, G_1^3) \end{aligned} \tag{244}$$

In particular for *S*-wave eigenvalues (244) becomes

$$k^2 = N_0 - \frac{5}{8} N_2 - (\nu_0 + \tfrac{1}{2})^2 - \frac{N_2}{8(2\nu_0 - 1)(2\nu_0 + 3)} + \frac{3}{8} N_4 + \cdots \tag{245}$$

At a first glance one might think that these expansions, as well as (240), are completely useless or even invalid since the physical values of ν_0 are positive or negative half-integers—i.e. precisely those values for which successive terms of the series diverge. However, the corresponding expansions—such as (21)—for the Mathieu equation contain exactly the same phenomenon, which is, in fact, characteristic of any even-power potential (see also Section 2.2). In the case of the Mathieu equation the regions of validity and convergence properties of the corresponding expansions have been examined in great detail by Meixner and Schäfke—so for analogous details we refer to their book[2]. Their results show, however, that for half-integral values of ν_0', the terms which do not contain the factor $\nu_0 - \nu_0'$ in the denominator constitute a valid approximation of that solution or eigenvalue. Where these expansions break down or become impracticable, different series are required, such as expansions for large $|G_i|$ which are presently under investigation[40].

4.5 *The S-matrix*

We now want to derive the *S*-matrix which has as its poles the eigenvalues (244). In the scattering region, of course, the quantum number ν_0 becomes meaningless as a half-integer. Consequently we consider it now a parameter defined by the solution of the secular equation for ν_0 or $\nu_0(\nu_0 + 1)$. To distinguish the present scattering case more clearly from that of the eigenvalue problem, we rename ν_0 as ν. Then

$$(\nu + \tfrac{1}{2})^2 = G_1 - G_0 - k^2 - \Delta G_1$$

or

$$\frac{\nu}{\bar{\nu}} = -\frac{1}{2} \pm \mathrm{i}\sqrt{(G_0 - G_1 + k^2 + \Delta G_1)} \tag{246}$$

The corresponding scattering solution may be written

$$\psi(r) = \sqrt{\left(\frac{2}{\pi}\right)} \frac{\Gamma(\nu + \frac{3}{2})}{\Gamma(\nu + \mu + 1)} \mathrm{e}^{-\mu\pi\mathrm{i}} (\sinh r)^{1/2} \cdot$$

$$\cdot \left[\mathcal{Q}_\nu^\mu(u) + \sum_{i=1}^{\infty} \sum_{\substack{j=-i \\ \neq 0}}^{i} P_{2i}(\nu, 2j) \mathcal{Q}_{\nu+2j}^\mu(u) \right] \tag{247}$$

Here the constant factor is determined by the condition that ψ has the asymptotic behaviour

$$\psi(r) \simeq \mathrm{e}^{\mathrm{i}kr} \quad \text{for} \quad r \to \infty \tag{248}$$

The other part follows from the eigenvalue solution (240) by replacing $\mathscr{P}$ by $\mathscr{Q}$ and ν_0 by ν. It is easy to convince oneself that the replacement of $\mathscr{P}$ by $\mathscr{Q}$ in (240) yields again a solution of our equation and in fact that solution which for $r \to \infty$ has the required asymptotic behaviour (248). Inserting (247) into the definition (144) of the Jost function $f(l, -k) \equiv f(L, -k)$ we obtain

$$f(l, -k) = (-\mathrm{i}k)^l \frac{\Gamma(\nu + \frac{3}{2})}{\Gamma(\nu + l + \frac{3}{2})} \cdot \left[1 + \sum_{i=1}^{\infty} \sum_{\substack{j=-i \\ j \neq 0}}^{i} P_{2i}(\nu, 2j) \cdot \frac{\Gamma(\nu - l + 2j + \frac{1}{2})\Gamma(\nu + l + \frac{3}{2})}{\Gamma(\nu + l + 2j + \frac{3}{2})\Gamma(\nu - l + \frac{1}{2})}\right] \tag{249}$$

where l is defined by

$$\mu = L + \tfrac{1}{2} = -(l + \tfrac{1}{2}) \tag{250}$$

In deriving (249) we used the relations

$$\mathscr{Q}_\nu{}^\mu(u) = \frac{\mathrm{e}^{\mu\pi\mathrm{i}}}{2^{\nu+1}} \cdot \frac{\Gamma(\nu + \mu + 1)\Gamma(\frac{1}{2})}{\Gamma(\nu + \frac{3}{2})} (u^2 - 1)^{\mu/2} \cdot u^{-\nu-\mu-1} \cdot {}_2F_1\left(\frac{\nu + \mu + 2}{2}, \frac{\nu + \mu + 1}{2}; \nu + \frac{3}{2}; \frac{1}{u^2}\right) \tag{251}$$

$${}_2F_1(a, b; c; 1) = \Gamma(c)\Gamma(c - a - b)/[\Gamma(c - a)\Gamma(c - b)] \tag{252}$$

In particular for S-waves and a potential $+p(p-1)/\sinh^2 r$, (249) reduces to

$$f(0, -k) = \Gamma(1 - \mathrm{i}k)/\Gamma(p - \mathrm{i}k) \tag{253}$$

The S-matrix therefore agrees again with our earlier expression (223). For many potentials, e.g. when the coefficients N_{2i} are independent of k, the factor involving the sums in (249) can be shown to be an even function of k. The S-matrix then assumes the simple form

$$S = \mathrm{e}^{2\mathrm{i}\delta(l,k)} = \frac{f(l, k)}{f(l, -k)} = \mathrm{e}^{\mathrm{i}\pi l} \frac{\Gamma(\nu + l + \frac{3}{2})\Gamma(\bar{\nu} + \frac{3}{2})}{\Gamma(\nu + \frac{3}{2})\Gamma(\bar{\nu} + l + \frac{3}{2})} \tag{254}$$

where $\bar{\nu}(k) = \nu(-k) = \nu^*(k)$. This formula therefore represents an analytic expression for the S-matrix of almost any type of Schrödinger potential which can be expanded in even powers of r. There are many consequences of this result.

4.6 *Regge poles*

Considering l again as a complex parameter, we see that the poles of S are given by

$$\nu + l + \tfrac{3}{2} = -n \tag{255}$$

$$\bar{\nu} + \tfrac{3}{2} = -n' \tag{256}$$

$n, n' = 0, 1, 2, \ldots$. The first of these is identical with our earlier eigenvalue condition $\nu = \nu_0$, remembering, of course, (250). The relation (256) when solved for l gives rise to a second family of Regge poles. We have not yet explored the full significance of these poles. However, one important relationship between the two families follows from the property $\bar{\nu}(k) = \nu(-k)$: At $k = 0$ we have $\bar{\nu} = \nu$ and hence

$$l = \alpha_n(0) = n' - n \tag{257}$$

i.e., the threshold points of the first family of trajectories are integer points along the real axis. It is possible that the trajectories (254) are related to the daughter trajectories discussed recently in the literature[41] or even to the extinct ghosts of Gross and Kayser[42].

4.7 *A field-theoretic example*

We leave potential theory now and turn our eye to a field-theoretic example[37] in order to show that the perturbation formalism is by no means restricted to space-dependent potentials. Moreover, this problem represents another example in which the harmonic oscillator plays a dominant role.

We consider a scalar field which interacts with a non-recoiling nucleon, the latter being localized at the origin of a spatial coordinate system. The Hamiltonian may be written[43]

$$H = H_0 + \lambda H_I$$

$$H_0 = m_0 + \int d\mathbf{k}\,\omega(\mathbf{k})a^*(\mathbf{k})a(\mathbf{k})$$

$$H_I = \frac{1}{(2\pi)^{3/2}} \int \frac{d\mathbf{k} f(k^2)}{\sqrt{[2\omega(k^2)]}} [a(\mathbf{k}) + a^*(-\mathbf{k})] \tag{258}$$

Here m_0 is the bare mass of the nucleon, and $a(\mathbf{k})$ and $a^*(\mathbf{k})$ are the destruction and creation operators for mesons of mass μ having energy $\omega(k) = \sqrt{(k^2 + \mu^2)}$. a and a^* are assumed to satisfy the familiar commutation relations

$$\left.\begin{aligned}[a(\mathbf{k}), a^*(\mathbf{k}')] &= \delta(\mathbf{k} - \mathbf{k}') \\ [a^*(\mathbf{k}), a^*(\mathbf{k}')] &= [a(\mathbf{k}), a(\mathbf{k}')] = 0\end{aligned}\right\} \tag{259}$$

as well as

$$[\psi^*, a] = [\psi^*, a^*] = 0 \tag{260}$$

where $|\psi^*\rangle \equiv \psi^*|0\rangle$ represents the 1-particle nucleon state with no surrounding meson cloud. The form-factor f in (258) is assumed to fall off rapidly enough so that all integrals we encounter are finite.

We now wish to determine the physical mass m of the nucleon, i.e. the eigenvalue of the equation

$$H|\Psi\rangle = m|\Psi\rangle \tag{261}$$

In solving this equation we consider H_0 as an unperturbed harmonic oscillator and H_I as an anharmonic perturbation. Then

$$H_0|\psi^*\rangle = m_0|\psi^*\rangle \tag{262}$$

and m may be written

$$m = m_0 + \lambda\delta \tag{263}$$

λ being the meson-nucleon coupling constant. Thus, by (261) and (263)

$$(H_0 - m_0)|\Psi\rangle = \lambda(\delta - H_I)|\Psi\rangle \tag{264}$$

The zeroth approximation to $|\Psi\rangle$ is therefore

$$|\Psi^{(0)}\rangle = \psi^*|0\rangle \tag{265}$$

Next we consider the expression

$$\mathscr{R}^{(0)} = \lambda(\delta - H_I)\psi^*|0\rangle \tag{266}$$

which is left uncompensated by the approximation (265). Clearly

$$\mathscr{R}^{(0)} = \lambda\delta\psi^*|0\rangle - \frac{\lambda}{(2\pi)^{3/2}}\int \frac{\mathrm{d}\mathbf{k} f(\mathbf{k}^2)}{\sqrt{[2\omega(k)]}}\, a^*(-\mathbf{k})\psi^*|0\rangle \tag{267}$$

However

$$(H_0 - m_0)a^*(\mathbf{k}_1)\psi^*|0\rangle = \omega(\mathbf{k}_1)a^*(\mathbf{k}_1)\psi^*|0\rangle \tag{268}$$

or more generally

$$(H_0 - m_0)|\mathbf{k}_1, \mathbf{k}_2, \ldots, \mathbf{k}_n\rangle = \sum_{j=1}^{n} \omega(\mathbf{k}_j)|\mathbf{k}_1, \mathbf{k}_2 \ldots, \mathbf{k}_n\rangle \tag{269}$$

where

$$|\mathbf{k}_1, \mathbf{k}_2, \ldots, \mathbf{k}_n\rangle = \frac{1}{\sqrt{(n!)}}\, a^*(\mathbf{k}_1)a^*(\mathbf{k}_2) \ldots a^*(\mathbf{k}_n)\psi^*|0\rangle \tag{270}$$

Thus a 1-meson state in (267) may be removed by adding a corresponding term divided by the energy of the meson to the zeroth approximation (265). Then the next contribution $|\Psi^{(1)}\rangle$ to $|\Psi\rangle$ becomes

$$|\Psi^{(1)}\rangle = -\lambda \int \frac{d\mathbf{k} f(k^2)}{\sqrt{[2(2\pi)^3\omega^3(k)]}} a^*(-\mathbf{k})\psi^*|0\rangle \tag{271}$$

Of course, for

$$|\Psi\rangle = |\Psi^{(0)}\rangle + |\Psi^{(1)}\rangle \tag{272}$$

to represent the solution of (261) to 0(1) in λ, we must also have

$$\lambda\delta\psi^*|0\rangle = 0 \tag{273}$$

i.e.

$$\delta = 0 \quad \text{to} \quad 0(0) \text{ in } \lambda$$

Next we have

$$\begin{aligned}
&\lambda(\delta - H_I)a^*(\mathbf{k}_1)\psi^*|0\rangle \\
&\quad = \lambda\delta a^*(\mathbf{k}_1)\psi^*|0\rangle - \frac{\lambda}{(2\pi)^{3/2}} \cdot \frac{f(k_1{}^2)}{\sqrt{[2\omega(k_1)]}} \psi^*|0\rangle \\
&\qquad - \frac{\lambda}{(2\pi)^{3/2}} \cdot \int \frac{d\mathbf{k} f(k^2)}{\sqrt{[2\omega(k)]}} a^*(-\mathbf{k})a^*(\mathbf{k}_1)\psi^*|0\rangle
\end{aligned} \tag{274}$$

The approximation (274) therefore leaves uncompensated the terms

$$\begin{aligned}
\mathscr{R}^{(1)} &= \lambda(\delta - H_I)|\Psi^{(1)}\rangle \\
&= -\frac{\lambda^2\delta}{(2\pi)^{3/2}} \int \frac{d\mathbf{k} f(k^2)}{\sqrt{[2\omega(k)]}\,\omega(k)} a^*(-\mathbf{k})\psi^*|0\rangle \\
&\quad + \frac{\lambda^2}{(2\pi)^3} \int \frac{d\mathbf{k}[f(k^2)]^2}{2\omega^2(k)} \psi^*|0\rangle \\
&\quad + \frac{\lambda^2}{(2\pi)^3} \int \frac{d\mathbf{k} f(k^2)}{\sqrt{[2\omega(k)]}\,\omega(k)} \int \frac{d\mathbf{k}' f(k'^2)}{\sqrt{[2\omega(k')]}} a^*(-\mathbf{k})a^*(-\mathbf{k}')\psi^*|0\rangle
\end{aligned} \tag{275}$$

The first and third terms of this expression may again be removed by adding corresponding terms divided by appropriate energy factors to the previous approximation (272). Thus $|\Psi^{(2)}\rangle$ becomes

$$\begin{aligned}
|\Psi^{(2)}\rangle &= -\frac{\lambda^2\delta}{(2\pi)^{3/2}} \int \frac{d\mathbf{k} f(k^2)}{\sqrt{[2\omega(k)]}\,\omega^2(k)} a^*(-\mathbf{k})\psi^*|0\rangle \\
&\quad + \frac{1}{2!}\left\{-\int \frac{\lambda\, d\mathbf{k} f(k^2)}{\sqrt{[2(2\pi)^3\omega^3(k)]}} a^*(-\mathbf{k})\right\}^2 \psi^*|0\rangle
\end{aligned} \tag{276}$$

The corresponding subsidiary equation for the eigenvalue is

$$\left\{\lambda\delta + \frac{\lambda^2}{(2\pi)^3}\int \frac{d\mathbf{k}[f(k^2)]^2}{2\omega^2(k)}\right\}\psi^*|0\rangle = 0 \tag{277}$$

We do not wish to overload this section with long and tedious calculations of higher contributions. The general procedure is always the same. It is found that no more terms arise to alter (277)—which is therefore exact. Similarly the first term in (276) is found to be cancelled out by a corresponding term arising from the next contribution. Adding various contributions we then have

$$\begin{aligned}|\Psi\rangle &= |\Psi^{(0)}\rangle + |\Psi^{(1)}\rangle + |\Psi^{(2)}\rangle + \cdots \\ &= \sum_{n=0}^{\infty}\frac{1}{n!}\left\{-\lambda\int\frac{d\mathbf{k}f(k^2)}{\sqrt{[2(2\pi)^3\omega^3(k)]}}\,a^*(-\mathbf{k})\right\}^n\psi^*|0\rangle\end{aligned} \tag{278}$$

Also, since (277) is exact,

$$m = m_0 - \frac{\lambda^2}{(2\pi)^3}\int\frac{d\mathbf{k}[f(k^2)]^2}{2\omega^2(k)} \tag{279}$$

These results agree with exact expressions as given by Schweber[43], thus showing the usefulness of our procedure.

References

1. Arscott, F. M., *Periodic Differential Equations*, Pergamon Press, London, 1964.
2. Meixner, J., and Schäfke, F. W., *Mathieusche Funktionen und Sphäroidfunktionen*, Springer, Berlin, 1954.
3. Aly, H. H., and Müller, H. J. W., *J. Math. Phys.*, **7,** 1 (1966), Appendix.
4. Dingle, R. B., and Müller, H. J. W., *J. Reine Angew. Math.*, **211,** 11 (1962); **216,** 123 (1964).
5. Müller, H. J. W., *J. Reine Angew. Math.*, **211,** 179 (1962).
6. Aly, H. H., and Müller, H. J. W., *J. Math. Phys.*, **8,** 367 (1967).
7. Spector, R. M., *J. Math. Phys.*, **5,** 1185 (1964).
8. Bertocchi, L., Fubini, S., and Furlan, G., *Nuovo Cimento*, **35,** 599 (1965).
9. Müller, H. J. W., *Ann. Physik* (*Leipz.*) (1968).
10. Müller, H. J. W., unpublished.
11. Müller, H. J. W., *Z. Physik*, **198,** 59 (1967).
12. Lovelace, C., and Masson, D., *Nuovo Cimento*, **26,** 472 (1962).
13. Mandelstam, S., *Ann. Phys.* (*N.Y.*), **19,** 254 (1962).
14. Bethe, H. A., and Kinoshita, T., *Phys. Rev.*, **128,** 1418 (1962).
15. Müller, H. J. W., *Ann. Physik* (*Leipz.*), **15,** 395 (1965).
16. Müller, H. J. W., *Physica*, **31,** 688 (1965).
17. Müller, H. J. W., and Schilcher, K., *J. Math. Phys.* (1968).

18. Erdélyi, A., Magnus, W., Oberhettinger, F., and Tricomi, F. G., *Higher Transcendental Functions*, Vol. III, New York, 1955.
19. Müller, H. J. W., *J. Reine Angew. Math.*, **212,** 26 (1963).
20. Müller, H. J. W., *Math. Nachr.*, **31,** 89 (1966); **32,** 49, 157 (1966).
21. Müller, H. J. W., *Ann. Physik* (*Leipz.*), **16,** 255 (1965).
22. Fivel, D. I., and Klein, A., *J. Math. Phys.*, **4,** 274 (1960).
23. Singh, V., *Phys. Rev.*, **127,** 632 (1962).
24. Ahmadzadeh, A., Burke, P. G., and Tate, C., *Phys. Rev.*, **131,** 1315 (1963).
25. Weinberg, S., *Phys. Rev.*, **131,** 440 (1963).
26. Warburton, A. E. A., *Nuovo Cimento*, **41,** 360 (1966).
27. Finkelstein, J., *Phys. Rev.*, **145,** 1185 (1966).
28. Müller, H. J. W., *Z. Physik*, **205,** 145 (1967).
29. Balázs, L. A. P., *Phys. Rev.*, **137B,** 1510 (1965).
30. Nambu, Y., and Sugawara, M., *Phys. Rev. Letters*, **10,** 304 (1963).
31. Serber, R., *Phys. Rev. Letters*, **10,** 357 (1963); *Rev. Mod. Phys.*, **36,** 649 (1964).
32. Aly, H. H., Lurié, D., and Rosendorff, S., *Phys. Lett.*, **7,** 198 (1963).
33. Müller, H. J. W., unpublished.
34. Müller, H. J. W., *Z. Physik*, **186,** 79 (1965).
35. Müller, H. J. W., *Z. Physik*, **183,** 402 (1965).
36. Fivel, D. I., *Phys. Rev.*, **142,** 1219 (1966).
37. Müller, H. J. W., *Z. Physik*, **205,** 149 (1967).
38. Aly, H. H., Müller, H. J. W., and Schilcher, K., *Nucl. Phys.*, **3B,** 401 (1967).
39. Aly, H. H., and Müller, H. J. W., *Nucl. Phys.*, **76,** 241 (1966).
40. Müller, H. J. W., to be published. Note added in proof: We have meanwhile succeeded in solving the radial Schrödinger equation (101) for the Gauss potential $V(r) = -g^2 \exp(-a^2r^2)$ when g^2 is large. The asymptotic expansion for the discrete eigenvalues is found to be (where $q = 4n + 3$, $n = 0, 1, 2, \ldots$):

$$k^2 = -g^2 + ga(2l + q) - \frac{a^2}{2^4}[3(q^2 + 1) + 4(3q - 1)l + 8l^2]$$
$$- \frac{a^3}{3\cdot 2^8 g}[q(11q^2 + 1) + 2(33q^2 - 6q + 1)l + 24(5q - 1)l^2 + 64l^3]$$
$$- \frac{a^4}{3\cdot 2^{15} g^2}[4(85q^4 + 2q^2 - 423) + l(2720q^3 - 71q^2 + 32q + 2976)$$
$$+ 32l^2(252q^2 - 12q + 64) + 256l^3(41q - 9)$$
$$+ 4096l^4] + 0\left(\frac{1}{g^3}\right).$$

41. Freedman, D. Z., Jones, C. E., and Wang, Y.-M., *Phys. Rev.*, **155,** 1645 (1967).
42. Gross, D. J., and Kayser, B. J., *Phys. Rev.*, **152,** 1441 (1966).
43. Schweber, S. S., *An Introduction to Relativistic Quantum Field Theory*, Harper International, New York, 1961.

Theory of Symmetry Violation and Gravitation

J. W. MOFFAT

1 Introduction

A theory of the violations of internal symmetry groups and the gravitational field is developed on the basis of a non-Riemannian displacement field. Divergence equations for the vector and axial vector current densities in weak interactions are derived. These equations show that if the groups U(3) and U(3) X U(3) are considered in a non-Riemannian space-time, then the group generators are not constants of the motion. Field equations for the non-symmetric fields are found, and in a weak field approximation vector and axial vector current sum rules can be obtained from the divergence equations.

In spite of the successes of the applications of groups like U(3) and U(6) to the classification of particles[1,2], and the successes of the current algebra programme, based on Gell-Mann's[3] U(3) X U(3) algebra, there exists a general feeling of dissatisfaction with the present situation in particle physics. Apart from the obvious lack of an overall dynamical theory of strong interactions, very little has been learned about the origin of the violation of the internal symmetries. The purely group theoretical statements about these violations do not lead to a deeper understanding. It is hoped that the symmetry violations may eventually be understood from first principle, in the sense that we understand the breaking of SU(2) by the electromagnetic field. The attempts to understand the symmetry violations as a spontaneous breakdown, analogous to the properties of a ferromagnet, have not as yet led to any significant advances. The hopes to combine the internal symmetries and the inhomogeneous Lorentz transformations (Poincaré group) have been thwarted, and the deeper connexion between the internal symmetries and space-time remains a mystery[4,5,6].

In the following, we shall study the problem of whether or not there exists a space-time structure that leads, in a natural way, to the violation of groups like U(3) and U(3) X U(3). Does the space-time structure of the universe violate the internal symmetries?

The innovation of general relativity removed the necessity for maintaining the 'inertial system' as a basic concept in physics. The theory remains the most satisfactory explanation of the phenomenon of the equality of inertial and gravitational masses. However, as a fundamental theory, it has had little influence on the course of modern physics, because of the weakness of the gravitational field in the realm of particle physics.

The inertial system is replaced in the general theory of relativity by the infinitesimal displacement field $\Gamma^{\lambda}_{\mu\nu}$ of Weyl and Levi-Cività. It was recognized by Einstein[7], in his search for a generalized theory of gravitation, that the gravitational theory is a quite special case of a general non-Riemannian displacement field. We shall investigate under what circumstances the generators of the groups U(3) and U(3) X U(3), determined by the three-dimensional integrals of the 4th components of the vector and axial vector current densities, are constants of the motion when we parallel displace these vectors in a non-Euclidean space-time. If we look for the non-conservation of the current densities due to the effects of space-time, then we discover that in a purely gravitational field the currents are conserved and the symmetry of U(3) X U(3) is not violated. However, in the presence of a non-Riemannian geometry, associated with a non-symmetrical connexion $\Gamma^{\lambda}_{\mu\nu}$, the currents are no longer conserved in the usual sense, and the generators of the group U(3) X U(3) are not constants of the motion.

The total Hamiltonian of the physical system is written[3] as

$$H = \bar{H} + H_8 + H_0$$

where $\bar{H}$ is invariant under the full algebra and leaves both the weak vector and axial vector currents conserved, while H_8 and H_0 describe the eightfold breaking and the violation of the conservation of the axial vector current, respectively. At present there is no knowledge of the dynamical nature of the terms H_8 and H_0; the only information we have of these terms is of a group theoretical nature.

If, indeed, the gravitational field and the fields that violate the internal symmetries of elementary particles have a common origin in the structure of space-time, then we may hope that a unified theory of these fields can be developed, and that the symmetry violations can be calculated, from first principle, on the basis of a general set of field equations incorporating the broken symmetry groups.

It is a common feature of theoretical physics that the more unified the fundamental premises become the more complex the machinery of the theory; the complexities of non-Riemannian geometry are sufficiently formidable to cause one to be tentative about introducing such a formalism into particle physics at the present time. However, there have been no new ideas forthcoming on the basis of our concepts associated with the conventional Minkowski space-time, which can provide hope for a deeper understanding of the dynamical aspects of particle symmetries; it is possible that we are required to delve deeper into the properties of space-time in order to throw new light on the problem.

In Section 2, we give a brief review of the theory of weak interactions and current algebras. The affine displacement field is studied in Section 3, and a new formalism is developed that incorporates the group properties of U(3) in terms of a nonet of affine displacement fields. In Section 4, a fundamental nonet of tensors $g^i_{\mu\nu}$ is introduced, and the covariant divergence of current densities is studied. This leads to the basic vector and axial vector current density divergence equations of the theory. A set of field equations, which determine the gravitational and symmetry violating fields, is obtained in Section 5 from a variational principle. Finally, in Section 6, the weak field approximation is discussed, and it is pointed out that the current algebra sum rules[8,9] can be obtained from the vector and axial vector density divergence equations of the theory.

2 Current Densities and the Algebra of U(3) X U(3)

The weak interactions can be described in the local approximation by the phenomenological Lagrangian in a Minkowski space-time

$$\frac{G}{\sqrt{2}}\mathscr{J}_\mu{}^+\mathscr{J}_\mu \tag{1}$$

where

$$\mathscr{J}_\mu = \mathscr{J}_\mu{}^l\,(\text{leptons}) + \mathscr{J}^h_\mu\,(\text{hadrons}) \tag{2}$$

The hadron current density $\mathscr{J}^h_\mu(x)$ is a member of a U(3) nonet of current densities $\mathscr{J}_\mu{}^i(x)(i = 0, 1, 2, \ldots, 8)$ and it can be written

$$\mathscr{J}^h_\mu(x) = \cos\theta(\mathscr{J}_\mu{}^1 + \mathrm{i}\mathscr{J}_\mu{}^2) + \sin\theta(\mathscr{J}_\mu{}^4 + \mathrm{i}\mathscr{J}_\mu{}^5) \tag{3}$$

where θ is the Cabibbo angle[10]. The nonet of currents $\mathscr{J}_\mu{}^i(x)$ has mixed parity and we can decompose it into the sum of a vector current $\mathscr{V}_\mu{}^i(x)$ and an axial vector current $\mathscr{A}_\mu{}^i(x)$:

$$\mathscr{J}_\mu{}^i(x) = \mathscr{V}_\mu{}^i(x) + \mathscr{A}_\mu{}^i(x) \tag{4}$$

The linear combination (3) is determined, in part, by the requirements of universality of the weak interactions. The $\mathscr{J}_\mu^1 + i\mathscr{J}_\mu^2$ and $\mathscr{J}_\mu^4 + i\mathscr{J}_\mu^5$ are the $\Delta Q = +1$ members of the nonet with $\Delta S = 0$ and $\Delta S = +1$, respectively.

It is known that the electromagnetic current of the hadrons is given by

$$\mathscr{V}_\mu^{\text{em}} = e(\mathscr{V}_\mu^{\,3} + \frac{1}{\sqrt{3}}\mathscr{V}_\mu^{\,8}) \tag{5}$$

Gell-Mann has postulated that the charges associated with the currents[3]

$$F^i(t) = -i\int \mathscr{V}_4^{\,i}(x)\,d^3x \tag{6}$$

$$F_5^i(t) = -i\int \mathscr{A}_4^{\,i}(x)\,d^3x \tag{7}$$

generate the minimal algebra of U(3) X U(3) under equal-time commutation

$$[F^i(t), F^j(t)] = i\,f_{ijk}F^k(t) \tag{8}$$

$$[F^i(t), F_5^j(t)] = i\,f_{ijk}F_5^k(t) \tag{9}$$

$$[F_5^i(t), F_5^j(t)] = i\,f_{ijk}F^k(t) \tag{10}$$

The components F^1, F^2 and F^3 are the three components of isotopic spin which are conserved by the strong interactions; the components F^4, F^5, F^6, F^7 and all the components of the axial charge F_5^i vary with time. The component F^8 is conserved and describes the hypercharge $F^8 = \sqrt{3}/2\,Y$.

The current algebras have been successfully used to obtain sum rules relating different weak interaction processes, and also relating weak and strong processes. The pivotal point of these applications has been the PCAC (partially conserved axial current) hypothesis, which relates the divergence of the $\Delta S = 0$ axial currents to the interpolating pion field[11]

$$\partial_\mu \mathscr{A}_\mu^{\,i}(x) = a\pi^i(x), \quad (i = 1, 2, 3) \tag{11}$$

Here a is the Goldberger–Trieman constant and $\pi^i(x)$ is the renormalized pion field operator.

By taking matrix elements of the equal-time commutators of currents between physical hadron states one obtains sum rules, e.g. the Adler–Weisberger[8,9] sum rule for the renormalization of the β-decay coupling constant is derived by taking matrix elements of the commutator of the axial charges between proton states, and then using (11) as a definition of the pion field.

3 The Infinitesimal Displacement Field Γ

Let us now discuss the concept of the parallel displacement of a contravariant vector. When a *nonet* of vectors $A^{\sigma i}$ $(i = 0, 1, 2, \ldots 8)$ is displaced parallel to itself by an infinitesimal distance $\mathrm{d}x^{\mu}$ the change in its components is to be given by[12]

$$\delta A^{\sigma i} = -h_{ijk}\Gamma^{\sigma j}_{\mu\nu}\,\mathrm{d}x^{\mu}A^{\nu k}, \quad (i, j, k = 0, 1, 2, \ldots, 8) \tag{12}$$

where $h_{ijk} = (1 - \alpha)f_{ijk} + \alpha d_{ijk}$. The f_{ijk} and d_{ijk} are the familiar structure constants of U(3) and α determines the relative strength of the F and D terms[3]. From the way the coefficients $\Gamma^{\sigma i}_{\mu\nu}$ enter the expression (12), it is clear that it would be unjustified at the beginning to specialize the $\Gamma^{\sigma i}_{\mu\nu}$ to be symmetric in the lower suffixes, and, therefore, we shall consider $\Gamma^{\sigma i}_{\mu\nu}$ to be a non-symmetric affinity.

When we displace a nonet of vectors $A^{\sigma i}$ parallel to itself around a closed, infinitesimally small cycle, we get the total displacement

$$\Delta A^{\sigma i} = h_{ijk}R^{\sigma j}_{\mu\nu\rho}A^{\mu k}s^{\nu\rho} \tag{13}$$

where $s^{\nu\rho}$ is the infinitesimal surface element, and the nonet of coefficients $R^{\sigma i}_{\mu\nu\rho}$ is the Riemann curvature tensor.

Let us use the Einstein notation and decompose $\Gamma^{\sigma i}_{\mu\nu}$ into its symmetric and antisymmetric parts[7]

$$\Gamma^{\sigma i}_{\mu\nu} = \underline{\Gamma}^{\sigma i}_{\mu\nu} + \underset{\vee}{\Gamma}^{\sigma i}_{\mu\nu} \tag{14}$$

where

$$\underline{\Gamma}^{\sigma i}_{\mu\nu} = \tfrac{1}{2}(\Gamma^{\sigma i}_{\mu\nu} + \Gamma^{\sigma i}_{\nu\mu})$$

$$\underset{\vee}{\Gamma}^{\sigma i}_{\mu\nu} = \tfrac{1}{2}(\Gamma^{\sigma i}_{\mu\nu} - \Gamma^{\sigma i}_{\nu\mu}) \tag{15}$$

If we put $\underset{\vee}{\Gamma}^{\sigma i}_{\mu\nu} = 0$, then the affine connexion reduces to the Christoffel symbols $\{^{\sigma}_{\mu\nu}\}^{i}$ of general relativity and the space-time structure is that of the pure gravitational field. The 'graviton' will be associated with the irreducible *singlet* representation of U(3), and therefore the Christoffel symbols $\{^{\sigma}_{\mu\nu}\}^{i} = 0$ for $i = 1, 2, \ldots, 8$. We shall demand that in general for the symmetric connexion $\underline{\Gamma}^{\sigma i}_{\mu\nu} \neq 0$ for $i = 0, 1, 2, \ldots, 8$. Moreover, we shall also consider the situation in which $\underset{\vee}{\Gamma}^{\sigma i}_{\mu\nu} \neq 0$ for $i = 0, 1, 2, \ldots, 8$, i.e. the skew tensors $\underset{\vee}{\Gamma}^{\sigma i}_{\mu\nu}$ correspond to a nonet of symmetry-breaking fields. If we just treated the case of U(3), then it would be sufficient to have only $\underset{\vee}{\Gamma}{}^{\sigma 3}_{\mu\nu}$ and $\underset{\vee}{\Gamma}{}^{\sigma 8}_{\mu\nu}$ as non-vanishing components, and we should only concern ourselves with the vector current density $\mathscr{V}^{\mu i}(x)$.

4 The Tensor $g_{\mu\nu}{}^{i}$ and the Covariant Divergence of the Current Densities

We shall associate with space-time a nonet of fundamental tensors $g_{\mu\nu}{}^{i}$ defined by

$$g_{\mu\nu}{}^{i} = g_{\underline{\mu\nu}}{}^{i} + \underset{\vee}{g}_{\mu\nu}{}^{i} \tag{16}$$

where we choose from the beginning $g_{\underline{\mu\nu}}{}^{i} = 0$ for $i = 1, 2, 3, \ldots, 8$, while $\underset{\vee}{g}_{\mu\nu}{}^{i} \neq 0$ for $i = 0, 1, 2, \ldots, 8$. A contravariant tensor $g^{\mu\nu i}$ can also be correlated to $g_{\mu\nu}{}^{i}$ according to the equation

$$g_{\mu\sigma}{}^{i} g^{\sigma\nu j} = \delta_{\mu}{}^{\nu}\delta_{ij} \tag{17}$$

The skew tensor of third rank $\underset{\vee}{\Gamma}{}^{\lambda i}_{\mu\nu}$ will be defined with mixed parity

$$\underset{\vee}{\Gamma}{}^{\lambda i}_{\mu\nu} = \underset{\vee}{U}{}^{\lambda i}_{\mu\nu} + \underset{\vee}{W}{}^{\lambda i}_{\mu\nu} \tag{18}$$

where $\underset{\vee}{U}{}^{\lambda i}_{\mu\nu}$ is a tensor and $\underset{\vee}{W}{}^{\lambda i}_{\mu\nu}$ is a pseudotensor with respect to parity transformations. Similarly for the skew tensor of second rank $\underset{\vee}{g}_{\mu\nu}{}^{i}$, we have

$$\underset{\vee}{g}_{\mu\nu}{}^{i} = \underset{\vee}{b}_{\mu\nu}{}^{i} + \underset{\vee}{c}_{\mu\nu}{}^{i} \tag{19}$$

where $\underset{\vee}{b}_{\mu\nu}{}^{i}$ is a tensor and $\underset{\vee}{c}_{\mu\nu}{}^{i}$ is a pseudotensor.

Let us now study the covariant derivatives of the nonet of weak current densities $\mathscr{J}^{\mu i} = \mathscr{V}^{\mu i} + \mathscr{A}^{\mu i}$. We have

$$\mathscr{J}^{\mu i};\nu = \partial_{\nu}\mathscr{J}^{\mu i} + h_{ijk}\Gamma^{\mu j}_{\sigma\nu}\mathscr{J}^{\sigma k} - h_{ijk}\Gamma^{\sigma j}_{\sigma\nu}\mathscr{J}^{\mu k} \tag{20}$$

Upon contracting the indices μ and ν, we get

$$\mathscr{J}^{\mu i};\mu = \partial_{\mu}\mathscr{J}^{\mu i} + h_{ijk}\Gamma^{j}{}_{\mu}\mathscr{J}^{\mu k} \tag{21}$$

where $\Gamma_{\mu}{}^{i}$ denotes the quantity

$$\Gamma_{\mu}{}^{i} = \Gamma^{\sigma i}_{\mu\sigma} - \Gamma^{\sigma i}_{\sigma\mu} \quad (\equiv 2\underset{\vee}{\Gamma}{}^{\sigma i}_{\mu\sigma}) \tag{22}$$

We shall postulate that the weak current satisfies the *covariant* conservation law

$$\mathscr{J}^{\mu i};\mu = 0 \tag{23}$$

From (21), we then have

$$\partial_\mu \mathscr{J}^{\mu i} = -h_{ijk}\Gamma^j_\mu \mathscr{J}^{\mu k} \tag{24}$$

or, splitting (24) according to parity*

$$\dot{F}^i(t) = \int \partial_\mu \mathscr{V}^{\mu i}\, d^3x = -h_{ijk}\int (U^j_\mu \mathscr{V}^{\mu k} + W^j_\mu \mathscr{A}^{\mu k})\, d^3x \tag{25}$$

and

$$\dot{F}^i_5(t) = \int \partial_\mu \mathscr{A}^{\mu i} d^3x = -h_{ijk}\int (U^j_\mu \mathscr{A}^{\mu k} + W_\mu{}^j \mathscr{V}^{\mu k}) d^3x \tag{26}$$

Here,

$$U_\mu{}^i = U_\mu{}^\sigma{}_\sigma{}^i - U_\sigma{}^\sigma{}_\mu{}^i$$
$$W_\mu{}^i = W_\mu{}^\sigma{}_\sigma{}^i - W_\sigma{}^\sigma{}_\mu{}^i \tag{27}$$

and $U_\mu{}^i$ and $W_\mu{}^i$ denote nonets of vector and axial vector fields, respectively.

If we put $\Gamma^{\sigma i}_{\underset{\nu}{\mu\nu}} = 0$ (and therefore $\Gamma_\mu{}^i = 0$) ($i = 0, 1, 2, \ldots, 8$) it follows from (24) that

$$\partial_\mu \mathscr{J}^{\mu i} = 0 \tag{28}$$

and the weak currents are conserved in the ordinary sense; it follows that $\dot{F}^i(t) = \dot{F}_5{}^i(t) = 0$ and the 18 generators F^i and $F_5{}^i$ are constants of the motion. Thus in the presence of a *pure* gravitational field the groups U(3) and U(3) X U(3) are symmetry groups of the infinitesimal unitary transformations. But from (25) and (26), we learn that in the presence of a non-Riemannian geometry, governed by a non-symmetric affine connexion $\Gamma^{\lambda i}_{\mu\nu}$, the generators F^i and $F_5{}^i$ are not constants of the motion (except F^3 and F^8 which are postulated to be conserved), and the symmetry of the group U(3) X U(3) is violated.

5 Field Equations

It is our hope that the nonet of skew symmetric fields $\Gamma^{\lambda i}_{\underset{\nu}{\mu\nu}}$ describes the fields that violate the symmetry of the groups U(3) and U(3) X U(3). We must now decide upon equations that will determine the unknown quantities $\Gamma^{\lambda i}_{\mu\nu}$ and Γ^i_μ.

In order to arrive at a suitable set of equations, we shall resort to a variational principle[7,12]. Let us consider the Palatini method, in which the $g^i_{\mu\nu}$ and the $\Gamma^{\lambda i}_{\mu\nu}$ are varied independently. On variation we get

$$\delta \int \mathscr{g}^{\mu\nu i} R^i_{\mu\nu}\, d^4x = \int (\delta \mathscr{g}^{\mu\nu i} R^i_{\mu\nu} + \mathscr{g}^{\mu\nu i}\delta R^i_{\mu\nu})\, d^4x = 0 \tag{29}$$

* Here we use the notation $\dot{F}^i(t) = \partial_t F^i(t)$.

where $\mathcal{g}^{\mu\nu i} = (\sqrt{(-g)}g^{\mu\nu})^i$ is a tensor density. The two parts of (29) vanish separately

$$\int \delta \mathcal{g}^{\mu\nu i} R^i_{\mu\nu}\, \mathrm{d}^4x = 0$$

$$\int \mathcal{g}^{\mu\nu i} \delta R^i_{\mu\nu}\, \mathrm{d}^4x = 0 \tag{30}$$

It is necessary to study $\delta R^i_{\mu\nu}$. In order to evaluate this let us consider the contractions of the Riemann curvature tensor

$$R^{\sigma i}_{\mu\nu\rho} = -\partial_\rho \Gamma^{\sigma i}_{\mu\nu} + \partial_\nu \Gamma^{\sigma i}_{\mu\rho} + h_{ijk}\Gamma^{\sigma j}_{\alpha\nu}\Gamma^{\alpha k}_{\mu\rho} - h_{ijk}\Gamma^{\sigma j}_{\alpha\rho}\Gamma^{\alpha k}_{\mu\nu} \tag{31}$$

We have the Einstein tensor

$$\begin{aligned} R^i_{\mu\nu} &= R^{\alpha i}_{\mu\nu\alpha} \\ &= -\partial_\alpha \Gamma^{\alpha i}_{\mu\nu} + \partial_\nu \Gamma^{\alpha i}_{\mu\alpha} + h_{ijk}\Gamma^{\beta j}_{\alpha\nu}\Gamma^{\alpha k}_{\mu\beta} - h_{ijk}\Gamma^{\beta j}_{\alpha\beta}\Gamma^{\alpha k}_{\mu\nu} \end{aligned} \tag{32}$$

and

$$R^{\alpha i}_{\alpha\nu\rho} = -\partial_\rho \Gamma^{\alpha i}_{\alpha\nu} + \partial_\nu \Gamma^{\alpha i}_{\alpha\rho} \tag{33}$$

It can easily be shown that

$$\delta R^i_{\mu\nu} = -(\delta\Gamma^{\alpha i}_{\mu\nu})_{;\alpha} + (\delta\Gamma^{\alpha i}_{\mu\alpha})_{;\nu} + 2h_{ijk}\Gamma^{\alpha j}_{\underset{\nu}{\beta\nu}}\,\delta\Gamma^{\beta k}_{\mu\alpha} \tag{34}$$

From (29) and (34), we arrive at the field equations

$$\partial_\alpha g^i_{\mu\nu} - h_{ijk} g^j_{\sigma\nu} {}^*\Gamma^{\sigma k}_{\mu\alpha} - h_{ijk} g^j_{\mu\sigma} {}^*\Gamma^{\sigma k}_{\alpha\nu} = 0 \tag{35}$$

$$\partial_\nu \mathcal{g}^{\mu\nu i}_{\ \ \nu} = 0 \tag{36}$$

$$R^i_{\mu\nu} = 0 \tag{37}$$

where ${}^*\Gamma^{\sigma i}_{\mu\nu}$ is the Schrödinger affinity[12]

$${}^*\Gamma^{\sigma i}_{\mu\nu} = \Gamma^{\sigma i}_{\mu\nu} + \tfrac{1}{3}\delta^\sigma_\mu \Gamma^i_\nu \tag{38}$$

This affinity satisfies the symmetry condition

$${}^*\Gamma^{\alpha i}_{\mu\alpha} = {}^*\Gamma^{\alpha i}_{\alpha\mu} \tag{39}$$

Equations (35) to (37) are the field equations of the theory, equal in number to the number of unknown quantities. Once we have solved for the Γ^i_μ, we can calculate the expressions $\partial_\mu \mathscr{V}^{\mu i}$ and $\partial_\mu \mathscr{A}^{\mu i}$ from (25) and (26). The current densities $\mathscr{V}^i_\mu$ and $\mathscr{A}^i_\mu$ are then postulated to satisfy the equal-time commutation relations of Gell-Mann[3]:

$$[\mathscr{V}_4^i(\mathbf{x}, t), \mathscr{V}_4^j(\mathbf{x}', t)] = -f_{ijk}\mathscr{V}_4^k(x)\delta(\mathbf{x} - \mathbf{x}') \tag{40}$$

$$[\mathscr{V}_4^i(\mathbf{x}, t), \mathscr{A}_4^j(\mathbf{x}', t)] = -f_{ijk}\mathscr{A}_4^k(x)\delta(\mathbf{x} - \mathbf{x}') \tag{41}$$

$$[\mathscr{A}_4^i(\mathbf{x}, t), \mathscr{A}_4^j(\mathbf{x}', t)] = -f_{ijk}\mathscr{V}_4^k(x)\delta(\mathbf{x} - \mathbf{x}') \tag{42}$$

The essential physics is contained in the equations presented so far although, as we shall see in the following section, the divergence equations (25) and (26) *already contain a lot of the information* which can be deduced from the equal-time commutation relations (40) to (42), and, therefore, we need not postulate these commutation relations in an *ad hoc* manner, but adopt just the fundamental divergence equations (25) and (26).

6 Weak Field Approximation and Sum Rules

In order to make an immediate connexion with particle physics, we shall resort to the weak field approximation. We assume that the $\underset{-}{\Gamma^{\lambda i}_{\mu\nu}}$ and $\underset{\nu}{\Gamma^{\lambda i}_{\mu\nu}}$ can be expanded in series

$$\underset{-}{\Gamma^{\lambda i}_{\mu\nu}} = \underset{1\,-}{\Gamma^{\lambda i}_{\mu\nu}} + \underset{2\,-}{\Gamma^{\lambda i}_{\mu\nu}} + \cdots \tag{43}$$

$$\underset{\nu}{\Gamma^{\lambda i}_{\mu\nu}} = \underset{1\,\nu}{\Gamma^{\lambda i}_{\mu\nu}} + \underset{2\,\nu}{\Gamma^{\lambda i}_{\mu\nu}} + \cdots \tag{44}$$

and similarly for the $\underset{-}{g_{\mu\nu}^{\;i}}$ and $\underset{\nu}{g_{\mu\nu}^{\;i}}$. We shall restrict ourselves to weak gravitational and symmetry violating fields of the first order. In particular

$$\underset{-}{g_{\mu\nu}^{\;i}} = -\delta_{0i}\delta_{\mu\nu} + \delta_{0i}h_{\mu\nu} \tag{45}$$

where $\delta_{\mu\nu}$ is the Kronecker tensor $\delta_{\mu\nu} = 1$ if $\mu = \nu$ and $\delta_{\mu\nu} = 0$ if $\mu \neq \nu$. Also, $\delta_{0i} = 1$ if $i = 0$ and $\delta_{0i} = 0$ if $i \neq 0$; the $h_{\mu\nu}$ are small quantities of first order.

Ideally, we should attempt to calculate the fundamental quantities $\Gamma^{\lambda i}_{\mu\nu}$ and Γ^i_μ from the field equations, but in order to see how the theory can lead to immediate physical results, we shall proceed in a semi-phenomenological fashion. From the divergence equations (25) and (26), to first order in the gravitational and symmetry violating fields, we can recover the currently fashionable sum rules.

The divergence of the vector and axial vector current densities are given in the theory by

$$-\partial_\mu \mathscr{V}^{\mu i} = h_{ijk} U^j_\mu \mathscr{V}^{\mu k} + h_{ijk} W^j_\mu \mathscr{A}^{\mu k} \tag{46}$$

and

$$-\partial_\mu \mathscr{A}^{\mu i} = h_{ijk} U^j_\mu \mathscr{A}^{\mu k} + h_{ijk} W^j_\mu \mathscr{V}^{\mu k} \tag{47}$$

Let us restrict the internal symmetry indices such that $i, j, k = 1, 2, 3$, and let us identify the U and W with vector and axial vector boson field operators, respectively. In particular, we shall make the physical identification to first order*

$$U^i_\mu = eA^i_\mu + B^i_\mu \tag{48}$$

where A^i_μ is an isotopic spin vector with the components 1 and 2 equal to zero and corresponds to the electromagnetic potential four-vector; B^i_μ is an isotopic vector boson field operator. We shall also adopt the further identification

$$-\varepsilon_{ijk} B^j_\mu \mathscr{A}_\mu{}^k = a\pi^i \tag{49}$$

where π^i is the interpolating pion field and

$$a = M_N M_\pi{}^2 g_A / g_r K^{NN\pi}(0) \tag{50}$$

Here g_r is the renormalized pion-nucleon coupling constant.

Our divergence equations now read to first order

$$-\partial_\mu \mathscr{V}_\mu{}^i = e\varepsilon_{ijk} A^j_\mu \mathscr{V}_\mu{}^k + \varepsilon_{ijk} B^j_\mu \mathscr{V}_\mu{}^k + \varepsilon_{ijk} W^j_\mu \mathscr{A}_\mu{}^k \tag{51}$$

and

$$\partial_\mu \mathscr{A}_\mu{}^i = a\pi^i - e\varepsilon_{ijk} A^j_\mu \mathscr{A}_\mu{}^k - \varepsilon_{ijk} W^j_\mu \mathscr{V}_\mu{}^k \tag{52}$$

By using a method due to Veltman[13], we can now calculate the well-known sum rules, e.g. the Cabibbo–Radicati sum rule[14] and the sum rules[15] for processes $\gamma + N \to N' + B$ can be derived from the vector divergence equation (51). The Adler–Weisberger sum rule[8,9] can be obtained from (52). Let us, for the sake of completeness, consider the derivation of the A–W sum rule.

The process of interest is

$$W + N \to W + N' \tag{53}$$

The S-matrix element is

$$S = \langle N', W_{\text{out}} | W_{\text{in}}, N \rangle \tag{54}$$

* It is understood that for convenience relevant coupling constants have been absorbed in the B and W field operators.

By using the reduction technique, we get

$$S = i\lambda e^i_\mu \int d^4x \, e^{-iq\cdot x}\langle N, W_{\text{out}}|\mathscr{A}^i_\mu(x)|N\rangle = i\lambda e^i_\mu M^i_\mu \tag{55}$$

where the e^i_μ denote the polarizations of the axial vector particle W, and we have used the equation (λ is a real coupling constant):

$$(\Box - M_W^2)W^i_\mu(x) = \lambda\mathscr{A}^i_\mu(x) \tag{56}$$

where M_W is the mass of the axial vector boson W. From (52), it follows that

$$q_\mu M^i_\mu = a\int d^4x \, e^{-iq\cdot x}\langle N', W_{\text{out}}|\pi^i(x)|N\rangle + i\lambda\varepsilon_{ijk}\frac{e^j_\mu}{(2q_0')^{1/2}}\int d^4x \, e^{i(q'-q)x}\langle N'|\mathscr{V}^k_\mu(x)|N\rangle \tag{57}$$

The last term in (57) just corresponds to the equal-time commutator of the axial charges. As pointed out by Veltman[13], the rest of the calculation follows the work of Adler and Weisberger[8,9], and leads to the sum rule

$$1 - \frac{1}{g_A{}^2} = \frac{4M_N{}^2}{[g_r K^{NN\pi}(0)]^2}\frac{1}{\pi}\int_{M_N+M_\pi}^{\infty}\frac{dE\,E}{E^2 - M_N{}^2}[\sigma_p^{\pi+}(E) - \sigma_p^{\pi-}(E)] \tag{58}$$

The derivation of the sum rules does not depend on the explicit use of the equal-time commutation relations (40) to (42), but only on the form of the divergence equations (46) and (47), which were derived from the basic postulates of the theory.

7 Concluding Remarks

Ever since Einstein discovered the theory of general relativity in 1915, there have been many attempts to generalize the theory so as to include the electromagnetic field in a unified scheme. These attempts culminated in the non-symmetric theory of Einstein, and Einstein and Straus[7]. In this theory it was hoped that the skew part of the tensor $g_{\mu\nu}$ described the electromagnetic field much in the same way as the metric tensor $g_{\mu\nu}$ of general relativity described the gravitational field.

The theory of symmetry violations and gravitation that we have described is based on the non-symmetric affine connexion as in Einstein's theory, but the physical interpretation is quite different. It is possible that with the combination of the internal symmetry group U(3) and the affine displacement field within the quantity $\Gamma^{\lambda i}_{\mu\nu}$, we have already affected as much of a unification of these two physical phenomena as is possible at present. The fundamental features of the theory only come to the foreground when we consider the exact solutions of the field equations, because it is then that we possess a real possibility of calculating the symmetry violations from the basic principles of the theory. In spite of the difficulties involved in finding exact solutions, we should not be deterred from thinking further about theories of this kind. The approach of conventional field theory has offered little or no explanation of the problems associated with our understanding of the rest masses of elementary particles and the mass differences of these particles within the framework of internal symmetries.

References

1. *The Eightfold Way* (Ed. M. Gell-Mann and Y. Ne'eman), W. A. Benjamin, Inc., New York—Amsterdam, 1964.
2. *Symmetry Groups in Nuclear and Particle Physics* (Ed. F. J. Dyson), W. A. Benjamin, Inc., New York—Amsterdam, 1966.
3. Gell-Mann, M., *Phys. Rev.*, **125**, 1067 (1962); *Physics*, **1**, 63 (1964). For reviews of this subject, see Moffat, J. W., in *Elementary Particle Theories* (Ed. P. Urban), Springer-Verlag, Vienna—New York, 1966, p. 113; Moffat, J. W., in *Symmetry Principles and Fundamental Particles*, W. H. Freeman and Co., San Francisco, Calif., 1967, p. 317.
4. McGlinn, W. D., *Phys. Rev. Letters*, **12**, 467 (1964).
5. Coleman, S., *Phys. Rev.*, **138B**, 1262 (1965).
6. O'Raifeartaigh, L., *Phys. Rev.*, **139B**, 1052 (1965).
7. Einstein, A., *The Meaning of Relativity*, Methuen and Co. Ltd., London, Appendix II, 5th ed., 1951 and 6th ed., 1956.
8. Adler, S. L., *Phys. Rev. Letters*, **14**, 1051 (1965); *Phys. Rev.*, **140B**, 736 (1965).
9. Weisberger, W. I., *Phys. Rev. Letters*, **14**, 1047 (1965).
10. Cabibbo, N., *Phys. Rev. Letters*, **10**, 531 (1963).
11. Gell-Mann, M., and Levy, M., *Nuovo Cimento*, **26**, 705 (1960); Nambu, Y., *Phys. Rev. Letters*, **4**, 380 (1960); Adler, S. L., *Phys. Rev.*, **137B**, 1022 (1965).
12. Schrödinger, E., *Space-Time Structure*, Cambridge University Press, 1954.
13. Veltman, M., *Phys. Rev. Letters*, **17**, 553 (1966).
14. Cabibbo, N., and Radicati, L. A., *Phys. Lett.*, **19**, 697 (1966).
15. Bég, M. A. B., *Phys. Rev. Letters*, **17**, 333 (1966).

Author Index

Subject Index